HORSES

A Guide to Selection, Care, and Enjoyment

HORSES

A Guide to Selection, Care, and Enjoyment

J. WARREN EVANS

UNIVERSITY OF CALIFORNIA, DAVIS

W. H. FREEMAN AND COMPANY
San Francisco

Project Editor: Pearl C. Vapnek
Manuscript Editor: Linda Purrington
Interior Designer: Marie Carluccio
Cover Designer: Sharon Helen Smith
Production Coordinator: Linda Jupiter
Illustration Coordinator: Audre W. Loverde
Compositor: Graphic Typesetting Service
Printer and Binder: The Maple-Vail Book Manufacturing Group

A number of the tables and figures in this book first appeared in
The Horse by J. W. Evans, A. Borton, H. F. Hintz, and L. D. Van Vleck.
W. H. Freeman and Company. Copyright © 1977.

Library of Congress Cataloging in Publication Data

Evans, James Warren, 1938–
    Horses : a guide to selection, care, and enjoyment.

    Bibliography: p.
    Includes index.
    1. Horses. I. Title.
SF285.E933   636.1   80–29070
ISBN 0–7167–1253–9

2 3 4 5 6 7 8 9 10   MP   0 8 9 8 7 6 5 4 3 2

Contents

Preface   xv

Introduction: Owning a Horse   1

    Ownership Obligations   1

    Ownership Costs   2

    Basic Facilities and Care   4

    Benefits of Ownership   5

PART I   **SELECTING AND BUYING A HORSE**   7

    1   **Considering Basic Factors**   9

        Seeking Help and Advice   9

        Locating a Horse to Buy   10

            Friends / Advertisements / Dealers and Trainers / Farms / Auction Sales

Cost Factors   12

Breed / Sex / Training / Age / Size / Color /
Conformation / Market Value

Gaits   14

Natural Gaits / Artificial Gaits

Common Defects in Gaits   23

Training   29

Age   30

Teeth and Age

Temperament and Behavior   38

Temperament Types / Agonistic Behavior /
Epimeletic and Et-Epimeletic Behaviors /
Investigative Behavior / Grooming Behavior /
Elimination and Ingestion / Play / Mimicry /
Learning / Sleeping and Waking /
Vices (Bad Habits)

Color   53

Head Marks / Leg Marks / Body Color

Sex   56

Pedigree   57

2   **Evaluating Conformation
and Soundness**   59

Conformation   59

Head / Neck / Chest / Withers / Shoulders /
Forelegs / Side View / Hindquarters

Blemishes and Unsoundnesses   83

Head / Shoulders

Leg Unsoundnesses   86

Forelegs / Feet / Hindlegs

General Unsoundnesses   98

Hernia / Heaves or Wind Broke / Roaring /
Thick Wind / Photosensitization

3 Choosing the Breed 101

Light Horse Breeds 102

American Quarter Horse / Arabian / Thoroughbred / Standardbred / Morgan / American Saddlebred / Tennessee Walking Horse / Gotland Horse / Racking Horse / Missouri Fox Trotting Horse / Morab / Paso Fino and Peruvian Paso / Galiceño / American Bashkir Curly

Registries with Color Requirements 121

American Albino / American Buckskin / Palomino / Paint, Pinto, and Spotted Horses / Appaloosa

"Native" Horses 128

Spanish-Barb / Chickasaw / Feral Horses in America

Draft Horses 131

Percheron / Suffolk / Clydesdale / Belgian / Shire

Heavy Harness or Coach Horses 137

Pony Breeds 138

Welsh or Welsh Mountain Pony / Connemara / Pony of the Americas / Shetland / Other Ponies

Miniature Horses 142

PART II EXERCISING OWNERSHIP RESPONSIBILITIES 143

4 Providing Physical Facilities 145

Planning 146

Identifying Needs

Codes and Regulations 147

Site Selection 147

Barns 149

Functional Design / Layout / Barn Exterior / Barn Interior / Portable Barns and Equipment

Fences 162

Breeding Facilities   163

Bedding   164

Fire Safety   165

5   **Feeding**   168

Anatomy and Functions
of the Digestive Tract   168

Common Feeds   173

Roughages / Grains and Grain By-Products /
Supplements

Nutrient Requirements   184

Water / Carbohydrates and Fats /
Maintenance and Special Requirements /
Protein / Vitamins / Minerals

Formulating and Balancing Rations   195

Rules of Thumb / Rations / Costs /
Commercial Feeds

Feeding Management   201

Group Feeding / Feeding Facilities / Feeding
Times / Age and Pregnancy / Ration Changes /
Overweight and Appetite / Other Factors

6   **Handling**   206

Catching and Haltering   206

Leading   208

Restraint   211

Tying / Hobbles / Twitches / War Bridles /
Chemicals

Safety   218

Actions / Clothes / Catching and Releasing /
Leading / Tying / Handling / Bridling / Saddling /
Mounting / Riding

Exercise   223

Exercise Methods

Conditioning   226
  Program / Evaluation of Condition and Stress
Transportation   229
  Trailers / Loading

7   **Grooming**   233
Basic Grooming   233
Baths   235
Blanketing   236
Trimming   237
Manes   238
Tails   240

8   **Caring for the Feet**   242
Hoof Handling   242
Daily Cleaning and Inspection   247
Trimming   248
Anatomy and Physiology   251
  Anatomy / Physiology
Shoeing   256
  Equipment / Shoes / Horseshoe Nails /
  Shoeing Procedure / Evaluation
Corrective Trimming and Shoeing   273
  Foals / Mature Horses

9   **Recognizing Illness
and Giving First Aid**   279
Recognizing Illness   279
  Respiration / Temperature / Heart Rate /
  Skin and Mucous Membranes / Excreta
Recognizing Lameness   282
Veterinary Assistance   284

First Aid   284

Cuts and Tears / Punctures, Burns,
and Abrasions / Wound Treatment—"Don'ts" /
Other Emergency Situations

Dental Care   298

10   Knowing About Parasites and Diseases   301

Parasites   301

Internal Parasites / External Parasites

Diseases   321

Common Diseases / Metabolic
and Other Diseases / Disease Prevention

PART III   USING AND ENJOYING THE HORSE   331

11   Riding and Driving   333

Riding Equipment   333

Western / Forward / Endurance and Competitive
Trail Riding / Bits and Bitting / Martingales /
Care, Storage, and Cleaning

Riding   355

Saddling / Bridling / Western Riding /
Forward Riding / Saddle Seat / Sidesaddle

Driving   371

Harnessing / Driving

12   Showing and Racing   375

Horse Shows   375

Halter Classes / Equitation Division /
Western Division / Jumper Division /
Hunter Division / Saddle Horse Division /
Roadster Division / Three-Day Events /
Gymkhana

Dressage   391

United States Dressage Foundation

Racing   398

> Types of Races / Supervision of Racing /
> Parimutuel Betting / Harness Racing /
> Quarter Horse Racing / Thoroughbred Racing

13   Rodeo, Trail, and Sports Events   411

Rodeo   411

> Rodeo Associations / Standard Rodeo Events

Cutting Horses   425

Fox Hunting   426

Social, Endurance, and Competitive
Trail Riding   427

Packing and Outfitting   430

> Equipment / Saddling / Packing

Polo   444

PART IV   BREEDING THE HORSE   447

14   Breeding the Stallion   449

Reproductive Tract   449

> Scrotum / Testes / Epididymis / Vas Deferens /
> Penis / Prepuce / Accessory Sex Glands

Semen Production and Characteristics   454

> Endocrinology / Characteristics

Semen Collection, Processing,
and Evaluation   455

> Collection / Care of AV Equipment /
> Processing / Evaluation

Breeding Soundness Evaluation   463

Infertility   465

Training Young Stallions   467

Castration   467

> Age / Procedure for Castration / Cryptorchids

15   Breeding the Mare   469

Reproductive Tract   469

Ovaries / Fallopian Tubes / Uterus / Cervix /
Vagina / Vulva / Mammary Gland

Physiology of Reproduction   475

Estrous Cycle / Behavioral Changes/
Hormonal Changes / Variability in the Estrous
Cycle / Foal Heat / Estrous Cycle Manipulation /
Records

Breeding   516

Mare Preparation / Stallion Preparation /
Breeding Area / Natural Breeding /
Artificial Insemination / Infertility Problems /
Embryo Transfer

Gestation   528

Length of Gestation Period / Hormonal Changes /
Placentation / Fetal Growth /
Pregnancy Examination / Abortions

16   Care During and After Pregnancy   538

Foaling   538

Pregnant Mare Care Immediately Before Foaling /
Signs of Approaching Parturition / Foaling /
Dystocia / Inducing Parturition /
Postnatal Foal and Mare Care

Foal Diseases   549

Orphan Foals   550

Weaning   551

Training the Foal   552

Congenital Defects in Foals   554

Malformations / Incompatible Blood Groups /
Combined Immunodeficiency Disease /
Other Immunity Failures

17   Understanding Basic Genetics   558

Basic Genetics   558

Genes and Chromosomes / Sex Determination /
Hybrid Crosses / Genetic Infertility /
Parentage Disputes / Coat Color Inheritance

Breeding Superior Horses   573

Qualitative Inheritance /
Quantitative Inheritance / Locating Animals
with Superior Genotypes / Linebreeding
and Inbreeding Relationships /
Stallion Selection to Complement the Mare

APPENDIXES   579

3-1   Addresses of Breed Associations and Registries,
Business Associations, and Magazines   581

5-1   Ration Formulation   587

12-1   Racing Tables   607

12-2   Glossary of Racing Terms   628

13-1   Glossary of Rodeo Terms   634

14-1   Semen Evaluation Procedures   636

15-1   Preparation of Cream-Gel
Semen Extender   646

15-2   Instructions for MIP-Test Kit
Pregnancy Determination   647

BIBLIOGRAPHY   649

INDEX   673

Preface

The horse has served humanity for centuries, first as a source of food and later as a vehicle of draft power. Today, the horse still serves as draft power in the underdeveloped regions of the world. In the technologically advanced nations, the horse serves primarily as a source of pleasure. This book was written to provide horsepeople with an introduction to the basic principles of horse care and management so that their enjoyment of the horse may be increased.

Aimed at the introductory equitation, horse-management, and horse-production courses at the undergraduate level, this book is also suitable for the general reader. The text assumes no knowledge of biology, chemistry, nutrition, physiology, medicine, or genetics; and every effort has been made to keep the technical vocabulary to a minimum and the exposition of concepts accessible to the lay reader.

The introduction to the book provides information on the moral and legal obligations, costs, and benefits of horse ownership. The book is organized around four central themes. The first concerns the selection and purchase of a horse. Guidance is given on locating a suitable horse and on considering basic factors, such as cost, age, temperament, color, and sex. The text offers biological bases for making purchase decisions. For example, an understanding of horse behavior can avert the purchase of a horse with bad habits and later can help prevent the horse from acquiring such habits. Evaluating conformation and detecting unsoundnesses, which affect performance, form the substance of the second chapter in this part. Common unsoundnesses and blemishes are defined, described, and related to a horse's serviceability. The concluding chapter of this part details the numerous breeds and their characteristics by categories such as light horse, draft horse, and pony.

The second part, on ownership responsibilities, discusses physical facilities, day-to-day tasks, such as feeding, handling, exercise, grooming, foot care, first aid, parasite control, disease prevention, and veterinary aid. The proper planning and construction of the stable site, barns, fences, breeding facilities, and bedding are outlined. The important topic of fire safety is discussed in the context of physical facilities. The related issue of accident prevention is paramount in the discussion of proper methods of restraint, routine handling, and basic training.

Using and enjoying the horse comprise the third major theme. This part encompasses riding and driving, horse shows and races, rodeo, trail, and other sports events. Basic procedures and equipment for riding, driving, racing, and other equestrian activities are described in detail.

The final section of the book concerns reproduction and genetics. One chapter is devoted to breeding the stallion, and another to breeding the mare. The material on reproduction is unique in that step-by-step details are provided for successfully breeding and foaling the mare. Pregnancy care and various problems conerning foals are discussed in a separate chapter. The part concludes with a chapter on basic genetic principles for selecting and breeding superior horses.

For the more serious student of horse husbandry, numerous appendixes contain detailed, technical information on ration formulation, semen evaluation procedures, and pregnancy determination. There are also glossaries of racing and rodeo terms and addresses of breed associations and horse magazines. An extensive bibliography, organized by chapter, concludes the book.

If the reader increases his or her enjoyment of the horse and if the general care of horses is improved, I will have achieved my objectives in writing this book.

March 1981   *J. Warren Evans*

HORSES

A Guide to Selection, Care, and Enjoyment

Introduction: Owning a Horse

We humans have been closely associated with the horse since its domestication in the Near East between 4500 and 2500 B.C. Even before domestication, the horse was a major source of food for us. By 1000 B.C., domestication had spread to Europe, Asia, and North Africa. The horse was first used as a draft animal in the Near East. After the development of machinery powered by fossil fuels, the horse was no longer used as a source of power in economically developed countries. In America, the horse population declined in the early 1900s, but horses regained their popularity in the late 1960s. Horses are now mainly kept for recreational purposes. Because few people had contact with horses between 1900 and 1960, horse management skills were not passed to the next generation. Today, many families are having their first experiences with a horse.

OWNERSHIP OBLIGATIONS

Uninitiated horse owners, lured by the horse's esthetic and athletic appeal, soon realize that owning a horse is not like the romantic stories about horses that they read as children. Owning a horse involves a considerable amount of time, dedication, and hard work. Owners must be prepared to spend more time caring for the horse than riding it. A horse requires care every day of the year when it is kept confined. It must be fed, watered, groomed, have its feet cleaned, and exercised. The owner must muck out and clean stalls daily and must take care of the tack and equipment.

OWNERSHIP COSTS

In addition to the labor responsibilities, the horse owner must be financially prepared to take care of the financial obligations of horse ownership. Before purchasing a horse, the prospective owner must consider economic facts. The initial cost of buying a horse depends on several factors. The main factors determining the horse's value are its training, age, breed, and temperament. Other factors, such as color, conformation, manners, and soundness, also affect decisions. All these factors are discussed at length in Chapters 1 to 3 (Part I, "Selecting and Buying a Horse").

Under confinement conditions, a horse consumes about 20 pounds of hay (9.07 kg) per day or approximately 3½ tons (3,175 kg) per year. Depending on the horse and its intended use, you may need to feed grain, as well. You must bed the horse's stall with straw, wood shavings, or another suitable material. Bedding needs to be completely changed at least once a week. You will need the services of a farrier every 6 to 8 weeks to trim or shoe the feet. You will need veterinary services when an injury occurs, and you will need to carry out a preventive medicine program to prevent common diseases and to control internal parasites. Basic, necessary equipment for taking care of the horse must be periodically replaced. Tack is a necessary expense, as are appropriate clothes for riding and caring for the horse. If the horse is to be transported for trail rides or horse shows, you will need a trailer and a car or truck to tow the trailer. Special equipment for towing the trailer such as trailer hitches and mirrors are added expenses. If the horse requires training, you must hire a trainer or some additional help to complete the training process.

Owning horses entails certain risks and possible extensive liability. Therefore, people involved with horses need added insurance. (Many horsepeople fail to recognize the need for insurance or may have been misled to believe that additional insurance is not necessary.)

A limited number of insurance companies offer horse mortality insurance. Large capital investments make it especially prudent to buy such insurance protection. You should discuss with the underwriter the conditions governing such policies. Humane destruction is sometimes necessary, but you must adhere to strict procedures to collect on the policy. The humane factor is the only consideration: A horse may not be destroyed because of injuries that merely prevent the performance function for which the horse is kept. Permission from the underwriter must be obtained, and this is usually given on the advice of a veterinarian.

In addition to the full mortality coverage, some companies offer a more limited peril policy covering fire, lightning, and transportation losses. The peril policy covers only loss from collision and/or overturn of the transporting vehicle. Losses are not covered for loading or unloading, van fits (horse going berserk), or injuries received during sudden braking of the vehicle to prevent an accident. (To cover the losses just described, a special accident policy must be obtained.)

Liability insurance must be purchased to protect the horseperson from potential liability for injury to people and damage to property. There are some basic misconceptions about liability coverage for horses. The standard homeowner's or tenant's policy provides coverage only for a pleasure horse ridden on the owner's property, adjacent to the owner's property, or in an area designated for pleasure riding. The homeowner's policy does not provide protection once business pursuits begin. Business pursuits begin once a horse owner does anything other than pleasure riding, such as taking a horse to a public horse show and showing it, using the horse in a parade, renting the horse to someone else, or participating in a gymkhana or rodeo. This also means that the homeowner's policy does not cover race horses, brood mares, and breeding stallions (including shares in syndicated stallions). You need a special, private, saddle animal liability policy to provide liability coverage for off-premises events. Professional horsepeople engaged in the business of teaching people to ride, training horses, and boarding or breeding horses need a comprehensive general liability policy as well as an umbrella policy. The umbrella policy is an excess limit policy that picks up not only the horseperson's liability exposures as a business but also the vehicles that the person owns. The comprehensive general liability and umbrella policies have an exclusion clause that leaves one basic area that the professional horseperson must attend to: Any animal in your care, custody, and control is excluded. Therefore, the professional needs a horse legal liability policy to cover horses that are the property of others and are in the insured's care, custody, and control for breeding, boarding, training, showing, or trucking purposes. This policy pays for all legal obligations arising out of the insured's legal liability for all risks of physical loss, injury, or humane destruction necessary to the horse in the care, custody, and control of the insured.

The professional horseperson who has farm and ranch employees may be subject to worker's compensation law requirements and may have to carry insurance to meet any liability covered by the law. Liability under worker's compensation law is without regard to negligence, and the employer must pay according to the law whether or not he or she was at fault.

There are also other forms of insurance. Most have to do with the reproductive status of a mare or stallion. Fertility insurance on stallions has become quite common with the advent of high price syndications. In utero policies are very popular with purchasers of mares in foal. Almost all breeding contracts state that the breeding fee is earned in full once the original owner who bred the mare has sold it. The purchase price of a mare in foal is determined in part by the foal in her, and the purchaser is interested in protecting the investment. There are two basic conditions of in utero insurance: The policy does not indemnify the owner if abortion occurs as the result of multiple fetuses or if a fetus is not in evidence for an autopsy report.

BASIC FACILITIES AND CARE

A horse must be stabled. Various types of facilities can be used to keep the horse under confinement conditions. Under most circumstances, a horse will be comfortable in an outside paddock that has a shelter to provide shade in hot environments or protection from wind in very cold climates. Details of stall construction and design are discussed in Chapter 4. Space must be available for tack and feed storage. An important consideration is adequate facilities for exercise: riding or exercising rings, or access to riding trails.

Many owners prefer to keep their horses at commercial stables. These stables range from a rented stall to elaborate equestrian complexes. When you select a commercial stable, you should consider several factors. Safety and cleanliness are very important. A clean stable is part of the disease and parasite control program. Manure should be removed from the paddocks and stalls at regular intervals (at least weekly) and stored in a compost pile. Removing trash and garbage helps control flies that carry diseases and parasites. No smoking should be permitted in the stable area, because it creates a fire hazard. Dogs should not be permitted unless the horses are accustomed to their presence. Equipment storage areas should be enclosed so that horses will not be injured by running into or stepping on the equipment. Fences should be safe for horses and kept in good repair. The stable area should be fenced to contain horses that accidentally get loose from owners. This can prevent accidents caused by the horses running out onto highways or streets and can prevent damage to neighboring property. Gates should open and close easily but should be constructed so that horses cannot open them.

Make inquiries about the disease and parasite control program for the stable. Are all horses required to have the standard immunizations? Are flies controlled? Are stalls disinfected before a new horse is brought in? Is there a quarantine area in case a horse becomes sick?

Also examine the barn and other buildings for their general condition and maintenance. Is the electric wiring protected from horses and rodents chewing on them? Are the buildings clean, and have fire safety precautions been taken? Stalls should be the appropriate size for your horse and properly maintained. A source of clean water is essential. Clean and properly designed feed mangers are important for the horse's safety and to prevent digestive upsets. A good, absorbent bedding material helps maintain a clean, dry stall.

The nutritional program is important to the well-being of the horse. The hays and grains fed to the horse must be of good quality. They should be stored in a clean area to which the horse cannot gain access. If the horse is kept at pasture, the pasture should be well maintained to provide quality forage. Overcrowded pastures decrease their quality and may be unsafe for your horse. Good stable managers keep horses that do not get along together in different pastures.

A well-managed and -designed stable provides appropriate facilities to ride in, to exercise the horses in, and to care for the horses. Areas to exercise horses during inclement weather are useful during certain periods of the year. Riding rings and access to riding trails should be conveniently located. The tack room should be secure and dry and have adequate room for storage of your saddles and riding equipment. Rest rooms, drinking fountains, and telephones are convenient for riders.

Evaluate the knowledge and attitudes of the owners, boarders, and employees. Is the owner a knowledgeable horseperson? What is his or her reputation with the local farriers, feed store managers, tack shop owners, and local horsepeople? Is he or she a good businessperson? What is the attitude of the stable employees toward the boarders? Are there enough employees working at the stable? What is the behavior of the boarders? Do they drink alcoholic beverages on the premises and use foul language? Are the boarders good horsepeople or do they abuse their horses? Is a trainer available to help you train your horse or to improve your skills with horses? If so, does the trainer have a good reputation? A veterinarian should be on call to the stable if his or her services are needed.

Examination of the horse often reveals if it is well cared for. It should not be too fat or too thin. Its coat should be in good condition and well groomed, and its feet should be properly trimmed or shod at all times.

BENEFITS OF OWNERSHIP

Horse owners and communities benefit from horses in several ways. Ownership and enjoyment of horses are not the exclusive privilege of the wealthy. In fact, 60 percent of all horse owners in the United States are among families with a below-average income.

Dealing with horses helps children develop a sense of responsibility. The urbanization and mechanization of agriculture has diminished the need for children to perform farm-oriented chores or to take care of animals. Taking care of, exercising, and riding one's own horse can play an important role in a child's training. Children also develop the discipline necessary to work with others by participating in mounted drill programs, horse-judging team contests, and U.S. Pony Club team competitions.

Riding is a sport in which the entire family can and often does participate together. A survey has found that there are 2.9 riders per horse-owning family and that 80 percent of families that own horses have at least two people who ride. About 40 percent of horse-owning families own four or more horses. Trail rides frequently attract several members of the same family. Father-and-son teams in team roping competition are quite popular. Riding horses provides relaxation and offers psychological benefits. And riding horses indirectly benefits those who never ride because horses help develop new trails and parklands used by backpackers and others.

The horse owners play a role in the economics of our nation by contributing to the tax base of the community, state, and nation. Taxes collected as a result of the horse industry amount to billions of dollars. Most of the obvious tax revenue is generated from parimutuel wagering and from real estate taxes on property devoted to the horse industry. The hidden sources are just as important. Taxes collected from the wages and salaries of people directly employed by the horse industry or by allied industries are important. Sales of horse-related clothing, tack, equipment, and so on are a source of tax revenue. Real estate taxes paid on land occupied by feed and equipment manufacturers and retailers, as well as transportation taxes on feed, equipment, and horses themselves, further contribute to the economics of the horse industry.

Riding horses improves the rider's physical fitness. Because the improvement in physical fitness is painless, riders seldom realize that it is occurring. Unlike jogging, horseback riding is seldom boring. Thus the horse, quietly but efficiently, helps solve one of the most challenging problems we face today—the deterioration of physical fitness.

Handicapped people can now enjoy riding horses. A horse can serve as an emotional outlet for them, and they gain a feeling of freedom from restraints imposed by their handicaps. Many handicapped children find it hard to relate to adults and other normal children, but they do not find it hard to relate to a pony. The pony or horse provides an exciting new and different learning situation for them.

Riding by the physically handicapped on horses is now an accepted form of physical therapy to rehabilitate damaged muscles. Equestrian therapy for the handicapped is developing all over the world. Starting in Europe, riding programs for the disabled have spread to Australia, New Zealand, Rhodesia, the United States, Canada, Bermuda, and South Africa.

Horseback riding has also become a recognized form of therapy for regaining mental health. Some benefits derived by patients are the conquering of fear and replacing it with self-confidence. Riding develops affection and companionship and a sense of mastery over a living creature. It often helps overcome a feeling of inferiority and requires the patient to make decisions and act on them. Riding improves their coordination and rhythm while providing a relaxing and stimulating form of exercise that creates a new interest.

I
SELECTING
AND BUYING
A HORSE

1
Considering
Basic Factors

SEEKING HELP AND ADVICE

The novice or less experienced horseperson buying a horse can get help from a variety of people in buying a suitable horse. Most novices consult a friend who is a horseperson before they talk to anyone else about buying a horse. The friend should be an accomplished horseperson, or the advice given may be based on myths and "folktales." Local 4-H Club horse leaders are an excellent source of information and help and have often already helped new club members buy horses for projects. Most of the horses they have to find for their members are of the type suitable for novices. Some farriers are excellent horsepeople and sources of advice, while others only know how to shoe a horse. A local farrier with a good reputation may help in locating a horse, because he or she is in constant contact with local horse owners. Members of local horse clubs may be willing to help a novice find a first, suitable horse. Ranchers can also help locate a horse, because most keep horses. A local veterinarian who specializes in horses can also be of help and should evaluate the horse for soundness and health. Professional help can be obtained from Cooperative Extension horse specialists, college animal science department faculty members who teach horse-related courses, and professional horse trainers.

The novice may want a trainer to work with a horse to correct deficiencies in its training or to help improve the novice's equitation or to prevent gross mistakes in training or riding. Disappointment usually awaits beginners who expect to become accomplished horsepeople without experienced horses and without the aid of a professional trainer. If novices want to break and train their own horses, they need a professional to watch and to direct their efforts. Failure to do so usually results in injury to the novice or in an improperly trained horse. The trainer should be carefully selected. Most trainers are specialists—they usually train horses for specific types of performance. Select a trainer who specializes in the type of horse you want. The trainer should be the most reputable, well-known trainer you can afford. One who has had years of training and riding experience is usually the most desirable. Previous clients are the best references to check the trainer's reputation. Observe the trainers in action with their current students or horses. They should be kind, considerate, and in command of the situation at all times.

Financial arrangements should be discussed before you hire the trainer. If the trainer is to train the horse or help school the horse and rider, agree on a monthly rate. The rate reflects the trainer's talents and ability to train horses and/or riders as well as his or her investment in facilities and equipment. There may be extra charges for grooming and for special training equipment. Some trainers require that the horse or rider be under their supervision for a minimum period of time. You should inquire as to how long each training session is and how regularly the sessions are given. You should also be able to observe the training sessions for the horse, so that you will know the hand and leg aids used to ride the horse. It is better if the owner is trained at the same time as the horse so that the horse is not confused when the owner starts to ride it. The trainer should limit the number of horses being trained. Horses seem to learn best from short (20- to 30-minute), intense training sessions 6 days a week. An assistant trainer may do most of the riding but the trainer should be riding and observing the horse at regular intervals. You should discuss the feeding program for the trainer's stable. The general condition and care of the horses in his or her custody should be evaluated. You should also discuss financial obligations for farriers and veterinary services as well as for expenses encountered if the trainer is to show the horse.

The trainer may want to inquire about the horse owner before agreeing to train a horse, select a horse to buy, or give the owner riding lessons.

LOCATING A HORSE TO BUY

Horses and ponies can be purchased by a variety of methods; each has its advantages and disadvantages. Regardless of the way in which a horse is bought and sold, the buyer must be aware that dishonest people can calm horses and mask lameness with drugs. Also, many horses are sold by word of mouth, and facts about the horse become distorted and inaccurate after they have been passed through several people. However, information that comes from someone who is familiar with the horse can be useful.

Friends

Friends can be a source of horses. But purchasing a horse from a friend can often strain relationships, because horses change their behavior in new surroundings and when they are ridden and cared for by new people.

Advertisements

Classified ads in newspapers or horse magazines can be an aid in locating a horse. These horses usually have some type of problem that makes the horse difficult to sell to people who are familiar with it. Problems are indicated by phrases such as "gentle but spirited" or "for experienced rider."

Dealers and Trainers

Local horse dealers keep a variety of horses. Most of the horses have some fault that the dealer can hide or has corrected so that he or she can sell the horse. Some horse dealers have the reputation of not being very honest. Honest horse dealers try to match a horse to the purchaser's needs and experience. A major problem horse dealers encounter is dishonesty by the person buying a horse. They try to impress the dealer or trainer by pretending to be much more advanced horsepeople than they really are. The end result is usually disappointment, because the rider is overmounted. Many professional trainers keep horses that are for sale. They may belong to the trainer or to the trainer's clients. Horses should be purchased only from the trainers who have a reputation for honesty. Trainers will make an effort to find an appropriate horse for someone who will take riding lessons from them and who will want the trainer to school the horse. They do expect to be compensated for their time and effort. A favorite mount belonging to the stable where you are receiving instruction can be a safe purchase, because you are familiar with the horse.

Farms

Breeding farms are good places to look for horses. A breeding farm offers the opportunity to look at horses typical of the breed desired. Breeding farms must sell horses that meet clients' needs, or they will have difficulty staying in business. Most breeders are interested in promoting their breeds and want to be sure that a client is satisfied with the horse. Satisfied clients are their best means of advertisement.

Auction Sales

Auction sales can result in bargains or headaches. There are several kinds of auctions: local weekly, breed association, breeder, and specialty performance. Generally speaking, it takes an experienced horseperson to evaluate a horse being sold at auction. There is something wrong with most horses sold at auction, particularly the local weekly or monthly auction for all kinds of horses. Breed association auctions tend to be more reputable, and at some auctions only select horses for a given breed are sold. An auction where a breeder sells horses once a year can be a good source of horses if the breeder has a reputation for quality. Some auctions specialize in specific types, such as race, cutting, roping, or polo horses. Quality Thoroughbred yearlings are sold at international markets such as those held at Keeneland, Saratoga, Del Mar, Hialeah, Deauville (France), Ballsbridge (Ireland), and Newmarket (England). International markets are also held for other breeds, such as the Arabian, Standardbred, and Quarter Horse.

COST FACTORS

Many factors affect a horse's price. Before the selection process starts, the potential buyers must decide how much money that they can afford to pay for a horse and then try to find a suitable horse for that price. Be realistic in setting standards for a certain amount of money. The subsequent discussion on selection details many of these factors. It is usually advisable to buy the best horse one can afford, because the costs of keeping a horse are the same regardless of its quality and value.

Breed

The breed and breeding of the horse influences its value. Grade horses are suitable for many purposes and are usually sold for less money than horses registered with a breed association. This is not always the situation, however, because some performance contests do not require a horse to be registered. Some grade horses are better performers and command the most money for that particular type of horse. Horses of fashionable, popular, or proven families within a breed registry usually command higher prices. Some of these may be of lower quality than other horses selling for the same price. The value of certain breeds varies depending on the locality. This is particularly true for breeds such as the Tennessee Walking Horse, Thoroughbred, and Quarter Horse. (Breeds are discussed more fully in Chapter 3.)

Sex

Sex of the horse influences the cost. A gelding, the most popular horse for riding and most performance contests, is generally more dependable and costs less than a mare or stallion. The better bred the horse is, the greater the price spread between the sexes. Well-bred stallions suitable for stud duty are quite expensive. Mares that will eventually be bred are worth more than geldings.

Training

Some of the most significant influences on price are the level and type of training and the horse's accomplishments. The horse's athletic ability increases its value. Many horses command very high prices because of their past or potential performances or training for certain types of performance.

Age

The horse's age is a prime factor. Young, untrained horses sell for less because of the future investment in training and time that the owner must bear. There is always the possibility of injury or failure to reach the desired level of training. Horses that are 5 to 10 years old are usually higher priced than those younger or older. At this age, the horse is at its peak in training and performance. The younger horse requires more delicate and skillful handling than the older, more experienced horse. Therefore, the less experienced the horse, the more experienced the rider needs to be. An inexperienced rider can gain much experience by working with an older, well-trained horse.

Size

Size influences price. Ponies usually are cheaper because their utility will be limited to a few years before the child becomes too large (unless the family has several children). Small horses are priced about the same as the average-size horses unless they are smaller than the minimal acceptable standards for the breed. Horses are measured in hands. A hand is equivalent to 4 inches (10.2 cm). The point of measurement is the distance from the highest point of the withers to the ground. Therefore a 15-hand horse is 60 inches high (152.4 cm) at the withers.

Color

Some horsepeople are interested in certain colors and will pay a premium for color. These horses are used for show, parade, or breeding purposes. If you will accept only a certain color for a specific purpose, the number of horses available for sale that meet your criteria will usually be very limited.

Conformation

Conformation of the horse and its soundness affect its value. Well-conformed horses almost always sell for more money compared with equivalently trained horses. Horses that are unsound for specific uses have a lower utility value. Blemishes detract from a horse's appearance and may decrease the value of horses used for show purposes. (Conformation and soundness are discussed fully in Chapter 2.)

Market Value

To help buyers form an opinion on value or costs for horses sold at auction, some breed associations and magazines publish the selling prices of the horses sold at various auctions. By consulting these sources and attending a few auctions, you can determine the current market values for certain types of horses.

GAITS

A gait is the horse's manner of moving its legs during progression. As a species, the horse is more versatile in selecting gaits than any other quadruped and uses several gaits unique to the species.

An understanding of gaits is important if one is to be able to detect lameness, to train a performance horse, or to cue a horse for a specific purpose. There are several gaits that a horse can use to move. Some are natural, while others are "artificial or learned" gaits. The walk, trot, canter, and run seem to be natural gaits for most horses. The other gaits are artificial gaits—most horses must be trained to execute them.

Several terms and concepts are used to describe the various gaits. The left side of the horse is the *near side,* and the right side is the *off side.* Today, these two terms are not used as commonly as earlier. A foot (or two feet simultaneously) striking the ground constitutes a *beat.* The beats may be evenly spaced in time, or they may not be. The *support sequence* is the sequence of the various combinations of feet that are touching the ground. A *step* is the distance between imprints of the two forelegs or two hindlegs. A *stride* is the distance between successive imprints of the same foot. During every stride, each leg goes through two phases: (1) the *stance* or *weight-bearing phase* and (2) the *swing or non-weight-bearing phase.* When horses are racing as trotters, pacers, or runners, the time required for the swing phase is almost constant, at about 0.32 seconds. Time for the swing phase appears to be independent of how fast the horse is racing; therefore, it takes about the same amount of time to prepare a leg for the weight-bearing phase regardless of speed or gait. Maintaining a constant swing during a race is important in keeping the horse sound. If the horse is unable to prepare its leg to begin the weight-bearing phase within the normal swing time, its leg structures are exposed to excessive concussive forces. When the leg is properly prepared for striking the ground, it is retracted so that it strikes the ground at a relative forward speed of zero. If the necessary swing time is longer, the leg must strike the ground before being properly prepared and traveling at zero speed relative to the ground. The foot is thus pounded into the ground with excessive shearing force. This predisposes the horse to many pathological problems resulting from excess concussion, trauma, and strain. A horse's ability to maintain the swing time for a leg is affected by the amount of time that all legs are off the ground (airborne phase), the stance phase for all legs, and the overlap phase (when two or more feet are on the ground simultaneously). In examining the strides (Figure 1-1) of Secretariat (1973 Triple Crown winner of Thoroughbred racing) and Riva Ridge (1972 Kentucky Derby and Preakness Stakes winner) during the running of the 1973 Marlboro Cup, in which both horses broke the world record for 1⅛ miles (1.8 km), we can determmine the effect of the overlap and stance phases. Secretariat ran faster because he spent less time with his legs in the stance and overlap phases; that is, his legs completed their ground contact quicker and more time was spent in the airborne phase.

Natural Gaits

There are six natural gaits of the horse: walk, trot, pace, canter, run, and back. Other gaits are said to be artificial.

Walk

There are several forms of the walk, but all show an even four-beat gait (Figure 1-2). The sequence of hoof beats is (1) left hind, (2) left fore, (3) right hind, and (4) right fore. Therefore, the sequence of beats is lateral, in that both feet on one side strike the ground before the feet on the opposite side strike the ground.

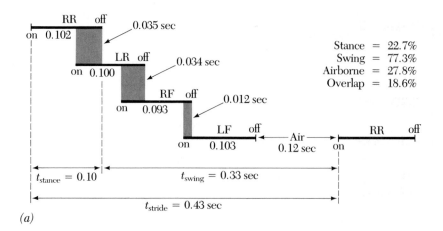

(a)

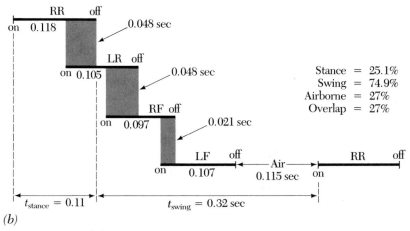

(b)

FIGURE 1-1

Comparison of the stride timing of *(a)* Secretariat and *(b)* Riva Ridge, showing the times at which each leg is placed on the ground or removed from the ground, the duration of the corresponding stance phases, the swing times, and the stride times. Read the succession of footfalls from left to right. The shaded regions show leg overlap. The percentages refer to percentage of the total stride time. (From Pratt and O'Conner, 1978.)

FIGURE 1-2 *(facing page)*

Sequence of beats for the gaits. Numbers 1–16 indicate the rhythm. LH = left hind. RH = right hind. LF = left fore. RF = right fore.

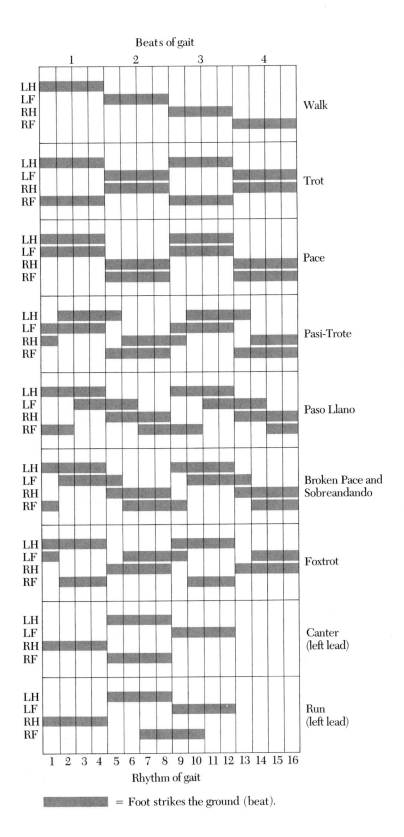

Beats of gait

1   2   3   4

Walk

Trot

Pace

Pasi-Trote

Paso Llano

Broken Pace and
Sobreandando

Foxtrot

Canter
(left lead)

Run
(left lead)

1 2 3 4 5 6 7 8 9 10 11 12 13 14 15 16

Rhythm of gait

= Foot strikes the ground (beat).

Trot

The trot is a two-beat gait in which the paired diagonal feet strike and leave the ground simultaneously (Figure 1-2). Between each beat, there is a period of suspension in which all four feet are off the ground. There are three classifications of the trot. An *ordinary* trot serves as the basis of comparison. An *extended* trot is when the horse is trotting at speed and the length of the stride is extended. A racing Standardbred typifies the extended trot, although western pleasure and dressage horses are also expected to be able to extend their trot in the show ring. A *collected* trot is when the horse slows down and uses extreme flexion of the knees and hocks. The Hackney often characterizes the collected trot.

Pace

The pace is a two-beat gait in which the lateral limbs strike the ground simultaneously (Figure 1-2). There is a lateral base of support, and a period of suspension with all four feet off the ground occurs between each beat. Because the horse is shifting its weight from side to side, there is a rolling motion to the gait.

Canter or Lope

This gait is a three-beat gait in which the first and third beats are made by two legs striking the ground independently and the second beat is made by two limbs striking the ground simultaneously (Figure 1-2). The legs that strike the ground independently are called the *lead limbs,* and each bears the entire weight of the horse for a short period of time. Therefore, the lead limbs are more subject to fatigue than the other two legs. In the left lead, the sequence of beats is (1) right hindleg, (2) left hindleg and right foreleg, and (3) left foreleg. A period of suspension follows the beat of the left foreleg. If the horse is being circled to the left, it should be in the left lead to maintain its balance. If turned to the right, the horse should change leads to the right lead. The changes of leads occur during the period of suspension (Figure 1-3). When changing to the right lead from the left lead, the left hindleg strikes the ground after the suspension period. The paired diagonals—right hindleg and left foreleg—are the second beat, and the right foreleg is the third beat. Frequently, untrained horses switch leads behind but fail to do so in front. This is called a *cross-legged canter,* and it is very rough to ride.

Horses change leads at will when they are not under saddle. They must be trained to start in a desired lead and to change leads at the appropriate time. They are signaled to take or change leads by specific cues (Figure 1-4).

FIGURE 1-3

Correct change of leads in a canter. In the first sequence, the horse is in a right (off) foreleg lead with a left (near) hindleg lead. At S, the horse is in a period of suspension and properly changes leads at this time, coming down with right (off) hindleg lead and a left (near) foreleg lead, as shown by the second sequence of hoofbeats. (From Adams, 1974.)

To make a horse take the left lead from the trot while moving to the left, use your right leg to place pressure behind the girth, which should shift the hindquarters to the left. Turn the horse's head slightly to the left with direct action of the left rein. Increasing right leg pressure should then cue the horse into the left lead. The cues for taking the right lead while moving to the right are opposite to those described for the left lead.

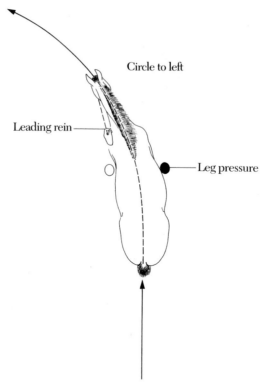

FIGURE 1-4

Cues to taking a left lead. Note the natural bend in the horse's body. (From *Western Horse Behavior and Training*, by R. W. Miller. Copyright © 1974 by R. W. Miller. Reprinted by permission of Doubleday and Company, Inc.)

Once the horse has learned to take the correct leads while moving in the circle, it should learn to take the leads on demand from the walk or trot while traveling straight forward. This requires that the horse first be trained to canter easily from a walk and two-track. To cue the horse for a right lead, start a two-track to the right, maintain left leg pressure and pressure by the left direct rein, and cue for a lope or canter (Figure 1-5). After the horse takes the leads on demand, it can be taught the flying change of leads. Teach the horse to make simple lead changes by slowing from the canter to a trot for a couple of strides and cueing it for a change of leads (Figure 1-6). After it executes these changes, teach the flying change by cantering in a large circle and giving the cues at the point where you want the lead change to occur. The cues for a change to the right lead are a shift of the rider's weight slightly forward of

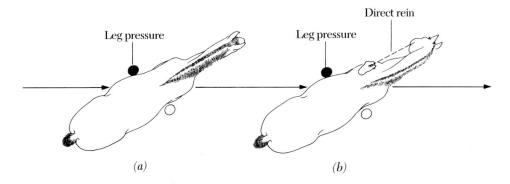

FIGURE 1-5

Taking a right lead on demand. *(a)* Step 1: Two-track to right. *(b)* Step 2: Cue for lope. (From *Western Horse Behavior and Training*, by R. W. Miller. Copyright © 1974 by R. W. Miller. Reprinted by permission of Doubleday and Company, Inc.)

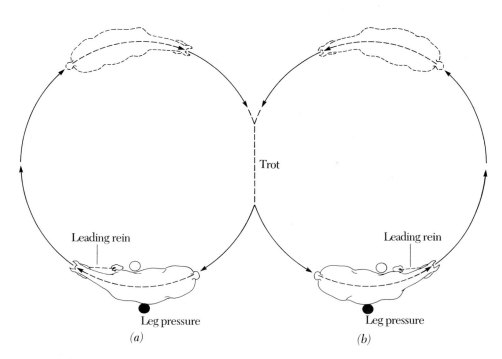

FIGURE 1-6

Cues for change of leads. *(a)* Lope: Right lead. *(b)* Lope: Left lead. Lope large circles, then drop to a trot to change directions and leads. Note use of leading rein to give the body of the horse a natural bend for the lead. Outside leg pressure encourages the horse into the correct lead. Vary the number of circles from one to several before changing direction. (From *Western Horse Behavior and Training*, by R. W. Miller. Copyright © 1974 by R. W. Miller. Reprinted by permission of Doubleday and Company, Inc.)

the horse's center of gravity, a squeeze with the left leg, and reining the horse to the right.

Run

The run is a four-beat gait (Figure 1-2). The gait is similar to the canter except that the paired diagonals do not land at the same time. The hindleg hits just before the foreleg. The lead limbs bear the full weight of the horse. In the left lead, the sequence of beats is (1) right hindleg, (2) left hindleg, (3) right foreleg, and (4) left foreleg. A period of suspension follows the four beats. If the horse changes leads, it will do so during the period of suspension. Because the legs bearing the weight of the horse become fatigued, the horse should change leads periodically when it is running. On the racetrack, the horse should change to the left lead as it goes around the turn and should change to the right lead along the straight portions to prevent excessive fatigue to the two lead legs.

Back

A horse backs by trotting in reverse, using a two-beat gait in which the diagonal pairs of legs work together.

Artificial Gaits
Flat-Foot Walk

The ordinary walk is the most common form of the walk and is used by all horses. The flat-foot walk is the natural walk of the Tennessee Walking Horse. It is slightly faster than the ordinary walk.

Running Walk

The running walk is the fast walk of the Tennessee Walking Horse. It is faster than the ordinary or flat-foot walk. The hindfoot also oversteps the hoof print of the forefoot by as much as 50 inches (1.27 m) but normally about 12 to 18 inches (30 to 46 cm). The horse travels with a gliding motion, because it extends its hindleg forward to overstep the forefoot print. However, the regular four-beat rhythm is maintained. When executing the gait, the head moves up and down in rhythm with the legs. If the horse is relaxed while performing the gait, its ears flop and its teeth click in rhythm with the legs. The gait is easy to ride and not strenuous for the horse.

Rack

The rack is an even four-beat gait but is executed faster than the walk or running walk. The forelegs are brought upward to produce a flashy effect. The hind feet do not overreach the prints of the forefeet as far as compared with the running walk. In the show ring, the rack is popular because of the speed and animation, but it is difficult for the horse to perform for extended periods. The excessive leg movement increases the amount of concussion and trauma to the forelegs.

Trot Variations

There are two variations of the trot (two-beat sequence), each with four irregular beats. The foxtrot is one of the accepted slow gaits for a five-gaited horse. It is similar to the trot except the hindfoot strikes the ground just before the paired diagonal forefoot. The paired diagonal feet leave the ground at the same time. When executing the *pasi-trote*, the Peruvian Paso horse allows the forefoot to strike the ground before the paired diagonal hindleg.

Pace Variations

A variation of the pace is the "stepping" or "slow" pace, which is the preferred slow gait for five-gaited show horses (Figure 1-2). It is the same as the *sobreandando* performed by Peruvian Pasos. The hindleg strikes the ground before the lateral foreleg so that the gait has four irregular-rhythm beats. In the takeoff, both legs leave the ground simultaneously.

The *paso llano* gait is also a broken pace, but the time between the hindfeet and the lateral forefeet striking the ground is longer than the sobreandando (Figure 1-2).

COMMON DEFECTS IN GAIT

There are several defects in the way that a horse may move its feet and/or legs while executing the gaits. Some defects are responsible for limb interference and may be severe enough to cause injury. Other defects are not serious but prevent the horse from giving its top performance. Some defects in the way of going are related to conformation, while others are related to injuries or to improper shoeing and trimming of the feet.

Cross firing occurs when the inside of the hindfoot strikes the diagonal foreleg (Figure 1-7). It is generally seen in pacers that have long backs. A long

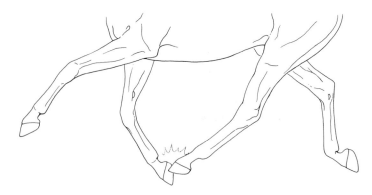

FIGURE 1-7
Cross firing: hindleg strikes opposite foreleg.

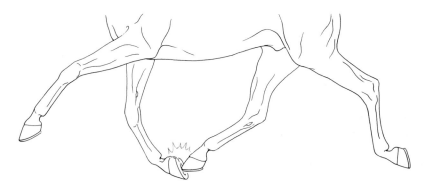

FIGURE 1-8
Forging: toe of hindfoot strikes sole of forefoot.

back causes the horse to twist or swing its hindquarters. As the hindleg moves forward, it swings inward and strikes the diagonal foreleg, which is extended back under the horse's body.

Forging occurs when the toe of the hindfoot strikes the sole area of the forefoot on the same side (Figure 1-8). It usually occurs as the forefoot is leaving the ground. Overreaching is similar to forging except the hindfoot comes forward more quickly and hits the heel of the forefoot before the forefoot leaves the ground. Both faults occur because the hindfoot breaks over and moves forward too soon relative to the forefoot breaking over and leaving the ground (Figure 1-9). In many instances, forging and overreaching can be corrected by speeding up the breaking over and forward movement of the forefeet

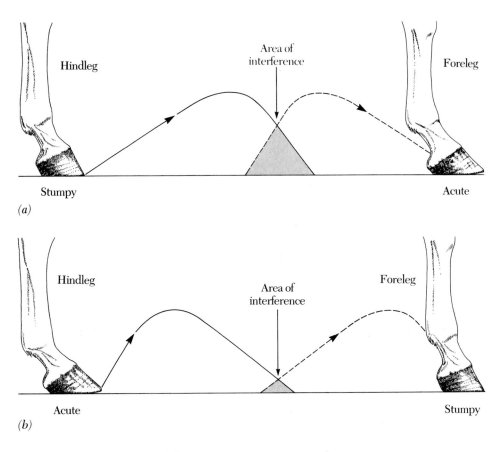

FIGURE 1-9
Corrective trimming for forging. *(a)* Incorrect. *(b)* Correct.

and retarding the hindfeet. This is usually accomplished by leaving the heels a little longer on the forefeet (increasing the angle of the hoof wall) and shortening the heels of the hindfeet. This changes the foot flight arc (Figure 1-10) and decreases the possibility of forging and overreaching. If necessary, the toes of the forefeet can be rolled to increase the ease of the forefeet breaking over (Figure 1-11). Also, trailers on the hindfeet will produce drag when the foot hits the ground and shorten the stride. Using heavier shoes on the forefeet than on the hindfeet carries the forefeet further forward because of inertia and increases distance between the feet.

Interference occurs when one foreleg strikes the opposite foreleg. If the contact is very slight, it is called *brushing*. Interference is usually the result of poor conformation—toe-wide or base-narrow and toe-wide. The line of foot flight of horses that handle their forefeet in one of three basic ways is illustrated

26

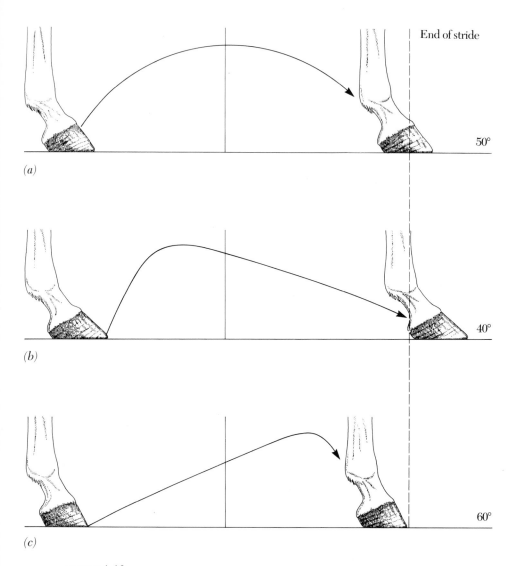

End of stride

50°

(a)

40°

(b)

60°

(c)

FIGURE 1-10

Effect of angle of hoof wall on foot flight arc. *(a)* Normal angle. *(b)* Acute angle.
(c) Too much angle.

in Figure 1-12. At the left, the horse stands straight, breaks over straight, and has an absolutely straight line of flight. The horse in the center is toe-wide (toes out). The toe-wide conformation fault causes the forefoot to break over the inside part of the hoof wall, swing in an inward arc, and land on the inside part of the hoof wall. This type of movement is called *winging, winging inward,*

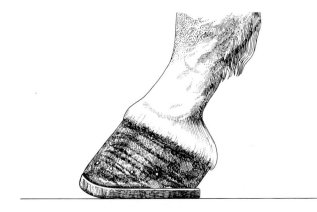

FIGURE 1-11
Rolled toe.

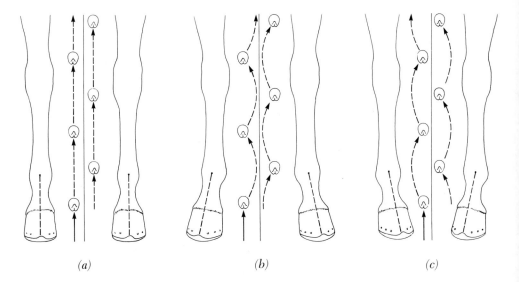

(a)   *(b)*   *(c)*

FIGURE 1-12
Lines of flight for the feet. *(a)* Normal. *(b)* Toe-wide. *(c)* Toe-narrow. (From Harrison, 1968.)

or *dishing*. It may be prevented by corrective shoeing. The base-narrow, toe-wide conformation causes the forefoot to break over the outside part of the hoof wall and then wing inward. This fault almost always causes interference. The horse at the right in Figure 1-12 is tow-narrow (toes in); thus its line of flight is more likely to be outward, and it will be a *paddler*. Although the

FIGURE 1-13

Paddling accompanies toe-in conformation.
(From Adams, 1974.)

way a horse stands in front usually provides a clue as to the anticipated line of flight of the front feet, this is not always a sure sign.

Paddling, or winging outward (Figure 1-13), is an outward swing of the forefeet. The feet break over the outside aspect of the hoof wall, move in an outward arc, and then land on the outside aspect of the hoof wall. Horses that are toe-narrow usually paddle.

Pounding is heavy contact with the ground and results in excess concussion trauma to the forelegs. Faults in conformation that shift the horse's center of gravity forward tend to make the horse's legs pound the ground.

Rolling is excessive movement of the shoulders from side to side. Horses that are wide between the forelegs and lack muscle development in that area tend to roll their shoulders. The toe-narrow fault in conformation also can cause rolling.

Scalping (Figure 1-14) occurs when the top of the hindfoot hits the toe of the forefoot right after the forefoot breaks over. As the area of impact moves up the hindleg, other terms are used to describe the defect. Speedy cutting occurs when the impact area is in the pastern and fetlock areas. Shin hitting occurs on the cannon bone, and if the hock is hit the fault is referred to as hock hitting. These defects in way of going are usually observed in racing trotters. As the speed of travel increases, the area of impact usually moves higher up the hindleg. Several faults in conformation predispose a horse to these defects in leg movement: short backs and long legs, leg weariness or hindlegs set too far under the body, short front and long back legs, and too long toes on forefeet.

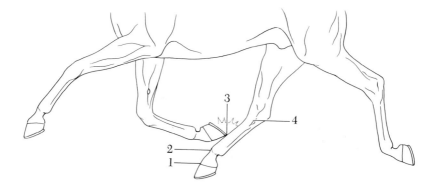

FIGURE 1-14
(1) Scalping. (2) Speedy cutting. (3) Shin hitting. (4) Hock hitting. (From Harrison, 1968.)

A trappy gait is a short, quick, choppy stride. Horses that have short and steep pasterns and straight shoulders tend to have a trappy gait. It may indicate unsoundness, as in the case of navicular disease.

Winding is a twisting of the moving foreleg around in front of the supporting foreleg. The horse appears to be walking a rope. Horses that are wide-chested tend to walk in this manner. It increases the likelihood of interference and stumbling.

TRAINING

Consideration of the type and degree of training of the horse by the potential buyer is critical. Beginning riders or horsemen should select horses that are well trained for the purpose that the horse is being bought. Older horses that are well schooled make excellent horses for beginners because their behavior and performance will probably remain the same for a long period of time. The beginner can make mistakes without spoiling the horse. Such a horse can be found at training stables or obtained from friends who have learned to ride it.

A beginner or intermediate rider should avoid buying a horse that is untrained or has only limited training. It takes years of experience before one is qualified to successfully continue a horse's training. Beginners have trouble knowing when to discipline a horse so that it associates the discipline with the mistake it made, knowing the fine line between discipline and cruelty, and knowing the fine line between kindness and permissiveness.

Intermediate riders may be successful in finishing the training required of a horse that has good basic training but needs additional experience and refinement of performance. The advice and help of a professional horseperson can keep the rider from making mistakes that would prevent the horse from reaching its performance potential.

A common mistake made by potential buyers is buying a horse that is trained beyond the buyer's ability to ride it. Well-trained horses are sensitive to specific cues. Such a horse soon reverts to a lower level of training because it is constantly waiting for the right cues applied at the right time. When the right cues are not used appropriately, the horse becomes confused and does not know what to do.

Previous training should be compatible with the techniques used by the potential buyer. It is time consuming and very difficult to retrain a horse to recognize a different set of cues.

Accomplished riders may wish to buy a horse and complete its training. They usually prefer to train their horses, because they do not have to buy someone else's mistakes. To train a green horse, the rider must have patience and training experience. Most important, the advanced rider must have sufficient time to train the horse. Irregular training schedules are a major cause of horses failing to become well trained and reach their performance potential. Riders who lack time, patience, and training experience should select horses that are well trained or at least have their basic training.

AGE

The desired use for a horse determines how old the horse should be. Generally, the best age at which to buy a horse is between 4 and 12 years of age. Before 4 years of age, the horse is not fully mature and is not ready for hard work and long hours of riding. Beyond 12 years of age, the usefulness of the horse is limited. There are exceptions to these ages. The average horse lives about 24 years, but for most purposes the horse has little value after 16 years of age. However, there are many horses that are serviceable in their early twenties. The Lippizaners of the Spanish Riding School frequently perform until the age of 24 and occasionally until they are 30. In some instances, an older horse or pony may be more desirable. Old age need not be considered if the horse is sound. A young child will probably become too large to ride a pony before an older pony must be retired because of age. A well-trained older horse serves its purpose for a person just learning how to ride. The inexperienced rider can learn a great deal about riding from an older, experienced horse. After 1 or 2 years, the rider will be ready for a different horse. Experienced riders buy horses at the peak of their physical development, because they will be exposed to strenuous exercise.

Most people make a mistake by buying a horse that is too young. The experienced horseperson can successfully train and handle a young horse. inexperienced horseperson usually spoils the young horse, so it never reaches its performance potential. Parents often buy a weanling or yearling because they want their children to grow up with the horse, the rationale being that the children and the horse can learn together. The result is a poorly trained horse and numerous accidental injuries to the children and/or horse.

Because the value of a horse depends largely on its age, familiarity with indications of age are extremely useful if the horse's registration certificate is not available or if it is unregistered. The teeth furnish the best index to the age of a horse, yet other general characteristics play an important role in determining the horse's age. In estimating the age of a young horse, size is a principal factor to be considered. In older horses, the sides of the face are more depressed, the poll is more prominent, the hollows above the eyes are deeper, the backbone becomes more prominent and starts to sag downward, the joints are more angular, and white hairs appear around the temples, eyes, nostrils, and elsewhere.

A horse that has been on the racetrack has a number tattooed inside its upper lip. Thoroughbred horses are tattooed according to a code. The letter indicates the year of birth (beginning with A in 1946 and J in 1980), and the four numbers are the last four numbers of the registration number. American Quarter Horses and Standardbreds do not have coded numbers.

Young horses are referred to according to their age. The young horse is referred to as a *foal* until it is weaned. The male horse is called a *colt* until it is 3 years of age, when it is called a *stallion*. Young female horses are called *fillies* until they are 3 years old, at which time they are referred to as *mares*. Some horsepeople refer to horses that are 6 months to 1 year of age as weanlings, 1 to 2 years as yearlings, and 2 years of age as 2-year-olds.

Teeth and Age

The order of appearance of the teeth and the way they are worn down are considered the most important and accurate cues for estimating the horse's age. The method is not absolutely accurate, because the nature and quality of feed, environmental factors, heredity, and disease materially affect the wear of the teeth. A horse fed soft, succulent feed usually has a younger-looking mouth. Horses fed hard feed or kept in dry, sandy areas of pasture may have mouths that appear 2 to 3 years older than normal. Consequently, one must consider all of these factors when examining the horse's mouth to estimate its age.

Up to 5 years, the age of the horse is commonly estimated by observing the appearance of the temporary and permanent incisors (see Figure 1-15). Therefore one must be familiar with the ages at which the temporary incisors appear and when they are replaced by the permanent incisors. Eruption of the temporary premolars, the replacement of the temporary premolars by permanent premolars, and eruption of the molars are fairly accurate indications of age but are used infrequently. The ages at which various teeth erupt are

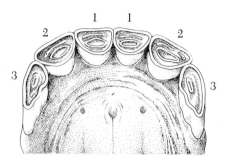

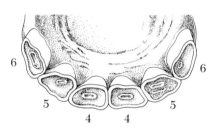

FIGURE 1-15

Positions and names of horse's incisor teeth.
(1) Upper central. (2) Upper intermediate.
(3) Upper corner. (4) Lower central.
(5) Lower intermediate. (6) Lower corner.

TABLE 1-1
The Average Times When Teeth Erupt

Tooth	Eruption
Deciduous	
1st incisor (or centrals)	Birth or first week
2nd incisor (or intermediates)	4 to 6 weeks
3rd incisor (or corners)	6 to 9 months
Canine (or bridle)	
1st premolar	Birth or first 2 weeks
2nd premolar	for all premolars
3rd premolar	
Permanent	
1st incisor (or centrals)	2½ years
2nd incisor (or intermediates)	3½ years
3rd incisor (or corners)	4½ years
Canine (or bridle)	4 to 5 years
1st premolar (or wolf tooth)	5 to 6 months
2nd premolar	2½ years
3rd premolar	3 years
4th premolar	4 years
1st molar	9 to 12 months
2nd molar	2 years
3rd molar	3½ to 4 years

SOURCE: Getty (1953).

given in Table 1-1. The ages of 6 to 12 years are determined by the wear of the dental tables. Beyond 12 years of age, the length and angle of the incisor teeth and Galvayne's groove are used to estimate the age.

To estimate the age of the horse, the horseperson must be able to distinguish between a temporary and a permanent incisor tooth and between the presence and absence of a cup. The temporary incisor tooth has a more defined neck where the crown and root meet at the gum line, is whiter in color, is more rounded, and is not as long as a permanent incisor.

The appearance of the temporary incisors allows one to estimate certain ages (Figure 1-16). The two central incisors on the lower jaw and/or two on the upper jaw are present at birth or appear shortly thereafter. The intermediate incisors on the upper and lower jaws appear at 4 to 6 weeks of age. The corner incisors appear at 6 to 9 months. A rule of thumb is that the incisors appear at 8 days, 8 weeks, and 8 months. The temporary premolars are not commonly used for estimating age because they are present at birth or appear during the first couple of weeks. Therefore, the young horse has 24 teeth: 12 incisors and 12 premolars. Some young horses may have wolf teeth, which appear in front of the temporary premolars. A wolf tooth is a vestigial tooth and has a very shallow root. It may not fall out prior to eruption of the second permanent premolar tooth (see section on care of teeth in Chapter 7).

At 2½ years of age, the central incisors are replaced by permanent incisors. The permanent intermediate incisors erupt at 3½ years of age; the permanent corner incisors, at 4½ years. The canine teeth, commonly called *bridle teeth,* erupt in the interdental spaces at 4 to 5 years of age in male horses. They only erupt in about 20 to 25 percent of all mares. Table 1-1 shows the ages at which the permanent premolars and molars erupt. At 5 years of age, the horse is said to have a "full mouth." The mature male horse has 40 permanent teeth, and the mare has 36 to 40 teeth, depending on the number of canines present.

The cup in a tooth is shown in Figure 1-17. An outside layer of enamel covers the tooth. On the top of the tooth (table surface), the enamel extends inward, forming a pit commonly called the *cup.* The cup is filled with a variable amount of central cement. The inside of the tooth is filled with dentine. In the innermost part of the tooth, the pulp cavity contains the blood and nerve supply to the tooth. From the front of the tooth, the pulp cavity extends upward between the cup and front surface. The teeth of the horse continue to grow during its life but are continually being worn down. When the horse eats, a rotary grinding motion wears the teeth down. The table surface is worn down in a characteristic manner. First the enamel on the table surface is worn off, leaving two rings of enamel around the outside edge and around the cup. As the wear continues, the table surface is worn down so that only the bottom of the cup remains. At this point, the cup is said to have "disappeared," even though the enamel from the bottom surface of the cup is still visible. Later, the enamel may completely disappear. When the cup disappears, the table surface has worn to the level of the pulp cavity. The pulp cavity is called the

1 month

3 months

9 months

2 years 6 months

3 years 6 months

4 years 6 months

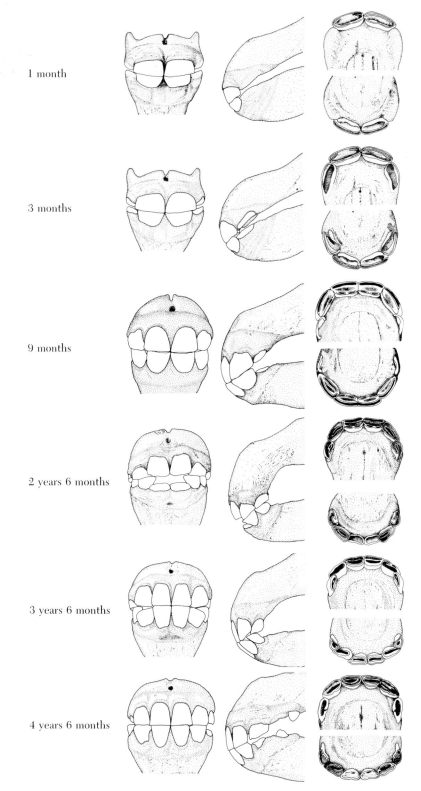

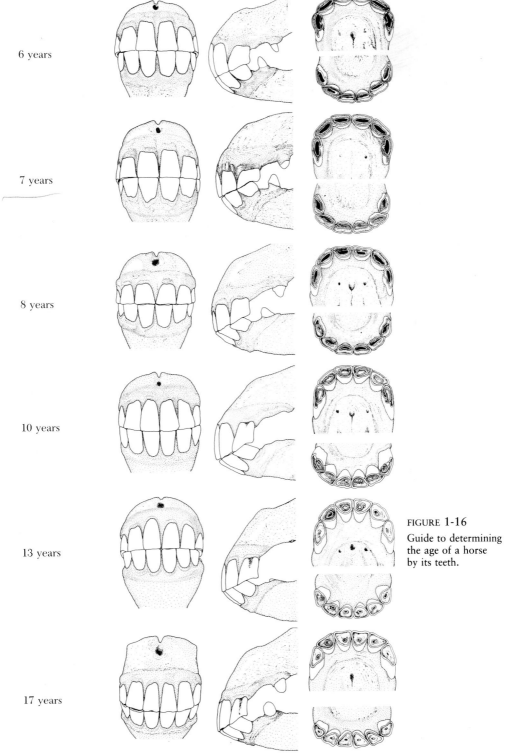

6 years

7 years

8 years

10 years

13 years

17 years

FIGURE 1-16
Guide to determining
the age of a horse
by its teeth.

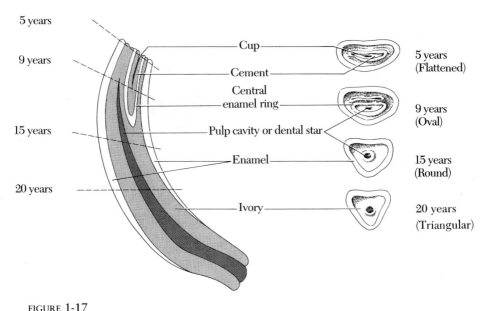

FIGURE 1-17

Anatomy of the horse's incisor tooth. A permanent middle incisor tooth at different ages and stages of wear. Changes in shape from oval to angular are shown in the cross-sectional views as wear progresses toward the root. (From Bradley, 1971.)

dental star, and appears at 8 to 10 years. It is first seen as a long, thin yellow line between the cup and the front of the tooth. As wear continues and the enamel on the bottom of the cup becomes dark, very small, and round, the dental star moves toward the center of the cup.

Between the ages of 6 and 12, the age can be determined by the presence or absence of the cups in the incisor teeth (Figure 1-16). Each year, the cups disappear from two of the incisors. At 6 years, they are absent from the lower central incisors. At 7, they are absent from the lower intermediate incisors, and at 8 they are absent from the lower corner incisors. Before age 9, the cups are present in all the upper incisors. At 9, the cups are absent from the upper central incisors, and at 10, they are absent from the upper intermediate incisors. When the horse is 11 or 12, they are absent from the upper corner incisors. At this age, the horse is said to be "smooth-mouthed."

Beyond 12 years of age, it is difficult to estimate the age of a horse and accuracy is decreased. In older horses, three indicators must be used: the incisive arcade, the table surface shape, and Galvayne's groove. When the incisors are viewed in profile, the incisive arcade—the angle formed between the upper and lower incisors—becomes more acute with age (Figure 1-18). In horses 20 years or older, the incisor teeth stick almost straight out. With advancing age, the table surface changes from a rectangular shape to a tri-

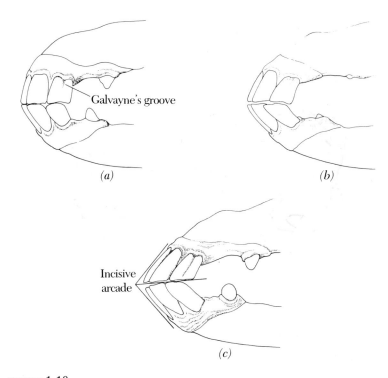

FIGURE 1-18
Profile of the upper corner incisor, incisive arcades, and Galvayne's groove. (a) 9 to 10 years—Galvayne's groove appears at gum margin. (b) 15 years—groove extends halfway down labial surface of incisor. (c) 20 years—groove extends to entire length of labial surface. In profile, the incisive arcade (the angle between upper and lower incisors) becomes more acute with age. (Courtesy American Association of Equine Practitioners.)

angular shape (Figure 1-17). By the time a horse is smooth mouthed, the table surfaces of the incisor teeth are very triangular in shape.

Galvayne's groove appears as a longitudinal depression at the gum line on the surface of the upper corner incisor at 9 to 10 years of age (Figure 1-18). The groove is recognized as a dark line. In some instances, it is hard to see but it is always a darker color than the enamel. As the horse gets older, Galvayne's groove grows down the tooth. At 15 years, the groove is about halfway down the tooth. At 20 years, the groove extends the full length of the tooth. It then starts to disappear at the gum line. At 25 years, it is absent from the upper half of the tooth and completely disappears by 30 years of age.

To examine the teeth, first stand to one side of the horse and separate the lips. Determine the incisive arcade, number of temporary and permanent teeth, and presence or absence of Galvayne's groove. If necessary, then examine the table surfaces. Place one hand on the horse's nose and grasp the tongue with the other hand. Remove your hand from the nose and use it to open the horse's mouth or to push the lips out of the way to look at the table surfaces of the upper and lower incisors.

TEMPERAMENT AND BEHAVIOR

The general temperament of the horse is an important quality. A good-tempered horse is obedient, intelligent, easy to train, courageous, confident, and calm. It is free of vices (bad habits) and easy to handle. Before buying the horse, spend considerable time observing and handling the horse to evaluate its temperament. If possible, obtain the opinion of the previous owner, trainer, dealer, and breeder. A horse that has had several owners may have an undesirable temperament, particularly if it has had rough or inconsiderate treatment.

Compatibility of your personality with the horse's temperament should be considered. In most circumstances, opposite types of personalities are the most compatible. A nervous person may be just the stimulation needed for a placid horse to perform. And nervous horses usually are not so excitable when they are ridden and handled by someone who has a quiet, easygoing manner. Nervous people tend to make quick movements, which excite a nervous horse.

Hunters, jumpers, and stock horses need willingness and courage to perform successfully. Children's ponies or horses need to be gentle and obedient. A sluggish horse may not be obedient, and an inexperienced rider will not enjoy riding it because he or she will have difficulty getting the horse to move or preventing it from going where it pleases.

Temperament Types

Horses are like humans in that they all have personalities and temperaments. There are at least six general temperament types of the horse: quiet, interested, nervous, extremely nervous, stubborn, and treacherous.

The quiet horse that is sluggish and has no interest in its surroundings is usually safe for the inexperienced horseperson. Because of its temperament, such a horse seldom advances to a highly trained state. It does not seem to care, and lacks the spirit and style for a well-schooled performance. The temperament of quiet horses makes them good for rental stables and for teaching beginners how to ride, because they are unconcerned about how a person moves or works around them.

The most desirable temperament is a horse that is interested in its surroundings and pays attention to what happens around it. When such a horse hears an unexpected noise or sees unexpected sudden movement, it responds by pricking its ears but does not shy or try to escape. These horses are usually easy to train and are willing performers. They have sufficient spirit to have animation and style in their performance. They are safe horses for beginners under the supervision of a knowledgeable horseperson or for the intermediate rider. When these horses are handled, they are sensitive to the handler's requests and respond calmly.

Nervous horses are easily excitable and shy away from strange objecs, movement, and noise. They are safe horses for knowledgeable horsepeople. Nervous horses respond to training and are usually capable of reaching a highly trained state. One problem is that they become excitable and flighty, and snort when they are exposed to new surroundings or unfamiliar objects. As these horses get older and gain experience, their nervous dispositions improve.

The extremely shy and nervous horse is safe for only the very experienced horseperson who understands horse behavior. These horses shy at the least provocation, such as the slamming of a stall door, movement of a dog or cat, seeing a moving shadow, or any unexpected movement or noise. Frequently they try to escape without regard for the handler, even trying to run over or knock down the handler. They flee without concern for their own safety and run into fences, equipment, and buildings.

Horses with stubborn temperaments are difficult to train to be obedient. They are slow to learn and require tact and patience by the trainer. When they get nervous or tired or are asked to perform quickly, they become sullen and refuse to work. Some of these horses reach a given stage of training and then seem to forget everything they learned. The trainer must start the training procedure all over at the beginning level.

A few horses have a treacherous temper and are very resentful. The most dangerous kind acts without apparent cause and when least expected. They strike, bite, or kick. However, one seldom encounters a true "man hater," which will pin its ears back against its head and attack a person with its mouth open and striking with its forefeet. It is capable of biting off the face or part of a limb of a person as well as of breaking bones when it strikes and kicks. The treacherous horse cannot be trained and is usually put to sleep.

The behavior of a horse when it is around other horses should be carefully evaluated. Horses are usually stabled adjacent to other horses and kept together in larger paddocks. The horse that continually kicks and bites other horses or kicks the stall wall separating it from another horse is difficult to keep. On trail rides, it is a nuisance because it must be kept separated from the other horses to prevent accidents and injuries.

Agonistic Behavior

To be able to accurately evaluate a horse's temperament, we must understand normal and abnormal horse behavior. One of the best ways to judge a horse's "personality," agonistic behavior, relates to facial expressions, postures, vocalizations, and locomotor patterns that signify dominance, submission, or some intermediate status or signify an escape pattern. Agonistic behavior causes many problems for the horse owner because it is highly variable and can be very pronounced.

Dominance Hierarchies

When horses are kept together, a dominance hierarchy is formed. The dominance hierarchy is established by aggression. Once established, only the threat of aggression is necessary to maintain the hierarchy.

The introduction of two strange horses to each other usually results in a basic pattern. They approach each other with their heads held high, and they may "toss" their heads. The necks are arched, and the ears are pointed forward. They encounter each other face-to-face by smelling or exhaling at each other's nostrils. During the face-to-face encounter, they frequently squeal, rear up, and threaten to strike with their forelegs. They may continue to encounter by smelling each other's neck, withers, flank, genitals, and rump. The process may be interrupted by one horse laying its ears back, pivoting the hindquarter toward the other horse, and kicking out with one or both hindlegs. A squeal may or may not be emitted when the horse kicks.

Horses that are familiar to each other but have been separated greet each other by briefly smelling at the nostrils. Some very aggressive horses meet a strange horse by immediately laying their ears back, running at the other horse with their necks held low, heads extended, and mouths open, trying to bite. At the last moment, they may pivot their hindquarters and kick.

To establish the dominance, a dominant animal threatens aggression or may fight the subdominant horse. It may be a traumatic experience, in which one or both horses are injured, or there may be little if any physical contact. Once the dominance hierarchy is established, only threats of aggression are necessary to maintain it although fights occasionally occur. Threats of aggression are visual gestures that may be accompanied by sound. They are enforced by physical contact if necessary. The mildest threat is laying back the ears and looking at the subordinate horse. The most common threat is the biting threat. The horse lays its ears back, exposes its teeth, opens its mouth, and nips in the direction of the other horse. About 30 percent of the bites are completed. Kicking threats are used less frequently than biting threats. In a mild threat, the horse may make a gesture to kick with one hindleg in a restrained manner. Serious kick threats are usually made by a squeal and a gesture to kick with both hindlegs.

Because submissive horses seldom challenge a dominant horse, the horses learn to get along together without very many injuries. From a management standpoint, the submissive horse must have adequate room for retreat and escape. When a new horse is introduced, the group should be observed until the dominance hierarchy is reestablished. If serious fighting occurs, the new horse may have to be removed. Putting two boss mares together usually causes serious fighting.

Most dominance hierarchies are linear, but some groups of horses have a complex hierarchy. If a complex hierarchy exists, one horse that is low in the order may be dominant over a horse that is at the top.

Young foals exhibit submissive behavior to older horses that allows the foals to move about in the herd. The foal extends its head, directs its ears sideways, and moves its lower jaw up and down. The foal also receives some protection from aggression by other horses when it is at its dam's side. Her position in the dominance hierarchy determines the amount of protection for the foal.

Social attachments occur between members of a group of horses. The social bonds may be at the family level or between horses that are not family members. A mare's reciprocal attachment with her foal begins about 2 hours after birth. The attachment continues to strengthen for a couple of weeks and then begins to weaken. By 6 months of age, the foal can be weaned with a minimum of stress if two or more foals are weaned together. Horses that are kept together for prolonged periods exhibit a preference to stay together. In a herd, some individuals form pair bonds. Some bonds are very strong and others very weak. When horses that have formed strong pair bonds are separated, they display considerable anxiety and call for each other by whinnying. This can cause problems when one is to be shown or ridden without the other present.

Epimeletic and Et-Epimeletic Behaviors

Epimeletic behavior is behavior involved with giving care or attention. This is a very common type of behavior in horses because they seek care and attention from each other. Epimeletic behavior is displayed several ways.

During the fly season, horses stand head to tail and mutually fight flies for each other. They also practice mutual grooming, nibbling each other with the incisor teeth. The nibble areas include the neck, base of the neck, withers, back, and croup. Each horse seems to have a regular grooming partner. Mutual grooming is most frequent during the spring, when the winter coat is being shed, and during the summer, when horses stand under a shade. Licking another horse is usually limited to a mare licking the foal after parturition for about 30 minutes.

Et-epimeletic behavior is the signaling for care and attention. It is used by all age groups, and is most frequently observed when horses are separated from each other. A young foal becomes very nervous when separated from its dam and will nicker or whinny for her. Mares also call for their foals by whinnying, and they quietly nicker to a foal to indicate protection. Mature horses accustomed to being kept together become upset and call for each other when separated. If a strong pair bond has been formed, it may take 3 to 4 days for the horses to settle down after they are separated. During this time, they will repeatedly whinny for each other and be very nervous. At first, they may be excited enough to run over or through fences and other objects just to get together. Therefore, one must be careful about separating horses so that injuries and accidents do not occur.

Horses exhibit varying degrees of contractual behavior. During inclement weather, they may huddle together. They will form a group when danger exists. Although this type of behavior is not well developed, it does exist.

In addition to behavior resulting from interactions with other horses, other types of behavior exist. Some of these types of behavior are useful in determining whether the horse is ill, eating properly, or normal. For example, horses become conditioned to being fed or exercised at a regular time. At feeding time or a time when the horse is turned out for free exercise, the horse starts nickering. Some horses wait until they see their owner or handler. At other times, they may pay no attention to the person.

Investigative Behavior

A horse's first reaction to a strange object in its surroundings is to raise its head and investigate the object. If the object appears to be a threat, the neck remains raised (sometimes with a pronounced arch), ears directed forward, nostrils dilated, and tail slightly elevated. The horse may move about nervously with hesitant steps, while obviously prepared to run. It emits loud, explosive blows of air (snorts) through its nostrils. The horse may advance toward the object for further investigation or may decide to escape. Once it has decided that the object is not a threat, it resumes its prior activities but may continue to display a certain caution.

Other horses may respond to the cues given by the first horse to see the object. They usually respond in a similar manner. If they are kept together in a large paddock or at pasture, they may move together to investigate it at a closer distance. If one horse decides to flee, the others usually follow.

Investigative behavior is the horse's inspection of its environment by use of the senses and movement. Horses are very curious about their environment and will use all the necessary senses to investigate it. When a horse is placed in new surroundings, it will investigate everything very thoroughly by use of its sight, smell, touch, hearing, and taste senses. Until the investigation is completed, the horse may be nervous and overreact to sudden noises or movements. When frightened, it may run into a fence or other object that it would usually avoid. If a new object is placed in its paddock, it normally cannot resist smelling, touching, tasting, and listening to the object. If not allowed to investigate the object, the horse will be apprehensive about it. Some horses must smell or touch new tack before it can be used on them without their becoming nervous. Investigative behavior can lead to accidents or other problems around a stable. While investigating a stall, a horse may learn to operate a latch and open the stall door. Horses can learn how to turn on water faucets and operate light switches because they play with objects during the investigation.

Grooming Behavior

Horses maintain their hair coats by several methods. Mutual grooming has been discussed but horses also groom themselves. They enjoy rolling in dry, fine soil such as sand and dust or in wet areas. Before rolling in a dry spot, they paw the area. After pawing, they lie down and roll on their backs. After they get up, they shake their whole body as if to dust themselves off. In fact, they do this after lying down for any reason.

Insects that irritate them cause them to rapidly contract the superficial muscles on their trunks and forelegs. The muscle contraction causes rapid skin movements. They use their heads to displace insects on the foreleg, shoulder, and barrel areas. Insects on the belly are displaced with the hindleg, and the tail is used on the hindquarters. Itching is relieved by rubbing on a fixed object or by scratching itself. The head is used to scratch or bite the forelegs, sides, croup, and hindquarters. The hindfoot is used to scratch the head and neck.

Elimination and Ingestion

Eliminative behavior is the behavior associated with urination and defecation. All horses have the same characteristic stance while urinating. The neck is lowered and extended; the tail is raised; the hindlegs are spread apart and extended posteriorly. The flaccid penis is extended by the male. The mare contracts and relaxes the lips of the vulva, exposing the clitoris several times after each urination. The interval between urinations is usually 4 to 6 hours.

Most horses will urinate in a stall or trailer. However, some are reluctant to do so, and they should be allowed out of the trailer every 3 to 4 hours to urinate. If they are not taken out of the trailer, they start nervously moving around and may start kicking the trailer. They do not like to urinate on hard surfaces because the urine splatters on their legs. If they are kept in stalls or small paddocks, most horses will urinate in the same area because the area becomes soft and the urine does not splatter as much.

Defecation occurs about every 2 to 3 hours. To defecate, the horse raises its tail and may also hold the tail off to one side. Ponies almost always stop moving to defecate, while horses will defecate when they are moving about. Stallions prefer to defecate in a small area when kept in a stall or small paddock. They usually back up to the pile and then defecate. They avoid walking on the manure pile if they can. Mares and geldings tend to have no particular place to defecate and scatter the feces in the paddock. Therefore, it is harder to keep the bedding in their stalls clean.

In pastures, a stallion smells the urination and defecation area and walks on or backs up to the area to urinate or defecate. Thus they waste very little of the pasture areas. Because of the mares' and geldings' behavior, however, the defecation area may become larger. Horses refuse to eat grass growing in these areas, so the manure must be scattered over the pasture with a rake to prevent the formation of defecation and urination areas.

Horses usually defecate when they get nervous. Rope horses commonly defecate on entering a roping chute. Some horses defecate immediately on entering a trailer or immediately after being unloaded.

Horses practice coprophagy—the eating of feces. Young foals will eat the feces of mares they are housed with. Adult horses will eat their own feces or the feces of other horses.

Ingestive behavior is the taking of food or water into the digestive tract. Foals may be stimulated to start eating solid food when they are only a few days old. They start by nibbling on the feed and rapidly learn to eat it. If given the opportunity, horses prefer to eat small amounts of feed throughout the day. Horses kept at pasture are able to eat in this manner. Horses kept in stalls and small paddocks eat at the convenience of the owner or handler. Very few horses eat small amounts of their feed at one time so that it will last for a 24-hour period, unless the horse is fed ad libitum (freely; food is always available). If fed ad libitum, they usually overeat and may get digestive disturbances. Preventing overeating is an important aspect of horse management.

Horses graze by taking a bite of grass, taking one or two steps, and taking another bite or two of grass. Therefore, they are moving about most of the time they are grazing, and they prefer to graze over a large area. If sufficient grass is available in the pasture, they eat the top part of the grass and leave the bottom portion. When the pasture has been overgrazed, the horse eats the grass down to the ground surface. Under these circumstances, they do not graze the defecation areas unless the area has been harrowed routinely to prevent the accumulation of feces.

Play

Horses enjoy playing. One of their favorite forms of play is running, either alone or in a group. When running alone, they frequently buck, slide to a stop in a corner, and jump and shy at familiar objects. When running in a group, they frequently push and nip other horses. They may chase each other. Two horses may be observed playfully nipping at and swinging their heads toward each other while rearing on their hindlegs and pawing at each other. This type of activity is also observed in young colts.

Horses also enjoy playing with objects they find in their paddock. They pick up small sticks and toss them about the paddock. They pick up and shake rags and pieces of paper. And they play with balls or other objects suspended in a stall, to relieve boredom. They will play with a latch on a stall door or gate for long periods of time. They can even learn to open the latch.

Mimicry

Allelomimetic behavior or mimicry is common among horses. They learn to copy the behaviors of other horses at a very young age. A young foal learns to chew on its tongue by watching its dam or another horse chew on its tongue. A foal can become addicted to sucking air by observing a wind sucker. Wood chewing is rapidly learned from other horses. A horse that is difficult to catch because it runs away each time it is approached will be watched by other horses, which thus learn to run away. When one horse in a group decides to run and play, other horses usually join it. Therefore it is important to separate horses with vices from other horses before they learn to copy the undesired behavior.

Learning

The horse can be taught many different tasks. It usually excels at those tasks that are athletic in nature but also has a good capacity for pattern discrimination learning. A horse first learns how to learn, and several factors influence this ability. The ability to learn how to learn is evident during pattern discrimination learning. It takes the horse numerous trials to learn to discriminate between the first two patterns. After it learns the correct response for a few patterns, fewer trials are needed to learn additional patterns. The horse can remember the correct response for several months. Once a horse has learned a specific type of athletic performance, it will remember it for years. After several years, the response may not be perfect but the horse can then be retrained to reach a high level of performance much more quickly than it took for the initial training. More difficult maneuvers or tasks are forgotten at a faster rate than easier tasks.

Early experiences affect the horse's ability to learn at a later time. Young foals can be taught to be led and handled. Early contact with humans under the proper environmental conditions result in a docile foal. Docility enhances later learning ability because the foal is not nervous about being handled. The docile horse pays attention to what it is being taught.

Sleeping and Waking

The way horses sleep and wake normally can help horse owners recognize abnormal behavior in illness. There are four recognizable stages in the sleep and waking cycle. Each stage has characteristic body and brain functions. The horse spends an average of 19 hours and 13 minutes awake. During this period, the horse is alert and active. It spends an average of 1 hour and 55 minutes in a drowsy state. Therefore it is awake for an average of 21 hours and 8 minutes each day.

There are two types of sleep. Slow-wave sleep is characterized by slow, regular brain waves, as monitored by an electroencephalograph. The slow, regular brain waves indicate that the mind is not functioning, and slow-wave sleep has been called "sleep of the mind." During slow-wave sleep, the muscles are not fully relaxed. The horse may enter slow-wave sleep while lying down in the sternoabdominal position or while standing up. The horse can sleep standing up because it can lock its knees and hocks by using the stay apparatus, which is a series of tendons controlling leg flexion. The horse spends about 2 hours and 5 minutes per day in slow-wave sleep. The time spent in slow-wave sleep is broken up into about 33 periods lasting 3.5 minutes each.

Between each period, the horse wakes up or enters a period of rapid eye movement (REM) sleep. REM sleep is characterized by irregular, rapid brain waves similar to the awake state. However, REM sleep is a deeper sleep than slow-wave sleep. The mind is very active but there is almost a complete loss of muscle tone. During REM sleep, the horse usually is lying on its side with its neck and head extended and resting on the ground. During the time it drifts back into slow-wave sleep, it may go back to the sternoabdominal position, or it may continue to lie on its side. It is not known if the horse must be on its side to go into REM sleep. There are about 9 periods of REM sleep that last about 5 minutes each, so that the horse spends an average of 47 minutes in REM sleep.

Changes occur in body functions as the horse sleeps. As the horse enters the drowsy state, the heart rate, respiration rate, and muscle tone decrease. They continue to decrease as the horse enters slow-wave sleep. During REM sleep, there is a slight increase in respiration rate and heart rate, but the rates are less than during the drowsy state. Muscle tone continues to decrease and is minimal during REM sleep. The eyelids indicate the horse's state of alertness. They are partially closed during the drowsy state and are barely separated during slow-wave sleep. During REM sleep, they are fully closed.

The horse spends an average of 22 hours standing and 2 hours lying down each day. About 80 percent of the time spent standing is during the daylight hours. There are usually about 4 to 5 periods of recumbency each day.

The horse lies down by going through the same maneuvers each time. The hindlegs are brought under the body and the forelegs are directed back under the horse so that all four feet are close together. The head and neck are lowered. At this time, the horse may move around in a small area to decide exactly where it wants to lie down. The forelegs are flexed and the head is elevated as the horse touches the ground with its knees and sternum. Next, the hindquarters are lowered. Once down, the horse may lay on either side with its neck and head extended outward and resting on the ground. The legs are extended. The horse may rest in an asymmetrical sternoabdominal posture (right or left). When resting in this position, the legs are flexed and the limbs on the side the horse is lying on are underneath the horse. The horse rises from a lateral recumbency position to the sternal position by first extending its forelegs in front of itself. The forelegs are used to raise the forequarters off the ground. When the forequarters are raised, the flexed hindlegs are extended to raise the hindquarters.

Vices (Bad Habits)

Abnormal horse behavior can have a variety of causes. Some atypical behavior may indicate illness or injury. Other types are referred to as *vices*. Usually, a vice affects a horse's usefulness, dependability, or health. Vices such as charging, striking, kicking, biting, and bucking are aggressive vices directed against a handler and may cause injury to the handler. Some of the same vices may be directed against other horses and cause them injury. Flight responses such as rearing, balking, shying, halter pulling, and running away can injure the rider or handler. Stall vices may cause injury to the horse, may cause poor performance, or may waste energy. These vices include stall kicking, wood chewing, cribbing, weaving, stall walking, bolting food, pawing, tail rubbing, and eating bedding or dirt. Bad habits are difficult to eliminate and require a lot of thought and effort on the part of the owner or manager.

Abnormal behavior as a result of illness or injury is discussed in Chapter 9. Horses should be observed at least once or twice a day to determine if they are injured or sick.

Stall vices are sometimes called "nuisance vices." They usually arise out of boredom from lack of exercise or insufficient activity out of a stall. Horses that receive enough exercise seldom develop stable vices. Providing sufficient amounts of bulky food for the horse to nibble on all day gives the horse something to do, and horses prefer to eat frequently. Boredom can be eliminated by providing toys such as plastic jugs hung from the ceiling. A horse may toss or carry about short lengths of plastic hose or a small stick. Occasionally a horse will carry a small rock around in its mouth to play with. Stable vices are contagious because horses mimic each other.

Other causes of vices are poor management, mishandling, or bad treatment at some previous time. Some people make a mistake of teaching a foal to do cute little things (such as nibbling tidbits from a hand) that cause the horse to lose respect for the horseperson when it grows up (such as biting).

The prospective horse should always be checked out for vices. Many of the vices can be detected by observing the horse in its stall or paddock, handling the horse, or riding it. When the horse is approached, it should not lay back its ears and threaten you with its mouth. By feeling all over the horse, you can detect any unusually sensitive areas. A tendency to kick can be detected when the horse lays back its ears and shifts its weight off the hindfoot nearest you.

Biting

Biting is a vice that is acquired as a result of incompetent handling. If a horse is fed sugar cubes or other treats by hand, it learns to expect these treats when handled. It soon begins nuzzling the hands and nosing the pockets of the handler each time he or she is within reach. The nuzzling develops into a nip, because the horse is disappointed. Before long, the horse starts to bite very hard, which can cause injuries. To prevent development of this vice, all treats should be placed in the feed manger.

Rubbing a horse's nose while petting it also teaches it to bite. The horse should be petted on other areas of the body and only occasionally on the nose.

Once a vice is learned, it is difficult to eliminate. Some horses respond to being slapped quite hard on the side of the mouth or face. An effective way of retraining the confirmed biter is to use a war bridle (Chapter 6). Each time the horse attempts to bite, one or two quick jerks on the bridle punish the horse quite severely. This is continued until the horse associates the punishment with biting. If it persists, a wooden gag can be used. A block of hard wood, about 5 inches long (12.7 cm) and 1½ inches square (9.68 cm²), with a chain run through a hole drilled through the center long ways is attached to the halter. The gag is adjusted so that it fits in the area of the bars. Each time the horse bites, the edges place pressure on the gums. It usually takes a few lessons to overcome most biting.

Bolting

Sometimes a horse bolts grain (gulps it down without chewing). The vice is undesirable because the whole grain passes through the digestive tract without being digested, leading to digestive disturbances such as colic. There are several methods to prevent bolting. The grain can be spread out in a large feeder with a flat bottom so the horse cannot get very much in its mouth with each bite. Large stones can be placed in the feed box so the horse must move the stones around to get to the grain, which slows down the rate of eating. The grain can be mixed with chopped hay to help induce chewing. Hay can be fed first, and grain can be fed after most of the hay ration has been consumed. The hay reduces the horse's appetite, so the horse is not inclined to eat as fast.

Attacking

Charging or savaging is a deliberate attack on a person. Not many horses are actually vicious, but they must be controlled. Stallions are prone to becoming unmanageable unless they are handled by a competent horseman. The vicious horse must be taught to obey, or to suffer accordingly. However, it must become confident in the handler, so the training must not inflict pain. The most effective means is to humiliate the horse by either pulling it to its knees with a "running W" and holding it there until it becomes calm or by laying it on the ground with a throwing harness and holding it down by restraining its head. This causes the horse no physical pain but does humiliate it.

Cribbing

Cribbing is a vice in which a horse sets its upper incisor teeth against an object, arches its neck, pulls backward, and swallows large quantities of air. The vice leads to colic and other digestive disturbances caused by the excessive amount of air in the digestive tract. Most cribbers are hard to keep because of the digestive disturbances and the fact that they would rather crib than eat. Moreover, it wears off the front edges of the incisor teeth. Wind sucking is similar to cribbing, but here the horse does not set its incisor teeth on an edge of an object.

There are several ways to prevent a horse from cribbing. However, once the vice is acquired, it is impossible to eliminate the habit if the opportunity to crib presents itself. A surgical procedure has been developed that stops some horses from cribbing. The most common preventive device is a neck strap that is placed rather tightly around the throat. When the neck is arched, the strap places pressure on the trachea, and the discomfort overrides the urge to crib. A spring-loaded strap that causes sharp, pointed nails to prick the horse when it arches its neck is more effective on those with a strong drive to crib. The straps cause the horse no discomfort when it is not trying to crib.

Fighting

Fighting is aggressive behavior by dominant individuals. Once the dominance hierarchy is established, there are seldom fights within the group. Sometimes two very dominant horses fight quite often because one is constantly challenging the other. In this situation, it is advisable to separate the horses before they injure each other.

Halter Pulling

Halter pulling can injure the horse's neck muscles—and the handler, if the horse lunges forward after pulling back. After the horse pulls back one or two times and manages to break loose, it usually forms the habit. A confirmed halter puller breaks halters and lead ropes as it pulls back, and it can then escape. And, of course, a horse running loose around a stable can cause accidents. Three hitches are commonly used to overcome the habit. As soon as you discover that the horse is testing its halter when it is tied, tie a strong rope that the horse cannot break around the horse's throatlatch. Tie the rope with a bow-line knot, so that it cannot slip and choke the horse. Run the rope through the ring in the halter and tie it to an immovable object. After the horse has been tied this way, it soon learns that it cannot break away and stops trying to do so. At this time, a lead rope can be used to tie the horse. However, if a horse has escaped several times and the habit is formed, more severe measures must be taken. A rope can be run around the horse's chest and placed just behind the withers. Secure the rope with a bow-line knot, and run the free end between the horse's forelegs and through the halter ring and tie it to an immovable structure. When the horse pulls back, it cannot tolerate the pressure being applied behind the withers. Some horsepeople prefer to place the rope around the horse's back in the area of the flanks. Form a small, stationary loop in one end and run the free end through it. Run the free end between the front legs to the halter ring and tie it so the horse cannot escape. As soon as the horse feels pressure around its body, it moves forward. After a few struggles and attempts to break away, it learns not to pull back.

Kicking

The horse that habitually kicks at other horses or people is dangerous to own or to handle. The vice usually results from incompetent handling. As soon as a horse makes an attempt to kick, it must be corrected. If the horse is continually corrected in the formative stages, the vice can be eliminated. The confirmed kicker is difficult to correct, however. Handle such horses cautiously, so that you are never where the horse can kick you. A horse that kicks other horses should be tied up away from them and should be ridden at the back of a group of riders.

Many horses that kick acquired the vice because they were fed too much grain and were not exercised. At first they playfully kicked the wall. Later, the kicking became a habit. Stall kicking can cause serious injury to the hindlegs and hocks. Some horses kick the walls of their stable for no reason other than just to kick. Other horses kick the stable wall at feeding time to display their impatience. Some horses only kick at night; a light in their stall will stop the kicking. An uncongenial horse kept in an adjacent stall will provoke some horses to kick. Often, placing the horse in another stall solves the problem. However, it is difficult to correct a confirmed stall kicker. Padding the stall wall stops some horses, because the lack of resistance or noise when a pad is kicked removes the desire. A rubber ball or stick attached to the fetlock with an elastic band is an effective corrective device for some horses. When the horse kicks, the ball or stick sharply strikes the leg. Some horses simply cannot be corrected.

Pawing

Pawing the stall floor is more a nuisance than a serious vice. It wastes energy and digs holes into the floor that must be filled periodically.

Rearing

Rearing is one of the most dangerous vices that a horse can have. When the horse rears up, the flailing forelegs can cause serious injuries to the handler, especially to the head. Such horses must be corrected with a lead shank or a whip because of the seriousness of the vice. A horse that rears while being ridden should not be ridden by beginning or intermediate riders. An experienced trainer can usually find the cause and correct the vice.

Shying

Shying at unfamiliar objects makes a horse dangerous to ride for even the experienced rider. It is a long and difficult task to overcome the habit. The horse must continually be taken to new surroundings and over new trails until it learns that nothing will harm it.

Stall Walking

Stall walking is a nervous habit that wastes energy. Stallions often walk constantly or pace circles around the stall or paddock. Frequently they start walking to work off excess energy and to calm their nervousness. Decreasing the energy content of the ration and giving regular exercise periods may eliminate the vice. Once horses start to stall walk, it is difficult to get them to stop. Placing a companion such as a goat, pony, or dog in the stall with them can be effective.

Striking

Striking with the forefeet is a dangerous vice, because the handler is always vulnerable to injury. While leading, grooming, and saddling such a horse, always remain alert and try to stay at the horse's side, not in front of it. Each time the horse attempts to strike, punish it with a war bridle or whip.

Tail Rubbing

Tail rubbing is usually started by horses whose tail areas are irritated by internal parasites, especially pinworms, or by a skin problem. Once the vice is acquired, the horse may continue to rub its tail even after the conditions that caused the initial irritation have been corrected. Tail rubbing is a common vice of Saddlebred horses that wear tail sets. To prevent the vice, a 2-inch by

12-inch board (5.1 by 30.5 cm) is used to construct a shelf that runs around the entire stall at a height just below the point of the horse's buttock. This tail board makes it impossible for the horse to rub its tail.

Horse's tails should be wrapped or protected with a leather tail guard when horses are shipped long distances in a van. If not, their tails will involuntarily be rubbed on the butt chain.

Weaving

Weaving is a rhythmical shifting of the weight of the forehand from one forefoot to the other. This is a nervous habit that takes a tremendous amount of energy. It results from too much food and insufficient regular exercise. It is almost impossible to eliminate the habit once acquired. A stable companion or suspended play toy may be beneficial. Turning the horse out in a large pasture also helps. The vice is rapidly learned by other horses, so the weaver should be isolated from their view.

Wood Chewing

Wood chewing is one of the most common vices. All horse buildings, paddocks, and fences must be designed so the horse cannot chew them. It takes only a few hours for a horse to chew through a 2- by 12-inch board (5.1 by 30.5 cm). Some horses chew wood every chance they get, while others chew wood only when the weather changes. The habit is dangerous to the horse because it can swallow splinters of wood. Also, it causes abnormal wear of the incisor teeth. The vice is expensive for the stable owner because of the need to continually replace badly chewed boards. Chemicals such as creosote can be applied to wood to discourage the horse from chewing it.

Other Vices

Many other vices can make it unpleasant to work with a particular horse. Some horses do not like to be groomed and will not stand still. Others do not like to be saddled and will jump around when the saddle is placed on their backs. A horse may inhale and hold the air while it is being cinched. After the horse is led around or ridden for a few minutes, the cinch can be tightened. Bucking is a serious vice that will get worse unless the horse is corrected by an experienced rider. Horses that are difficult to catch are a nuisance. All these vices detract from a horse's value and usefulness. If possible, one should avoid buying a horse that has one or several of them.

COLOR

The color of a horse normally is not a consideration in selecting most horses. However, it may be important if the horse is to be used for certain purposes. If a breed registry has certain qualifications for registration that are based on the color or color pattern, the color pattern is very important. The registration associations of Appaloosas, Paints, Pintos, Buckskins, and Palominos are based almost solely on color. Other breed associations, such as the American Quarter Horse Association, may disqualify for registration horses that have certain types of color patterns. (See Chapter 3 on breeds and breed associations.) Breeders must be knowledgeable about the inheritance of these color patterns, or they may produce individuals that cannot be registered. The color or color patterns of a horse may make it more or less desirable for certain types of performance classes where a horse or rider's ability is judged, because some colors or color patterns seem to attract a judge's attention.

Finally, the color and color pattern of a horse serves as a basis for identification because it is one of the most conspicuous traits of a horse. (See Chapter 17 for information on the genetic inheritance of color and pattern.)

Head Marks

The head markings of a horse usually consist of the presence of white hairs in specific areas. The most common head markings are illustrated in Figure 1-19. We lack knowledge concerning the inheritance of head markings. It is not clear whether they are due to dominant or recessive genes. The chin spot is believed to be due to recessive genes.

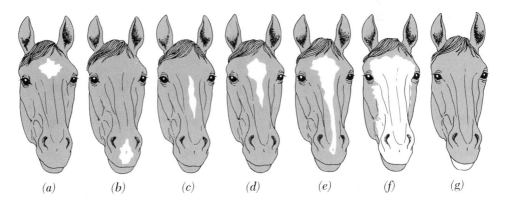

(a) *(b)* *(c)* *(d)* *(e)* *(f)* *(g)*

FIGURE 1-19

Head marks. *(a)* Irregular star. *(b)* Snip. *(c)* Narrow snip. *(d)* Star and connected narrow stripe. *(e)* Star and connected snip. *(f)* Bald face. *(g)* Chin spot. (Courtesy Jockey Club.)

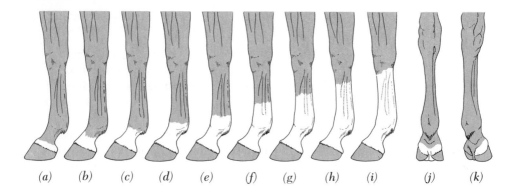

FIGURE 1-20

Leg marks. *(a)* Coronet. *(b)* Half pastern. *(c)* Pastern. *(d)* Pastern and part of ankle. *(e)* Ankle. *(f)* White to above ankle. *(g)* Half stocking. *(h)* Three-quarters stocking. *(i)* Full stocking. *(j)* Heel. *(k)* Outside of heel.

Leg Marks

The common leg marks are shown in Figure 1-20. Very little is known about their inheritance. The distal spot is a small spot of dark hair that is located just above the coronary band in a white area. The marks are carried by a dominant gene.

Body Color

There are several basic body coat color patterns and several variations of these patterns. The inheritance of some coat color patterns is known, while the knowledge of the inheritance of others is only theoretical. It is difficult to predict the coat color of foals because the mode of inheritance is complex.

The black horse is a uniform black color on the body, mane, and tail. The skin color is black. The uniform black color pattern is referred to as "jet black" if the coat color does not fade when the horse is kept out in the sun for several days. The black coat color of other horses fades in the sun to a "blackish bay" color pattern.

The bay color pattern is characterized by a black mane and tail, black points (black hair below knees and hocks, black muzzle, and black tips on the ears), and a reddish body. The color of the body may vary from a light to a dark reddish color. The dark reddish bay color is called a "blood bay." The basic color of the bay horse is black, but the black color is restricted to the mane, tail, and peripheral parts of the body. Some bay horses may have zebra markings—a dark stripe down the back and extending down each shoulder and transverse bars on the forearm.

The seal-brown color is another modification of the black coat color. It is recognized by the brown hairs located in the flank areas, on the muzzle, under the eyes, and on the tips of the ears. A dark seal-brown horse may be difficult to distinguish from a black horse that fades in the sun. Those that have a lighter-colored body area are easier to recognize and are normally called "brown" horses.

The chestnut or sorrel horse is another basic color pattern. The skin color is brown, and the hairs are red. The lighter-colored horses are called "sorrel" horses whereas the darker pattern is the "chestnut." The very dark red chestnut horse is called a "liver chestnut."

White horses are born and remain white throughout their lives. They have pure white hair, pink skin, and blue eyes. The white horse has a dominant gene for the white pattern. There are two other white coat color patterns that are modifications of the chestnut and black coat color patterns. The cremello pattern is an off-white or cream-colored body and blue eyes. Some cremellos have a lighter mane and tail. The outlines of face and leg markings are evident. The cremello pattern is a modification of the basic chestnut pattern. Perlino horses are modifications of the black coat color pattern. Their bodies are an off-white or pearl-white color. The mane and tail is a light rust color.

Grulla horses have black manes, tails, and peripheral parts like bay horses except the body is a dilution of the black hairs on the body to a sooty black.

The dun and buckskin color patterns are further modifications of the bay coat color pattern. The dun color is a modification of the dark bay color, whereas the buckskin is a modification of the light bay color. Both color patterns are characterized by the black mane, tail, and legs. They may or may not have the dark stripe down the back, on the shoulders, and across the forearm. The buckskin color pattern has a light yellow body color. The body color of the dun is darker than the buckskin and may be described as a dingy yellow. The mane and tail may not be as black in some dun horses.

The palomino color pattern is characterized by a yellow body color and a lighter mane and tail. The mane and tail may be almost white or flaxen.

The coat color of gray horses is characterized by white hairs mingled with hairs of the basic color. As the horse gets older, more white hairs appear in the coat. The foal may be born a solid color, or it may have a few white hairs in the coat. The skin is one of the two basic colors. Colored hairs are continuously being replaced with white hairs, so that older gray horses are almost white. When black horses with the gray gene have a higher proportion of black hairs than white hairs, the horse is referred to as an "iron-" or "steel-gray" horse. Red grays are modifications of the bay pattern, and chestnut grays are modifications of the sorrel and chestnut colors. Unless one knows the pedigree or history of a gray or roan horse, one cannot tell them apart.

The roan pattern is a mixture of white and colored hairs. The roan color is present at birth and does not change when the horse gets older. A blue roan is a mixture of white and black hairs. The roan pattern superimposed upon the basic bay pattern is called a "red roan." Strawberry roans are sorrel or chestnut horses with white hair mixed in with the colored hairs.

Paint and pinto horses have four color patterns. The main distinguishing character of these patterns is the white spotting. In the tobiano pattern, the white spotting crosses over the top of the horse's back and extends downward. The white extends from the belly and legs toward the back in the overo pattern. Horses with black pigmented skin and coat color are referred to as "piebald" whereas brown pigmented horses are called "skewbalds." There can be piebald tobianos, piebald overos, skewbald tobianos, and skewbald overos.

Appaloosa horses have a variety of spotting patterns, but they must meet three minimum requirements if they are not spotted: mottling of the skin, striped hooves, and an unpigmented sclera. The mottled or particolored skin appears particularly on the nostrils, muzzle, and genitalia. The striped hooves have alternate vertical stripes of white (unpigmented) and dark (pigmented; black or brown) colors. White around the cornea of the eye is a white sclera.

There are many coat color patterns, and all have at least one of the three characteristics previously described. All of the spotting patterns are of two basic patterns. The leopard color pattern is a white coat with dark spots scattered over the horse's body. The other basic pattern is the blanket pattern in which a white blanket, usually containing dark spots, crosses over the horse's croup, loin, and/or back. There is considerable variation in the distribution of white and of spots from the classical basic pattern. In the mottled pattern, unpigmented hairs are grouped together into small, uneven areas of white that are not always continuous with another. Pigmented spots may be found in the white areas. The pattern is usually located over the dorsal area. Speckled patterns are characterized by small clusters or specks distributed rather evenly throughout the pigmented area involved. Some horses have dark spots distributed over a colored coat pattern that contains no white areas. The appaloosa may also have a white blanket that contains no spots.

The genetics of coat color inheritance is complex, and our knowledge is largely based on theory. The present concepts of coat color inheritance are discussed in Chapter 17, on genetics.

SEX

The sex of a horse influences its value and behavior and thus its desirability for certain potential uses. The sex to choose is controversial. Some horsepeople prefer a mare because of her potential as a brood mare if she has good conformation and becomes well trained. Others do not use mares because of their behavior when they are in estrus. Before deciding to buy a mare to show in performance classes, it is advisable to determine her behavior during estrus. Some mares do not show the characteristic signs of estrous behavior unless they are actively teased by a stallion. Very seldom do their estrous cycles influence the performance of such mares. Other mares display estrous behavior when they are exposed to any strange horse. Their behavior does not allow them to be shown when they are in estrus. Some mares develop an erratic and sour disposition during estrus and are not attentive to the rider. If the mare will never be bred, she can be spayed (ovaries removed) to eliminate the mare's estrous cycle. The surgical procedure is simple, and the mare can be worked 10 to 12 days after surgery.

Only accomplished horsepeople should select stallions to be ridden or worked unless they are to be used for breeding. Their behavior usually causes problems and makes them difficult to handle unless the handler is competent. Because of their aggressive sexual behavior, they must be kept separate from other horses. Their paddocks and stalls must be well made because most stallions will kick or run into them periodically to see if they can escape. It is difficult to haul them with other horses unless mentholatum is rubbed in their noses. Otherwise, they smell the other horses and become aggressive. Some stallions are well behaved and have good manners because they have been competently trained and handled. However, if they are not continually corrected, they soon become difficult to handle. They should never be trusted, because even the most docile ones occasionally exhibit erratic behavior for a few moments. During these few moments, injuries can occur. Very few stallions consistently give a good performance when shown or worked.

Geldings—castrated male horses—are the most popular horses to ride and use because of their attitude. Their behavior is predictable and usually stays the same. They can be kept together or with other horses. Some geldings will mount and copulate with estrous mares, but most do not display any sexual behavior. Horses are gelded or castrated between birth and 2 years of age, but older stallions can be castrated. Colts that have poor conformation and/or a poor pedigree are gelded as soon as their testicles descend into the scrotum. The testicles may be descended at birth and, if not, usually descend by 10 months of age. Well-conformed colts may be given a performance test and are castrated if they do not perform or train as expected. Older stallions that do not produce quality foals should be castrated. Cryptorchids (only one or neither testicle descended into the scrotum) should be heavily discriminated against. They are referred to as "ridgelings" after they are 15 months of age. Ridgelings display intense sexual behavior and retain their stallion attitude for several months after castration. They are difficult to castrate, because a flank incision must be made.

PEDIGREE

One of the decisions that the potential buyer must make is whether or not to buy a registered or an unregistered (grade) horse. Many grade horses are excellent horses and are suitable for many purposes. A large percentage of pleasure riding horses are unregistered, as are many horses used by riding stables and by ranches. Some horse shows or specific classes in a horse show do not require a horse to be registered. Some high-quality grade horses that have excellent performance records command a high price.

Breed associations have been formed to issue registration certificates and to maintain records of lineage and performance. The stud books published by the breed associations give name, registered number, sire, and dam for all horses registered each year. Buying a registered horse has several advantages. If the horse has a performance record with a breed association, the record can be obtained. This record is important in evaluating the horse for future performances or for possible value (prepotency) as a breeding animal. The horse's ancestors can also be evaluated in terms of their performance ability or ability to produce quality foals. For example, the following records can be obtained for horses registered with the American Quarter Horse Association: racing activity, arena performance, dam's produce, and get of sire. In addition, the Quarter Race Breeding Analysis computes statistics on sire's records for percentage of starters that attained register of merit status, percentage of starters that attained AAA racing rating (high on a scale of AAA+, AAA, AA, A, and B), percentage winners per starters, and other pertinent information. The Jockey Club Statistical Bureau keeps racing records on Thoroughbred horses. For a fee, it will provide a five-cross pedigree by tracing the horse's ancestors in the *American Stud Book*, which is the record of foals registered by the Jockey Club. It uses other records to provide race records, produce records, crop analysis of sire, and domestic family histories. Another computerized record service provides a breeder's statistical report for all the racehorses that are racing and are bred by an individual, a monthly brood mare statistical report for all the foals of a mare that are racing, and a weekly stallion statistical report for all the horses a stallion has sired that are racing. Their reports are obtained from the data published by the *Daily Racing Form*.

Racing data for Quarter Horses and Thoroughbreds are published in the *American Racing Manual* and the *Chart Book*. The *American Racing Manual* is published yearly and contains the records of horses, owners, trainers, jockeys, breeders, stallions, yearling sales, stake race purse distribution, and many other statistical analyses. The *Chart Book* is published monthly and contains the official charts of all races run at recognized North American tracks during the month of issue. The quality of a horse can be determined from these records, and a horse's pedigree can be a valuable aid for selecting young horses before they have had a chance to prove their performance ability.

2
Evaluating Conformation and Soundness

CONFORMATION

The light horse is used primarily for athletic purposes, whereas the draft horse is used for pulling power. The conformation of a horse is related to the performance potential in that the relationships of the parts determine the way in which a horse moves. Different conformation types of horses are suited to different types of performance. Variations of desired conformation are personal preferences in many respects, but there is a basic conformation that is desired for the light horse. Conformation determines the ease, freedom, and direction of leg movement. Limb interference can result from faults in the basic conformation of a horse. Faults in conformation predispose a horse to certain unsoundnesses, which are discussed later in this chapter. Potential buyers can save much time by avoiding horses whose potentials are limited by faults in conformation.

To understand the basic conformation of a horse, one should be familiar with the nomenclature of the parts of the body and the bones supporting it

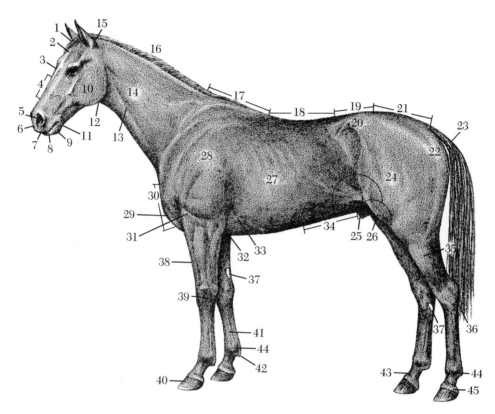

FIGURE 2-1

General conformation and points of the horse. (1) Forelock. (2) Forehead. (3) Face. (4) Bridge of nose. (5) Nostril. (6) Muzzle. (7) Upper lip. (8) Lower lip. (9) Under lip. (10) Cheek, jaw, or jowl. (11) Chin groove. (12) Throatlatch. (13) Jugular groove. (14) Neck. (15) Poll. (16) Crest. (17) Withers. (18) Back. (19) Loin. (20) Point of hip. (21) Croup. (22) Buttock. (23) Dock. (24) Thigh. (25) Flank. (26) Stifle. (27) Barrel. (28) Shoulder. (29) Point of shoulder. (30) Chest. (31) Arm. (32) Elbow. (33) Girth. (34) Abdomen. (35) Gaskin. (36) Hock. (37) Chestnut. (38) Forearm. (39) Knee. (40) Hoof. (41) Cannon. (42) Ergot. (43) Pastern. (44) Fetlock. (45) Coronet.

(Figures 2-1 and 2-2). One should develop a logical procedure for examining conformation, to be followed for each horse that is evaluated.

The overall balance and symmetry of the horse should be evaluated from a distance (Figure 2-3). The measurements of Secretariat, winner of racing's Triple Crown in 1973, illustrate the unusual proportions in relation to his head, neck, shoulder, hip, and waist; heart girth to the legs; and length to height. For the standard horse, the length of the head is about the same as the length of the neck, shoulder, waist, and hip; the heart girth is about the same length as the legs; and the height is about the same length as the horse's length from shoulder to hip. Some horses look as if they were made up of parts from several different-sized horses, so that they are out of balance. From a distance,

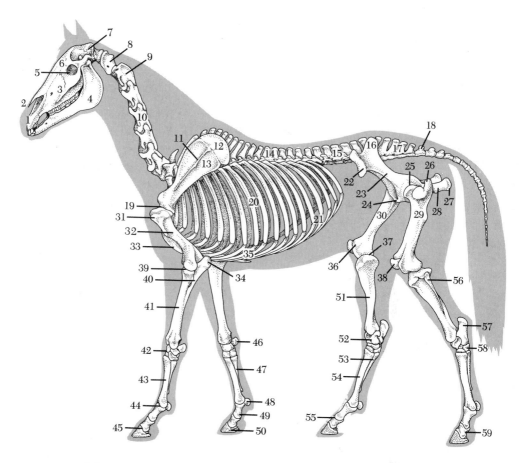

FIGURE 2-2

Skeletal system of the horse. (1) Incisive bone (premaxillary). (2) Nasal bone. (3) Maxillary bone. (4) Mandible. (5) Orbit. (6) Frontal bone. (7) Temporal fossa. (8) Atlas (first cervical vertebra). (9) Axis (second cervical vertebra). (10) Cervical vertebrae (there are seven of these, including the atlas and axis). (11) Scapular spine. (12) Scapular cartilage. (13) Scapula. (14) Thoracic vertebrae (there are usually eighteen of these). (15) Lumbar vertebrae (there are usually six of these). (16) Tuber sacrale. (17) Sacral vertebrae (sacrum) (there are usually five vertebrae fused together). (18) Coccygeal vertebrae. (19) Shoulder joint. (20) Ribs (forming wall of thorax: there are usually eighteen ribs). (21) Costal arch (line of last rib and costal cartilages). (22) Tuber coxae. (23) Ilium. (24) Pubis. (25) Hip joint. (26) Femur, greater trochanter. (27) Tuber ischii. (28) Ischium. (29) Femur, third trochanter. (30) Femur. (31) Humeral tuberosity, lateral. (32) Humerus. (33) Sternum. (34) Olecranon. (35) Costal cartilages. (36) Femoral trochlea. (37) Stifle joint. (38) Patella. (39) Elbow joint. (40) Ulna. (41) Radius. (42) Carpus. (43) Metacarpus. (44) Fetlock joint. (45) Coffin joint. (46) Accessory carpal bone (pisiform). (47) Small metacarpal bone (splint bone). (48) Proximal sesamoid bone. (49) First phalanx. (50) Distal phalanx (third phalanx). (51) Tibia. (52) Talus (tibial tarsal bone) (astragalus). (53) Small metatarsal bone (splint bone). (54) Metatarsus. (55) Pastern joint. (56) Fibula. (57) Calcaneus (fibular tarsal bone). (58) Tarsus. (59) Middle phalanx (second phalanx).

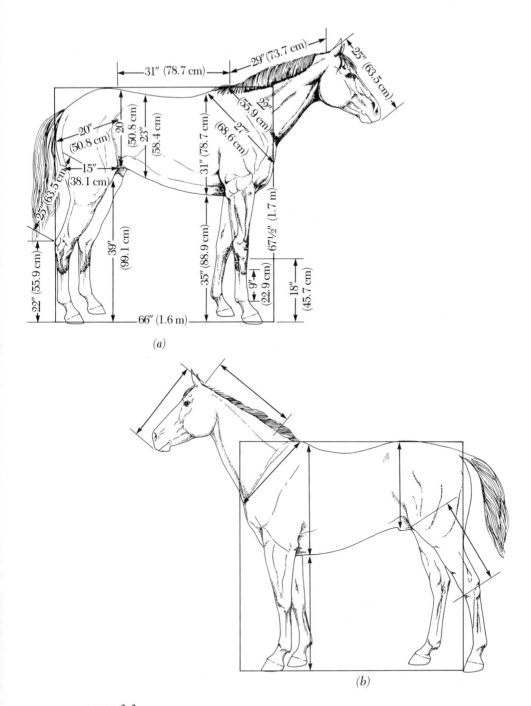

FIGURE 2-3

(a) Secretariat's conformation versus *(b)* standard conformation. (Reprinted with permission from the February 1977 issue of *Horseman*. Copyright © 1977. All rights reserved.)

observe the horse walking and trotting both toward and away from you. This gives you the opportunity to observe faults in the horse's way of going. Then observe the horse up close for a detailed examination.

Head

Generally the horse's head is considered for its esthetic value. The exact shape of the ideal head varies depending on the breed, but you should evaluate certain functional aspects: The head shape influences the visual field, balance, and respiratory function.

The ears should be clean-cut and alert. The ear should curve around and come to a fine point. Well-formed ears may indicate good breeding in grade horses. Ears spaced too far apart, droopy, or placed too close together are displeasing. Large ears are distracting and make the head look out of proportion. Alert ears mean the horse is attentive to its surroundings. Ear movement can indicate the horse's temperament. Horses with sour dispositions tend to pin their ears back. They usually do this before they bite, kick, or buck.

The head should be generally triangular from front and side views (Figure 2-4). From the side, the jaws should be deep, and the head should taper down

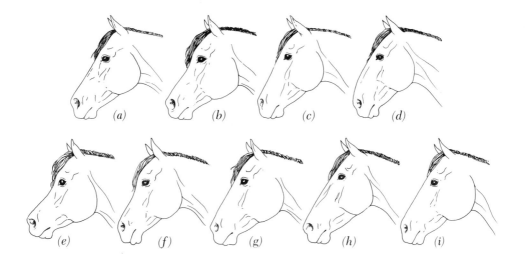

FIGURE 2-4

Profiles of horse's heads. *(a)* Good head, straight profile, Thoroughbred type. *(b)* Good head, Arab type, prominent forehead but with dished profile below eyes, deep through jowl, small muzzle, eye typically low-set. *(c)* Plain head, prominent forehead but no dish, and with "receding forehead." *(d)* Common head—Roman-nosed, same type as *(c)* but with arched nasal bone. *(e)* Pony type of head—dished face (between eyes as well as the whole profile). *(f)* "Camel" head, also called "elk-nosed." *(g)* Head as triangular as *(b)* but lacking the dish below the eyes. *(h)* "Jug head," almost the same width from jowl to muzzle, high-set eye. *(i)* "Platterjaw." (From Edwards, 1973.)

to a reasonably-sized muzzle. If the muzzle is too large, the horse's field of vision immediately in front will be limited. The head should not be too long or too short. Adequate length between the eyes and nostrils is important, because inspired air is heated or cooled before entering the lungs by the turbinates in the head. The turbinates are soft, bony structures that are well supplied with blood. Excessive head length tends to shift the horse's center of gravity forward. The heavy head is held out on the end of the neck, which acts as a long lever. Horses with excessively long heads also have a larger blind spot on the ground in front of them, due to interference with the visual field.

There should be adequate width between the jaws—three to four fingers—to provide adequate space for the trachea, which crosses the area.

The face shape from a side view varies from breed to breed. Arabian horses have the dish-shaped face that improves frontal vision. The convex shape, commonly called the Roman nose, impairs frontal vision. A straight profile is preferred for most breeds.

The head should be wide between the eyes and should taper down to the muzzle, giving a triangular impression. The eyes should be large, clear, and set out on the sides of the head. When the eyes are wide apart and set out on two corners of the triangle, the horse has the widest field of view. The "pig-eyed" horse has small eyes that are set too far back into the head. This impairs vision because the horse is "looking out of a tube" and finds it difficult to see behind. Impaired vision makes the horse nervous when something approaches it from behind because it cannot see what is approaching as well as can a horse with large, prominent eyes. Horses with slanted or "almond-shaped" eyes cannot see above and below their eyes as well as can horses with large, round eyes.

The horse should have large nostrils, and when the horse is working, the nostrils should flare out so that they do not restrict air flow into the respiratory system. Oxygen supply is a major limiting factor for horses working at speed over distances greater than a mile. If air flow is limited by nostril size, the horse is unable to perform to its potential.

It is important to examine the horse's teeth. The upper and lower incisors, as well as the lips, should meet evenly. If the upper jaw is too long, giving the appearance of a parrot beak, the horse has an overshot jaw. The opposite condition, in which the upper jaw is too short, is undershot jaw. Horses with these conformation faults have difficulty eating and require more careful management. The faults are inherited, so such horses should not be chosen as breeding stock.

Excessive wear on the front teeth may indicate the horse has a vice. Horses that crib wear off the front edges of their incisor teeth. Playing with a chain used to lock a gate also wears off the incisors excessively.

Examine the head to determine if melanomas or warts are present that will be irritated by the bridle or halter. Melanomas are tumors of the pigment-forming cells in the skin of gray horses. They may be located across the forehead and behind the jaw on the throatlatch area.

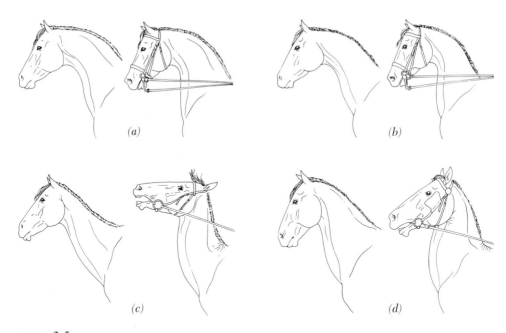

FIGURE 2-5

Various types of necks and their attitudes when the horse is collected. *(a)* Arched neck. *(b)* Swan neck. *(c)* Ewe neck. *(d)* "Close-coupled" and upside-down neck. (From Edwards, 1973.)

The throatlatch area should not be too thick, and the angle should be wide and open (Figure 2-5). The trachea passes through the throatlatch area, and if the angle is too sharp it will be compressed when the head is flexed at the poll. Horses with a jaw that is too large ("platter jaw") have the same difficulty when flexing. Because of the discomfort from flexing, these horses do not make good performance horses, which must flex for a proper headset. Periodically, they throw their heads upward or stick their noses outward to relieve the discomfort. The classic Arabian horse typifies the open throatlatch and it can flex at the poll without discomfort.

Neck

The neck should be moderately long and smooth (Figure 2-5). Proper length and weight contribute to the horse's balance and style. Flat, long, smooth muscles increase the ease and freedom of movement of the forelegs. The neck should grade smoothly into the shoulder and chest, but a definite line of demarcation should be evident. The swan-type neck is usually not discriminated against. The swan-type neck is a long, smooth neck that forms an S-shaped curve. This allows the horse to flex at the poll. The neck muscles are longer than usual and contribute to a smooth gait with a long stride because rotation of the shoulder is improved. Because of the flexation, the head is carried closer to the body, and the center of gravity or balance point is shifted posteriorly. This lightens the forehand of the horse and reduces concussion and trauma to the forelegs. In contrast, a short and very thin neck that is placed high on the chest causes the horse to have shorter strides. Lack of muscle length inhibits the movement of the forelegs and makes the horse rough to ride. Most jumping horses have long necks, which enable them to raise their forelegs up high when they are folded.

The "upside-down" neck is characterized by excess fat deposits and heavy muscling on the underside of the neck. Horses with this type of conformation encounter two problems. Flexing at the poll causes discomfort, or the horse may not be able to flex sufficiently at the poll for the desired headset. The center of gravity is shifted forward so that the horse is predisposed to common unsoundness of the foreleg, such as ring bone, side bone, and navicular disease.

A cresty neck is the opposite of the upside-down neck. Excess fat deposits are located on the crest of the neck. The excess fat can cause the crest of the neck to break or fall over to one side of the neck. The excessive weight on the forehand leads to the same problems as with the upside-down neck. Both types of horses are not well suited for performances that require fine control with reins, such as is required of the reined cow horse.

The "ewe-necked" horse has a sagging top line and bottom line neck profile. It has difficulty flexing at the poll. When such horses are stopped with a bit, they throw their heads upward. Therefore they are unsuitable for performance classes, western pleasure, and reining.

Chest

Observed from the front, the chest should not be too wide or too narrow (Figure 2-6). Either extreme is a fault. If the chest is too wide between the forelegs, the horse will have a rolling or waddling type of gait and will usually paddle with its feet. A too narrow chest causes leg interference when the horse travels, because the horse tends to criss-cross its front legs. One cause of a narrow chest in front is that the shoulders may be set too far forward on the body. The American Quarter Horse usually has a definite V-shape between its forelegs, because of its heavy muscles, but this V should not be too wide. The Thoroughbred does not have as much muscle, so it does not have as definite a V-shape.

Withers

The spiny processes of the thoracic vertebrae, located at the base of the neck and between the scapulas, form the horse's withers. The withers serve as a base of attachment for the muscles supporting the neck. They should be high enough

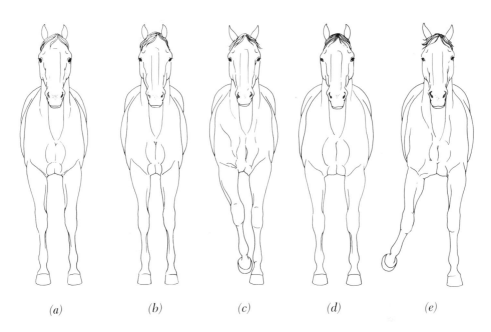

(a)   (b)   (c)   (d)   (e)

FIGURE 2-6
Front view of horse. *(a)* The proper width between legs allows correct straight forward movement of the legs. *(b)* The horse is too narrow, which may cause limb interference, as in *(c)*. *(d)* The horse is too wide in front. *(e)* Some wide-fronted horses may paddle. (From Edwards, 1973.)

to prevent a saddle from slipping to the side of the horse. If the withers are too high, the horse can easily be injured by a saddle. Withers that are too low are called "mutton withers" and predispose a horse to being clumsy and to forge. They restrict the rotational movement of the shoulder. There is usually a rolling motion to the way such horses move, so that the rider is shifted diagonally back and forth across the saddle. The saddle also tends to be displaced forward and can injure the horse. Mutton withers frequently accompany heavy front ends, shifting the center of gravity forward. This reduces the horse's athletic potential and usually shortens the length of stride. The withers should not be lower than the croup. This fault also shifts the center of gravity forward and makes it difficult for the horse to pick up its forehand and make maneuvers such as roll backs.

Shoulders

The ideal shoulder on a horse is long and sloping. Shoulder length is measured from the shoulder point to the top of the scapula. The ideal shoulder slope is 45° (Figure 2-7). The angle is measured at the intersection of (1) a horizontal

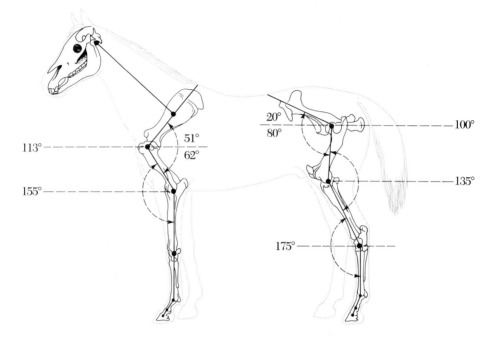

FIGURE 2-7
Articular angles for a horse.

line at the level of the shoulder point and (2) a line bisecting the scapula or shoulder blade. To judge the shoulder slope, look at the scapula, not at the shoulder point and the top of the withers. Some horses' shoulders are placed too far forward on their bodies, which gives the impression of long, sloping shoulders when in fact the shoulder is short and has a steep angle. A 50- to 52-degree slope is acceptable. The junction of shoulder and arm is one of the main concussion-absorbing mechanisms in the horse's foreleg. If the angle is too steep, the wear and tear on the leg increases as a result of the increased trauma. Horses with steep shoulders are predisposed to the common unsoundnesses caused by poor concussion absorption. The horse is usually rough to ride, because the concussion travels all the way up the horse's leg.

Shoulder length increases the stride length. A long shoulder means the muscles are working on a longer lever, giving more power and strength to the swing of the foreleg. Longer strides enhance the horse's running ability, because a horse's forelegs strike the ground a fewer number of times in each mile. Therefore, horses with short shoulders are more prone to lameness over a period of years. In addition to absorbing concussion inefficiently, short, steep shoulders make a horse look out of proportion. They are usually accompanied by short, steep pasterns, so that the foreleg looks too straight vertically. The rotational movement of shoulders that have too steep an angle is retarded, so that the foreleg moves with less freedom and ease.

A long scapula is desirable because it increases the area for attachment of the muscles that tie the foreleg to the spine. These muscles form a sling-type arrangement suspending the horse's body between the horse's forelegs without a bony attachment (Figure 2-8). The suspension aids in absorbing concussion. Shoulder muscles should be flat and smooth, not thick and bulky.

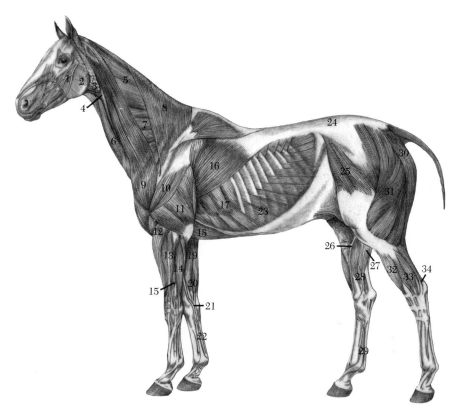

FIGURE 2-8

Muscular system of the horse. (1) Facial nerve. (2) Masseter muscle. (3) Parotid salivary gland. (4) Jugular vein. (5) Splenius muscle. (6) Sternocephalic muscle. (7) Serratus ventralis muscle. (8) Trapezius muscle. (9) Brachiocephalic muscle. (10) Deltoid muscle. (11) Triceps muscle. (12) Radial carpal extensor muscle (extensor carpi radialis muscle). (13) Common digital extensor muscle. (14) Ulnaris lateralis muscle. (15) Lateral digital extensor muscle. (16) Latissimus dorsi muscle. (17) Serratus ventralis muscle. (18) Pectoral muscle. (19) Radial carpal flexor muscle (flexor carpi radialis muscle). (20) Ulnar carpal flexor muscle (flexor carpi ulnaris muscle). (21) Cephalic vein. (22) Digital flexor tendons. (23) External abdominal oblique muscle. (24) Gluteal muscles. (25) Tensor fasciae latae muscle. (26) Saphenous vein. (27) Gastrocnemius muscle. (28) Long digital flexor muscle. (29) Digital flexor tendons. (30) Semitendinosus muscle. (31) Biceps femoris muscle. (32) Long digital extensor muscle. (33) Lateral digital extensor muscle. (34) Achilles' tendon (also called the "hamstring"), consisting of tendons of gastrocnemius, biceps femoris, and superficial digital flexor muscles attaching to calcaneus.

Forelegs

Arms

The humerus, also called the arm, should be short in relation to shoulder length. Excess arm length tends to shorten the stride, because the forearm cannot be carried as far forward. Excess length places the foreleg too far under the horse's body, causing the horse to carry excessive weight on the forehand. The arm can also be too short, decreasing the stride length.

The arms should operate in planes that parallel the body plane. If the elbow is set too close to the body—"tied in at the elbow"—the whole foreleg is rotated outward so that the horse stands toe-wide or "splay-footed." The proper space between elbow and body is about the thickness of a human hand. If the elbow is pointed outward too much, the foreleg is rotated in, and the horse stands toe-narrow or "pigeon-toed." The toe-narrow conformation fault is serious and is discussed more fully later.

The forearm should be long, with long, lean muscles that tie in close to the knee. This type of conformation allows long strides, particularly when the horse has a relatively short cannon bone. Long muscles work on a longer lever, which makes it easier to lift the foreleg and thus contributes to ease and freedom of movement.

Knees

A wide, flat, smooth, clean-cut knee is desirable. A large knee increases the supporting surface so that there is less stress per unit of area. The knee should be centered on the radius, and the cannon bone should be centered on the knee. A common conformation fault of the knee is the "offset" or "bench" knee. The cannon bones are placed to the lateral side of the knee (Figure 2-9), causing an increased stress on the medial side of the knee, particularly on the medial splint bone. When "bench-kneed" horses are subjected to hard work, such as training, before the splint bone fuses to the cannon, the horse will tend to get splints. Splints result from irritation of the interosseous ligament attaching the splint bone to the cannon. After the splint bone fuses to the cannon, the splint bone cannot move, so no irritation occurs.

The "open-knee" fault in conformation (Figure 2-9) occurs when the knee joint bones are not large enough relative to the cannon and radius bones. A depression about the size of an index finger crosses the middle of the front surface of the knee. A horse is not criticized for having open knees until after it is 3 years of age, when the knee bones stop growing. Because the bones are small relative to the horse's size, they are subject to injury during exercise. Injuries result from greater than normal concussion per unit of cross-sectional area. Excess concussion can stimulate excess bone growth, leading to unsoundness.

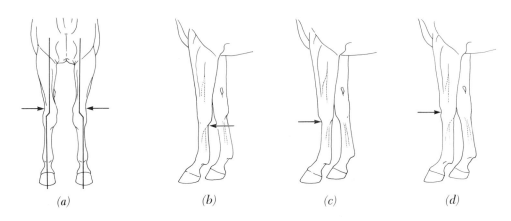

FIGURE 2-9
Common conformation faults of the knee. *(a)* "Bench" or "offset" knee. *(b)* Tied in at knee. *(c)* Chopped at knee. *(d)* Open knee.

Cannon Bones

The cannon bone should be short, compared to the forearm, so that the knees and hocks are closer to the ground than in horses with long cannons. A short cannon increases the horse's stability and increases the length of stride. The cannon area should look refined and have thin skin. The outline of the cannon bone, tendons, ligaments, and blood vessels should be prominent. The cannon bone is round but should appear flat from the side view because the tendons are located posterior to the cannon. If the cannon bone is too small for the horse, it will look set too far under the knee when viewed from the side (Figure 2-9). This conformation fault is described as "chopped" at the knee. "Tied" or "constricted" at the knee describes tendons that are too small for the horse (Figure 2-9). Immediately below the posterior aspect of the knee, the cannon area is too small and is constricted. This fault predisposes a horse to problems such as bowed tendons when it is stressed during exercise.

Pasterns

Moderately long, sloping pasterns are desirable (Figure 2-10). If the pasterns are too long, the fetlock descends too much, which places excess strain on the deep flexor tendon, sesamoids, and suspensory ligament, and predisposes the horse to bowed tendons. Well-conformed horses are usually easy riding because the excess movement of the pasterns absorbs the concussion, whereas the horse with upright pasterns is rough riding because the concussion is inadequately absorbed. Upright pasterns predispose the horse to the common pathological

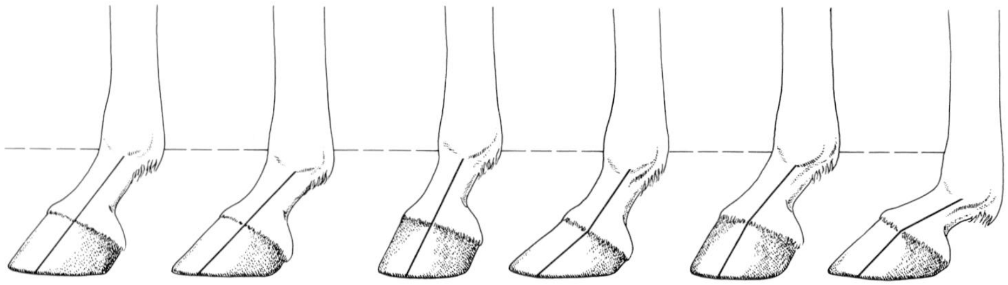

FIGURE 2-10

Slope and length of the pasterns. *(a)* Proper slope and length. *(b)* Excessive slope. *(c)* Not enough slope. *d)* Broken angle caused by excessive length of the toe. *(e)* Broken angle caused by long heels. *(f)* Broken angle caused by weak pastern.

problems resulting from the excess concussion—ring bone, side bone, and navicular unsoundnesses. The pastern angle should be the same as the shoulder angle. Upright shoulders usually are accompanied by upright pasterns. If a horse has an upright shoulder but weak pasterns with too much slope, it is subject to tendon strain.

Length and slope of the pasterns affect the foot flight arc (Figure 2-11). The ideal foot flight arc is smooth, like a quarter of a circle. Pasterns with too much slope along with a long toe do not allow a smooth arc. As the foot breaks over the toe, it suddenly goes very high in the air and then coasts back to the ground. This type of forward movement stresses the tendons. With upright pasterns, the toes are usually shorter and the heels longer. The foot breaks over the toe very easily, with a gliding motion, comes to a peak, and then abruptly comes down, striking the ground with excessive concussion. Pathological problems often develop as a result of the increased concussion.

Feet

Observe the size and shape of the horse's feet carefully. Unsound feet usually mean the horse has no use value except possibly for breeding purposes. Feet should be large but still in proportion to the horse's body. The feet are one of the main concussion-absorbing mechanisms, and if they are too small, they suffer increased wear and tear. Very few horses have feet that are too large, but one of the traits that has been selected for by breeders is small feet. Today, too many horses have small feet and develop foot unsoundness. The slope of the pasterns and the front of the hoof wall should be continuous (Figure 2-10). Improper foot care, leading to long toes, long heels, or short heels, causes the

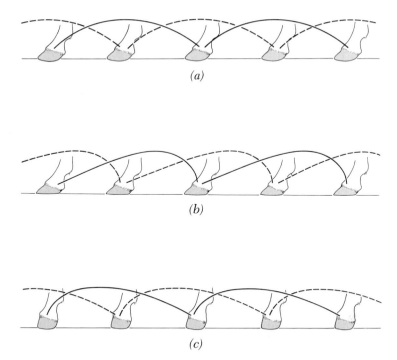

FIGURE 2-11

Foot flight arcs. *(a)* Correct arc: balanced feet. *(b)* Incorrect arc: long toes, flat heels. *(c)* Incorrect arc: short toes, high heels.

hoof wall angle to deviate from the pastern angle. These deviations stress tendons and suspensory ligaments. The hoof wall should be thick, free of defects, and open at the heel. The frog should be well developed and in light contact with the ground. Shallow heels indicate undeveloped digital cushions, and navicular disease may develop. The foot is discussed in detail in Chapter 8.

Forelegs—Front View

Viewed from the front, the forelegs should be straight (Figure 2-12). If two lines were drawn from the shoulder point to the ground, they should be parallel, bisect the forelegs, and strike the ground at 90° angles. Any deviations of the legs from these lines are conformation faults. The common faults are base-wide, base-narrow, toe-wide, toe-narrow, knock-knees, and bowlegs. Viewed from the side, a line should bisect the foreleg, touch the bulb of the heel, and strike the ground at a 90° angle. A deviation of the foreleg from this line is a conformation fault. Commonly observed faults are buck-knees, calf-knees, camped under, and camped out.

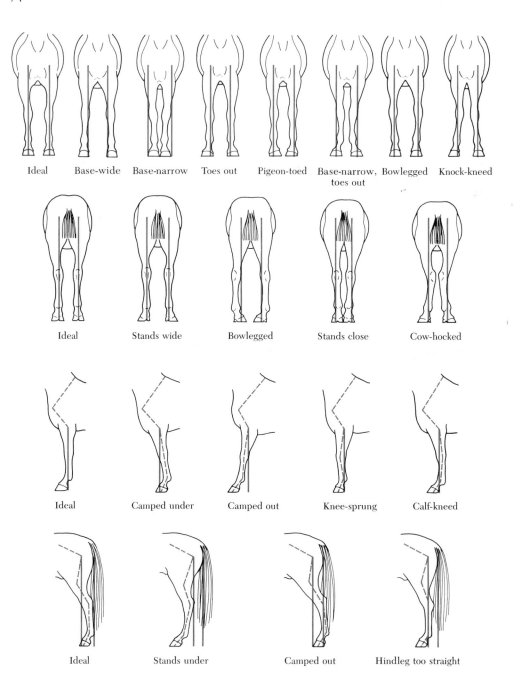

Ideal | Base-wide | Base-narrow | Toes out | Pigeon-toed | Base-narrow, toes out | Bowlegged | Knock-kneed

Ideal | Stands wide | Bowlegged | Stands close | Cow-hocked

Ideal | Camped under | Camped out | Knee-sprung | Calf-kneed

Ideal | Stands under | Camped out | Hindleg too straight

FIGURE 2-12

Conformation defects in the front and rear legs and viewed from the front, rear, and side. (From *4-H Horse Judging Guide,* Cooperative Extension, Pennsylvania State University, University Park, Pennsylvania.)

Base-Wide

The base-wide fault occurs when the distance between the center of the hooves is greater than the distance between shoulder points. Horses that have narrow breasts tend to be base-wide. The base-wide condition causes excess trauma to the medial (inside) side of the horse's leg, predisposing it to medial ring bone, medial side bone, medial splints, and navicular disease. When the horse travels, the foot breaks over the medial side of the toe, wings inward, travels back outward, and strikes the ground on the medial part of the hoof wall. If the horse is barefooted, the medial side of the hoof wall is worn down at a faster rate than the lateral side because the foot breaks over and lands on the medial side. Some horses are exceptions and may paddle when they travel.

If the horse is toe-wide and base-wide, the faults in way of going are exaggerated. Such horses have a greater tendency to wing inward and to interfere with the opposite foreleg. Often the horse just brushes the medial aspect of the opposite coronary band; this defect is recognizable by the hair being worn off in that area. In exaggerated cases, the opposite cannon bone may be struck during each step.

Base-Narrow

The base-narrow fault in conformation is characterized by the distance between the centers of the hooves being less than the distance between the shoulder points. When the horse travels, the foot may break over the center of the toe or it may break over the lateral side of the toe. If the foot breaks over the center of the toe, it almost always lands on the lateral side of the hoof wall first. The excess concussion on the lateral (outside) side of the foot and leg leads to concussion-related unsoundnesses on that side. If the horse is not shod, the lateral hoof wall wears off at a faster rate than the medial side.

Toe-Narrow

Frequently, the horse is toe-narrow as well as base-narrow. The foot breaks over the lateral side of the toe and lands on the lateral side of the hoof wall, predisposing the horse to unsoundness on the lateral side of the foot and leg. Such horses usually paddle when they travel. If the horse is toe-wide and base-narrow, the foot still breaks over the lateral side of the toe and lands on the lateral side of the hoof wall. This fault in conformation places the greatest strain on the fetlock and is considered the worst type of foreleg conformation. Since the foot breaks over the lateral side of the toe and lands on the lateral side of the hoof wall, the foot almost always wings inward. Because the forelegs are close together, there is almost always leg interference.

The toe-narrow fault shifts the center of gravity forward. When the horse is ridden, the rider feels top-heavy, particularly when turning or working in circles. The horse does not respond to the bit as quickly as it should. If the condition is extreme, the horse shifts its weight from side to side in a rolling motion.

Toe-Wide

The toe-wide or splay-foot fault causes the foot to break over the medial side of the toe, wing inward, swing back outward, and land on the medial side of the hoof wall. The excess wear and trauma occurring on the medial side of the foot and leg predisposes the horse to the unsoundnesses resulting from excess concussion. The toe-wide fault does not affect a horse's balance or its base of support. Such horses normally can perform quite well if there is no leg interference.

Knock-Knees and Bowlegs

Knock-knee and bowleg faults should be discriminated against because of the excess strain placed on the knee and lack of mobility of the leg.

Leg Rotation

Some horses have a straight leg down to the knee and then a rotation of the leg toward the inside or outside. Other horses have straight forelegs to the fetlock and then pasterns set so that they are rotated inward or outward. Both faults result in a horse that stands toe-wide or toe-narrow.

Buck-Knees

The buck-knee fault exists when the knees are in front of the line that bisects the foreleg. Other terms used to describe the condition are "goat knees," "over the knee," and "knee-sprung." A slight buck-knee is not too serious and seldom causes problems. Buck-kneed horses are clumsy and have a tendency to stumble, particularly if the toe is too long. The fault places an excess strain on the flexor tendons and on the sesamoid bone. Therefore, such horses tend to have bowed tendons and sesamoiditis. If the condition is severe, their knees may shake and tremble when they stand.

Calf-Knees

One of the worst faults in foreleg conformation is the calf-knee. The knee is behind the line that should bisect the leg and is in a compromised position when the foot strikes the ground. This causes excess strain on the knee, which breaks easily, particularly when muscles are tired and the knee goes further backward. Such horses tend to chip their knee bones on the front edges as their knees move backward when their feet strike the ground.

Camped Out or Under

The "camped out in front" fault occurs when the forelegs are not placed square under the horse. They are placed on an angle so that the feet are too far forward. This condition is usually observed in horses that are foundered or have navicular disease. If it is their normal stance, they are usually poor athletes because they are unable to move their forequarters very easily.

The opposite condition is "camped under in front." When the legs are directed backward under the horse, it carries too much weight on the forehand. Such horses have shorter strides and a tendency to stumble because of a low foot flight arc.

Side View

After viewing the front legs, view the horse from the side to evaluate the body. The body should be properly balanced. The back and loin should be short and well muscled. The chest should be deep and full. Long, well-sprung ribs are desirable.

Back and Loin Muscles

Muscling in the horse's back and loin is important to support the weight of a rider (Figure 2-13). The loin has no support other than the vertebral column. Lack of muscling usually leads to a swayback or sagging back. Excessive length in the back and loin contributes to the swayback condition and affects the horse's gait. Long-backed horses have a rolling motion n their hindquarters. They have a tendency to cross-fire—the hindfoot strikes the opposite foreleg. This is a problem in racing pacers. The roach-backed horse (with a convex back) is discriminated against because the back lacks flexibility. These horses have a short stride. They tend to overreach and to forge because the distance between the forelegs and hindlegs is shortened.

Chest and Ribs

The chest contains the lungs and heart. A deep, full chest provides adequate room for these vital organs. Horses that lack depth through the chest usually do not have endurance capacity. Well-sprung ribs serve to protect the vital

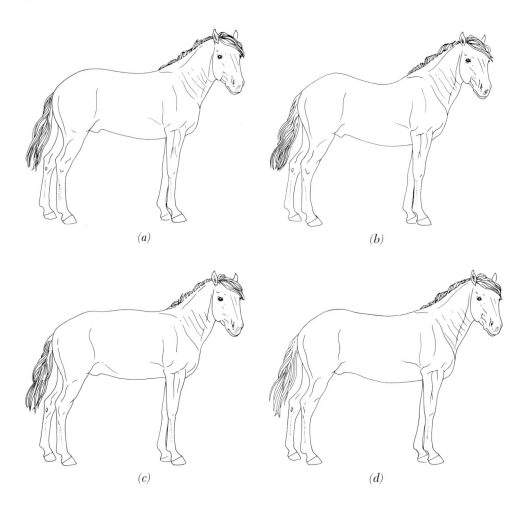

FIGURE 2-13
Side view of horse's back to evaluate conformation. *(a)* Good top line. *(b)* Sway back. *(c)* Roach back. *(d)* Long.

organs and as a point of attachment for muscles of the foreleg. Horses with flat ribs are referred to as being "slab-sided" and lack chest capacity (Figure 2-14). This fault usually makes horses hard to keep. They require more feed to maintain body condition. Flat ribs are often associated with legs that are placed too far under the horse's body, so such horses are heavy on the forehand. The ribs can be so arched that the horse becomes "barrel-chested." It is difficult to keep a saddle on such a horse, and pressure from the saddle is not distributed over a sufficient surface area of the chest. The result can be sore spots on the back and sides. Look under the chest to determine if the horse has a prominent sternum. A prominent sternum is irritated by the girth.

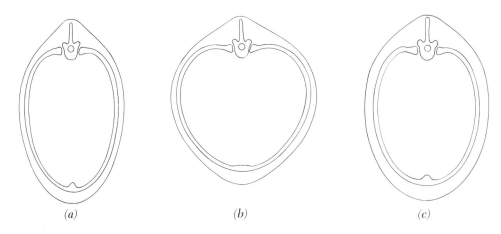

(a)   *(b)*   *(c)*

FIGURE 2-14

Cross-section of the thorax. *(a)* "Slab-side." *(b)* "Round" cylinder. *(c)* Ideal shape: large heart and lung capacity inside. (From Green, 1969.)

Coupling

The junction of the loin with the hindquarters is the coupling. The coupling should be heavily muscled and wide to provide a strong support for the loin. The coupling should be deep so that the horse is not cut up (lacks depth) in the flank. Depth in this area contributes to a large capacity for the digestive tract and gives the horse a long underline. Lack of depth in the flank should not be confused with horses in excellent physical shape. Racehorses and endurance horses lack depth in the flank when they are well trained and fit. Horses lacking depth in the flank are sometimes described as "wasp-waisted" or "hound-gutted."

Hindquarters

The hindquarters are the propulsion mechanism of the horse. The croup should be long, fairly flat, muscular, broad, and have the tail set well on it (Figure 2-15). A long distance from the hip point to the hock, coupled with a desirable type of croup, allows speed and athletic ability.

The croup angle affects the movement of the hindlegs. A flat croup causes a long, flowing stride, and the foot stays close to the ground. A steep croup causes the hindleg to be driven into the ground. The steep croup is desirable for horses that need a quick start and will run for a short distance. Therefore, Arabian breeders want a fairly flat croup on their horses, while the calf roper wants a steep croup. If the croup is too flat, the hindlegs are carried too far behind the horse and limit its athletic potential. Racing Thoroughbred horses

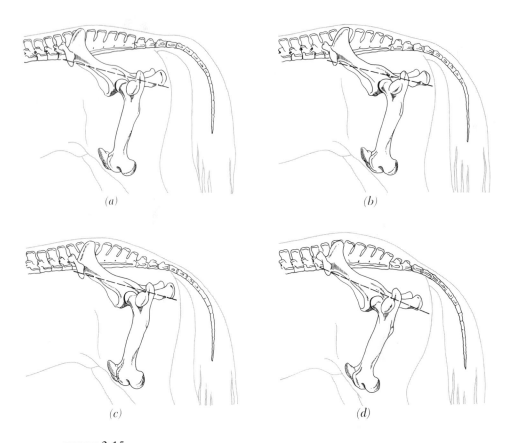

FIGURE 2-15

Relationship between slope of pelvis and the sacrum of Arabians. *(a)* The so-called level croup desired in the Arabian horse. However, only the sacrum is level (and even it has some slope) and the pelvis itself has just as much slant as the "sloped" and "steep" (for an Arabian) croups shown in *(b)* and *(c)*. In *(b)* the actual position of sacrum is the same as in *(a)* but the crest of the ilium is higher and somewhat farther back. This also has the effect of making the back appear longer. When exaggerated in height, the crest of the ilium is very prominent and is termed the "jumping hump," because it is commonly seen in hard-conditioned jumpers and hunters. In *(c)* the sacrum itself is straight; it slopes upward in *(a)* and *(b)*, while the topline thereof, which of course makes the outline of the croup, slants down, making a "steep croup," although the pelvis itself has the identical slope of *(a)* and *(b)*. Therefore, this kind of slope would have no effect on speed or power compared to the other two, regardless of "raciness" a somewhat sloped croup may imply. Because of the angle of the sacrum, the tail is set on much lower, further exaggerating the impression of sloped croup. On the other hand, *(d)* shows the goose rump, common to draft horses. Here, the pelvis itself is very slanted, and the sacrum follows suit, also slanting downward, with tail very low set on. (From Edwards, 1973.)

should be intermediate. The length of the croup affects the length of the stride. Short croups are associated with short strides, and long croups with long strides.

The degree of muscling in the hindquarters depends on the breed of horse

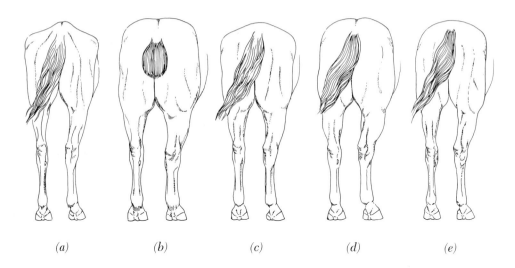

(a)   (b)   (c)   (d)   (e)

FIGURE 2-16

Muscling of the hindquarters. *(a)* Peaked appearance of hips on emaciated horse, also showing wasted condition of thigh muscles. *(b)* The broad "double" hindquarters of a draft horse. *(c)* "Rafter-hipped," "flat-hipped," or "ragged-hipped"—too flat across the top and with hip bones extending beyond the thigh muscles. *(d)* The "pear shape" desired in a Quarter Horse, with thigh muscles bulging well beyond the point of hip. *(e)* Normal hindquarters, the croup well rounded from this rear view, and thighs forming a square, rather than "pear" shape. "Rounded hips" refers to this rear view of the hips and is not to be confused with a "round croup" (apple rump) as viewed from the side. (From Edwards, 1973.)

and its purpose (Figure 2-16). Horses that need power need larger muscles, and horses that work at speed need long, slim muscles. The width at the stifle should be greater than the width at the hip point when viewed from behind. Lack of muscling makes the hip points wider, and such a horse is referred to as being "rafter-hipped." Lack of croup muscling is described as "goose-rumped," and such a horse lacks power and endurance. Thigh muscles need to be powerful. Stifle muscles should extend well down toward the hock, so that the muscles are long. The gaskin should be heavily muscled to the inside and outside of the leg, and the muscles should be long. The gaskin muscling largely controls the course of the leg in forward movement, provided that the bone structure is well aligned. Excess muscling on one side or another causes the leg to move toward the side with the excess muscle.

The length and set of the femur, tibia, and cannon bones affect the horse's way of going and its speed. The femur is set to be inclined forward, downward, and slightly outward. The femur must be turned outward slightly to allow the stifle joint a full range of movement without interference from the abdomen. The tibia should be long for maximum development of speed. Tibia length increases the area for attachment of muscles that provide driving power for the hindquarters. A short tibia decreases the length of the stride.

Hocks

A good hock joint is wide and deep, well defined, and strong. The skin is fine textured and loose fitting, but the prominence and depressions are well marked. The angle or set of the hock is important. The angle when viewed from the side should be about 175° (Figure 2-8). If the angle is too large, the leg is said to "lack set." A straight hock increases trauma to the hock bones. Such a horse is predisposed to bog spavins, thoroughpins, and upward fixation of the patella. Too small an angle at the hock results in the sickle-hock type of conformation. This places an excess strain on the plantar ligament, which attaches just below the hock in the posterior surface of the cannon bone. The excess strain causes inflammation and enlargement of the plantar ligament at the point of attachment, which is referred to as a "curb." The sickle-hock conformation reduces stride length, and such horses tend to forge and overreach. Horses that must perform a lot of sliding stops are not discriminated against for a slight sickle hock, because the horse can get its hocks up under itself and can work.

The hind cannon bones should be short, and they are larger than the front cannon bones.

Hindleg Faults

The set of the hindlegs should be viewed from the side and from behind (Figure 2-12). A line drawn from the point of the buttocks should touch the back of the hock, parallel the cannon, and strike the ground perpendicularly about 3 to 4 inches (76 to 102 mm) behind the heel. Several conformation faults can be observed from this view. The cannons of the sickle-hocked horse do not parallel the line but are directed forward at an angle to the line. Horses may be camped out or in behind. When the horse is camped out, the hindleg is posterior to the line; when camped under, the leg is anterior. The camped-out-behind fault prevents the horse from getting its legs under itself to move. Such horses cannot collect themselves. They tend to jab their legs into the ground and are unable to lift their bodies sufficiently to be good hunters and jumpers. If the horse is camped under behind, the distance between the fore- and hindlegs is decreased, and limb interference may occur. The fault may be associated with sickle hocks or a roached back.

From behind, lines dropped from the point of the buttocks to the hocks should bisect the hocks, cannons, and heels in half. Faults include base-wide, base-narrow, cow hocks, and bow-legs. Very seldom does a horse have the base-wide conformation behind. The base-narrow fault is common and increases the opportunity for hindleg interference. Excess muscling on the outside of the leg usually causes the horse to be base-narrow. Excess strain is placed on the outer side of the leg.

The cow-hocked horse stands with hocks inside the two parallel lines, the hock points rotated inward, and base-wide and toe-wide feet. Cow hocks place an excess strain on the inside of a horse's leg, predisposing it to bone spavin.

Power and movement in the hindquarters are lost, because the limbs move upward and outward, not straight ahead. Whenever a movement requires a straightforward thrust of the hindleg, the cow-hocked horse is limited. To a rider, the horse feels weak behind. The horse resents binding its joints and consequently develops bad habits such as rhythmic irregularity in gait, a high head carriage, and head tossing.

The horse is bowlegged behind when the hocks are wider than the two parallel lines. The points of the hock turn outward. As the horse moves, the hocks tend to rotate. The horse seems stiff in the hindquarters because it lacks flexibility. Any straightforward impulsion is impaired. It has weak hindquarters and fatigues quite easily.

One type of hindleg conformation that is not discriminated against and is preferable for many types of horses that are well-trained athletes is a hock turned inward so that the hindfeet are slightly toe-wide. The cannons are parallel and straight. This allows the horse to work off its hocks and to be collected, permitting maximum extension of the hindleg for long strides. The horse usually develops a good sliding stop.

BLEMISHES AND UNSOUNDNESSES

Before the horse is purchased, it should be evaluated for soundness. If possible, a veterinarian should make the examination. The veterinarian may have to use specialized equipment such as x-ray machines to detect or verify some unsoundnesses that are not visually apparent. The word *soundness* is a general term that can mean anything from a slight injury to a defect that renders the horse useless. A horse may be sound for one purpose but unsound for another intended use; for example, a crippled horse may be sound for breeding purposes but not for riding. Horsepeople refer to various defects as *blemishes* or *unsoundnesses*. A blemish is an injury or imperfection that affects the horse's value but not its serviceability. Injuries such as rope burns and insignificant wire cuts are normally classified as blemishes. An unsoundness is an injury or abnormality that affects serviceability. Most unsoundnesses drastically reduce the horse's value. In many instances, it is difficult to definitely classify a defect as a blemish or as an unsoundness. If the horse is sound for certain uses, a defect is a blemish. For another use, the horse may be unsound and of little value. At the time of an injury, the horse may be unsound but once the wound is healed it is a blemish.

It is important for a horseperson to be able to recognize the common blemishes and unsoundnesses, know their causes, and understand how they affect a horse's serviceability. When selecting breeding animals, it is extremely important to determine whether an unsoundness was caused by poor management such as accidental injuries, nutrition, or strenuous training or whether it was caused by faulty conformation. If caused by faulty conformation, do not use the horse for breeding.

Head

There are several unsoundnesses of the horse's head. Most of these unsoundnesses are quite serious.

Blindness

Check for blindness in one or both eyes. The first indication that a horse may be blind is the way it walks. Blind horses hesitate before taking a step, and their ears move around more than normally. To verify blindness, move your hand close to the eye. If the horse can see, it will make a reflex movement.

Periodic Ophthalmia

Commonly called "moon blindness," periodic ophthalmia is characterized by a cloudy or inflamed condition of the eye. The eye goes through a series of recoveries followed by relapses. Each series becomes worse, eventually ending in blindness. At one time, the disease was believed to be the result of riboflavin (one of the B vitamins) deficiency. However, there are several other causes of moon blindness, such as leptospirosis, the parasite *Filaria equina* within the eye, reaction to systemic parasitism, or antigens developed as a result of repeated streptococcal infections. Moon blindness should not be confused with night blindness.

Night Blindness

Night blindness is faulty vision during the twilight hours and is caused by a vitamin A deficiency. Night blindness can be difficult to detect. It is usually evident if the horse is forced to move about duringtwilight hours in strange surroundings. It stumbles over unfamiliar objects or may bump into things as it moves around.

Poll Evil

Poll evil is an inflamed condition resulting from a bruise in the poll region. The swelling usually contains pus. A veterinarian should treat it, because healing is very slow and difficult. It may break out again after it appears to be cured. The bruise is caused by someone hitting the horse between the ears with a whip or club or by the horse accidentally bumping its head in a barn or trailer. It may also be caused by the horse hitting its head on the ceiling of a stall when the horse rears up.

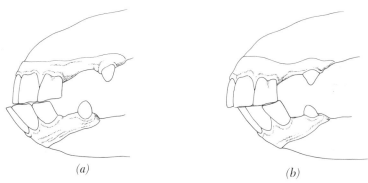

(a)   (b)

FIGURE 2-17
Side view of how teeth meet for (a) undershot jaws and (b) overshot jaws.

Overshot and Undershot Jaw

The overshot jaw, commonly called "parrot mouth," occurs when the lower jaw is shorter than the upper jaw (Figure 2-17). The opposite condition is undershot jaw. Both conditions are hereditary, and the overshot jaw is believed to be carried on a dominant gene. With either unsoundness, the horse has difficulty eating and requires extra care.

Wobbler

Wobbler describes various conditions that affect a horse's coordination by having their effect on the spinal cord. A so-called true wobbler is due to spinal-cord compression in the neck. The symptoms may appear at any time from birth to 3 or 4 years of age. In the early stages, it is difficult to recognize. It begins with hindlimb incoordination. Both hindlimbs are affected. There is a gradual change in that the incoordination becomes worse. The signs of incoordination are a dragging of the toes of the hindfeet, and errors in the rate, range, force, and direction of movement. The symptoms may increase in severity until the horse falls down if it tries to make a sudden change in the direction of movement. It is a serious unsoundness because true wobblers do not improve. They are not safe for riding purposes because of the incoordination. The condition may be genetic so it is doubtful if they should be used for breeding purposes.

Shoulders

Today you see very few horses with unsound shoulders. However, two conditions are worth mentioning.

Fistulous Withers

Fistulous withers are inflamed by bruising. In the past, the condition was caused by pressure from a horse collar, but today it frequently occurs as a result of a badly fitting saddle. The saddle either has a narrow tree and is being used on a wide-backed horse, or it has a wide gullet and is being used on a horse with high withers.

Sweeny

Opinions vary regarding the true definition of a "sweeny," but the term actually applies to any group of atrophied muscles, regardless of location. Most people confine its usage to the shoulder muscles. Muscle atrophy or degeneration results either from disuse or from loss of nerve supply. In "shoulder sweeny," the nerve crossing the spine of the shoulder blade has been damaged. It is usually easy to see, but in the early stages after the injury it may be difficult to see if no lameness is present. However, the horse's performance capability is limited when it needs the atrophied group of muscles. There is no known treatment, but nerve regeneration may occur. Nerve regeneration usually takes a long time—at least 6 months. For cosmetic value, some horse traders inject irritants into the area to fill it with scar tissue so they can hide the unsoundness.

LEG UNSOUNDNESSES

Forelegs

Bucked Shins

Bucked shins are a temporary unsoundness that occurs primarily in racing 2-year-olds and some 3-year-olds and other horses that are trained hard (Figure 2-18). It occurs infrequently in older horses. Shins are usually bucked during the first few weeks of training when the horse is overworked. It is a very painful inflammation of the periosteum (bone covering) along the greater part of the front surface of the front cannon bone. The horse becomes quite lame and is extremely sensitive when the slightest pressure is applied to the shin. Bucked shins are caused by constant pressure or trauma to the periosteum, resulting from concussion during fast workouts or riding for too long a period before the horse is in good physical condition. The horse usually recovers after 30 to 60 days' rest. If it becomes lame again after the rest, the lameness may be caused by "saucer" fractures of the front cannon bone.

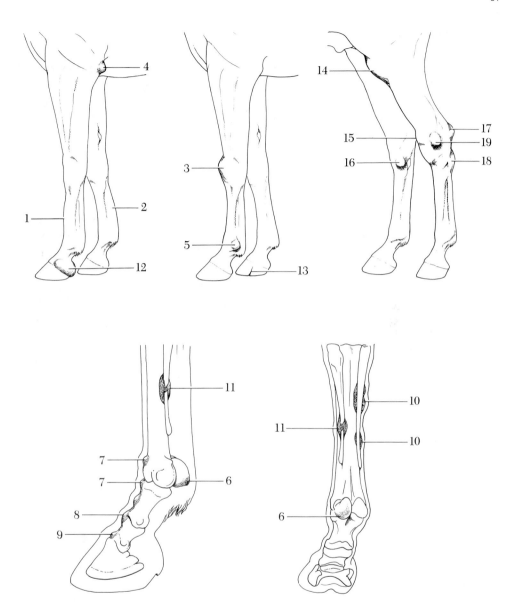

FIGURE 2-18

Unsoundnesses and blemishes. (1) Bucked shins. (2) Bowed tendon. (3) Popped knee. (4) Shoe boil. (5) Wind puff. (6) Sesamoiditis. (7) Osselet. (8) High ring bone. (9) Low ring bone. (10) Splint. (11) Broken splint bone. (12) Side bone. (13) Quarter crack. (14) Stifled. (15) Bog spavin. (16) Bone spavin. (17) Capped hock. (18) Curb. (19) Thoroughpin.

Bowed Tendons

A bowed tendon is a result of a severe strain and/or wear and tear of the deep flexor and/or superficial flexor tendon (Figure 2-18). The tendon sheath separates from the tendon (attachments are torn) so that there is a hemorrhage inside the tendon sheath. There may also be some tearing of the tendon fibers.

Bowed tendons are commonly seen in the front legs of horses that work at speed. They usually occur when the horse is fatigued and its entire body weight is on one leg. Contributing or predisposing factors also include:

1. Long, weak pasterns (the fetlock joint moves too far downward and stretches the tendon).

2. Long toes (this condition retards the breaking over the foot and consequently places a strain on the tendon).

3. Overexertion.

4. Fatigue.

5. Improper shoeing.

6. Muddy tracks.

7. Weak tendons.

The condition primarily occurs on forelegs, but it occasionally occurs on a hindleg. It shows up as a thickened enlargement on the tendons that occupy the posterior space in the cannon regions—between knee and fetlock joints or between hock and fetlock joints. Only about 20 percent of these horses are sound for heavy work after they recover. Therefore, they are usually unsound for use as hunters, jumpers, cutting horses, or rope horses. They usually rebow the tendon when they are worked hard.

Ring Bone

Ring bone is new bone growth on the first, second, or third phalanges (Figure 2-18). It may occur on any foot, but is most common on the forefeet. It is caused by an excessive pulling on the ligaments of the coffin and/or pastern joints, trauma, a direct blow to the phalanges, concussion from constant work on hard surfaces or hard work over the years, or poor conformation, such as straight pasterns and toe-in or toe-out conditions. The periosteum (bone-forming tissue) is disturbed, and new bone growth results. Ring bone is classified in two ways: (1) position on bones and (2) joint involvement. High ring bone appears on the lower end of the first phalanx and/or the upper end of the second phalanx. Low ring bone appears on the lower end of the second phalanx and/or the upper end of the third phalanx. The excess bone growth may be on either side of the bones or all the way around. Articular ring bone involves the joint surface of the pastern or coffin joints. Periarticular ring bone is around the joint but does not involve the joint surface. The following terminology is used to describe ring bone: periarticular, high ring bone; periarticular, low ring bone; articular, high ring bone; articular, low ring bone. Horses with ring bone often do not show very many symptoms. If there is no lameness, the enlargement may go unnoticed. In articular ring bone, there is always lameness, but in periarticular ring bone there is usually none. If the ring bone is high periarticular, the horse may never be unsound unless the ring bone becomes articular. Low ring bone is always regarded as unfavorable, because the animal travels on its heels.

Shoe Boil or Capped Elbow

Shoe boil or capped elbow is a soft, flabby swelling caused by an irritation of the elbow (Figure 2-18). There are two common causes: (1) injury from a long heel on a front shoe or the heel calk of the shoe and (2) injury from contact with the surface the horse is lying on. It should be treated in the early stages by applying tincture of iodine daily and by using a shoe-boil boot or roll. The roll is strapped around the pastern and prevents the heel of the foot or shoe from pressing on the elbow while the horse is lying down. Long heels should not be left on front shoes.

Splints

Splints are abnormal or new bony growths on the cannon or splint bones (Figure 2-18). They are caused by a disturbance of the fibrous interosseous ligament between the second and third metacarpal bones or between the third and fourth metacarpal bones; that is, disturbance of the ligaments between the cannon and splint bones. This disturbance can be caused by excessive strain or by a blow to that area. Strain causes an increased blood supply to the area, and the area is stimulated to form new bone. The most common cause of splints is overworking young horses, which places too much strain on medial splint bones. Hard stops, fast turns, jumping on hard ground, and galloping on hard surfaces are particularly straining. Gaited horses with high action where the foot strikes the ground very hard (rack) are predisposed to splints. Horses with "bench knees" tend to develop splints because of the extra strain on the inner (medial) splint bone. Similar problems occur in all crooked legs: toe-in, toe-out, straight pasterns, and so on. Young horses in training should wear splint boots to protect their legs when they do turns and roll backs. Splints usually cause lameness only in the early stages. At first, the horse may walk all right but trots lame. A little later, it is also lame when it walks. There may be no swelling, but careful inspection will reveal a hot, painful spot in the splint area. In time, if not treated, the splint calcifies and a permanent hard lump appears.

At that time, lameness disappears, and the lump is considered a blemish rather than an unsoundness. However, if it is the result of "bench knees" or other conformation faults or if it interferes with a tendon or ligament, it is always an unsoundness. Do not confuse splints with a broken splint bone. They are not the same, but bony knots may be present in both. A broken splint bone is an unsoundness if it interferes with a ligament, tendon, or the carpal joint.

Wind Puffs

Wind puffs are sometimes referred to as windgalls or "puffs" (Figure 2-18). They are enlargements of fluid sacs located immediately around the pastern or fetlock joints—front or back feet. They result from too hard or too fast work on hard surfaces or just heavy work. Very few old horses escape them. The size of a wind puff may be reduced by applying cold packs followed by a liniment, but the puff will reappear when the horse is exercised. Almost all roping horses, as well as other horses that have been used very hard, have wind puffs. Usually, they are not serious, and no permanent benefit results from treatment. They indicate that the horse has been worked hard for a period of time in its life.

Osselets

Osselets are inflammatory conditions around a fetlock joint (Figure 2-18). They result from strain or pressure on immature bones. Concussion is often a major cause, so horses with upright pasterns are predisposed to osselets. Osselets are easy to recognize because there is a fairly well-defined swelling on an area of the fetlock joint. If you touch it, it feels like putty or mush, due to hemorrhage, and fluid collects beneath the periosteum. The osselet may be warm or hot. The horse usually travels with a short, choppy stride and shows evidence of pain when the fetlock joint is flexed. An osselet usually affects the horse for several years, because it is an arthritic condition.

Popped Knees

Popped knees are inflammatory conditions of the knees (Figure 2-18). Popped knees usually have a sudden onset. They are caused by sprained or strained ligaments that hold the knee bones in position or by damage to the joint capsule, followed by increased fluid within the capsule. Popped knees are most common in racehorses during hard training but may occur in other horses that are worked too hard. Trauma can lead to popped knees, as when a horse paws the stall or trailer and hits its knees. Horses hit their knees when they eat out of wide feed troughs or hang their heads over stall doors while stomping flies. Bad conformation, such as bench-knees or calf-knees, predisposes horses to the condition. Horses that have popped knees rarely regain a degree of soundness that allows them to race or work hard again. Often popped knees are associated with articular bone growth.

Sesamoiditis

Sesamoiditis is an inflammation of the proximal sesamoid bones that are located at the back of the fetlock joint (Figure 2-18). It results from an unusual strain to the fetlock area and is common in racehorses, hunters, and jumpers. Sesamoiditis can be detected when the horse is moving. The horse will not allow the fetlock to descend to a normal level and shows that it is in pain. On closer examination, swelling is evident. If pressure is applied to the sesamoid bones, the horse will flinch. In chronic cases, the periosteum is disturbed and there is new bone growth.

Mud Fever

Mud fever is an inflammation of the superficial layers of skin, accompanied by an exudation (discharge). It is caused by the action of mud—exactly how, no one seems to know. It usually affects the limbs from the fetlock upward when they have been repeatedly splashed with mud and is common in horses kept in a muddy paddock or pasture. The skin becomes thick, tender, and covered with scabs. When the scabs come off, the hair falls off with it. Mud fever is the reason for the old saying, "Don't wash off mud but allow it to dry and then brush it off." The problem seems to be aggravated by washing the legs off.

Feet
Contracted Heels

Contracted heels usually occur in the front feet but also occur in the hindfeet. The condition is characterized by drawing in or contracting of the hoof wall at the heels. Improper shoeing causes and/or aggravates it. Contracted heels can be caused by lack of frog pressure if the frog is cut out each time the horse is shod. Another common cause is lengthening of the toe (long toes), because the slight expansion of the hoof is decreased or obliterated when the foot strikes the ground. The retraction force in the foot is greater, and the heels start to contract inward. In true contracted feet, the sole becomes abnormally dished. This puts pressure on the third phalanx and thus causes lameness. In some cases, the condition can be corrected in 1 or 2 years. A Chadwick spring can be used, the heels of the shoe can be slippered, or the hoof wall can be weakened with a rasp.

Corns or Bruised Sole

Corns are bruises of the soft tissue underlying the sole of the foot, marked by a reddish discoloration of the sole immediately below the affected area. They can be caused by fast work on rough, hard surfaces; flat soles; stepping on small, hard objects; and/or poor shoeing. If the shoes are left on the feet too long, the heels of the shoe are forced inside the wall and press on the sole at the angle of the wall and the bar. Usually, the working horse must rest until the bruise heals; in some cases, the condition may be chronic. Chronic lameness is usually caused by new bone growth on the third phalanx.

Laminitis or Founder

Founder is an inflammation of the laminae of the foot. It is characterized by a passive congestion of the laminae with blood. Severe pain results from the inflammation, which presses on the sensitive laminae. It can have several causes:

1. Overeating can cause laminitis. The toxin histidine formed in grain during digestion is decarboxylated to histamine and is thought to cause laminitis. The toxin is not present in cooked grain. Wheat and barley have more of it than other cereal grains. It can be transmitted via stomach contents of affected animals. The symptoms occur 12 to 18 hours after eating the grain.

2. Overwork or road founder is caused by concussion from a hard surface. If a horse is not used to exercise, riding it for several hours may cause founder.

3. Water founder is caused by an overheated horse drinking large amounts of cold water. The cause in this case is not known. An overheated or hot horse should always be cooled out before it drinks.

4. Postparturient laminitis is caused by a uterine infection, and the fetal membranes may or may not be retained.

5. Grass founder is common among overfat horses that are grazed on lush pastures. The cause is not known but may be histamine.

6. Miscellaneous causes of founder include viral respiratory injections, drugs, and overeating beet tops.

All the feet are susceptible to laminitis, but the forefeet are more susceptible than hindfeet. If all four feet are involved, a standing horse carries its hindfeet well up under it and carries its forefeet posteriorly. It has a very narrow base of support. If only the forefeet are involved, the hindfeet are carried well up under the body and the forefeet are carried forward. The horse is obviously in pain and is reluctant to move. Heat is present over the sole, wall, and coronary band. You may observe muscle tremors, increased respiration, and fever. These are usually due to anxiety and pain. If your horse has acute founder, stand it in a cool mud hole or wrap gunny sacks around the fetlocks and feet and keep them wet with a hose. Call the vet.

In chronic cases or after a severe case, several characteristics enable recognition of a foundered horse. The sole is dropped and flat and is usually very flakey. The hoof wall grows very rapidly. Because the toe grows faster than the heel, the feet may develop a long, curled-up toe.

A very heavy ring forms around the wall of the foot for the rest of the horse's life. A horse that has been foundered and cured almost always has rough rings around the hoof wall. (However, you can also detect different phases of nutritional status by looking at the rings on the feet. In this case, the rings merely indicate changes in the rate of hoof wall growth and should not be confused with founder.) In severe cases, separation of the laminae, rotation of the third phalanx, and penetration of the third phalanx through the sole may occur. Such horses are lame for the rest of their lives, as are the ones in which the third phalanx rotates but does not penetrate the sole.

Navicular Disease

Navicular disease is an inflammation of the small navicular bone of the front foot. It is actually a degenerative and erosive lesion of the fibrocartilage surface of the bone that begins as an irritation of the navicular bone and deep flexor tendon. It is impossible to determine the exact cause. Some predisposing conditions are hard work, upright pasterns, and small feet. Trimming the heels too low on horses with upright pasterns increases the pressure of the flexor tendon on the navicular bone. Affected horses usually go lame; have a short, stubby stride; and usually point the lame foot when standing.

Treatment consists of special shoeing to make it easier for the foot to break over and to reduce the anticoncussive activity of the deep flexor tendon against the navicular bone. Raising the heels and rolling the toe allows the foot to break over faster and reduces concussion to the navicular bone. To further reduce trauma, bar shoes or pads are used. This type of correction tends to contract the heels by reducing frog pressure. The tendency for the heels to contract can be reduced by slippering the heels of the shoe so that the wall slides outward as the foot strikes the ground. Wall expansion can be further aided by thinning the quarters of the wall with a rasp. If special shoeing does not help, the navicular nerve can be cut so the animal does not have any sensation in the foot. If you are buying a horse and want to see if the nerves have been cut, look for two small scars—½ to 1 inch long (13 to 25 mm)—above the bulbs of the front feet. It is my opinion that these horses are useless for hard work, because there is continual wear on the tendon, which may rupture.

Quarter Crack or Sand Crack

A vertical split in the wall of the hoof is called "quarter crack" or "sand crack." The crack may extend down from the coronet band or up from the bearing surface. It is located in the quarter part of the hoof (if it is located on the toe, it is called a "toe crack"). The condition results from the hoof becoming too dry and brittle, an injury to the coronet band, long hoof walls that start to split, or not having the hoof wall surface level when shod. When it first starts, a crack can usually be stopped by burning a moon-shaped crescent at the tip of the crack or by filing a groove across the tip. Cracks can be prevented by keeping the hooves flexible with plenty of moisture and hoof dressing and by making sure the feet are trimmed or shod properly at regular intervals.

Severe toe, quarter, and heel cracks are frequent. Each crack requires individual corrective treatment and shoeing. The general principle is to use a toe clip on each side of the crack to prevent wall expansion and to lower the wall under the crack so the wall does not bear weight on the shoe. Plastics are often used to seal the crack and to keep the crack from expanding and contracting during foot action.

Quittor

A quittor is a deep-seated sore that drains at the coronet. It causes severe lameness. The infection may arise from many causes—puncture wounds, corns, and so on. It may be a chronic inflammation of a collateral cartilage of the third phalanx that drains through the coronary band.

Scratches or Grease Heel

Scratches (grease heel) are mangelike inflammations of the posterior surfaces of the fetlocks and are usually found in the rear legs. The condition is well named, because there is greasy, foul-smelling secretion. It is usually caused by filthy conditions or muddy paddocks. The skin becomes chapped, cracked, and infected. Grease heel can be corrected by clipping the hair very close, cleaning with soap and water, and applying an antiseptic.

Side Bones

Side bones are ossified lateral cartilages located immediately above and toward the rear quarter of the hoof head. They usually occur in the front feet. One or both sides of the foot, as well as one or both feet, may be affected. Lameness may or may not be present. Side bones usually develop from concussion—running or working horses on a hard surface, injury to the cartilage from a blow, or poor conformation that increases concussion. Toed-in or base-narrow conformation causes lateral side bones; toed-out or base-wide causes medial side bones. The horse may be lame during ossification, but rarely after.

Thrush

Thrush is disease of the frog caused by many organisms, of which *Spherophorus necrophorus* is the most common. It is found in any one or all of the four feet. It is brought on by unsanitary conditions—the affected horse is usually standing in mud or on urine- or feces-soaked ground or bedding. Thrush is also the result of failure to clean the feet regularly. There is a very offensive characteristic odor and a blackish discharge. The horse becomes lame if the sensitive structures of the foot become involved. Thrush is easy to cure by trimming away the affected part of the frog and by applying an antiseptic.

Gravel

Actually, the word *gravel* is a misnomer for this condition. Foreign material such as rock gravel is not necessarily involved. Most gravels are nothing more than abscesses that occur between the sensitive sole, frog, or laminae and their insensitive or horny counterparts. Usually the infections form as a result of bruises. The blood is not sterile, and an abscess forms. Dead tissue can act as the aggravating foreign material. In some cases, the white line cracks in a dry foot, and an infection is formed. The infection is the cause of "gravel." When the infection shows up at the coronary band, it is not a piece of foreign material working out. Instead, pus from the infection follows the path of least resistance, which is up the laminae. Therefore, the pus migrates to the coronary band and forces its way out. There is a drastic improvement as soon as the pressure is relieved at the site of pus formation.

Seedy Foot or Seedy Toe

Seedy foot is a separation of sensitive and insensitive laminae, usually at the point of the toes. It is a dead growth of the sole and feels almost like sawdust. While it can be cut out, it grows back the same way every time, leaving a place for infection to enter the foot. In addition to causing unsoundness, the wall is weak and will not hold a shoe the way it should.

Hindlegs
Stifled

The horse's stifle corresponds to the human knee. A horse is said to be "stifled" where the patella of the stifle joint has been displaced. The patella usually moves upward and to the inside. It can be placed back in normal position. Older horses are normally not very sound once they are stifled. Young horses may recover and be sound after they get a little older and stronger.

Stringhalt

Stringhalt is an involuntary flexion of the hock during progression. It may affect one or both hind limbs. The true cause is unknown but has been blamed on nervous diseases or degeneration of nerves. Symptoms are quite variable. Some horses show mild flexion of the hock during walking, while others show a marked jerking of the foot toward the abdomen. Some horses show signs with each step, while others are spasmodic. The signs are usually exaggerated while turning and after the horse has had a rest. The important point is that the horse does not have good stability whenever it is making a turn. However, many stringhalted horses are serviceable and can gallop and jump perfectly satisfactorily. Unfortunately, the condition frequently gets worse with age.

Treatment is surgery to remove part of the tendon of the lateral digital extensor. Most cases then show some improvement.

Thoroughpin

Thoroughpin is a puffy condition in the hock web caused by trauma or tendon strain (Figure 2-18). Excess synovial fluid forms and accumulates in the tarsal (hock) synovial sheath, which encloses the deep digital flexor tendon of the hindleg. Normally there are no signs of pain, lameness, or heat. The condition can be determined by the puff's movement. When pressed, the puff will move to the opposite side of the leg. Thoroughpin can be treated by applying pressure and massaging, but these are not always successful. It is best to let a veterinarian treat the condition. If left untreated, the joint action is not influenced, and the horse is seldom unsound.

Bog Spavin

A bog spavin is a filling of the natural depression on the front face of the hock. It is a chronic distension of the joint capsule of the hock. In addition to the large swelling on the front surface of the hock joint, two smaller swellings are usually located on the back part of the hock joint, lower than the swellings of thoroughpin. The bog spavin is much larger than the blood spavin. Bog spavin is caused by faulty conformation (horses that are too straight in the hock joint suffer increased concussion and trauma to the hock joint) and by trauma (injury to the hock joint as a result of quick stops, quick turns, and so on).

A veterinarian can treat bog spavin that is caused by accidental trauma, such as the horse getting kicked by another horse. In this case, it is a blemish if no bone changes are present. Many horses with bog spavin work for years without lameness. However, it is extremely difficult to treat bog spavin that results from faulty conformation, because it recurs.

Blood Spavin

Blood spavin is an abnormally swollen vein on the front face of the hock, just above where a bog spavin is located. Lameness, heat, and swelling usually accompany the condition. Actually, there is no agreed-on definition for a blood spavin. Many people refer to a blood spavin as being the enlarged saphenous vein that crosses a bog spavin. No successful treatment is known.

Bone Spavin

A bone spavin is new bone growth on the inside (medial side) and upper (proximal) end of the third metatarsal or hind cannon bone and a couple of the bones of the hock (third and central tarsal bones). The condition causes arthritis of the hock joint. Bone spavins are caused by the same conditions as bog spavin except that the faulty conformation is sickle hocks or cow hocks. Sickle hocks and cow hocks stress the inside of the hock joint. Horses with narrow, thin hocks are predisposed to bone spavins. A very large bone spavin is usually referred to as a "jack spavin."

One symptom is pain when the horse flexes the hock joint; the horse therefore reduces the height of the foot flight arc. This causes the foot to land on the toe, so that the toe gets short and heels tend to get long. Also, the horse tends to drag its toes and rolls its hips as it moves. These signs are caused by incomplete flexion of the hock.

Treatment for bone spavin is surgery by a veterinarian, but the horse will not recover completely.

Blind or Occult Spavin

Blind spavin is a disease originating in the hock joint and causing typical spavin lameness. However, there may be no outward palpable or radiological signs. The damage is usually to the articular cartilages. The horse is usually lame all its life.

Capped Hock

Capped hock is an enlargement at the point of the hock (Figure 2-18). It is caused by bruising when the horse kicks a trailer or walls of the stall. It is unsightly but need not be considered serious unless it interferes with the horse's work. In the early stages of development, it can be reduced by daily painting with a tincture of iodine solution. Once the "cap" has set or become fibrous, about all that can be done is to give a series of blisters therapy treatments and hope resorption takes place. It is often impossible to cure because the horse is a confirmed stall kicker.

Curb

Curb is a condition in which there is a fullness of the rear of the leg just below the hock, caused by an enlarged plantar ligament (Figure 2-18). Anything that increases ligament stress can cause curb, which thickens the ligament. Sickle hocks and cow hocks are conformation faults that often lead to curb. Sliding a horse too far when stopping it or stopping it too suddenly in deep, soft ground strains the plantar ligament enough to "pop a curb." Firing and blistering are the usual treatments (firing is making a series of skin blisters with a hot needle over the area of lameness).

GENERAL UNSOUNDNESSES

Hernia and abnormal respiratory conditions are general unsoundnesses. They greatly lower the usefulness and value of a horse.

Hernia

A hernia is any protrusion of an internal organ through the wall of its containing cavity. Usually it involves part of the intestine passing through an opening in the abdominal muscle. Umbilical and scrotal or inguinal hernias are fairly common in young foals. They usually correct themselves; if they do not, surgery is delayed until after the foal is weaned.

Heaves or Wind Broke

Horses with the heaves have trouble forcing the air out of their lungs during or after exercise or after drinking cold water. The horse starts to expire, pauses, and then contracts its abdominal muscles forceably. The jerking of the ribs is characteristic, as is the formation of the so-called heave line in back of the ribs.

A sequence of events usually precedes heaves. The condition actually results from a persistent cough probably caused by prolonged feeding of poor-grade roughage or of dusty feed or by keeping horses in very dusty, confined conditions. The persistent coughing breaks down the alveolar walls in the lung, so that the lung loses elasticity. Thereafter, expiration is no longer passive, and the horse must force the air out.

A horse with the heaves is unsound for any strenuous or fast work. There is no satisfactory treatment. It is best to turn the horse out to pasture, only use it for light work, and keep down the dust.

Roaring

An animal that whistles or wheezes when its respiration is speeded up with exercise is said to be a "roarer." The whistling or roaring noise occurs only with inspiration, not expiration. Roaring can be caused by broken rings in the trachea or, as in most cases, by paralysis of the muscles that control vocal cord tension. Air passing through the sagging muscles makes them vibrate in such a way as to produce the noise. Once established, roaring usually gets worse. There are no known means of prevention, but about 70 percent of cases can be surgically corrected.

Thick Wind

Thick wind simply refers to any difficulty in breathing.

Photosensitization

During the summer, a condition known as *photosensitization* may be observed. Photosensitization is an allergic reaction to light, particularly sunlight. Certain plants sometimes develop chemicals in their leaves that, when digested and absorbed by the horse, get into its bloodstream and eventually reach the skin cells. These chemicals cause the skin to have a severe allergic reaction when hit by light. Usually, only unpigmented skin is involved, because dark skin contains melanin, a dark pigment that shields the skin against the sun's rays. When the horse is exposed to sunshine, the white or unpigmented areas become violently inflamed. Unpigmented hairless areas, being even more unprotected, are most severely affected. The owner is often puzzled by the pattern of the skin lesions, which so distinctly limits itself to the unpigmented skin. A sorrel or bay horse may have severe involvement of the blaze on its face, lips, and ankles on stockinged feet, while the rest of the body remains normal. In Paint horses, all the white areas may be severely inflamed, but the darker areas will be absolutely normal.

The skin looks as if it has been severely burned. It blisters, weeps, cracks, and may even slough off (die and fall away). In severe cases, the horse may be quite sick. In mild cases, the horse may not seem to notice the ugly lesions, but the owner surely will.

The feeds most likely to develop photosensitizing chemicals are the legumes, such as clover and alfalfa. Vetch has also been reported to cause the disease, as has puncture vine (*Tribulous terrestris,* often called "bull head" or "goat head"), which is common in the southwestern United States.

Common sense directs treatment of the horse. It should be promptly put in a dark stall and removed from the offending feed source. A laxative should be given to remove remaining feed from its digestive tract, and the diet should be changed. Although the disease is disfiguring and alarming, nearly all horses completely recover.

3
Choosing
the Breed

Many breeds of horses are found throughout the world. The various breeds were developed according to the needs and desires of horsepeople in each locality, or they developed naturally. The usual definition for a breed of horse is that it is a group of horses that have certain distinguishable characteristics, such as function, conformation, and color. Wayne Dinsmore (1935) of the Horse and Mule Association of America says, "A breed of livestock may be considered to exist when there are a substantial number of animals within a given species that possess certain common characteristics, differentiating them from others of the same species, with power to transmit said distinguishing characteristics with a high degree of certainty to their progeny. The distinguishing characteristics must have value or not enough breeders will participate in the production of such animals, nor will the work be continued with sufficient persistence to evolve a breed."

At some time in history, all the breeds of horses have come under human influence, and we have controlled the selection of breeding stock. It is interesting to trace the evolution of the breeds and see how a few horses have played a role in the development of almost all the breeds.

As early as the reign of King James I (1603–1625), Arabian horses were imported into England and were crossed with native light horses. During the reign of Charles II (1660–1685), the so-called Royal Mares were imported. Byerly Turk, Darley Arabian, and Godolphin Barb were brought to England between 1689 and 1728. These three stallions are the foundation sires of

Thoroughbred horses. The descendants of Byerly Turk through Herod established one of the sire lines. True Britain descended from Byerly Turk and was the sire of Justin Morgan, foundation sire of the Morgan horse breed. From the Morgan, several other breeds developed. Allen F-1 was a grandson of a Morgan stallion and is recognized as the foundation sire of the Tennessee Walking Horse. The Morgan horse also contributed to the development of the Standardbred, Quarter Horse, American Albino, and Palomino breeds. Eclipse can be traced from Darley Arabian, and is one of the sire lines of Thoroughbreds. Blaze, the foundation sire of the Hackney, and Messenger, the progenitor of the American Standardbred and American Saddlebred, were also descendants of Darley Arabian. The American Saddlebred stems from Denmark, who was a Thoroughbred and the sire of Gaines Denmark. Hambletonian 10, a great-grandson of Messenger, was one of the most influential sires of the Standardbred breed. Thoroughbreds have been very influential in establishing the American Quarter Horse breed.

LIGHT HORSE BREEDS

If you decide to buy or keep purebreds of a certain breed of horse, you should join that breed association. Each association has rules that govern the registration, ownership transfer, and activities of the breed. Various records must be kept and submitted to the association at specified times. Failure to adhere to the rules of the association may result in fines or restriction of one's activities associated with the breed. The rules and procedures for conducting business with each association can be obtained by writing each association (Appendix 3-1).

When you register a horse with a breed association, you must understand the type of registration certificate that has been issued. Some associations have more than one type of registration certificate. Some certificates restrict the horse's participation in various breed activities and the registration of foals the horse produces or sires. It is impossible to present here all the rules for the associations, and, furthermore, they change periodically. Therefore, you should obtain the rule books from the breed association you are interested in.

The breed that a person chooses to buy, own, or breed is determined primarily by his or her interest in certain types of horse activities. Some breeds are versatile, while others are suited for specific purposes. The more versatile breeds are more popular (Table 3-1).

American Quarter Horse

The American Quarter Horse originated in America during the colonial era, although an official breed association was not formed until 1941. (See Figures 3-1 and 3-2.) Racing was popular with the colonists. The races were short and

TABLE 3-1
Foal Registrations, 1960–1976

Breed	1960	1968	1973	1976	Total as of 1977	Registry Formed
American Quarter Horse	37,000	57,000	87,568	100,321	1,350,000	1914
Thoroughbred	12,901	22,700	26,760	29,500	~760,000	1894
Appaloosa	4,052	12,389	20,357	20,471	~275,000	1938
Arabian	1,610	6,980	12,266	14,962	~125,000	1908
Standardbred	7,100	10,200	11,393	13,184	~400,000	1938
Half-Arab and Anglo-Arab	2,200	9,800	13,222	9,052	178,400	1955
Tennessee Walking Horse	2,623	8,492	7,116	6,600	190,000	1935
Paint	0	2,390	4,331	6,203	~30,000	1962
Morgan	1,069	2,134	3,052	3,700	~67,000	1894
American Saddlebred	1,600	3,500	4,011	3,671	165,092	1891
Pinto	230	2,258	2,270	2,511	30,200	1956
Palomino	657	1,262	1,580	1,604	9,013	1936

FIGURE 3-1
A painting of the "ideal" American Quarter Horse by Orren Mixer.
(Courtesy American Quarter Horse Association.)

FIGURE 3-2

Sugar Vandy, American Quarter Horse Association stallion, is the sire of many outstanding cow horses and cutting horses. His get have been 1972 and 1977 California Reined Cow Horse Association snaffle bit futurity champions, and 1975, 1976, and 1977 world champion all-around stock horse. Sugar Vandy stands at stud at Dos Pinos Ranch, Davis, California, and is owned by Dick and Dianne Winger. (Photo by Baldwin, courtesy D. Winger.)

about a quarter mile in length. They were frequently run down village streets or along country lanes near the plantations. Horses that could successfully race for a quarter of a mile were known as "quarter pathers." As the frontier moved westward, the quarter pather moved with it to run on the newly established short racetracks. In addition to racing, these horses served several purposes. They were adopted by ranchers and cowboys to use with range cattle, because they had an inherent "cow sense." Therefore this type of horse was very popular in the Southwest during the late 1800s, and its popularity has continued to increase. The American Quarter Horse Association was founded in 1941 and has become the largest horse breed organization in the world.

Two types of registration certificates are issued by the American Quarter Horse Association (AQHA), so its *Official Stud Book* has two parts. The numbered section of the registry, which makes up the largest portion of registered American Quarter Horses, consists of horses registered from parents that are numbered American Quarter Horses. The second type of registration is the appendix listing of horses that have one parent that is a numbered American Quarter Horse and one parent that is a Thoroughbred registered either with the Jockey Club of New York or in the *Stud Book* of the Mexican Jockey Club. An "appendix" horse may receive a number by fulfilling the requirements for a "register of merit."

Many breeds and types of horses played a role in the synthesis of the Quarter Horse. Several famous sires such as Steeldust, Old Billy, Shiloh, Yellow Jacket, Peter McCue, Little Joe, Traveler, and Copperbottom were responsible for the early establishment of the breed. Later, sires such as Doc Bar, Old Sorrel, Three Bars, Top Deck, Leo, King, Easy Jet, and Go Man Go have left their mark on the breed. Wimpy, owned by the King Ranch, was the champion stallion at Fort Worth, Texas, in 1941 when the association was formed and was given the registration number 1.

Today there are many types of Quarter Horses because they are used for so many purposes. Because of the interest in racing created by the betting public and increased purses, some horses are bred primarily for speed. Others are used on ranches, so some breeders try to maintain the conformation and desire to work cattle. These types of horses are also shown in western event classes at horse shows. Other horse show classes, such as hunters and jumpers, have resulted in Quarter Horses that have a Thoroughbred conformation.

Arabian

The early history of the Arabian horse is obscure. Horses similar to the best Arabians were in North Africa more than 1,000 years before good horses were known in Arabia. They were apparently introduced to Arabia from Africa sometime between the first and sixth centuries. The Arabian proper is a descendant of the Kohl breed, and all Arabians descend from the five foundation mares known as Al-Khamesh ("The Five"). They were all owned by King Solomon about 1635 B.C. There are five great families of Arabian horses descending from Al-Khamesh: Keheilet Ajuz, with thirty-seven subfamilies; Maneghi, with four subfamilies; Hadban, with five subfamilies; Jelfon, with two subfamilies; and Homdani, with two subfamilies. From Keheilet Ajuz come the most distinguished families. Darley Arabian descends from Keheilet Ajuz. The Arabians were bred as purebreds and also used to cross on native mares to improve or develop breeds in many localities.

The Arabian (Figures 3-3 and 3-4) was first imported to America in 1730. Since that time, many horses have been imported from many countries. They have several distinguishing characteristics. Arabians are known for their endurance and tend to dominate endurance races and competitive trail rides.

106

FIGURE 3-3
Na Ibn Moniet, 98751, a purebred Arabian stallion and champion at halter, is owned by
Jarrell McCracken of Brentwood Farm. (Courtesy International Arabian Horse Association.)

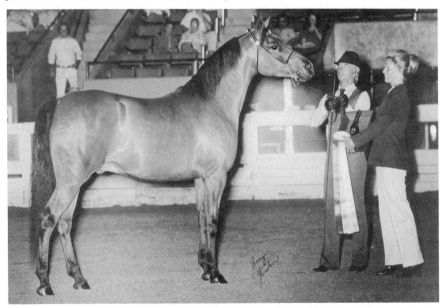

FIGURE 3-4
Santiago Gino was the 1977 U.S. national champion half-Arabian mare. (Photo by Johnston,
courtesy International Arabian Horse Association.)

Their heads have a characteristic dished face. They are usually 14 to 15.2 hands tall and weigh 900 to 1,000 pounds (408 to 454 kg).

The Arabian horse industry in Canada and the United States has three separate organizations not formally related, but all working closely together with many interlocking members. The first of these is the Arabian Horse Registry of America, Inc. This organization functions, as its name implies, strictly as a registry of all purebred Arabian horses in the United States. The registry was formed in 1908 and before that time, the Jockey Club in New York registered purebred Arabian horses to a limited degree. The International Arabian Horse Association was formed in 1950 with the moral and financial blessing of the Arabian Horse Registry, because they did not wish to go into promotional work or horse show activities or to establish member clubs. At that time, the Arabian Horse Registry was also the registering authority for all of Canada. In about 1960 or 1961, the Canadian Arabian Horse Registry was formed to operate in the same manner as the Arabian Horse Registry of America; that is, to register all purebred Arabian horses in Canada.

Shortly after the International Arabian Horse Association was founded, it was given the opportunity to acquire the Half-Arabian and Anglo-Arabian registries from a private party who had recently acquired them from the American Remount Association. These two registries then became part of International's operations, even though its primary members and promotional efforts are targeted toward the purebred Arabian horse but also including the Half-Arabian and Anglo-Arabian.

In the Half-Arabian Registry, the sire or dam must be a registered purebred Arabian registered either with the Arabian Horse Registry of America or the Canadian Arabian Horse Registry. The Anglo-Arab Registry registers foals by registered Thoroughbred stallions out of registered Arabian mares, foals by registered Arabian stallions out of registered Thoroughbred mares, foals by registered Thoroughbreds or registered Arabian stallions out of registered Anglo-Arab mares, and foals by registered Anglo-Arab stallions out of registered Anglo-Arab mares, registered Thoroughbred mares, or registered Arabian mares. However, such foals must be not more than 75 percent nor less than 25 percent Arabian blood.

In 1973, International inherited the activities of the Arabian Racing Association of America. Arabian racing has taken place in Nevada, Michigan, Kentucky, New Mexico, Florida, Georgia, and California. Arabian racing is actually in its infancy, but the interest among Arabian owners is increasing. International is prepared to keep records of the performance of Arabians on the racetrack.

Thoroughbred

The history of the Thoroughbred horse (Figure 3-5) has its early beginning in England. The early racing stock had endurance but not speed. They were native mares that had been crossed with horses brought into England by the Romans

FIGURE 3-5
Pretense ranks as one of the best Thoroughbred stallions in North America.
(Photo by Stackhouse, courtesy of Delaney's Stock Farm, Murrieta, California.)

and the French Normans. Oriental sires (Arabians, Turks, and Barbs) were used to try to improve the speed of the "Royal Mares," which are regarded by some to be the foundation mares of the breed. However, it was not until Byerly Turk, Darley Arabian, and Godolphin Barb were imported to England that the Thoroughbred that we know today was established as a refined running horse. Byerly Turk reached England in 1689 and was the first sire to have a significant effect on speed. Darley Arabian was imported in 1705 or 1706 and continued to improve the horses. Godolphin Barb was not in England until 1728 to 1730. One of his fillies, Post, was sent to America in 1733. Because of his popularity as a racing sire, a large number of his offspring were imported to America. All English and American Thoroughbreds trace their lineage in the male line from these three studs. The three foundation sires were Matchem, grandson of Godolphin Barb; Herod, great-great-grandson of Byerly Turk; and Eclipse, great-great-grandson of Darley Arabian. Bulle Rock was the first Thoroughbred imported to America, arriving in 1730. From that time, mares and stallions were imported from England to improve the running horse being developed in America. Since then, the Thoroughbred has continued to be bred and selected to improve its speed and endurance.

Thoroughbred horses were first registered in England in the *General Stud Book*. The first volume appeared in 1793. All English Thoroughbreds must trace back to horses registered in the *General Stud Book*. The *American Stud Book* was first published in 1873. The Jockey Club purchased the registry in 1894 and has continued to register American Thoroughbreds.

The Thoroughbred is one of the most versatile horse breeds and has influenced the development of many other breeds. They are primarily running horses but are also used for hunters, jumpers, dressage, three-day events, cow horses, and so on.

Many famous racing horses have captured the hearts of Americans. Eleven horses have won the Triple Crown (the Kentucky Derby, the Belmont Stakes, and the Preakness Stakes) for 3-year-olds. Citation won the Triple Crown in 1948, and no other horse won it for 25 years until 1973, when Secretariat won it. In 1977, Seattle Slew won it, and Affirmed won it in 1978. Other horses, such as Swaps, Round Table, Buckpasser, Man o' War, Forego, Kelso, and Dr. Fager, did not win the Triple Crown but compiled extensive race and/or sire records. The centers of racing have moved from colonial Virginia to California, Kentucky, Florida, and New York.

Most of the running Thoroughbreds are 15.1 to 16.2 hands high and weigh 900 to 1,200 pounds (408 to 544 kg). They have long, smooth muscles; short cannon bones; long forearms; long, sloping shoulders; and long distance from the hip to the hock.

Standardbred

The Standardbred breed (Figure 3-6) includes both trotters and pacers. They have a common history, and some Standardbreds have been successfully raced at the trot and pace. The name *Standardbred* was derived from the original standard for speed, which was adopted in 1879. For a horse to be registered, it had to trot a mile in 2:30 minutes or pace it in 2:25. Since then, the standard has been lowered to 2:20 for 2-year-olds and 2:15 for older horses. Speed at the trot and pace is their most distinctive characteristic. The breed is of American origin but traces to horses imported from England. The breed can be traced directly to Darley Arabian, and the most important stallions were Messenger and Bellfounder. The American influence came from the early colonial days. The colonists selected horses that would travel long distances at the trot or pace. In the early history of the breed, three families were the most popular, the descendants of Messenger, Justin Morgan (the Morgan), and Grand Bashaw (the Clays). Hambletonian 10 was the most influential early American-bred sire. He stood at stud 23 years and sired 1,333 colts. Hambletonian 10 was linebred to Messenger, who appears four times in the third and fourth

FIGURE 3-6
Adios, an all-time great Standardbred stallion that sired primarily pacers.
(Courtesy United States Trotting Association.)

generation removed. Four sons of Hambletonian 10 established sire lines. In the early 1900s, four male trotting lines were prominent: Peter the Great, Axworthy, McKinney, and Bingen. More recently, Adios has exercised the most powerful influence on the breed (see Figure 3-6). Currently Meadow Skipper is a leading sire (Figure 3-7).

Since the early history of the Standardbred, several associations have maintained records. In 1870, the National Trotting Association was organized, and in 1877 the American Trotting Association was founded. J. H. Wallace published the American Trotting Register in 1868. The register was sold to the American Trotting Association in 1891. The register was continued, and today all Standardbreds are registered in the *American Trotting Sires and Dams Book*. Today, the official registry is maintained by the U.S. Trotting Association, which was formed in 1939 and is now closed. All foals registered must have registered parents.

FIGURE 3-7
Meadow Skipper, one of the leading Standardbred sires.
(Courtesy United States Trotting Association.)

Morgan

The Morgan horse breed (Figure 3-8) traces its history to Justin Morgan, foundation sire of the breed. Justin Morgan was foaled in 1789. Originally named Figure, the man who brought him to Vermont from Springfield, Massachusetts, named him Justin Morgan, after himself. He was a small horse that stood slightly over 14 hands and weighed a little over 1,000 pounds (454 kg). He gained a reputation over the years by pulling a log that a draft horse could not move, carried a President on a parade ground, and outran the best quarter-mile racehorse in central Vermont. He produced three stallions—Sherman, Bulrush, and Woodbury—that established him as a sire. Black Hawk, a grandson of Justin Morgan, produced Ethan Allan 50, champion trotter of the world.

The Morgan has been a versatile breed since its beginning. Originally Morgans were developed as general-purpose horses for New England farms.

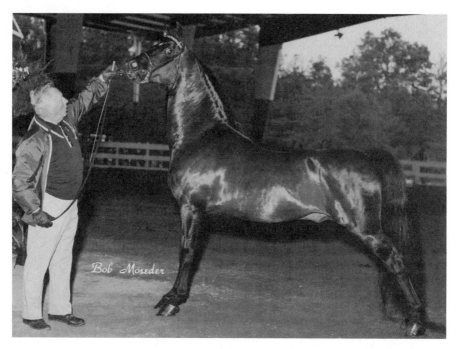

FIGURE 3-8

UVM Promise, world champion Morgan stallion in 1978 and world champion Park Saddle horse in 1979 and 1980, is owned by Mr. and Mrs. Epperson of Tennessee Valley Morgan Horse Farm. (Photo by Moseder, courtesy Mr. and Mrs. Epperson.)

Today they are used as working cow horses, trail horses, and youth horses and are shown under saddle, in harness, as jumpers, and as roadsters.

The Morgan Horse Registry was formed in 1894 by Colonel Battell. Since then, the organization has undergone two name changes. The Morgan Horse Club took over the registry in 1930 and is presently named the American Morgan Horse Association. In 1907, the U.S. Morgan Horse Farm was established in Middlebury, Vermont, to perpetuate the breed and was operated by the U.S. Department of Agriculture. The farm, a gift from Colonel Battell, was given to the University of Vermont in 1951.

The breed characteristics have changed little since Justin Morgan. They stand 14.2 to 15.2 hands and weigh 900 to 1,100 pounds (405 to 495 kg).

American Saddlebred

The American Saddlebred horse was developed in Kentucky by pioneers who needed a utility horse. The breed developed from the Thoroughbreds, Morgans, and Narragansett pacers and trotters, and was used for riding and driving. Some were used for plowing. The foundation sire for the breed is Denmark, a

Thoroughbred. Even though much of the development of the breed preceded Denmark, it was his mating to the "Stevenson Mare" that produced Gaines Denmark, through which most saddlers are traced. Nine other sires were important to the development of the breed and were recognized as Noted Deceased Sires in 1910. In 1891, a group of breeders met in Louisville, Kentucky, to form the National Saddle Horse Breeders' Association. The name was changed to American Saddle Horse Breeders' Association in 1899. Today, the average height is 15 to 16 hands, and the average weight is 1,000 to 1,200 pounds (454 to 544 kg). Saddlebreds have short backs, flat croups, sloping shoulders, arched necks, and long pasterns. They are used for riding, driving, hunting, jumping, parades, cow work, and show. They excel as show horses and are shown as three- and five-gaited and fine harness horses. The three-gaited horses are required to perform at the walk, trot, and canter, whereas the five-gaited horses are also shown at the slow gait and the rack. They are judged on their action, conformation, animation, manners, and soundness. The five-gaited horse (Figure 3-9) is shown with a full mane and tail, whereas the three-gaited horse (Figure 3-10) is shown with a clipped mane and tail. The

FIGURE 3-9

Wing Commander, an American Saddlebred horse that was six times winner of the world's grand championship five-gaited division at Kentucky State Fair.
(Photo by Horst, courtesy American Saddle Horse Breeders' Association.)

FIGURE 3-10
Blue Meadow Princess, world champion three-gaited American
Saddlebred horse. (Photo by Horst, courtesy American Saddle
Horse Breeders' Association.)

fine harness horse is shown at two gaits, the walk and trot. They are shown to
an appropriate four-wheeled vehicle and carry a full name and tail (Figure
3-11).

Tennessee Walking Horse

The Tennessee Walking Horse (Figure 3-12) originated in the middle Tennessee
bluegrass region. The breed had its beginning as the Southern Plantation Walking Horse. These horses were of mixed breeding, which included Narragansett
and Canadian pacers as well as trotters and pacers. Later, other breeds such as
the Thoroughbred, Standardbred, Morgan, and American Saddlebred were
used in the area to produce horses. The horses that developed were utility
animals but they had a natural ability for an easy, gliding gait. In 1885, a cross
between a stallion called Allendorf, from the Hambletonian family of trotters,

FIGURE 3-11

Main Glitter, world champion fine harness mare, owned by Cynthia Wood Stable, being shown by Bob Vesel. (Photo by Fallow, courtesy Cynthia Wood.)

FIGURE 3-12

Threat's Supreme, 1979 world champion Tennessee Walking Horse, owned by D. and S. Perry. (Photo by Dixon, courtesy Tennessee Walking Horse Breeders Association.)

and Maggie Marshall, a Morgan mare, produced Allan F-1. Later, Allan F-1 was recognized as the foundation sire for the Tennessee Walking Horse. Other bloodlines, such as the Copperbottoms, Gray Johns, Slashers, Hals, Brooks, and Bullett families, had already produced the Tennessee pacer prior to Allan F-1's arrival in middle Tennessee in 1903. However, it was the cross of Allan F-1 on the Tennessee pacer mares that produced the Tennessee Walking Horse as it is known today. Refinement of the style and conformation of the Tennessee Walking Horse was started soon after by the arrival of Giovanni. The Saddlebred stallion crossed quite well with the Allan F-1 mares to produce higher-quality horses.

In 1935, the Tennessee Walking Horse Breeders Association formed. When the stud book was established, 115 horses were selected as foundation stock. Almost every breed of horse in the United States was represented in the bloodlines of the foundation stock. The distinctive characteristic of the breed is the ability to do the running walk. The running walk is characterized as a four-beat gait in which all four feet strike the ground separately at regular intervals. The hindfoot overstrikes the footprint left by the lateral forefoot by 18 to 50 inches (45.7 to 127 cm). This gives the horse a smooth, gliding motion that is comfortable to ride.

Gotland Horse

The Gotland horse is one of the oldest breeds of horses, because it has a history for 10,000 years on the Swedish island of Gotland in the Baltic Sea. Descendants of the Tarpans, the Gotland horses are called Skogsruss by the Swedes. Throughout history, they have been used as calvary mounts, draft animals, and carriage horses. In 1880, the Swedish government placed the Gotlands under strict government protection and supervision. At that time, some oriental horses were used in a crossbreeding program to expand the genetic base. Gotlands were imported to the United States in 1957 from Sweden and are registered with the American Gotland Horse Association.

Gotlands are small horses ranging between 12 and 14 hands. Their size and temperament make them excellent youth horses.

Racking Horse

The Racking Horse Breeders' Association of America was formed in 1971 to register horses with the natural ability to perform the pleasure walk, slow rack, and fast rack. The rack is a four-beat gait in which the hindleg and its diagonal foreleg leave the ground simultaneously but the hindleg leg strikes the ground prior to its diagonal foreleg. The registry is open, and the foundation mares and stallions are yet to be selected.

Missouri Fox Trotting Horse

The origin of the Missouri Fox Trotter is obscure. The settlers that came from the mountains and plantations of Kentucky, Tennessee, and Virginia brought with them their saddle horses. Some of these horses could do the foxtrot, which was easy to ride, and the horse could travel long distances at the gait. The foxtrot gait is a broken trot in which the forefront strikes the ground prior to the diagonal hindfoot. The hindfoot glides forward of the print made by the lateral forefoot. Because the front of the horse is rising upward and the hind-quarters are being lowered, the gliding motion gives the rider a comfortable ride. The speed of the gait is 5 to 8 miles per hour (8 to 12.9 km).

In 1948, the Missouri Fox Trotting Breed Association was formed to preserve the type of horse that was developed from the settlers' foxtrotting horses after the infusion of Arabian, Morgan, American Saddlebred, Tennessee Walking, and Standardbred blood. Gradually sire lines developed, and the lines of the Diamonds, Copperbottoms, Brimmers, Red Bucks, Chiefs, Steel Dusts, and Cold Decks are recognized. However, the only requirement for registration is the ability to perform the foxtrot gait. In 1970, a rival association, the American Fox Trotting Horse Breed Association, was formed. Many foxtrotters are registered with both associations.

Morab

The Morab (Figure 3-13) is a new breed created in 1973 with the formation of the Morab Horse Registry of America by James Alan Miller. The first Morabs were probably bred in the 1800s by crossing the Arabian and

FIGURE 3-13
Tezya, the first horse registered as a Morab. (Courtesy Morab Horse Registry of America.)

Morgan breeds. Since then, many breeders have used the cross to produce good horses. The horses on which the Morab association and registry were established were selectively bred by Martha Doyle Fuller. The breed was perpetuated by Fuller beginning in 1955 by crossing Morgan mares with Arabian stallions to produce distinct characteristics in the resultant foals. They stand 14.3 to 15.2 hands high and weigh 950 to 1,200 pounds (428 to 540 kg).

Paso Fino and Peruvian Paso

The Pasos had the origin of their breed in the Spanish provinces of Cordova and Andalusia. When the Moors invaded Spain, they brought Arabs and Barbs to the area. These horses were crossed with the native horses, which had remnants of the Andalusian breed. One strain to develop was the Spanish Jennet, which was noted for its comfortable saddle gait and for its ability to pass the gait on to its offspring. In 1492, Columbus took some horses of this descent to Santo Domingo. From this area, the Spanish Jennet horse with the saddle gait spread through the Carribean, Mexico, and South America. From these horses, the Pasos developed. They have the common characteristic of the paso gait, a broken pace in which the paired lateral limbs leave the ground together but the hindfoot strikes the ground prior to the lateral forefoot. This eliminates the jarring effect of the pace and makes the gait comfortable to ride. They also exhibit termino, which is the outward swing of the foreleg.

Two registries register Paso Fino horses (Figure 3-14). In 1964, the American Paso Fino Pleasure Horse Association was formed; it registers purebred

FIGURE 3-14

A Paso Fino being shown on lines for conformation and gait.
(Courtesy Paso Fino Owners and Breeders Association.)

FIGURE 3-15

A Peruvian Paso, the 3-year-old stallion Osado. He was 1977 Pacific Coast champion. (Photo by Frederick, courtesy Peruvian Paso Horse Registry of North America.)

Paso Finos. The Paso Fino Owners and Breeders Association also promotes and breeds Paso Fino horses. Most horses registered by these associations have come from Puerto Rico and Colombia.

Peruvian Paso horses (Figure 3-15) developed from Paso horses brought to Peru by the Spanish conquistadors. They were used by riders on long journeys between haciendas on the coastal plains of Peru. Two U.S. registries register Pasos imported from Peru. The American Association of Owners and Breeders of Peruvian Paso Horses was formed in 1967. They register horses regardless of color. The Peruvian Paso Horse Registry of North America only registers horses of a single color, thus excluding roan, tobiano, and overo horses.

Galiceño

Galiceño horses (Figure 3-16) originated in the Galicia Province in Spain. They were brought to North America by Cortez in 1519 and were kept by natives in the coastal regions of Mexico. Even though purebred Galiceños never migrated north to join other American horses, they are the ancestors of countless

FIGURE 3-16

Canta, a champion Galiceño mare. She was ridden by Shareen Allen
of Kyle, Texas, to win the Highpoint Youth All-Around in 1977.
(Photo by Falkner, courtesy Galiceño Horse Breeders Association.)

Mustangs. In 1958, they were imported to the United States from Mexico, and
in 1959 the Galiceño Horse Breeders Association was formed.

The Galiceños are small horses that are usually 12 to 13.2 hands in height
and 625 to 700 pounds in weight (283 to 317 kg). They are used primarily
by young adults for most types of performance events. Many of the Galiceños
are recognized for their natural ability to perform the running walk.

American Bashkir Curly

Horses with curly coats are an ancient breed and have been depicted in art and
statuary in early China. In Russia, they have been raised for centuries on the
southern slopes of the Ural Mountains by the Bashkiri people, for whom they

are named. The Russian Lokai breed has individuals with curly coats, and some have appaloosa markings. The milk from the mares is used for milk, cream, butter, and cheese. It is also fermented into kumiss, an alcoholic drink. The older animals are used for meat, and their hides are used to make clothes. The horses are also used for draft and transportation purposes.

In 1898, Peter Damele and his father found three curly-haired horses in the Peter Hanson Mountain Range in central Nevada. It is unknown how the horses got there, but since then other curly-haired horses have been seen on the Damele range. In 1971, the American Bashkir Curly Registry was formed to prevent their extinction in the United States.

Several characteristics differentiate these horses from other breeds: long, silky, curly hair (in waves or ringlets); mane and tail hair completely shed each summer; absence of ergots in some individuals; four black hooves; and five lumbar vertebrae.

REGISTRIES WITH COLOR REQUIREMENTS

Within breeds, many coat colors and/or color patterns are attractive to horse owners. People who found particular colors appealing established registries for these horses. The Palomino Horse Association of California was the first color breed association. Shortly thereafter, registries for color patterns such as Appaloosas, Albinos, Paints, Pintos, and Buckskins were established. Some registries only require color for registration, while others have conformation standards. The color breeds do not breed true, with the possible exception of the overo color pattern.

American Albino

American Albino horses (Figure 3-17) as a breed originated on the White Horse Ranch at Naper, Nebraska, in 1937 under the ownership of Caleb and Ruth Thompson. The foundation sire was Old King, who was milk-white in color with pink skin and brown eyes. Foundation mares were Morgans. Coloring is the primary requisite for registration. Horses must have snow- or milk-white hairs with pink skin and brown, blue, or hazel eyes when born. They are not true albinos, because they do not have pink eyes. The dominant white gene results in snow-white horses. Cream-colored horses are horses with the gene(s) for dilution of the normal color patterns. To separate the dominant white from the cream-colored horses, the American Albino Association changed its name to White Horse Registry in 1980, and the cream-colored horses are registered with the Cream Horse Registry. They vary in height from 12.2 to 17 hands and weigh about 1,000 pounds (454 kg).

FIGURE 3-17
American albino ponies in a field. (Courtesy White Horse Registry.)

American Buckskin

The American Buckskin Registry Association was started in 1963 to register Buckskin, Dun, Grulla, and Red Dun horses. In 1971, the International Buckskin Horse Association was formed. The dilution gene(s) responsible for these color patterns was present in the Norwegian Duns and the Sorraia horses of Spain. The Spanish conquistadors brought horses with the dilution gene(s) to America and thus the descendants of the Buckskin-colored horses (see Figure 3-18).

The American Buckskin Registry Association registry describes four color patterns that are eligible for registration. Buckskin horses must have a body coat of some shade of yellow but not red. The points are black or dark brown. A dorsal stripe is not necessary. Dun horses have a yellow to nearly white body coat with dark points such as red, black, brown, and dark yellow. The mane and tail must be darker than the body. Grulla horses are smoky gray in color, with no white hairs mixed in. A dorsal stripe must be present. Red Duns have a dun-colored body coat (flesh-colored or red). The points are dark red or red mixed with black. A dorsal stripe is necessary. Palominos and bays with dorsal stripes are not eligible for registration. Horses with blue eyes or white spots

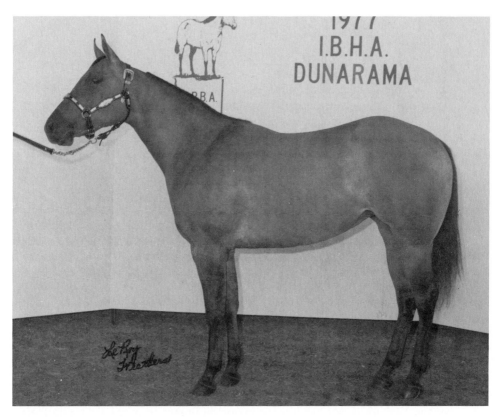

FIGURE 3-18
Reeds Skippa Oak, 1977 grand champion buckskin mare
at International Buckskin Horse Association Convention.
(Photo by Weathers, courtesy International Buckskin Horse Association.)

on their bodies are not eligible for registration. White markings are not allowed to extend above the hocks and knees except for face markings. Once eligibility for color is established, an inspector must approve the horse before registration.

The International Buckskin Horse Association describes seven color patterns that are eligible for registration: Buckskin, Dun, Grulla, Red Dun, Copper Dun, Claybank (a variation of dun), and Perlino Dun.

Palomino

The palomino color pattern has been responsible for many myths and legends. The gold-colored horse appears in ancient tapestries and paintings of Europe as well as in Japanese and Chinese art of past centuries. However, the exact history has never been recorded.

The Palomino Horse Association was organized in 1936 for the perpetuation and improvement of the palomino-colored horse through the recording of bloodlines and issuing of certificates of registration to qualifying horses. Many Palomino horses are also registered with other breed associations such as the American Quarter Horse Association and American Saddlebred. However, there are no purebred Arabian horses with the palomino color pattern.

The Palomino Horse Association defines the characteristics as:

1. Body color near that of an untarnished gold coin. The color may vary lighter or darker but must be natural. They must have a full white mane and natural white tail.

2. Dark or hazel eyes, both eyes the same color.

3. Good head; wide between the eyes; small, alert ears; standing on well-set legs; good withers; short, straight back; and a good, natural tail carriage.

4. Height 14.1 to 16 hands.

5. Weight approximately 1,100 pounds (495 kg).

6. White on face is permitted but shall not exceed a blaze, stripe, or star. White on legs is permitted but shall not extend above the knees and hocks.

Palomino horses are also registered by the Ysabella Saddle Horse Association. The Ysabella Association registers Palominos and sorrel horses with silver manes under their Gold Seal. Under their Silver Seal, they register white horses that have silver manes and tails and blue eyes. The Silver Seal Registry of the Ysabella Association is the major difference from the Palomino Horse Association. The Ysabella Association registers horses of the genotypes that can be used to produce the palomino color, whereas the Palomino Horse Association does not.

Paint, Pinto, and Spotted Horses

Three breed registries register horses with body markings of white and another color. These horses are described as being of the overo or tobiano pattern and of the piebald or skewbald color. The overo pattern (Figure 3-19) is due to an autosomal recessive genotype that causes a basic color with white body marks. The white does not cross the back between the withers and the tail. At least one and often all four legs are of the dark color. Head markings are often bald or apron- or even bonnet-faced. Irregular, rather scattered, and/or splashy white markings are on the body and are often referred to as "calico" patterns. The tail is usually of one color. The horse may be predominantly white or dark colored. The tobiano color pattern (Figure 3-20) is thought to be due to a dominant gene. The heads of tobianos are marked like those of a solid-colored horse and may have a star, a stripe, a snip, or a blaze. All four legs are white,

FIGURE 3-19

Far Ute Keno, palomino stallion with the overo color pattern. He is superior all-around horse, superior halter, superior bridle path hack, and superior western pleasure horse in the American Paint Horse Association. (Courtesy John Tiger and Tiger Stripe Farm.)

FIGURE 3-20

Dandy Diamond, APHA stallion, has the tobiano color pattern. He is 1971 APHA national halter champion and 1977 APHA national champion working cow horse. (Photo by Marge, courtesy Bill and Jan Neel, the owners.)

at least below the knees and hocks. They usually have the dark color in one or both flanks. The spots are regular and distinct, often coming in oval or round patterns that extend down over the neck and chest, giving the appearance of a shield. A piebald is a black horse with white spots; a skewbald is any color except black with white marks.

In 1962, the American Paint Horse Association was formed to register, promote, and keep records on the stock and quartertype Paint horse. The association was formed primarily because the American Quarter Horse Association does not register horses with white body markings. All registered Paint horses must be sired by a registered Paint, Quarter Horse, or Thoroughbred. Conformation is equally important as color for a horse to be registered. All Paint stallions must meet conformation standards after they are 2 years of age.

The Pinto Horse Association of America was formed in the mid-1950s to register Pinto horses of various types. Some of the most popular types are stock, pleasure, saddle, and hunter. Because of the growing interest in ponies, the Pinto Horse Association has developed a Pony Registry for Pintos under 14 hands.

The Morocco Spotted Horse Cooperative Association of America was organized at Adel, Iowa, in 1935. Originally named the Iowa Spotted Horse Association, its charter was reissued in 1939 under its present name. After a peak in 1949, the association became dormant in the early 1960s. In 1971, the records and files were transferred to an active group of breeders, and the organization was revived.

The primary foundation sires of the Morocco Spotted horses were Hackneys. One of the early imported piebald stallions was Stuntney Benedict 8660 (American Stud Book Number 1000).

Appaloosa

The appaloosa color pattern (Figure 3-21) dates back to prehistoric times. Drawings by early humans on the walls of caves in France depicted the pattern. Spotted horses later appeared in Chinese art dating from 500 B.C.and in Persian art of the fourteenth century.

During the exploration of western America, the Nez Percé were the only Indian tribe to have many appaloosas. They bred their horses to a distinctive type. The breed nearly disappeared after the Nez Percé War of 1877 because the horses belonging to the tribe were sold. The spotted horses derived their name from the Palouse River, which drains the area that was home for the Nez Percé Indians. This area became known as the Palouse Country, and the spotted horses in the area were called Palouse horses. "A Palouse" horse became slurred to "apalouse" and later "apalousie." Finally, the horses were given their present name—appaloosa.

Formed in 1938, the Appaloosa Horse Club has been very active. In 1948, it held its first national show. Today, the show draws over 2,000 Appaloosa

FIGURE 3-21

Britches, registered Appaloosa stallion. (Courtesy Appaloosa Horse Club.)

horses, making it one of the largest breed shows in the world. A racing program was developed in the early 1960s. Long-distance trail rides are also sponsored and are a part of an awards program. They are used for western events and cattle work as well as for hunters and jumpers.

In 1949, the stud book was closed to open registration of foundation stock. All Appaloosas are recognized by three color characteristics that separate them from other breeds. Their eyes are encircled with white (white sclera). The skin is mottled, with an irregular spotting of black and white. Hooves are striped vertically black and white. Mane and tail may be sparse, although some Appaloosas have a full mane and tail. There are six basic coat color patterns:

1. Black with white containing black spots over loin and hips.
2. Black with white over loin and hips.
3. Blue or red roan.
4. White with black spots over entire body.
5. Black with white spots over entire body.
6. Black with white spots over the loin and hips.

"NATIVE" HORSES

American Indians caught some of the horses that escaped from the early Spanish explorers. These horses were subsequently bred by the Indians to form a type or breed of horse. The appaloosa is a typical example of this type of horse and became quite popular while the other types or breeds have not become so popular. These horses as well as the ones that became feral horses are discussed in this section.

Spanish-Barb

The Spanish-Barb Breeders Association (SBBA) was formed in 1972 to perpetuate, through selective breeding and linebreeding, the few strains of true-blooded Spanish-Barbs bred for 50 to about 130 years by several dedicated people (see Figure 3-22). The SBBA foundation sires that represent the three major strains are (1) the Scarface horses, an isolated strain kept intact since the mid-1800s under the Romero family and presently owned by the Weldon McKinleys of New Mexico; (2) the Rawhide horses, representing a strain of Spanish-Barbs selectively bred for nearly half a century by Ilo Belsky of Nebraska; and (3) the

FIGURE 3-22
Spanish-Barb stallion, El Mano de Oro, of Ilo Belsky breeding. (Courtesy Spanish-Barb Breeders Association.)

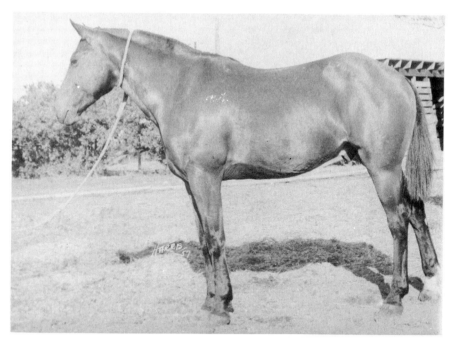

FIGURE 3-23
Chickasaw gelding. (Photo by Harris, courtesy Chickasaw Horse Association.)

"Buckshot" strain of isolated feral ancestry from the early 1920s. Standard height of the Spanish-Barb is 13.3 to 14.1 hands, and weight is 800 to 975 pounds (363 to 442 kg).

Chickasaw

After attacking de Soto's expedition in 1539, the Chickasaw Indians captured many of the horses that escaped. The Chickasaw horses were used by the colonial Americans because they were all-around utility horses that were closely coupled and well muscled (Figure 3-23). They were about 13 hands 2 inches in height. They were crossed with imported horses to produce some of the ancestors of the modern Quarter Horse. The Chickasaw Horse Association was formed to preserve the early Chickasaw-type horse.

Feral Horses in America

Feral horses are horses who have been (or their immediate ancestors have been) domesticated and freed. It is difficult to document how, when, and where the first horses escaped from or were stolen from the Spaniards and spread through-

FIGURE 3-24
The spread of the horse in America. (Adapted from Zarn, Keller, and Collins, 1977a.)

out America (Figure 3-24). They arrived with Columbus on his second voyage in 1493. During the early years of Spanish exploration, horses accompanied each voyage. Breeding farms established in the West Indies supplied horses for the explorers. Indian tribes in Texas and New Mexico were probably the first to obtain horses, and by 1780 most tribes had horses. Although no accurate estimates of feral horse numbers exist for the 1700s and 1800s, there were believed to be 2 to 5 million head. The greatest numbers occurred in the Southwest, especially in west-central Texas. Barbed wire and fencing reduced their numbers and shifted their range. At the end of the 1800s, most wild horse concentrations were found west of the Rocky Mountains. Presently, public land administered by the U.S. Bureau of Land Management and the U.S. Forest Service and located in California, Colorado, Idaho, Montana, Nevada, New Mexico, Utah, and Wyoming contains habitat for feral horses. In 1975, about 50,000 feral horses were present on these lands. Some of these horses have been feral for many generations, while others have been recently released.

Out of public concern for the plight of feral horses (now referred to as *mustangs*), Congress has passed two federal laws to protect them. Public Law 86-234 prohibits pollution of water holes for trapping them and the use of aircraft or motorized vehicles to capture or kill them. Public Law 92-195 places feral horses and burros roaming on natural resource lands under the jurisdiction of the secretaries of the interior and of agriculture for protection, management, and control. It provides a penalty for harassing, capturing, killing, or selling them, as well as for processing them into any commercial product. There is a program under which individuals can adopt a wild horse and can keep it. However, it cannot be sold.

DRAFT HORSES

Draft horses were once used as war horses and as beasts of burden and were later used for their pulling power. Thus they were developed to be heavy (1,500 to 2,000 pounds, or 680 to 907 kg); to have a low center of gravity; to be wide, deep, compact, and strong; and to have large bones. In the underdeveloped countries, they are still used for draft purposes when they are available.

Percheron

The Percheron horse (Figure 3-25) originated in the ancient province of Le Perche, near Normandy in France. Percheron horses in the seventeenth and eighteenth centuries were smaller and more active than the modern Percheron. They were 15 to 16 hands and were for general use. In the nineteenth century, the French government established the Haras de Pin as an official stud stable to improve the breeding industry. The horses at de Pin were selected with great care and the breed was improved. In 1823 or 1824, the stallion Jean-de-Blanc was foaled and became the most famous Percheron sire. All of the modern Percherons trace their bloodlines to Jean-de-Blanc.

The French Percheron Society was established in 1833 to promote the purity of the breed. The first registry of French Percherons was published in 1885. Because the Percheron was gaining popularity in America as early as 1850, the first stud book was published in America in 1876. At one time, America had more registered Percheron horses than any other breed.

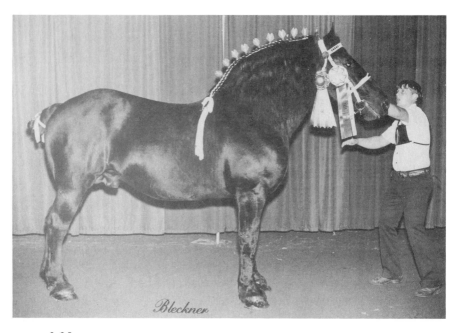

FIGURE 3-25

Percheron stallion, Radalta King, 251008, owned by Willard Wilder's Valley View Percherons. (Photo by Bleckner.)

The ideal Percheron has been described as a medium-sized, heavy-boned, up-headed horse. Stallions stand 16 to 17 hands and weigh 1,800 to 2,000 pounds (816 to 907 kg). Mares are 15 to 16 hands in height and weigh 1,500 to 1,600 pounds (680 to 726 kg). They are predominantly black or gray in color. The Percheron is more refined and has more balance than some of the other draft breeds. Thus they are frequently crossed with Thoroughbred horses to produce hunters and jumpers.

Suffolk

The Suffolk or Suffolk-Punch horse breed (Figure 3-26) was developed in East Anglia, the counties of Norwich and Suffolk, England. The first English stud book was published in 1880, although horsepeople in the district had kept private records earlier. Crisp's Horse of Ufford, foaled in 1768, is the foundation stallion of the breed, and practically all Suffolk horses today trace to him in the direct male line. They are recognized to be the only draft breed that breeds completely true to color—chestnut. Seven shades of the chestnut color are recognized. The horses are 16.1 hands in height, although some stallions are 17 hands.

FIGURE 3-26
Nortonean Ladyship, 19948, registered Suffolk mare bred by J. T. Thistleton-Smith,
Fakenham, Norfolk, England. Imported by Upwey Farms, South Woodstock, Vermont,
and owned by Mr. and Mrs. Lloyd B. Wescott, Clinton, New Jersey.
(Courtesy American Suffolk Horse Association.)

The Suffolk was bred exclusively for farm work, not for city dray work. They were popular during the 1930s, but the American Suffolk Association ceased to function in the 1950s. Then, in the early 1960s, when the draft horse made a recovery for show ring and pulling contest purposes, the association was revitalized.

Clydesdale

The native home of the Clydesdale (Figure 3-27) is southern Scotland, in the county of Lanark. Credit for establishing the breed is given to John Paterson. He imported a Flemish stallion, Locklyoch's Black Horse, who dramatically improved the local horses. Not much is known about their development until

FIGURE 3-27
An eight-horse hitch of Clydesdales. (Courtesy Flying U Rodeo Company.)

1780 when another stallion, Blaze, was imported from Ayrshire. In 1808, Glancer's dam was purchased. Most of the noted Clydesdale sires descended from Glancer.

The first imports to North America were to Canada in 1842. It was not until the 1870s that they were imported to the United States from Canada and Scotland.

The stallions weigh 1,700 to 2,000 pounds (771 to 907 kg) and stand about 16.2 hands. The mares weigh about 200 pounds (90.7 kg) less and are about 1 to 2 inches (25.4 to 50.8 mm) shorter. The horses are noted for the growth of long hair on the back of the cannons—"feathers." One problem with the Clydesdale is the difficulty in keeping its legs clean, dry, and free from disease.

The American Clydesdale Horse Association was formed in 1877 and maintains the stud book. Scottish Clydesdales were registered by the Scottish Clydesdale Horse Society, which was organized in 1878. Today, the Clydesdale Breeders Association of America maintains the stud book in America.

Belgian

Belgian draft horses (Figure 3-28) are native to Belgium. For centuries, the greater part of the Roman cavalry consisted of Belgian horses. No particular horses or individuals played a role in the development of the breed; instead,

FIGURE 3-28

Belgian colt. (Photo by Strandlund, courtesy M. Carlson.)

they were a product of their environment. They were improved as a breed after the government breeding stud at Tervueren was established in 1850. They were imported to America in 1866 but were not imported in large numbers until the 1880s.

Belgians became popular because of their size, strength, endurance, and action. They weigh about 2,000 pounds (907 kg) and are 16.2 to 17 hands tall. Records of Belgian horses have been kept since 1866 by the National Draft Horse Society of Belgium and since 1887 by the American Association of Importers and Breeders of Belgian Draft Horses. In 1937, the name was changed to Belgian Draft Horse Corporation of America.

Shire

Shire draft horses (Figure 3-29) developed in the lowlands of east-central England. They have been known as the Great Horse, War Horse, Cart Horse, Old English Black Horse, Giant Leicestershire, Strong Horse, and, lastly, Shire

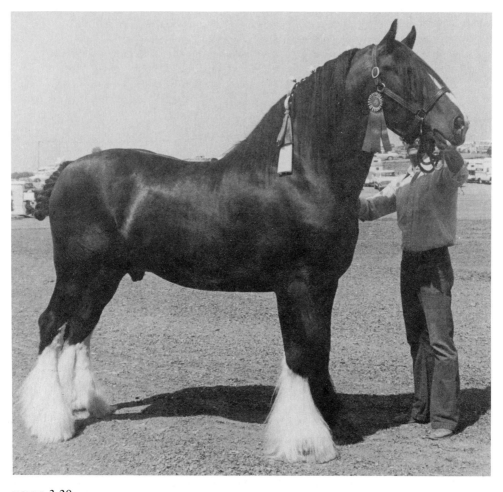

FIGURE 3-29
Ballasize Select, a Shire stallion. In 1977, he was grand stallion of all draft breeds
at the Clark county fair in Washington state. Owned by Ike and Kathy Bay, Hillsboro, Oregon.
(Courtesy Mrs. J. H. Erskine.)

horse. The early Shire horse was of mixed German, Flemish, and English origin.
Robert Bakewell, father of improved livestock husbandry in general, improved
the Shire breed during the middle 1700s. He imported large mares from Hol-
land, crossed them with English stallions, and continued to breed the horses
only if they met minimum criteria.

Shires were imported to Canada as artillery horses in 1836 and to the
United States in 1853 as draft horses for farming. They did not become popular
in America and were criticized for having straight shoulders and pasterns,

feathers, and a sluggish temperament. They weigh about 2,000 pounds (907 kg) and stand 16.2 to 17 hands.

Early records were maintained by the English Cart Horse Society, formed in 1878, and by the American Shire Association, formed in 1885.

HEAVY HARNESS OR COACH HORSES

There are several types of heavy harness horses: Hackney, French Coach, German Coach, Cleveland Bay, Yorkshire, Russian Orloff, and American Carriage Horse. These horses were primarily used to pull coaches. However, the Cleveland Bay (Figure 3-30) was used as a general utility horse, and the Hackney was used for park driving.

FIGURE 3-30

Choc Dorton Yeoman, Cleveland Bay stallion imported to America from England in 1975. (Courtesy Cleveland Bay Horse Society of America.)

FIGURE 3-31
Welsh pony. (Courtesy Welsh Pony Society of America.)

PONY BREEDS

Welsh or Welsh Mountain Pony

The original home of the Welsh pony (Figure 3-31) was in the rugged hill country of Wales. The severe winters required the development of sound, hardy, self-reliant ponies. They resemble a small Arabian in that they have fine features. Welsh ponies have been used to pull chariots, work in coal mines, work cattle, and serve as postmen's mounts. The ponies respond well to discipline and thus are ideal for young children.

Welsh ponies were imported to America in the 1880s. George E. Brown of Aurora, Illinois, imported a large number of them between 1884 and 1910. Through the efforts of George E. Brown and John Alexander, the Welsh Pony and Cob Society of America was formed in 1907. In 1946, "Cob" was dropped from the name. Welsh and English breeders formed their own registry in 1901. In America, their popularity declined during the Depression years but revived in the late 1950s and 1960s. Today, they are popular projects for 4-H, Vo-Ag, and Future Farmer club members. They are also crossed with other breeds to produce 13- to 14-hand ponies. These half-Welsh ponies can be registered with the registry. Because of the popularity of the crosses, the Cross-Bred Pony Registry was formed in 1959 to record the animals' pedigrees.

Connemara

The Connemaras are the largest of the pony breeds (Figure 3-32). The breed developed on the west coast of Ireland. The rugged ponies developed in this harsh environment. At various times during their early history, Spanish, Arab, and French horses were bred to Connemara mares or stallions. In 1606, King James I was given a dark dun Connemara stallion as a gift, which sired other ponies that he gave to King Philip of Spain.

FIGURE 3-32
Hideaway's Grey—Tifferary, a 3-year-old Connemara gelding. (Courtesy Hideaway Farms.)

In 1923, the Connemara Pony Breeders Society was formed in Ireland. By inspection, 9 stallions and 93 mares were admitted to the registry as foundation animals. Connemaras were first imported to America in 1951. By 1956, the interest was sufficient to form the American Connemara Pony Society.

Connemaras have been bred for utility. They are gentle and have good dispositions, so they are desirable mounts for children. There have been several outstanding Connemaras. Nugget was one of the great show jumpers, and Little Model had a distinguished dressage record.

Pony of the Americas

The Pony of the Americas (POA) is a small horse that was developed specifically for use by young people and families for better horsemanship, sportsmanship, and family fun (Figure 3-33). The breed was started by Les Boomhower of

FIGURE 3-33
Heather Hancock, a supreme champion
Pony of the Americas gelding.
(Courtesy Pony of the Americas Club.)

Mason City, Iowa. The breed was started as a cross of a Shetland stallion bred to an Appaloosa mare. The first foal was Black Hand No. 1. In 1955, the Pony of the Americas Club was formed.

POA horses must be 11.2 to 13 hands and must meet certain color markings. The horse must have Appaloosa coloring and characteristics and one registered POA parent to be eligible. In effect, the POA is a small horse with the Appaloosa color pattern. The POA blends features of Arabians, Quarter Horses, and Thoroughbreds as well as other breeds into the desired color pattern, size, and conformation.

Shetland

The Shetland pony (Figure 3-34) developed in the Shetland Islands, located 100 miles (161 km) south of Scotland and 350 miles (563 km) from the Arctic Circle. The ponies were present in this harsh environment when Norsemen arrived in the islands in about 850. The breed is characterized by small size, shaggy winter coat, and conformation. The maximum height is 11.2 hands, but most are about 10 hands. Before refinement through selection in some

FIGURE 3-34
Shetland pony. (Courtesy ANIMALS ANIMALS/Karen Tweedy-Holmes.)

countries, they had a draft horse conformation, so they were desirable for work in mines.

They are registered with the American Shetland Pony Club, which was formed in 1888.

Other Ponies

Several other breeds of ponies have resulted from crosses of the main pony breeds and various types of horses. The Americana was started in 1962 by crossing Shetland and Hackney ponies to produce a Saddlebred type of pony. The Welsh Pony and Tennessee Walking Horse were crossed to originate the American Walking Pony in the 1950s. The smaller horses failing to meet minimum height requirements of the Saddlebred, Tennessee Walking Horse, and Quarter Horse are registered with the American Saddlebred Pony, American Walking Pony, and American Quarter Pony Associations, respectively.

FIGURE 3-35
Little Hassler, mature miniature horse stallion, is 26½ inches tall.
(Courtesy J. R. Bridges, the owner.)

MINIATURE HORSES

There are no breeds of miniature horses, and they are rare. They are not dwarfs but are small-sized ponies and horses (Figure 3-35). The maximum height for registration by the International Miniature Horse Registry is 32 inches (813 cm). Tom Thumb measured 23 inches (58.4 cm), and Cactus measured 26 inches (66.0 cm) in height. They are kept by several breeders as pets and are used in circuses. The famous Falabella horses of Argentina were first bred by the Falabella family in 1868. They were primarily selected from Shetland ponies and have developed a reputation for excellent conformation. The smallest miniature has been 15 inches (38.1 cm) tall.

II
EXERCISING
OWNERSHIP
RESPONSIBILITIES

4
Providing Physical Facilities

Stable management is among the important activities of a good horseperson, and the building and facilities and their arrangement very much affect management efficiency. Carefully planned, well-designed facilities make horse care easier and add to the personal satisfaction and enjoyment of owning horses. In planning or inquiring about a facility for horses, you should (1) know what is needed, (2) know how to get it, and (3) keep within your budget.

Today most people in this country live in or near large cities. Because of this urban background or, more exactly, the lack of a rural background, many people who buy a horse for the first time lack experience in handling and caring for large animals. Many new horse owners have virtually no conception of appropriate facilities for their situation or whether, in fact, they should have any facilities. Consequently, to ensure that they are at least meeting the minimum standards for horse housing, they may overbuild their facilities, construct poorly conceived, inadequately designed, or inappropriately located buildings.

First, one must not forget the psychology of the horse. A horse is a cross between a large beaver and a gopher. It loves to eat everything available, including wood and other types of facilities, and it likes to dig by pawing. Never forget the horse's fondness for digging, pawing, and eating wood when you think of constructing a facility for a horse.

PLANNING

The most critical step in establishing a good horse operation is the planning that precedes land acquisition and building construction. Planning is just as important for remodeling a facility as for buying a facility someone else has built. If you plan to remodel, two criteria should be met: (1) the remodeled facility should be at least as good as you can build, and (2) it should not cost more than a new building.

Identifying Needs

The first step in the planning is determining the size and makeup of the horse population to be kept. Make a very specific list showing the number of each kind or type of animal and how each type of animal will be handled, managed, and housed. The list may be quite short (one or two riding horses) or quite extensive (a group including stallions, brood mares, and young stock such as are found on a breeding farm). It is extremely important to plan for expansion and future changes in the herd. A long-range plan for the herd allows better planning for facilities. Therefore, before establishing a new facility, set the ultimate goals of the entire operation. Do not forget about future expansion and needs. Identifying such needs is very critical, and you must be able to tell the architect or builder what the horses need. The mistakes should be made on paper, not in the building.

Once the type of operation, the number of animals, and the future expansion of the herd have been determined, the next step is to determine the type of facility needed to manage these horses. You must figure out the total amount of barn space needed under a roof to stable and care for the animals. Consider basic space needs for stalls, traffic lanes, and storage of feed, tack, and equipment. Include special features such as wash racks, trailer storage, breeding area, show stable for selling horses, riding arena, rest space, and other areas needed under the same roof.

Also figure out the amount of space needed under a roof for separate but associated buildings; for example, open shelters, riding arenas, sales barn, training barn, equipment storage, exercise area, hay and feed storage, office and lounge, and other areas, including living quarters if they are needed. Each person must decide whether certain areas should be placed under a roof or not. To make these decisions, consider the local climate; the location relative to homes, roads, and so on; and the requirements for the operation. For example, in a very cold, damp climate, a horse owner needs an enclosed, perhaps even heated, grooming area. In a warm southern climate, a more open arrangement may be adequate.

Determine the amount of space needed for open and fenced areas, such as lanes and roads, outside lots and corrals, exercise areas, parking space for cars and trailers and other vehicles and equipment, special events, and distance between buildings for a pleasing appearance, fire protection, and, most importantly, future expansion. If horses are to be boarded, an area for a loading ramp may be desirable. If horses are seldom trucked to or from the farm, the ramp may be unnecessary.

Carefully compare alternate housing arrangements and management methods and select those most suited to the situation. Ideas on horse facilities may be collected from publications, tours, visits to existing layouts, experienced horse owners, professional managers, farm advisers, and so on.

When planning, investigate available methods and then plan for inoffensive manure storage and disposal. Assess available fire protection and modify plans accordingly, adding protection and fire controls if necessary. It is also important to find out what building codes, zoning regulations, and other local restrictions may affect development, construction, or use of the facilities. Finally, review the budget allocation, placing priorities on the necessities, conveniences, overall desires, and frills, in that order.

Then determine the size needed for the complete layout. Exclude family housing, large pastures, and crop or other land not directly associated with the stabling and facilities complex. Then determine the size of the total layout, which is based on the land area covered by the necessary buildings and the directly related open and fenced areas. This provides the basis for establishing how much land to buy or to make sure is available within a complex that is already built.

CODES AND REGULATIONS

The next step in selecting the area on which to build is determining what effects various regulations and restrictions may have on the proposed stable. Society places certain limitations on ownership, development, and use of property. This step is just as important for land presently owned as for land being considered for purchase.

There are three types of codes and regulations. Generally speaking, building codes set construction standards, zoning prohibits the use of property for specific purposes, and sanitary regulations have to do with public health, related pollution, and pest control. Other restrictions may also limit the use of the property, including deed restrictions, easement, and covenants between property owners. Paying particular attention to the codes and regulations may save grief in the long run.

SITE SELECTION

Selecting a site for construction of a stable or a combination stable and riding arena involves some very elementary common sense. The most level site that can be obtained should be used for the construction site of the barn or stable.

The building site should be well drained, accessible, and have a slope of about 5 feet per 100 feet (1.5 m per 30.4 m) away from the building in all directions to assure good surface drainage. Topography of the site particularly affects the cost of site development, and site preparation should be completed before building construction is started. Sites on steep slopes or rocky terrain and sites requiring considerable landfill are costly to develop and may make compromises necessary. Make sure that the property is not in a flood plain or other areas where high water could cause problems.

The first thing that should be considered is water supply. An adequate all-year water supply of the quality needed must be available at the site, either from a public water system or from convenient groundwater and surface water developments. Any well needed should be drilled before any other construction begins. Also keep in mind the distance water lines must go and the electrical supply available.

If at all possible, the facilities should be constructed away from distracting noise, offensive odors, and heavy traffic. Existing nuisances may affect use and enjoyment of the site, and nuisances that horses and horsepeople create may affect neighbors.

In larger operations, wind breaks may be important. Consider the natural protection from the elements provided by wooded areas, knolls, hills, and ravines.

Existing buildings may be a determining factor in site selection but only if their size, location, physical condition, and inside arrangement fit into the overall plan for horse facilities. The proximity of the barn in relation to residences and management of labor and the evaluation of the site in relation to ultimate use of the barn (private use, commercial stable, breeding operation, and so on) are important.

Esthetics are important. Appearance, personal taste, pride of ownership, and available funds may affect site selection. However, the horse-housing facilities should not overshadow the residence.

Building layout will probably be a deciding factor in selecting the site, and the site chosen may affect both building layout and building style.

The size of the operation also affects site selection. In addition to room for the planned buildings, the site should provide space for other planned activities and areas. It should also provide for future building and paddock expansion; good traffic patterns; safe and convenient handling of animals, vehicles, equipment, and materials; and snow removal, if necessary.

Horse facilities definitely need a good, all-weather road to provide access in all kinds of weather. You must be able to get to the animals during inclement weather, and you may need to haul feed and/or horses into the facility during inclement weather.

The physical and social aspects of manure disposal can be a major problem, so be sure to follow local regulations on manure handling. It is essential to clean stalls, paddocks, and exercise areas of horses kept in confinement. Generally, spreading the manure and bedding on crop land is most satisfactory.

If the location does not include fields where manure can be spread 365 days a year, you will need a manure storage area. Keep all manure storage and disposal areas well removed from streams or land that slopes directly into streams, and never spread manure on frozen land.

In some areas, location of feed sources is important. Even saving $1 per ton (907 kg) of feed on hauling costs can make quite a difference over the years.

In most cases, it is less expensive to grade and prepare a site satisfactory in a majority of the factors discussed than it is to pick a level site that satisfies no other factors. However, it is expensive to hire construction personnel such as mountain climbers, alligator fighters, and scuba divers to prepare the site and construct the stable. If the site must be graded, all the top soil should be removed. The construction pad should be higher than the surrounding areas outside the building so that there will be good drainage. The base or pad must be well compacted by the grading machine or the site must be graded well in advance of actual construction to allow rain to compact the area. Also, rain indicates low areas on the site that can be filled before construction begins.

BARNS

The function of a horse barn is to protect horses from temperature extremes and injury, keep them dry, and eliminate drafts. It should be easy for the operator to work in. Adequate space in the barn should be provided for the well-being of the animals and the convenience and safety of the people caring for them. There must be a sufficient supply of fresh air in both winter and summer. A barn should have sufficient light and be free from odors.

Functional Design

First decide on the set of functions to be satisfied by the proposed building. Then choose a design that provides these functions. A common error is selecting a building that includes functions or features that are not really needed. This drastically increases the cost of the building. For example, a covered exercise area costs nearly as much to erect as a building filled with stalls. If a training program requires the use of the exercise area, choose a design incorporating that feature. But if an exercise area is not needed, do not build one.

Barn styles are determined by the distinctive shape of the roof. The three shapes most widely used are the shed, gabled, and offset gabled (Figure 4-1). The once-popular gambrel, gothic, and monitor roofs, very much in evidence on old barns, are now used only occasionally on new barns. The trend is toward one-floor structures with clear-spanned or post-beamed roofs (Figure 4-2). A one-floor structure eliminates or holds to a minimum costly overhead

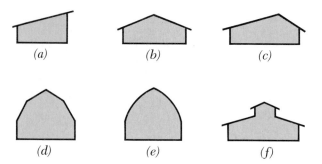

FIGURE 4-1

Barn styles. *(a)* Shed. *(b)* Gable. *(c)* Offset gable.
(d) Gambrel. *(e)* Gothic. *(f)* Monitor.
(Adapted from Blickle and others, 1971.)

storage. The use of one floor also permits lighter, less costly construction. However, it may be more economical to use overhead storage if ground space is at a premium. Clear-spanned roofs also add flexibility to buildings by making them more adaptable to other uses. The style chosen must first be utilitarian; esthetics are secondary.

Layout

There are many horse barn layouts. Basic designs incorporate one, two, three, or four rows of stalls (Figure 4-3). Most barns for housing one or two horses are of the one-row, porch-type construction. For more horses, stalls can be arranged around the sides of a large building, leaving an area in the center for riding or exercise (Figure 4-4). If the barn is smaller, a service alley may divide two rows of stalls. A large barn may have two such alleys (Figure 4-5). Alternately, the stalls can be grouped in the center with the exercise area around the outside (Figure 4-6). This is a popular racetrack layout. Simple facilities are shown in Figure 4-7. However, you should choose the layout that will work best for your situation. The best way to do this is to study many of the plans that are available, visit other farms, and discuss the situation with people involved with similar requirements. Be quite critical about the layout and the building plans, because there are marked differences in the amount of floor space used to house horses. An eight-horse barn may be 10 by 92 feet (3.05 by 28 m). In this case, almost all of the floor space is used to house horses. Another plan using a 50- by 24-foot (15.24- by 73.1-m) barn uses 80 percent of the area to house horses. Other plans use only a small percentage of space for horse housing.

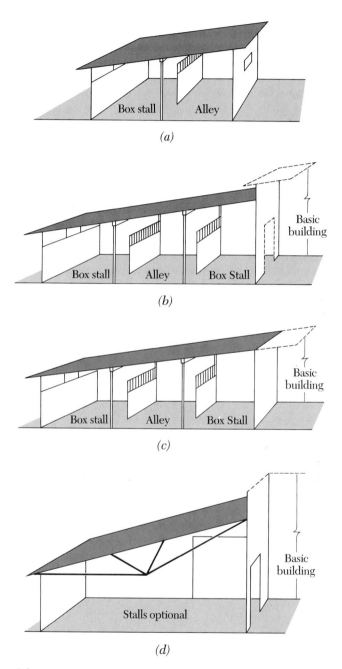

FIGURE 4-2

Shed roof options. *(a)* Free-standing building (post and beam). *(b)* Attached shed (post and beam). *(c)* Roof extension (post and beam). *(d)* One-slope truss (clear span). (Adapted from Blickle and others, 1971.)

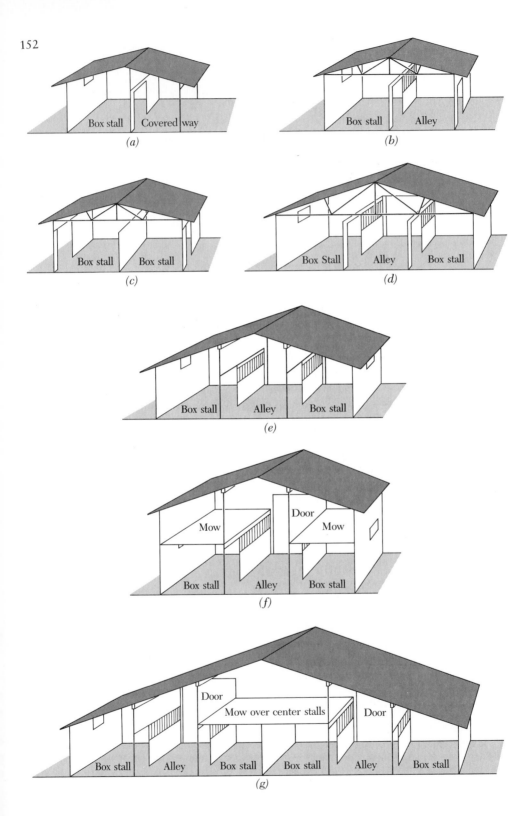

(a) Box stall | Covered way

(b) Box stall | Alley

(c) Box stall | Box stall

(d) Box Stall | Alley | Box stall

(e) Box stall | Alley | Box stall

(f) Mow | Door | Mow | Box stall | Alley | Box stall

(g) Door | Mow over center stalls | Door | Box stall | Alley | Box stall | Box stall | Alley | Box stall

FIGURE 4-3 *(facing page)*

Gable roof options. *(a)* One row of stalls serviced from covered way (post—beam—wall). *(b)* One row of stalls serviced from enclosed alley (clear-span truss). *(c)* Two rows of stalls serviced from outside (clear-span truss). *(d)* Two rows of stalls serviced from center alley (clear-span truss). *(e)* Two rows of stalls serviced from center alley (post and beam). *(f)* Two rows of stalls serviced from center alley; mow over stalls (post and beam). *(g)* Four rows of box stalls serviced from two alleys; two center rows of stalls back to back with mow above (post and beam). (Adapted from Blickle and others, 1971.)

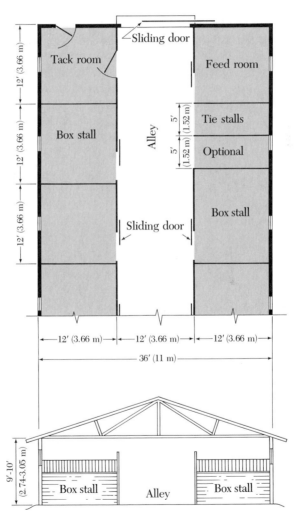

FIGURE 4-4

Barn layout for two rows of stalls and center alley for inside service. (Adapted from Blickle and others, 1971.)

154

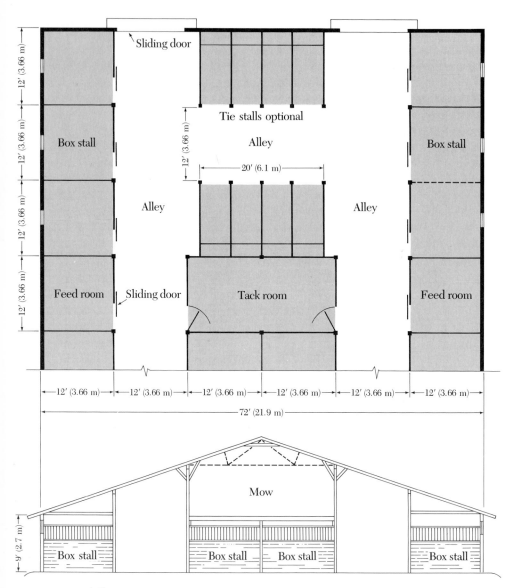

FIGURE 4-5

Barn layout for four rows of stalls and two alleys for inside service.
(Adapted from Blickle and others, 1971.)

There are some design tips to follow when preparing or examining a barn layout. Plenty of head room for the horse and the rider is a necessity if any riding is to be done in the area. Eight feet (2.4 m) of head room is the minimal acceptable height for an average horse. For a horse and rider, at least 12 feet

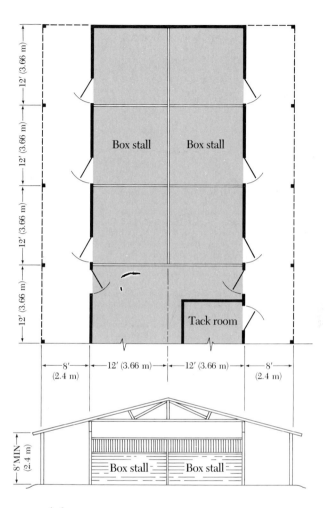

FIGURE 4-6

Barn layout for two rows of stalls and two alleys for outside service. (Adapted from Blickle and others, 1971.)

(3.66 m) are needed. Ceiling height in stalls should be no less than 9 feet (2.7 m). Riding arena ceilings should be at least 14 feet (4.3 m) high. If jumping events or training is to be held in the arena, the ceiling should be at least 18 feet (5.5 m) above the ground.

The foundation is important. Regardless of the type of construction, whether it is wood, block, or steel, design an adequate foundation, with the bottom below the frost line. Poles should be sunk at least 3 feet (0.9 m) in the ground for stable areas, with a concrete pad below the poles. For arenas and

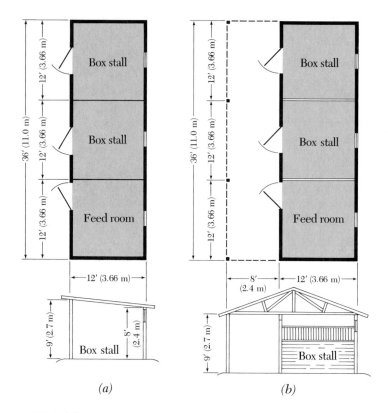

FIGURE 4-7

Barn layout to keep two horses, outside service. (a) Basic facility.
(b) One row of stalls, covered way. (Adapted from Blickle and others, 1971.)

other large buildings, poles should be 4 feet (1.2 m) in the ground, with a concrete pad below. Moreover, one-quarter of the pole should be in the ground. For pole construction, the foundation of the building should consist of treated tongue-and-groove boards around the outer base of the building. These boards should be buried at least 2 inches (5 cm) below the surface of the building pad. These considerations are particularly important for keeping the barn level so that the doors will work without difficulty and the walls do not crack.

Barn Exterior

The exterior walls can be of wood, block, steel, or aluminum. There are several points to remember in selecting the type of material. First of all, it should be durable—something that a horse cannot kick through or cannot eat through.

Consider appearance and adaptability to changes in operation that require modifying walls for such things as doors and windows. Outside wooden walls of 4-foot by 8-foot (1.2- by 2.4-m) plywood sheets or tongue-and-groove material can be stained or painted for an excellent appearance. When aluminum expands and contracts with changing temperatures, it crackles. The noise bothers some horses and some people. Also, when the wind blows, aluminum can make a bothersome noise.

The number, size, and types of windows depend on the operation and the owner's desires. If a window is to be opened so that the horse can hang its head outside, it must be carefully designed. Horses will chew the outside of the building. Also, wood sashes are not particularly good for horse barns. They are extremely difficult to protect from breakage by the horse and from chewing. However, a wood sash may be more attractive depending on the type of exterior walls. Aluminum sliding windows are basically maintenance free, require no painting, and provide no surface that the animal can chew on. They are easy to protect from horses with bars, metal screens, or pipe. Horses cannot put their heads out through these windows. However, sliding windows only allow half the ventilating area, for the size of the glass, that other windows allow.

There are at least four types of roofing to consider: aluminum, metal, asphalt shingles, or wood shakes. All four types are durable if properly installed. However, there are some minor characteristics to consider when selecting the roof material. Aluminum and metal roofs may leak around the fasteners. Many people now use screws rather than nails to hold aluminum and metal roof material to the frame. Then to expand facilities, the screws can be taken out and the sheets of aluminum or metal removed without damage. Also, skylights are usually easier to install on aluminum or metal roofs. Skylights of any kind should be put in the aisle, not above the stalls, because the light panels transmit heat. Skylights may also cause condensation problems, dark areas in arenas, and excess heat buildup in summer. However, they do cut down on the amount of electricity required to light the barn. With a shingle roof, condensation is not a problem. And the shingle roof system is warmer even without insulation.

Any door over 4 feet (1.2 m) wide should slide instead of hinge to eliminate warping and sticking. They should be sturdy and durable. They will be used often and should be very easy to handle. A Dutch door should have a metal covering over the top of the lower section to prevent chewing. Plan, construct, and use doors in such a way that they are not potentially dangerous to horses and people. And doors should be convenient. For example, double-aisle doors may be more convenient than a large single door, especially when children are to be opening them. One large door may be too heavy for a child to slide. If a horse is to be ridden through the door into the building, the door should be at least 12 feet (3.66 m) wide. Doors to accommodate trucks or other large equipment should be at least 16 feet (4.88 m) wide.

Barn Interior

The primary purpose of the barn is to modify the environment for the horse. Therefore, it should allow control of temperature, humidity, odors, light, velocity of air, and the amount of fresh air available.

Temperature and humidity are extremely important. A major problem in horse housing is moisture or condensation, which can be controlled in a properly constructed facility. Horses, like other animals, produce heat and moisture. The moisture (2 gal, or 7.6 liters, per horse daily) must be removed from the stable to prevent condensation and buildup of odors. Humid conditions contribute to respiratory diseases and cause odors to develop. Condensation on wood can accelerate the aging process by providing a good environment for growth of fungus and molds, which can deteriorate wood rapidly.

Ventilation removes the moisture and odors and controls the temperature. It is important to install a ventilation system for the stable when the building is being constructed. This can be done by providing, in addition to doors, eave openings, ridge ventilators, and adjustable windows or wall panels in both uninsulated and insulated buildings used for cool housing. Ventilation fans, fresh air inlets, adjustable windows or wall panels, and supplemental heat can be installed in insulated buildings that are used for warm housing. As a rule, horse owners in mild or moderate climates can obtain adequate housing for their animals with uninsulated buildings; however, wood or masonry side walls and tight sheeting under the roof are desirable for some insulation.

If you must insulate a barn, make sure that water does not condense readily on the insulated surface (normally it does not). The roof should always be insulated, while insulation in side walls is desirable but not essential. Minimum insulation is equivalent to about 3½ inches (8.89 cm) of fiberglass bats. Place a vapor barrier on the warm side to prevent or slow down passage of water vapor through the walls and ceilings. It must be installed between the insulation and the inside wall or ceiling lining. Many materials can be used as a vapor barrier; the most common is 4-mil polyethylene plastic.

Ventilating open-front buildings is rather simple. However, there must be good air movement during cold weather to prevent moisture from condensing on the underside of the roof and in hot weather to keep the inside temperature acceptable. Locate the building on a well-drained site with the open side away from prevailing winter winds. If drafts and snow blow-in are problems, close the front wall at each end and adjacent to solid partitions. Use spaced-board wind breaks at the end of the building. Install cross-partitions from floor to roof in long, open sheds if indoor drafts continue to be a problem.

The box stall is the most important component of the barn. The first consideration is size. A 10- × 10-foot (3.05- × 3.05-m) stall is the minimum, and a 10- × 12-foot (3.05- × 3.66-m) area is most desirable. Foaling stalls should be larger—12 × 12 feet (3.66 × 3.66 m). Stallion stalls should also be 12 × 12 feet or 14 × 14 feet (4.27 × 4.27 m). Ponies need an 8- × 8-foot

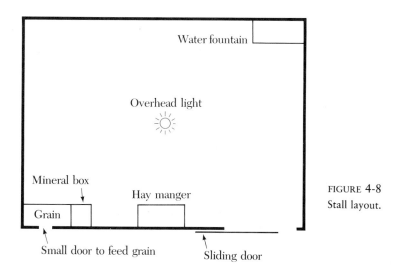

FIGURE 4-8
Stall layout.

(2.4- × 2.4-m) stall. The stalls must be sturdy and free from sharp, project-ing objects. Stall lining should be solid for 4 to 7 feet (1.22 to 2.13 m) above the ground. Bars (2¼ inches, or 5.7 cm, between bars); heavy, woven wires; or expanded metal can be used above that point. Pressure-treated lumber is more durable than other kinds of lumber. Boards should be placed on posts so that the posts take the pressure, not the nails.

Locate the feeders, waterers, and other stall equipment where the animal cannot injure itself (Figure 4-8). Most people usually prefer to put the water trough in one corner of the stall. The hay manger and feed box should be located across the front wall so that the grain or hay can be added to the stall without actually going into the stall.

The stalls should be lined with oak or some other durable material. The lining should be installed so that surfaces available for chewing are eliminated as much as possible. When installing linings, try to eliminate dead spaces between walls where rodents can nest.

The minimum ceiling height in stalls should be 9 feet (2.7 m). The stall floor is a major concern. In many instances, the existing soil can be used if it is acceptable. Clay can be added to stalls after construction. Concrete can be used if desired, but a drainage system should be built into each stall, and you will need to use deep bedding. Asphalt flooring has become quite popular and is very useful. The asphalt should be very porous; that is, the rocks used in the asphalt should be No. 1 and 2 grade. Resilient mats can be used to cover the stall floor. A specially compounded synthetic resinous material has been developed that, when applied over an asphalt, concrete, or wood base, results in a smooth or textured, resilient, uniform, durable surface for use in stalls and wash racks.

Stall doors are one of the most important features to consider about the stall because they are used constantly and it is easy for them to get sticky and difficult to open. They should be at least 4 feet (1.22 m) wide. Sliding doors are more desirable because they are easy to use and do not pose any danger to horses or people. Doors that swing into the aisle can be dangerous to people, and those that swing inward can injure horses. Sliding doors can be solid or can have an opening on the upper section. The latch should require at least two types of movement to open. Moreover, it should be easy for the manager to look down the alley way and to check that all the stall doors are latched. The latches should be simple enough so that they do not bind and do not require frequent adjustment.

Aisle floors of asphalt or concrete are less dusty than dirt, but the surface should be coarse, not smooth. Make sure that the surface is rough. If it is concrete, it should have a rough broom finish. The asphalt should be made with No. 1 and 2 grade stones. Recently, some people have successfully used indoor-outdoor carpets on their alley ways and grooming areas. Do not lay water lines under asphalt or concrete. The water lines should be laid outside the building and then brought individually into each stall.

The most important aspect of the tack room is security. Moreover, its size should fit the size of the stable. If possible, place the tack room between two stalls. Then small observation windows can be put in the walls adjacent to the stalls. These can be very useful when foaling mares, and so forth. The most common floor material in the tack room is concrete. The walls should be insulated if heat or air conditioning is to be installed. In damp areas, heat is desirable to keep the tack dry.

Grooming and washing areas depend on the type of facility and its primary function (Figure 4-9). It can consist of a single slab placed on the outside of a building where the water runs off or it can be very elaborate. Elaborate indoor wash areas usually have hot and cold water. They also have radiant heat lamps for drying purposes.

Barn lighting is extremely important. Lights should be shielded to prevent horses from breaking them. They should be evenly distributed in the alley way or aisle and in the riding arenas. If not, the horse will spook and try to jump the shadows. Fluorescent lights are cheaper because they produce $2\frac{1}{2}$ times the amount of light per watt of electricity. It is wise to be able to turn outside lights on and off from the house. Make sure there are sufficient outlets for clippers and other appliances. All electrical wire should be placed in conduits to protect it from rats and horses.

The feed room should be large enough to accommodate the size of the barn, especially if hay storage is not provided elsewhere in the building. The floor should be concrete. The feed bin should be rodent proof; in fact, the whole feed room can be made rodent proof by putting rat-proof wire mesh ($\frac{1}{4}$-inch $\times$ $\frac{1}{4}$-inch, or 6.4 mm, hardware cloth) behind all of the ceilings, floors, and walls. Organize the room so that it is easy to clean. This will also minimize the potential for rodent infestation. The door to the feed room should

FIGURE 4-9
Wash rack. (Photo by B. Evans.)

be large enough to allow feed bags, hay bales, and so on to be carried in. It should also be conveniently located so that it is easy to get the feed from the entrance to the feed room. In some operations, a door to the outside may be desirable for feed deliveries. If there are a large number of horses, you may want to explore the idea of having a bulk bin, which saves labor. It is much more convenient to store the hay on the ground level rather than to put it into an overhead loft. If there is an overhead loft, some type of mechanized system is desirable to get the hay into the loft. Normally it is cheaper to expand first-floor facilities than to build an overhead loft strong enough to support hay storage. One ton (907 kg) of hay requires 200 cubic feet (5.66 m³) of space. The following formula determines the space needed to store 125-pound (56.7-kg) hay bales:

$$\frac{\text{Number of horses} \times \text{lb/day} \times 365 \times 200}{2,000}, \text{ or}$$

$$\frac{\text{Number of horses} \times \text{kg/day} \times 365 \text{ days}}{907}$$

The space should hold hay for one year.

Each horse drinks 8 to 12 gallons of water per day (30.3 to 45.4 liters); therefore, the system must supply sufficient water for all the horses. It must also supply water for other necessary functions—wash rack, lawn, fire sprinkler system, and so on. Water should not move from one water fountain or trough to the next, because such systems spread diseases very rapidly. Decide whether to use automatic or individual waterers. This is a labor consideration. Decreasing labor requirements as much as possible means more free time to enjoy the horses. In commercial establishments, it also means money savings. Automatic water fountains should be easy to clean, have protected floats, and be designed so a horse cannot get its teeth stuck in them. They should have shut-off valves for emergency use.

Portable Barns and Equipment

Many portable barns and corrals are available. Some are inexpensive; others are not. They offer advantages such as fast construction, low maintenance, and movability. Both single stalls and shelters or barns with many stalls are available, as are portable round corrals for training young horses and a variety of other equipment.

FENCES

Fences require considerable planning. First decide what should be fenced and, second, decide what type of fence to use. Aluminum alloy tubular fencing is expensive but is maintenance free. If you choose a wire-mesh fence, install a board in the middle and on top, because a horse must be able to see the fence. Foals that run into wire fences may break their necks. The wire should cover the boards so the horses cannot chew on them. Make sure that the mesh is small enough (2-inch × 4-inch V mesh, or 5.1 × 10.2 cm) that a colt cannot stick a foot through. Wood fences should have three or four rails. Wood is not desirable because it must be kept painted, splinters, breaks, rots, and is chewed by horses. Chain-link fences are difficult to maintain because they are stretched out of shape by horses kicking and rubbing them. Precast concrete fences are strong, durable, maintenance free, attractive, and functional. Rubber-nylon fencing must be properly installed to prevent horses from stretching it and escaping through it. Do not use rubber- or plastic-covered cable as horses chew off the covering, and the cables act like knives if a horse gets caught in them.

A horse fence should be at least 4 feet 8 inches to 5 feet high (1.4 to 1.5 m). Horses like to play across fences, so barbwire should not be used along the top of the fence. A board fence covered with wire mesh or an electric fence wire is best.

Gates in fences must be strong and designed for safety, so that they do not sag and become difficult to open. For some reason, horses like to use the area around a gate as a rest area. Therefore, the gate should be made out of a material that they cannot chew up. Gates should be designed so that a horse cannot get its foot caught, because horses tend to paw at a gate when they become anxious. The gatepost must be strong and set in the ground so it will not lean. Latches must be designed so that a horse cannot learn to open them.

BREEDING FACILITIES

Facilities for breeding are discussed in the chapter on reproduction in the mare. In addition to the breeding area, palpation chutes, and teasing facilities, brood mare barns may be necessary. Usually sheds or open-front barns are sufficient to provide adequate protection for mares and foals.

Foals need creep feeders. In a stall, the creep feeder is designed so that the foal can eat the creep feed but the mare cannot (Figure 4-10). The bars across

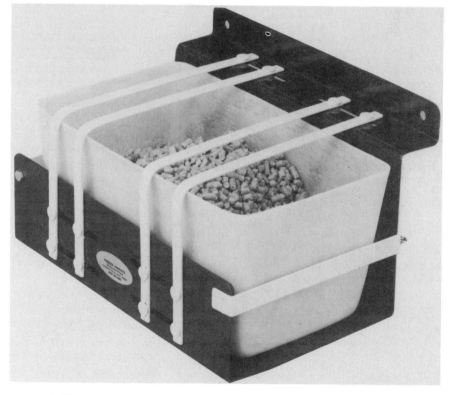

FIGURE 4-10

Foal creep feeder box (Gro-Foal feeders by Paddock Products, 339 Bentleyville Rd., Chagrin Falls, Ohio). Bars must be adjustable from 2¼ inches (5.71 cm) to fully open. Feed box is easily removed for cleaning.

the top are adjustable so they can be opened up more as the foal gets larger. In the pasture or paddock, the creep is fenced and must be close to the mare's feeding area. If it is not located properly, foals will not leave their dam to go to it. Creep feeders are dangerous because a foal can injure itself by trying to run in or out of it.

BEDDING

Bedding for horses should absorb urine (Table 4-1), be free from dust, be readily available, be inexpensive, provide a comfortable bed, and be easy to handle. Several materials are used, but barley and oat straw are the most common. The straw should be left in long lengths to keep it free from dust, even though the absorptive capacity of chopped straw is much higher. If it is left

TABLE 4-1
Water Absorption of Bedding Material

Material	Lb Water Absorbed per 100 Lb Dry Material
Straw	
Barley	210
Flax	260
Oat	
long	280
chopped	375
Rye	210
Wheat	
long	220
chopped	295
Wood Products	
Pinewood chips	300
Hardwood chips	150
Pinewood sawdust	250
Hardwood sawdust	150
Pine shavings	200
Hardwood shavings	150
Other	
Corn stover	250
Corn cobs	210
Hay, chopped	200
Peanut hulls	250
Peat moss	1,000
Sand	25

SOURCE: Ensminger (1969).

long, the stable personnel can quickly separate the clean areas from the dirty areas in the stall and thus less straw is used each day.

Wood shavings are preferred by many horse owners. To prevent respiratory problems, shavings must be free from sawdust. The softer woods have a greater absorption capacity. Shavings offer a definite advantage for people traveling to horse shows or sales because they can be purchased in bales.

Rice hulls and peanut hulls provide excellent bedding when they are available. However, they do not absorb water, and the stall floor soon becomes saturated with urine.

Sand is poor bedding. It absorbs very little urine, and horses that pick around in it for hay may develop sand colic. It is also quite heavy to handle.

The average horse produces about 8 tons (7,257 kg) of manure per year. Because manure contains 60 percent water, 3.2 tons (2,903 kg) of dry manure must be disposed of each year. Bedding is usually disposed of at the same site. To control internal parasites, pick up manure from stalls and paddocks at least once a week. To control flies and odors, pick it up daily. If it is not removed from the premises at the time it is picked up, the manure should be stored on a concrete slab. It will compost, and then it can be used for fertilizer. If it is to be spread on pastures, it should compost for at least 2 weeks to kill the larvae of internal parasites. The pile should be sprayed regularly to kill flies.

The location of the compost pit should be convenient to the stable area. For esthetics, it should be located out of sight. The pit should be accessible for easy loading into a truck when the compost pile is removed. If many horses are kept at the stable, a small tractor with a front-end loader is handy to haul manure.

FIRE SAFETY

The problem of stable fires is acute, but the attitude of most people toward them is similar to their attitude toward car wrecks: They always seem to happen to someone else—that is, until they happen to you. Between 1960 and 1970, there were at least 140 reported stable fires in the United States. A total of 1,400 horses were burned to death, and property damage amounted to $30 million. Therefore, preventing stable fires concerns all horsepeople.

Protecting a horse in a stall from the danger of fire is a totally different problem from fire prevention in a person's home. The horse is usually standing in some type of bedding that is kept dry. Oat straw develops a temperature of approximately 300°F (148.9°C) at the head of a fire approximately 1 minute after it ignites. Barley straw takes approximately 5 minutes to develop the same temperature. In other words, horses stand in material that develops as much heat and at the same rate as gasoline. All that is required to start a stable fire is a match or a spark.

A common problem is that there is usually only one door to the stall, so the horse cannot escape into the paddock. The size of the stall is approximately 10 × 10 feet or 12 × 12 feet (3.05 × 3.05 m or 3.66 × 3.66 m). Approximately 2 or 3 minutes are required for a straw fire in a stall to burn an area 10 feet (3.05 m) in diameter. By the time the fire covers an area 4 feet (1.22 m) in diameter, most horses have been injured. Their lungs are seared when the fire has covered an area 6 feet (1.83 m) in diameter. They start to suffocate when the area is 8 feet (2.44 m) in diameter and are dead by the time an area 10 feet (3.05 m) in diameter is burned.

If a horse is to survive, the fire must be extinguished in the stall within approximately 30 seconds. How does the horseperson cope with this special problem?

The most common answer is proper stable construction. Cement-block barns or similar types of construction are meant to save the building, not the horse in an individual stall. If fire is to be prevented from spreading through a barn, some construction features are mandatory. Fire-retardant paints effectively delay combustion of wood framework inside the barn. Every third or fourth stall should have partition walls that extend to the ceiling to help delay the spread of the fire through the barn.

Sprinkler systems can also be used, but most sprinklers were not designed to put out a fire under the circumstances that exist in a stall. Automatic fire-extinguishing systems must meet at least two requirements to be used in a stable. They must not suffocate the horses when activated, and they must react within seconds after a fire starts. Most automatic sprinklers are designed to throw a circular pattern instead of a square pattern, so the spray must be strong enough to reach the corners of the stall. If a fog-type system is used, the fog must suffocate the fire to extinguish it, so the horse will suffocate with the fire. Water can cause extensive smoke formation, which in turn will suffocate the horse or at least cause lung damage. This can be prevented if the sprinkler is activated before the fire covers an area 1 foot in diameter. A thermal lag system with sensors located above the stalls takes 4 to 5 minutes to activate, which is too long. Recent developments have resulted in sprinkler systems that begin to operate approximately 5 seconds after a fire starts.

Of course, not all fires start inside a stall. A layer of chaff and dust covers the floor of many barn lofts. Chaff and dust burn like gasoline. The loft and other storage area of a barn should be swept regularly to remove dust. Periodically, the inside of a barn should be hosed down with water to remove all dust and cobwebs. This is just as important as keeping the alleys clean.

An adequate number of exits from the barn are necessary so that fire does not block the only exit. All exits should open into enclosed areas. In case of fire, the horses can be turned loose but they must not flee the stable area. Fleeing horses may be killed by automobiles when they cross or travel down roads. They may also ruin other property such as lawns or crops.

The stall doors should open by sliding. When horses flee through emergency exits, they can catch their hips and other parts on doors that open into stalls. When the doors slide open to one side, someone can run through the stable, opening all the doors and letting the horses out. Also, sliding doors cannot close accidentally. Because time is important in emergencies, stall door latches should be strong and easy for people to open, but not for horses.

If a barn is constructed in this manner, all the stall doors can be opened by the rescuer on one trip through the alley. Horses that leave by themselves will be caught in the outside enclosed area. However, someone should prevent them from returning inside the barn because a scared horse likes to return to a familiar stall. Because of this trait, many horses will not leave their stalls unless they are led out. It may be necessary to cover their heads with a sack or something similar to prevent them from seeing the fire and smoke. They are easier to lead this way; if led any other way, they try to climb the back walls of their stalls.

All electricity in a barn should be turned off at night. The master switch should be located close to an entrance, so the lights can be turned on and one can see what is taking place inside the barn during an emergency. Flashlights and the bouncing, moving shadows and lighted areas they cause frighten horses. Do not use flashlights unless absolutely essential. Electrical wire in a barn must be protected from mice and rats, which will chew the insulation off, leaving two bare wires that can easily start an electrical fire. At the same location as the electrical switch, which should be at an exit, a fire extinguisher should be available in case of a small fire. At this location, there should also be a water hose that can reach the length of the barn. This hose should remain attached to a water hydrant so that one can quickly get water to all parts of a barn.

As another protective measure, each stall should have a halter and lead rope hung on a hook beside it. Then you need not waste time taking a halter off one horse and putting it on another while rescuing horses. Time is of the essence.

People working at the stables should obey a strict set of fire prevention rules. Smoking should not be allowed in or adjacent to buildings. For convenience, large receptacles for disposing of cigarettes should be placed at building entrances. All electrical equipment should be in safe working condition and should be inspected regularly. Appliances such as hot plates, coffee pots, and radios should not be left unattended when in use. They should be used only in a special area. Inflammable materials such as lighter fluid, solvents, or cleaning fluids should not be used in the stable area. The owner or manager should not tolerate rule infractions by employees or visitors. The rules must be rigidly enforced at all times.

5
Feeding

Nutrition is one of the most neglected aspects of horse care and management. Horses seem to suffer from two opposite extremes of malnutrition—either they are too fat or they are starving. An understanding of nutrient requirements, characteristics of common feeds, feeding management, and nutrient costs enables an owner to properly and economically feed horses.

ANATOMY AND FUNCTIONS OF THE DIGESTIVE TRACT

The digestive tract of the horse consists of the mouth, pharynx, esophagus, stomach, small intestine, large intestine, rectum, and anus (Figures 5-1 and 5-2). Two large glands, the liver and the pancreas, are associated with the digestive tract and release part of their secretions directly into the digestive tract. The functions of the digestive tract include prehension of food, mastication, digestion, absorption, and initial storage of the nutrients.

The lips of the horse's mouth are quite mobile and help in sucking and prehension of food. They are very sensitive tactile organs. The hard palate, the upper part of the mouth, is a bony plate covered with a mucous membrane, crossed by many ridges. It has a rich blood supply. The hard palate extends back to the soft palate. The tongue is an important tactile organ and is necessary for mastication and swallowing. It serves as a plunger to drive the food into

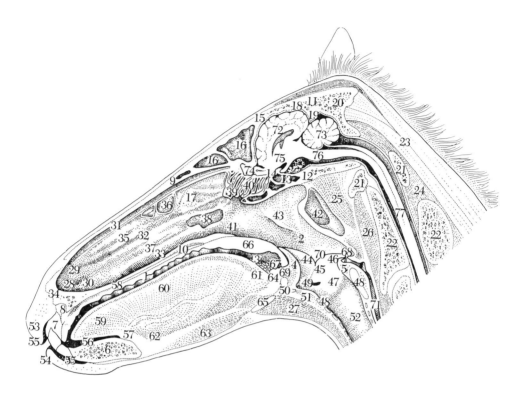

FIGURE 5-1

Sagittal section of the horse's head. (1) Oral cavity. (2) Nasopharynx. (3) Oropharynx.
(4,5) Laryngopharynx. (6) Incisive part of mandible with first incisor. (7) Incisive bone with
first incisor. (8) Incisive duct. (9) Nasal bone. (10) Osseous palate. (11) Interparietal bone.
(12) Sphenoid bone. (13) Sphenoid sinus. (14) Ethmoid bone. (15) Frontal bone.
(16, 17) Conchofrontal sinus, composed of (16) frontal sinus and (17) sinus of dorsal nasal cocha.
(18) Parietal bone. (19) Tentorium cerebelli osseum. (20) Occipital bone. (21) Atlas. (22) Axis.
(23) Funicular part of ligamentum nuchae. (24) Rectus capitis dorsalis. (25) Longus capitis.
(26) Longus colli. (27) Sternohyoideus and omohyoideus. (28) Nasal vestibule. (29) Alar fold.
(30) Basal fold. (31) Dorsal meatus. (32) Middle meatus. (33) Ventral nasal meatus. (34) Nasal
septum, almost entirely removed. (35) Dorsal nasal concha. (36) Conchal cells. (37) Ventral
nasal concha. (38) Sinus of ventral nasal concha. (39) Middle nasal concha, opened.
(40) Ethmoid conchae. (41) Choana. (42) Pharyngeal opening of auditory tube. (43) Right
gutteral pouch, opened. (44) Epiglottis. (45) Aryepiglottic fold. (46) Arytenoid cartilage.
(47) Corniculate process. (48) Cricoid cartilage. (49) Entrance to lateral laryngeal ventricle.
(50) Thyroid cartilage. (51) Cricothyroid ligament. (52) Trachea. (53) Upper lip. (54) Lower lip.
(55) Labial vestibule. (56) Sublingual floor of oral cavity. (57) Frenulum linguae. (58) Hard
palate with venous plexus and palatine ridges. (59) Apex. (60) Body of tongue. (61) Root of
tongue. (62) Genioglossus. (63) Geniohyoideus. (64) Hyoepiglotticus in glossoepiglottic fold.
(65) Basihyoid with lingual process. (66) Soft palate with glands and muscles. (67, 68) Rostral
and caudal boundaries of intrapharyngeal opening. (69) Free border of soft palate. (70) Caudal
end of the palatopharyngeal arch. (71) Esophagus. (72) Cerebrum. (73) Cerebellum.
(74) Olfactory bulb. (75) Optic chiasma. (76) Brain stem. (77) Spinal cord. (From Nickel and
others, 1973.)

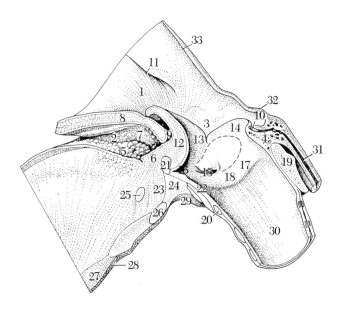

FIGURE 5-2

Sagittal section of the pharynx and larynx of the horse. (1) Nasopharynx. (2) Oropharynx. (3, 4) Laryngopharynx. (5) Root of tongue with tonsillar follicles (lingual tonsil). (6) Glossoepiglottic fold. (7) Palatoglossal arch. (8) Soft palate. (9, 10) Boundaries of intrapharyngeal opening. (9) Free border of soft palate. (10) Palatopharyngeal arch. (11) Pharyngeal opening of auditory tube. (12) Epiglottis. (13) Aryepiglottic fold. (14) Corniculate process of arytenoid cartilage. (15) Entrance to lateral laryngeal ventricle; the broken line marks the extent of the ventricle. (16) Median laryngeal ventricle. (17) Vocal process of arytenoid cartilage. (18) Vocal fold. (19, 20) Cricoid cartilage. (21) Thyroid cartilage. (22) Cricothyroid ligament. (23) Hyoepiglotticus. (24) Fat. (25) Hyoideus transversus. (26) Basihyoid and lingual process. (27) Geniohyoideus. (28) Mylohyoideus. (29) Sternohyoideus. (30) Trachea. (31) Esophagus. (32) Caudal pharyngeal constrictors. (33) Hyopharyngeus. (From Nickel and others, 1973.)

the pharynx. Salivary glands—the parotid, mandibular, and sublingual glands—produce saliva, which lubricates the food and which contains amylase, a minor enzyme that helps digest carbohydrate. Each day, as much as 10 to 12 liters (2.6 to 3.2 gal) of saliva are secreted. Teeth are discussed in Chapters 1 and 7. They grind food, which increases the efficiency of digestion.

The pharynx connects the mouth with the esophagus and connects the nasal cavity with the larynx and trachea leading to the lungs. Both air and food pass through the pharynx, which directs them to their proper destinations. There are two cavities in the pharynx: the nasopharynx (part of the respiratory channel) and the oropharynx (part of the digestive tract). These two cavities are separated by the soft palate. The epiglottis passively covers the laryngeal opening during swallowing to prevent food or liquids from passing into the lungs. During inspiration or expiration of air, it rests against the upper border of the soft palate to close off the back of the oral cavity. Because of this arrangement, a stomach tube can be passed through the nostrils into the esophagus and stomach. Care must be taken not to pass the tube into the trachea. If liquids such as worm medicines are passed into the lungs, the horse may suffocate, or die from aspiration pneumonia.

The esophagus extends from the pharynx and opens into the stomach. It lies close to the skin on the bottom and left side of the neck. You can see the horse swallow, because a muscular contraction moves the food. If a stomach tube is being passed, the operator should feel the tube as it moves down the esophagus. The esophagus joins the stomach at a very oblique angle. At the junction, there are a strong muscular sphincter and numerous mucosal folds. Because of the angle that the esophagus enters the stomach and the strong sphincter, it is almost impossible for stomach contents or gas to be forced from the stomach into the esophagus. The stomach usually ruptures before stomach contents are forced into the esophagus. Therefore the horse usually cannot vomit or belch. This one reason is why horses are susceptible to colic and ruptured stomachs.

The horse's stomach is small for the size of the animal (see Figure 5-3). Its capacity is about 8 to 15 liters (2.1 to 4 gal) and is about 10 percent of the digestive tract capacity. Because of its limited capacity, food does not stay in the stomach very long (less than 2 hours) before being forced into the small intestine. Even though frequency of feeding has been demonstrated not to affect the efficiency of digestion, the size of the stomach does suggest other feeding management considerations. At least two feedings per day help to prevent overdistension of the stomach, colic, and founder. In the natural state, the horse tends to eat small amounts continuously.

Some digestion takes place in the stomach as a result of gastric secretions and limited bacterial fermentation. Limited bacterial digestion of carbohydrates to lactic acid does occur.

The small intestine of the horse is composed of three parts that contain 30 percent of the digestive tract capacity: duodenum, jejunum, and ileum. The bile and pancreatic ducts open into the duodenum. Bile salts are secreted by the liver and are necessary to emulsify lipids (fats). The bile salts are secreted continuously into the digestive tract, because the horse has no gall bladder to store them. Pancreatic juice contains enzymes for protein and carbohydrate digestion. Intestinal cells also secrete digestive enzymes. The small intestine is the primary site of protein digestion and absorption of amino acids. About 60 to 70 percent of the protein eaten may be digested and absorbed before reaching the large intestine. Dietary protein content and quality is important for young horses.

Most soluble carbohydrates are digested in the small intestine. The end products are glucose and volatile fatty acids, which are absorbed and used for energy. The small intestine is also the primary site of fat digestion and absorption. Even though the horse has no gall bladder, it can still tolerate high-fat diets and can use dietary fat. Fats are a concentrated source of energy, and

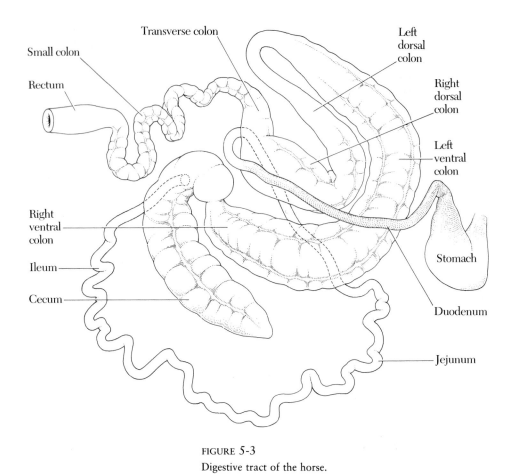

FIGURE 5-3
Digestive tract of the horse.

horses that have a high-energy requirement can efficiently use 10 to 15 percent fat added to the ration. Because of fat absorption in the small intestine, it is presumed that the fat-soluble vitamins (A, D, E, and K) are absorbed here, as are B vitamins. Most calcium, zinc, and magnesium is absorbed in the small intestine. And the same amount of phosphorus is absorbed here as in the large intestine.

The terminal end of the small intestine is the opening of the ileum into the large intestine. The opening extends into the large intestine because the ileum and the sphincter are very muscular. The muscular structure of the ileum allows the digesta to be forced into the large intestine even though the pressure may be higher there. Because the ileum prevents backflow of digesta and equalizes pressure between the large intestine and jejunum, colic can result when this sensitive equalization system malfunctions.

The large intestine is composed of the caecum and colon. It has a large capacity. The caecum, a pouch close to the ileum, may hold as much as 70 liters (18.5 gal) but normally contains about 30 liters (7.9 gal). The ascending or great colon is composed of the right ventral colon, sternal flexure, left ventral colon, pelvic flexure, left dorsal colon, diaphragmatic flexure, and right dorsal colon. From a practical viewpoint, the great colon is in the form of two U-shaped structures that lie more or less on top of each other.

The descending or small colon has a smaller diameter than the great colon. The small colon is shaped to form two rows of regular almost semispherical sacs that form feces into the characteristic balls.

The large intestine serves as a fermentation vat. The microbial flora are similar to those found in cattle. These bacteria break down cellulose into volatile fatty acids: acetic, propionic, and butyric. Some soluble carbohydrates also reach the large intestine and are converted to volatile fatty acids. Bacteria synthesize amino acids in the large intestine. The amino acids are absorbed here, but the importance of their synthesis and absorption has not been fully evaluated. The bacteria are able to use some nonprotein nitrogen for protein synthesis, or at least increase nitrogen retention. The caecum is the primary site of water absorption, but the large colon also plays an important role. B vitamins are synthesized by microflora in the large intestine, and some of them are absorbed.

The digestive tract ends in the rectum, which is 20 to 30 cm (8 to 12 inches) long, and the anus.

COMMON FEEDS

The common feeds that are fed to horses are usually classified into three categories—roughages, concentrates, and supplements. Roughages are high in fiber and relatively low in energy. They include pasture, hay, forage crops, and silage. Pastures and hays are the most common roughages. Concentrates are low in fiber and high in energy. They include grains and some grain by-products. Supplements are used to balance rations, to make up for deficiencies in protein, minerals, and vitamins. Protein supplements are of plant or animal origin, while many minerals are available in organic or inorganic substances. Each vitamin has several good natural sources; in some instances, the horse can synthesize them.

Roughages

Hay

The most common form of roughage given to the horse kept in confinement is hay. It is difficult to produce or to buy good-quality horse hay. There are many variables in hay production, and an understanding of some of the general factors affecting hay quality will help you select high-quality hay.

Plant species affect certain qualities of good hay. Legumes are higher in protein, energy, calcium, and vitamin A than grass hays. Red clover is dustier and not as green in color as alfalfa. A mixture of legume and grass plants may be desirable for certain types of horses or geographical areas. Weeds are undesirable, because they add woody material that is low in acceptability and digestibility to the horse, and they have bad tastes and odors.

Growing conditions such as rainfall affect hay quality. Drought conditions result in stunted growth and fewer leaves. Excessive moisture often encourages diseases that attack and decrease leaves. Adequate fertility ensures a higher nutrient content, more leaves, and less fiber.

Stage of plant growth at harvest time affects nutrient content. As plants grow from the vegetative stage to the reproductive stage, their protein content, digestibility, and palatability decline. The ratio of stem to leaf increases, so the plant has a higher fiber content. For maximum nutrient content and good tonnage per acre, legumes are harvested when a few flowers start to appear in the field. Grasses are harvested when seed heads begin to appear. Grain hays are cut when the grain is in the soft-dough stage of development.

Weather conditions affect field-cured hay. The two most important curing conditions that reduce quality are rain and too much sunlight. Rain beats the leaves from the legumes, leaches out soluble carbohydrates, and packs the hay so it does not dry properly. The moisture content of hay should be about 12 to 18 percent but may extend up to 25 percent. If the hay is baled or processed before the moisture content is reduced to the acceptable limit, it will become moldy and have a musty, moldy odor. Mold can be toxic to horses. Excessive sunlight bleaches the color out of the leaves, causing a loss of vitamin A. Therefore, hay should be dried as quickly as possible. If cured too slowly, hay begins to ferment and lose its nutrient content.

Harvesting procedures also affect quality. If hay is cut and placed in windrows, the stems should be crushed to aid drying. Crushing or crimping the stems may prevent the necessity of moving the windrow for proper drying. Excessive movement after the hay starts to dry shatters the leaves and mixes dirt into the hay. If hay is left in windrows too long, it dries too much, and the leaves fall off when the baler picks it up.

The preceding discussion and the following criteria can help you select high-quality hay. Examine the hay for stage of plant maturity by looking for flowers or seed pods. No mature seed or plants in full bloom should be present. Estimate the percentage of leaves present. Leaves are the part of the plant with the highest quality, so look for a high proportion of leaves relative to stems. Stems are low in digestibility and quality. Determine what plants are in the hay and the relative proportions of each. A bright green color indicates a minimum amount of bleaching and loss of vitamin A. Discolored hay may mean it has been rained on and the nutrients leached out. The hay should have a good, clean smell to it. Moldy and dusty smells are undesirable. No foreign material should be present. Examine it for sticks, stubble, weeds, dirt, bottles, paper, and so on. They may cause mouth injuries or otherwise harm the horse, and you are purchasing them instead of quality hay. Economics are based on nutrient content, hauling, and storage costs. If hay is purchased in large lots, have samples of it analyzed by a commercial laboratory to determine nutrient content.

Each kind of hay has certain characteristics that the horseperson should be familiar with. Many types of grass hays are fed to horses in the United States. However, only a few are in widespread use.

Timothy hay once was one of the most popular horse hays and still is in some areas. It is easy to harvest as clean, bright hay free from mildew and mold. In some areas, it is quite expensive, because it must be shipped great distances. Timothy must be harvested in the pre- or early-bloom stage; afterward, the protein content rapidly declines. The second cutting is the best cutting to feed, because later cuttings are much lower in quality, and the first cutting is usually full of weeds.

Bermudagrass hay is used in the southern United States. Common Bermudagrass does not grow tall enough for hay production, but certain varieties of Coastal Bermudagrass can be used. The same stand of grass can be cut four to five times per year for hay. It is as nutritious as timothy hay. Its value can be increased by growing it with a mixture of legumes.

Prairie hay or *wild native hays* are suitable for horses if they are processed right. However, very seldom can one buy good-quality prairie hay outside of the Midwest. Most is made from poor-quality grass that is quite mature. It usually has lots of weeds and trash baled in it, such as foxtails, which cause mouth wounds and irritations. The protein content may be around 2 to 4 percent. In the Midwest, bluestems are normally grown with other grasses for hay production. Bluestems are highly palatable provided they are cut before reaching full maturity and while still green.

Sudangrass hay (a sorghum) can be fed to horses and is popular in some areas. It must be cut in the early-bloom stage for maximum nutritional content. However, if cut earlier, it may contain large amounts of prussic acid, which is toxic to animals. The protein content is about 10 percent, and the energy content is about the same as for oat hay. Johnson grass, also related to the sorghums, makes good hay when cut at the early-bloom stage. Two or three cuttings can be made per year. It is usually grown in the southern United States.

Oat hay is an excellent feed for horses. The choice between alfalfa and oat hay depends on price per unit of energy or protein and on what kind of horse is being fed. Depending on its processing and the area in which it is grown, it can be low in protein and contain only marginal calcium, phosphorus, and carotene.

Barley hay is very similar to oat hay except it has barbed beards that cause mouth irritations and wounds. Do not feed it unless absolutely necessary. There are some smooth varieties that can be used for horse hay.

Alfalfa hay is one of the best to feed horses. Several characteristics of alfalfa make it excellent. First, it is highly palatable. Most horses really enjoy eating alfalfa hay. In fact, they usually overeat and may get colic unless intake is restricted. Another reason to restrict availability is that it is high in energy—120 percent more energy per unit of weight than oat hay. Therefore feed approximately 85 percent as much alfalfa as oat hay to get the same energy intake. It does not cause excessive sweating, as believed by some horsepeople. However, the high energy content may lead to overfeeding and to a nervous or fat horse. It is safe as the only hay fed. Many people feel that alfalfa needs to be fed or mixed with another hay. It is not necessary to mix it, but it can be mixed to give a mature, nonworking horse more bulk. If fed to lactating mares or growing horses, it should not be diluted with other hay, because they need the high energy and protein content. It is high in good-quality protein: as much as 18 to 19 percent crude protein. People used to think that alfalfa would burn a horse's kidneys out, because an increased urine output high in ammonia results. The excess protein in alfalfa is converted to energy compounds, and nitrogen must be eliminated as ammonia. Alfalfa is a good source of vitamins A and B, and, if sun cured, of vitamin D. The ratio of calcium to phosphorus is about 6:1 and must be considered when feeding young, immature horses. It is more laxative than grass hays but not enough to be of concern. Scientific evidence does not support the old idea that it is too laxative if cut before the late-bloom stage. It should be cut at early- to mid-bloom stage, when you can see a few blooms or flowers when you look across a field. At the late-bloom stage, when the field is solid with flowers, the nutrient content of the hay has decreased considerably—as much as 5 percent.

There are usually five to eight cuttings from an alfalfa field each year. The first cutting usually is full of trash—weeds, grass, paper, and so on. The second cutting is usually clean, and the stems are fine. The third cutting is good hay, but the fourth and fifth cuttings start to have more stems and fewer leaves. As the hay changes from fine stems to coarse stems, the nutrient content and palatability decrease.

Clover hays are nutritious and somewhat similar to alfalfa, because they are also legumes. Clover is difficult to cure unless mixed with grass hays, so pure clover hay is rare. There are five kinds of clover hays: red, common white, crimson, alsike, and ladino. White and ladino clovers are usually grown for pasture, not hay. The other three contain about 14 to 16 percent crude protein. Red clover causes "slobbers" in horses—excessive salivation that does not hurt the horse. However, the appearance of the horse may not be pleasing, and excess moisture in the feed box may result in moldy feed.

Hay can be processed several ways. The traditional way is in bales. Each "three-wire bale" should weigh 125 pounds (56.7 kg) or 16 bales to a ton. A "two-wire bale" is much lighter. It varies in weight, but 66.66 pounds (30.2 kg) are standard. Thirty bales make a ton (907 kg). Hay can also be "cubed"—run through a press that makes small cubes about 1½ inch square (38 cm²) of varying lengths. Cubes require less storage space and are easy to carry to horse shows. There is practically no waste when cubes are fed. However, a horse takes 3 to 4 days to learn to eat them. Less time is spent eating, so the horse has more time to chew on fences, buildings, and so on. And there is always a possibility of a cube lodging in the throat and causing the horse to choke.

Pasture

Pasture is used as a source of roughage for horses kept in larger paddocks or in pastures. Many horses are kept on native pastures of grasses, legumes, and forbs. There are great nutrient content variations in annual pastures depending on the type of pasture and growing season. Early in the growing season, the green pasture is high in moisture and low in energy content. In fact, horses can starve while eating pasture that is 1 to 2 feet high (30.4 to 61.0 cm). Later in the season, the moisture content is lower and the nutrient content is higher. As the pasture continues to mature and weather, it becomes deficient in protein, energy, and other nutrients. A supplementation program must be considered for pastures during low nutrient periods. Irrigated pastures vary in quality depending on the pasture species and management. Irrigated pastures may supply all the required nutrients for some classes of horses.

Horse owners are somewhat limited as to the kinds of pasture grasses and legumes that can be grown in each geographical area. Certain species grow profusely, and others grow little or not at all. The horse owner should consult with the local Cooperative Extension Office for specific recommendations for specific areas. Generally, the pasture should be a mixture of one or two grasses with one or two legumes. The most common warm-season perennials used are Bermudagrass, Johnson grass, bluestems, and native grasses. Common warm-season annuals include sorghum-sudan hybrids, sudans, and millets. Cool-season perennials used are fescue, orchardgrass, timothy, smooth bromegrass, perennial ryegrass, and bluegrass. Cool-season annuals include oats, barley, wheat, rye, and annual ryegrass. Legumes include alfalfa, clovers, and birdsfoot trefoil.

Before selecting grasses to plant, the horseperson must decide what the purpose of the pasture will be. It may be used primarily for exercise or to supply a significant portion of the daily nutrient requirements. Each grass has certain characteristics that make it desirable for the different uses. Kentucky bluegrass and white clover form a sod that is quite resistant to close cropping and trampling. Tall fescue is suited for areas that have a heavy traffic pattern, because it is most resistant to trampling.

Several management factors can improve forage production. Grasses respond to fertilization. The soil should be tested to determine the nutrient needs for optimum production and allow one to make meaningful decisions about fertilizers. The local county Cooperative Extension agent can be of help. The proper combination of fertilizers helps maintain the balance of grasses and legumes in a pasture. Pastures should be irrigated at intervals that do not allow plants to be stressed by lack of water. The amount of water will be determined by the kinds of plants and soil. Adequate drying time should be allowed to minimize trampling and plant injury.

Producing abundant, good-quality forage is possible only under proper grazing management. Pastures must be used in such a way that sufficient leaf area and root reserves are maintained. This is necessary for immediate regrowth. Minimum recommended grazing heights are bluegrass, 2 inches (5.1 cm); native meadow grass, 3 inches (7.6 cm); bromegrass, 3 inches (7.6 cm); orchardgrass, 3 inches (7.6 cm); tall fescue, 3 inches (7.6 cm); Bermudagrass, 2 inches (5.1 cm); and small grains, 4 inches (10.2 cm). The pasture should be grazed evenly. If horses refuse to graze a certain area, moving the water supply, salt, or the feed supplement may be necessary to encourage its use. At regular intervals, the pasture can be clipped to a recommended minimum height. Grazing forage evenly is encouraged when the pasture is dragged with a chain harrow to scatter manure. Grazing pastures with both cattle and horses increases the uniformity of grazing.

There are other considerations for good pasture management. Several small pastures are better than one large one. This allows rotational grazing and an opportunity to irrigate, clip, fertilize, and control weeds. Do not put the horses on the pasture before its growing season; give the forage a head start.

The amount of pasture required per horse can be estimated. The average mature horse consumes 2.5 percent of its body weight in dry matter per day. Therefore, a 1,000-pound (454-kg) horse consumes 25 pounds (11.3 kg) of dry matter per day or 750 pounds (340.2 kg) per month. Forage production varies from 1 to 10 tons (907 to 9,070 kg) or more per acre each month. The local Cooperative Extension agent can provide data for a given area. A horse uses about 60 percent of the forage production. After taking into account the growing season for the pasture, you can calculate the number of horses per month per acre. The most common cause of pasture deterioration is overcrowding and overgrazing, so a minimum of 2 acres per mature horse and 1 acre per pony or yearling is necessary.

Pastures should be free from pits, holes, stumps, and other hazards that could injure the horse. Moles and gophers should be controlled to prevent tunneling.

Horses should be wormed for internal parasites prior to turning them out on the pasture and at regular intervals thereafter. This prevents parasite buildup on the pasture.

Poisonous Plants

Poisonous plants can be a problem in pastures and in hays. Hundreds of plants are toxic to horses. You should be familiar with the ones that grow in your geographical area if you plan to pasture horses. Poisonous plants are frequently baled into hay along with the hay forage, so you must be able to recognize them when examining the hay prior to purchase or feeding. Horses also nibble on poisonous ornamental plants around stables or homes. All horsepeople should be familiar with the most common ones. If you suspect poisoning, call a veterinarian immediately.

Oleander is a common ornamental bush that is extremely poisonous to horses. They usually chew on the bush or prunings from it.

In hay fields, fiddleneck *(Amsinkia intermedia)*—also called tarweed or fireweed—and senecio are common. They contain toxic substances that cause liver cirrhosis. The effects may not be noticeable until after prolonged feeding of the contaminated hay. Clinical symptoms include weight loss, jaundice, incoordination, depression, aimless wandering, colic, rectal prolapse, diarrhea, and anemia. There is no known treatment.

Yellow star thistle is also toxic to horses. The pattern of poisoning is unknown. Horses must consume at least 600 pounds (272 kg) during a 1- to 2-month period to produce the symptoms. After horses acquire a taste for it, they actively seek out the plant. Some clinical symptoms are being unable to swallow, being unable to grab food, pushing the head into water to drink, wrinkled skin around the mouth and nostrils, a wooden expression on the face, lips pulled over the teeth like a purse-string suture, tongue running in and out of mouth, chewing movements with or without feed in the mouth, yawning, standing or walking around with head down, exhibiting depression, and not being able to close its mouth when attempting to graze. In most advanced cases, the horse starves to death. There is no treatment.

Avocado causes sensitive mastitis in mares and reduces milk flow. Damage usually is not permanent.

Sudangrass and hybrid sorghums are cyanogenic plants. Drought and freezing increase the hydrocyanic acid content in these plants, which if then ingested cause death. The presence of lathyrogenis precursors in sorghum plants may indicate that these substances are responsible for cystitis syndrome, a peculiar syndrome of horses that graze sundangrass pastures. Clinical signs include posterior ataxia, urine-stained hair and skin ulcerations in the perineal region, and frequent urinations. Reproductive consequences include abortion and fetal deformities.

Further information about poisonous plants can be obtained from the Cooperative Extension Service or by referring to Fowler (1963) and Kingsbury (1964).

Grains and Grain By-Products

The concentrate part of the ration contains the grains, which are higher in energy and lower in fiber than roughages. Many grains are fed to horses, the most common ones being oats, barley, and corn.

Oats

Oats are the most popular grain fed to horses. They are highly palatable, and very few horses refuse to eat oats. One characteristic that has made them popular with horsepeople is their fiber content—about 13 percent—which is higher than corn or barley. This means they have more bulk per nutrient content, and the horse must eat more to satisfy its nutrient requirements. Bulk makes it more difficult for the horse to overeat and get colic or founder. Therefore it is a safe grain to feed because of a wider margin for error by children, employees, or owners. Oats store quite well when they are harvested and dried properly. Kernels should be plump, heavy, and clean and have a low ratio of husks to kernels, bright color, and clean smell. The weight of a bushel of oats can help determine if they are heavy and have a low ratio of husks to kernels. The heavier the oats, the more nutrients they contain per unit of weight. U.S. No. 1 heavy oats weigh 36 pounds (16.3 kg) per bushel, whereas U.S. No. 4 oats weigh about 27 pounds (12.2 kg) per bushel. The standard weight is 32 pounds per bushel. To decrease the percentage of husk, lower the fiber content, and increase the nutrient content per unit of weight, the ends of the kernel may be clipped. Clipped oats may weigh 42 to 50 pounds (19.1 to 22.7 kg) per bushel. Oats should be bought according to the least cost per unit of energy, provided they are clean and stored properly. Cleanliness is important; oats should be cleaned to remove dirt, weeds, seed, and broken kernels. Dust may indicate old oats and thus a decreased vitamin content or improper storage or cleaning. Dusty feeds lead to respiratory problems. Oats can be checked for dust by filling a can with them and pouring them back into the bin. A musty smell indicates that the oats have been stored with too much moisture and will spoil. They should not taste sour or bitter.

They may be fed whole or processed. Processing includes crimping, rolling, or crushing the kernel. There has been considerable discussion of whether or not oats should be processed. Whole oats are easily eaten and digested by the horse. When the kernel coat is broken by processing, less chewing is required, and digestive juices have better access to the kernel. The increased digestibility is about 5 percent. With processing, the nutrients deteriorate faster, and 2 or 3 weeks after processing the loss exceeds the increased digestibility. Therefore the cost difference between whole and processed oats should not be more than 5 percent. It is usually more economical to buy whole oats, although a few whole kernels do pass through the digestive tract. If oats are processed, a slight crimp is sufficient.

Barley

Barley is comparable to oats as a horse feed except for some specific characteristics that influence how it is used. Generally, it is substituted for oats if its cost per unit of energy is less. Barley is lower in fiber than oats and is classified as a "heavy" feed. Barley has a greater energy density and weighs more per unit of volume (48 pounds per bushel, or 21.8 kg). The barley kernel is harder than the oat kernel, so it should be rolled before feeding. If the kernel is crushed or ground, it is too heavy and can cause colic unless mixed with a bulkier feed such as wheat bran. Barley is evaluated the same way as oats to determine if it is of good quality.

Corn

Corn is one of the most energy-dense feeds, because it contains a large amount of carbohydrate. Corn has a high energy content per unit of weight and a high weight per unit volume (56 pounds per bushel) (Table 5-1). Therefore, a given volume of corn contains approximately three times the amount of energy of an equal volume of oats. Because of its energy content, corn became known as "too hot" a feed for horses. However, if the horse is fed according to its energy needs, corn is an excellent feed. Horses that require a lot of energy for exercise and work are usually fed corn. It is one of the few grains that has any vitamin A value. Corn is low in protein and deficient in lysine, which is an essential amino acid.

Corn quality is judged by the moisture content and by the percentage of well-formed kernels. Very few damaged kernels should be present. Kernels should be well formed, plump, firm, and separated. There should be no insect damage or mold. Moisture content should be less than 14 percent. Because the kernel is high in starch, it ferments readily and becomes toxic. A musty odor indicates excess moisture.

Corn can be fed while still on the cob. This can be used as a management technique for horses that bolt their grain. However, older horses and horses with bad teeth have difficulty eating whole corn. Whole ears keep longer than shelled corn but cost more to store. Many shelled corn kernels pass through the digestive tract without being digested, so it is best to crack or roll it before feeding. Cracked corn may be preferred; it allows the digestive juices to enter

TABLE 5-1
Volume, Weight, and Energy of Common Grains

Grain	Weight per Bushel (pounds)	Weight per Quart (pounds)	Volume per Pound (quart)	Energy Content, Mcal per Pound	Mcal per Quart
Corn, whole	54	1.7	0.6	1.82	3.09
Oats	27–34	1.0	1.0	1.40	1.40
Milo	54–56	1.7	0.6	1.60	2.72
Wheat	60	1.9	0.5	1.76	2.72
Barley, whole	36–47	1.5	0.7	1.66	2.49

and increase the digestibility. Rolled or crushed particles are too small and ferment quickly in the digestive tract. This may lead to colic if the horse is being fed a high-concentrate diet. Whole-ear ground corn can be fed because the cob maintains a high-fiber and low-energy content.

Other Grains

In addition to the commonly used grains, others are also used in horse rations.

Milo is a high-energy grain that is fed to horses primarily in the Southwest. It is a very heavy feed and should be fed with a bulky feed to avoid digestive disturbances. Horses have difficulty chewing it unless it is cracked. It may not be very palatable to some horses.

Wheat is seldom fed except to very young horses. Most wheat is used for human consumption, and only the by-products of the milling process are available for horse feed. Wheat bran is the hard outer coating of the kernel. One popular way of feeding wheat bran is in the form of a mash. It is highly palatable to horses and is frequently used to add bulk. The content should be kept at 10 to 12 percent of the ration. It is also fed to increase the phosphorus content. Wheat bran has about 12 percent digestible protein. Wheat middlings are fine particles of the wheat kernel obtained during the milling process. Some wheat middlings or shorts are fed to horses, but they must be mixed with a bulky feed. They should be limited to 25 percent of the ration since it is a heavy feed. Excessive amounts of middlings can cause colic and founder since they tend to pack in the stomach.

Molasses is a popular component of mixed concentrates. It is a by-product of sugar refining. Horses like its flavor, it is a cheap source of energy, and it reduces dust in the feed. The proportion of molasses generally should not exceed 10 to 12 percent of the ration; 5 percent is the most common proportion. Excessive amounts make the feed sticky and difficult to handle.

Supplements

Ration supplements are usually intended to increase nutritive value. There are four common types of supplements: protein, vitamin, mineral, and a combination of the three.

Protein

Adding a protein supplement is usually indicated when a horse has a high protein requirement. As previously discussed, horses that are growing, lactating, or in later stages of pregnancy and performance horses in high-stress situations have higher protein requirements. Additional protein may be required when horses are being fed a poor-quality roughage.

Protein is expensive, and many rations contain excess protein. The additional protein does not harm the horse because the horse converts it to energy. Excess nitrogen, which is a product of the conversion process, must be eliminated from the horse's body via urine, in the form of urea. This increases urine formation, and the urine has a strong ammonia smell.

Excessive feeding of nonprotein nitrogen sources such as urea can result in harmful ammonia toxicosis. Horses are more tolerant of urea than are cattle, so the urea-containing feeds (5 percent urea) fed to cattle are safe for horses.

Several feeds are commonly used as protein supplements: alfalfa, linseed, soybean, and cottonseed meals; meat, bone, and blood meals; milk products; and dried brewer's yeast. Alfalfa meal and legume pellets are excellent sources of dietary protein for all horses. Of the three oilseed meals, soybean meal is the best protein source, particularly for young, growing horses. Soybean meal contains a better balance of the essential amino acids and has a higher lysine content compared with the other oilseed meals. Lysine is one of the limiting amino acids for growth. Cottonseed and linseed meals should not be the primary source of dietary protein unless they are fed in excess or supplemented with a lysine source. Modern methods of processing oilseeds extract a higher percentage of the oil. As a result, linseed meal is no longer fed to horses to enhance their coat because the oil content in the meal is lower. Instead, add 2 to 3 tablespoons (28.6 to 42.9 ml) of corn oil to the ration each day. Cottonseed meal contains gossypol, which is toxic to pigs. However, mature horses can eat about 1.5 pounds (0.68 kg) per day without problem. Dried skim milk is an excellent source of dietary protein, particularly for young, growing horses (weanlings and yearlings), because it contains a high level of lysine.

Vitamins

Vitamin supplementation is a very widespread mismanagement practice. If the horse receives sufficient vitamins from its ration or synthesizes them in its body, no vitamin supplement is needed. It is very uneconomical, and the horse derives no benefit from the excess vitamins. In fact, excess vitamins can greatly harm a horse. If a ration needs additional vitamins to balance it and to provide the daily requirements of the horse, several sources can be used. Supplementary sources of vitamin A are commercially synthesized vitamin A, fish liver oils, and liver meal. Carotene, a precursor of vitamin A, can be supplemented by adding dehydrated alfalfa meal, corn gluten meal, and good-quality hay. Fresh green forage such as pasture is an excellent source of carotene. Supplements high in B vitamins are dried brewer's yeast, animal liver meal, wheat germ, soybean oil meal, and dried legumes. The primary source of vitamin D is sun-cured hay and exposure to sunlight. Fish liver oils and irradiated yeast are excellent sources. Vitamin E is found in most feeds and seldom needs to be supplemented. Alfalfa pellets are a good source of vitamin E.

TABLE 5-2
Mineral Supplements

Supplements	Calcium (%)	Phosphorus (%)	Magnesium (%)
Bone meal, steamed	32.3	13.3	0.6
Calcium carbonate	36.7	0.5	0.3
Dicalcium phosphate	23.7	18.8	—
Limestone, ground	36.1	—	2.1
Monosodium phosphate	—	25.8	—
Monodicalcium phosphate	16.8	22.1	0.5
Phosphate, defluorinated	31.7	13.7	0.3
Sodium tripolyphosphate	—	25.1	—

SOURCE: National Research Council (1978).

Injectable vitamins should be used only on the recommendation of a veterinarian to treat a disease or a deficiency.

Minerals

Mineral supplements are frequently needed to balance the mineral content of rations (see Table 5-2). The most commonly deficient minerals are calcium, phosphorus, sodium chloride (salt), and iodine. Salt and iodine requirements can be met by keeping iodized salt available to the horse in the form of a salt block or granules. There are many sources of calcium and phosphorus. The choice of which supplement to use depends on availability and cost. Ground limestone and oyster shell flour are good sources of calcium. Monosodium and disodium phosphates supply only phosphorus. Both calcium and phosphorus are supplied by steamed bone meal, dicalcium phosphate, monocalcium phosphate, and spent bone block.

The remainder of the minerals are referred to as *trace* minerals and are found in most natural feeds unless feeds are grown in areas of known mineral deficiency. A trace-mineralized salt block can be used if necessary.

NUTRIENT REQUIREMENTS

For normal body functioning, horses need six nutrients: water, carbohydrates, fats, protein, minerals, and vitamins. In addition to understanding which nutrients are required, the horseperson must understand the factors that regulate the amounts of nutrients required. These factors are growth, maintenance, reproduction, lactation, and work.

Water

The horse must have a source of fresh, clean drinking water. It drinks about 10 to 12 gallons (37.8 to 45.4 liters) per day during normal environmental conditions. Several factors control the amount of water each horse needs. During hot weather or when subjected to hard work, the water requirement is higher and may even be doubled because of the water lost via sweating and respiration. Pregnant mares require about 10 percent more water than do nonpregnant mares. Lactating mares require 50 to 75 percent additional water to replace that secreted in the milk. Horses eating dry feed consume more water.

The horse should have water available for drinking at all times. Some horsepeople prefer to water the horse in the morning and afternoon by placing water in the stall. Make sure the horse drinks enough to meet its daily requirements. Automatic drinking fountains ensure a source of fresh water but may get stuck open, flooding the stall.

During hot weather or while working, the horse should be allowed to drink frequently. This helps prevent overheating and keeps the horse refreshed. To prevent water founder, do not allow horses that are hot after being heavily exercised to drink too much cold water at one time (see the section on founder in Chapter 2). The horse should be cooled out and allowed to take a few swallows every couple of minutes while it is being walked around.

Horsepeople disagree as to the proper time of watering in relationship to feeding. The horse can drink before, during, or after eating without any adverse effects on the horse or digestion.

Depriving a horse of water reduces its feed intake. After 3 to 4 days, the horse will eat very little if any feed. The feed reduction is accompanied by rapid weight loss. The weight loss is primarily due to dehydration and can be replaced in 3 to 4 days.

Carbohydrates and Fats

The primary function of carbohydrates and fats is to provide energy. The expression *feed energy* denotes the value of feed for its greatest function, which is to furnish energy for body processes. There are several ways to express the energy content of feeds but all rations are formulated for their digestive energy content or total digestible nutrient content. Gross energy—the total energy content of a feed—is determined by measuring the heat produced when the feed is ignited in a bomb calorimeter. Digestible energy is the gross energy of the feed minus the energy lost in the feces. Digestible energy (DE) is determined by digestion trials. Metabolizable energy (ME) is a more refined definition of the amount of energy retained in the horse's body. Energy lost in the urine and gaseous products of digestion is subtracted from the DE value to obtain the

ME value. Net energy (NE) is the energy actually used by the horse. The units to express these energy values are calories. A calorie is the amount of heat necessary to raise the temperature of 1 gram of water 1 degree Celsius; 1 kilocalorie (kcal) is 1,000 calories, and a megacalorie (Mcal) is 1,000 kilocalories.

An outdated method of expressing energy content of feed is total digestible nutrients (TDN). TDN is the sum of all the digestible organic nutrients retained in the horse's body. It is determined according to the same method as digestible energy but is expressed as weight or percent TDN. One pound of TDN is equivalent to 2,000 kcal (1 kg TDN = 4,410 kcal). Some feed tags still give the TDN content.

Maintenance and Special Requirements

The requirement for energy varies according to the physiological state of the horse and its activity (Appendix Tables 5-1 to 5-4). The basic requirement for energy is referred to as the *maintenance requirement*. This is the amount of energy required to maintain a constant body weight during normal activities of a nonworking horse. The maintenance energy requirement is related to body weight: 400-pound (181.4-kg) horses require 1.87 Mcal per 100 pounds (45.4 kg); 900-pound (408-kg) horses require 1.6 Mcal per 100 pounds (45.4 kg); 1,100-pound (499-kg) horses require 1.5 Mcal per 100 pounds (45.4 kg); and 1,300-pound (590-kg) horses require 1.42 Mcal per 100 pounds (45.4 kg).

Appendix Tables 5-1 to 5-4 must be considered as a starting point when calculating a ration and feeding a horse. At first, it is best to feed the horse an additional 10 to 15 percent DE. This is necessary because the table values are estimated and allow no margin of error. They do not take into account an activity factor, environmental influences (cold weather), parasitism, or individuality. Some horses are "hard keepers" and require more energy. Others are "easy keepers" and require less energy than normal. If the horse starts to get fat, the amount of feed can be reduced, and vice versa. The most common problem is obesity from overfeeding coupled with lack of exercise. The most noticeable effects of insufficient energy intake are loss of weight, loss of condition, and excessive fatigue.

Exercise

Exercise increases the DE requirement above the maintenance requirement. Table 5-3 gives the influence of various types of exercise on the energy requirement. When calculating the energy requirement for a horse being exercised, add the maintenance requirements (Appendix Tables 5-1 to 5-4) to the exercise requirement (Table 5-3).

TABLE 5-3
Estimated Energy Requirements for Various Types of Exercise

	Requirement	
Activity	(kcal/hr/kg body wt)	(kcal/hr/lb body wt)
Walking	0.5	0.2
Slow trotting, some cantering	5.1	2.3
Fast trotting, cantering, some jumping	12.5	5.7
Cantering, galloping, jumping	24.0	10.5
Strenuous effort	37.0	17.7

SOURCE: Hintz and others (1971).

Reproduction

Reproduction influences the energy requirement. Stallions need more energy during the breeding season because their exercise increases. Semen production does not significantly increase the energy requirement. Pregnancy significantly increases the energy requirement only during the last 90 to 120 days of pregnancy (Appendix Tables 5-1 to 5-4). Prior to the last trimester of pregnancy, fetal growth is slow, and only 50 percent of fetal growth occurs over the long period of time. When fed according to the recommendations, the mare's weight should increase 5 to 6 percent, and the foal weights should be normal. Under ideal management conditions, the products of conception are about 10 percent of the mare's weight. This means that mares convert 5 percent of their body tissues to fetal tissue. If a mare converts slightly more than 5 percent of her tissue to the products of conception (loses a little more than 5 percent of her body weight), the foal's birth weight will be maintained, but the mare loses too much condition, which decreases her milk production and affects the foal's growth rate. If the mare loses too much body weight, the birth weight will be decreased as well as subsequent milk production.

Lactation

The lactating mare has the highest energy requirement (Appendix Tables 5-1 to 5-4). She needs enough energy for good milk production and to regain the 5 percent of body weight she lost during gestation.

Growth

The energy requirement for growth has not been established, because optimal growth rates have not been defined. Some horses may grow too fast in terms of body weight increases, which in turn may influence soundness and longevity. The recommendations in Appendix Tables 5-1 to 5-4 are based on the requirements for daily gain, taking into account the size of the horse relative to its expected final mature weight. Insufficient energy intake results in poor growth rates and an unthrifty appearance.

Protein

Protein requirements of horses are expressed as crude protein (CP) or digestible protein (DP). Crude protein is a measurement of the amount of nitrogen contained in the feed. Not all the nitrogen is available to the horse in the form of protein. Digestible protein is the available amount of protein in the feed. The percentage of crude protein that is digestible varies from feed to feed and among classes of feed. About 70 percent of the CP in grass forage is DP, whereas it is about 85 percent for legumes and mixed rations.

Another factor that must be considered when protein requirements are discussed is protein quality. Protein is made of building blocks called *amino acids*. Some amino acids cannot be synthesized within the horse's body and are called *essential amino acids*. Nonessential amino acids can be synthesized from carbon skeletons resulting from carbohydrate and fat metabolism. The requirement for protein should really be expressed per amino acid but the necessary research has not been performed. Protein quality indicates the amino acid balance of a feed. Good-quality protein has a balance of amino acids about the same as found in the horse's body. Poor-quality protein is deficient in one or more essential amino acids. Protein quality is important because all the amino acids must be present and in the right amounts for protein synthesis to occur. If one amino acid is left out, synthesis cannot occur, and if one amino acid is not present in a sufficient amount, its deficiency limits the use of the other amino acids. To obtain the proper balance of amino acids, several feeds are usually combined to form a ration. This is important for young, growing horses.

Appendix Tables 5-1 to 5-4 list the requirements for protein. For horses kept on a maintenance ration, the CP should be kept at 8 to 10 percent (Table 5-4). Lower levels of CP are usually associated with low palatability of the ration. Exercise does not increase the protein requirement beyond what would be supplied by the increased feed intake to meet the energy needs. The same holds for the stallion. During the last 90 to 120 days of pregnancy, the mare's protein requirement increases to meet the needs of the rapidly developing fetus. About 50 percent of fetal development occurs during this period. The ration should contain 10.5 percent CP. Failure to supply adequate protein during gestation results in decreased birth weight and a loss in the mare's weight. During lactation, a ration containing 13.5 percent CP is required. Growing horses appear unthrifty, have a high ratio of feed to weight gain, and have poor growth if their protein needs are not met. The rations for weanlings and yearlings should contain 16 and 13 percent CP, respectively. It is important that they be fed high-quality protein. Lysine, an essential amino acid, appears to be the most limiting amino acid for growth. Dried skim milk and soybean and alfalfa meal are good sources of lysine. A mixture of feeds should be used for the young, growing horse.

Vitamins

Vitamins are needed in exceedingly small amounts. They serve as catalysts for bringing about many transformations and reactions in the body tissues. Most vitamins act as coenzymes, which are small but essential parts of an enzyme that catalyzes some body process. Some vitamins can be synthesized in the body, while others must be present in the diet.

Vitamins are divided into two general groups. Vitamins A, D, E, and K are soluble in lipids or lipid solvents and are referred to as fat-soluble vitamins. They are stored by the body in fat cells and in the liver. Thus, the horse can maintain a reserve supply of the fat-soluble vitamins. The vitamin B complex and vitamin C belong to the water-soluble group. They are not stored by the body and must be ingested or synthesized daily.

Vitamin A

The functions of vitamin A are not fully understood, but deficiency and toxicity symptoms are well documented. One of the first deficiency symptoms is impairment of vision. Lacrimation—excessive watering of the eyes—is observed. This is in response to the cornea becoming too dry. The cornea can be damaged, and blindness can result. Night blindness is also observed. The horse has difficulty seeing when adjusting from bright light to dim light. The skin becomes rough and dry, with a brittle hair coat. Skin lesions may develop that are difficult to heal. Poor hoof growth, digestive disturbances, and respiratory illness are observed.

Vitamin A requirements of horses can be met by carotene, a precursor of vitamin A or by synthetic vitamin A. Carotene is found in all fresh green forages and is converted to vitamin A in the wall of the intestine. Carotene as well as vitamin A is unstable in sunlight, oxygen, and high temperatures. About 90 percent of the carotene is lost when hay is left unprotected from sun and rain for 1 year. The carotene content also decreases when hay is properly stored, so that after 6 months of storage a significant percentage of the carotene has broken down. Fortunately, vitamin A is stored in the liver, and horses feeding on green forage can store 3 to 6 months' supply. Care must be taken not to oversupplement vitamin A, because toxicity causes bone fragility, hyperostosis, and exfoliated epithelium.

TABLE 5-4
Nutrient Concentrations in Diets for Horses and Ponies (100 percent dry matter basis)

	Maximum Amount of Dry Matter Intake per Day	Digestible Energy		Example Diet Proportions Hay Containing 2.2 Mcal/kg	
	% body wt	Mcal/kg	Mcal/lb	Concentrate[a]	Roughage
Mature horses and ponies at maintenance	1.5	2.2	1.0	0	100
Mares, last 90 days of gestation	2.0	2.5	1.1	25	75
Lactating mare, first 3 months	3.0	2.8	1.3	45	55
Lactating mare, 3 months to weaning	3.0	2.6	1.2	30	70
Creep feed	4.5	3.5	1.6	100	0
Foal, 3 months old	4.5	3.25	1.5	75	25
Weanling, 6 months old	4.5	3.1	1.4	65	35
Yearling, 12 months old	2	2.8	1.3	45	55
Long yearling, 18 months old	2	2.6	1.2	30	70
2-year-old, light training	2.5	2.6	1.3	30	70
Mature working horses					
light work[b]	2	2.5	1.1	25	75
moderate work[c]	2.25	2.9	1.3	50	50
intense work[d]	2.5	3.1	1.4	65	35

Note: Values are rounded to account for differences among Tables 5-1 to 5-8 and for greater practical application.
[a]Concentrate containing 3.6 Mcal/kg.
[b]Examples are horses used in western pleasure, bridle path hack, equitation, and so on.
[c]Examples are ranch work, roping, cutting, barrel racing, jumping, and so on.
[d]Examples are race training, polo, and so on.
SOURCE: National Research Council (1978).

It is estimated that 25 International Units (IU) per kg of body weight (1,136 IU/100 lb body weight) are adequate for maintenance, that 40 IU per kg (1,818 IU/100 lb) are adequate for growth, and that 50 IU per kg (2,272 IU/100 lb) are adequate for pregnancy and lactation.

Vitamin D

Vitamin D is not a dietary requirement. The action of ultraviolet rays in the sunlight converts a substance in the skin to vitamin D, which is then absorbed by the body. Exposure to sun for 30 minutes per day is adequate to meet the daily requirement. The other source of vitamin D is sun-cured hay.

The exact requirements for vitamin D have not been established. A deficiency symptom is rickets but it is seldom observed. Oversupplementation is

Hay Containing 2.0 Mcal/kg		Crude Protein (%)	Calcium (%)	Phosphorus (%)	Maximum Ca:P Ratio	Vitamin A Activity	
Concentrate[a]	Roughage					IU/kg	IU/lb
10	95	8.5	0.30	0.20	6:1	1,600	725
35	65	11.0	0.50	0.35	5:1	3,400	1,550
55	45	14.0	0.50	0.35	5:1	2,800	1,275
40	60	12.0	0.45	0.30	5:1	2,450	1,150
100	0	18.0	0.85	0.60	1.5:1		
80	20	18.0	0.85	0.60	2:1	2,000	900
70	30	16.0	0.70	0.50	3:1	2,000	900
55	45	13.5	0.55	0.40	3:1	2,000	900
40	60	11.0	0.45	0.35	3:1	2,000	900
40	60	10.0	0.45	0.35	4:1	2,000	900
35	65	8.5	0.30	0.20	6:1	1,600	725
60	40	8.5	0.30	0.20	6:1	1,600	725
70	30	8.5	0.30	0.20	6:1	1,600	725

a common problem. Ten times the daily requirement of 2,500 to 5,000 IU (6.6 IU/kg or 300 IU/100 lb) for several months can lead to toxicity symptoms. Excessive calcium deposits develop in soft tissues such as heart, muscle, and arteries. Once deposited, the calcium cannot be removed. Excess amounts of the vitamin can be stored in the liver, so it is very seldom necessary to add vitamin D to the ration.

Vitamin E

Vitamin E is a fat-soluble vitamin about which little is known. It is found in green, growing forage; good-quality hay; cereal grains; and wheat germ oil. It has been synthesized, and alpha tocopherol is used when a supplement is required. There have been many claims about the benefits of vitamin E but few are based on scientific evidence. When combined with selenium, it has been an effective treatment for white muscle disease (degenerative skeletal muscles). It has been associated with muscle development, oxygen transport, red blood cell stability, and fertility.

Vitamin K

Vitamin K is synthesized by the bacteria in the digestive tract. Its major function is involved with blood clotting.

Vitamin C

Vitamin C is synthesized in the liver and is not a dietary essential.

B Vitamins

The 10 vitamins known as the B complex are water-soluble. Their functions are better understood than those of the fat-soluble group. The B complex vitamins are synthesized by bacteria in the digestive tract. The only conditions in which supplementation may be necessary are heavy stress and heavy training and performance or when the horse is fed very poor-quality roughages.

Most of the information about the B complex has been confined to thiamin, riboflavin, and vitamin B_{12}. Thiamin is involved with carbohydrate metabolism. A deficiency is manifested by loss of appetite. Three mg of thiamin per kg feed (approximately 1 mg/lb) meets the requirement. Vitamin B_{12} has received more attention than the other B vitamins because of its involvement with the synthesis of red blood cells. Considerable amounts of money have been spent on vitamin B_{12} injections to raise red blood cell counts, to no avail.

Minerals

Minerals are inorganic compounds that serve as components of body tissues and as catalytic compounds that trigger reactions in the body. The major minerals in the body are salt (sodium and chloride), calcium, and phosphorus and are the ones most often added to a ration. The minerals required in very small quantities are referred to as *macrominerals* (magnesium and potassium) or as *trace minerals* (iodine, cobalt, sulfur, copper, iron, manganese, fluorine, zinc, and selenium).

The intake of the major minerals must be closely monitored and controlled. Several factors control the intake. The soil content in a given area influences the mineral content of the forage. There are iodine-, magnesium-, and selenium-deficient soil areas in the United States. Horses consuming forage grown in those areas may need additional sources of the deficient mineral. The ratio of hay to grain and the type of hay fed is important because plants differ in mineral content. Quality of the feed is important, because improperly cured and stored feed has a lower mineral content. Failure to supply adequate amounts of minerals leads to deficiency symptoms, and excessive consumption can lead to pathological physiological problems.

Sodium and Chloride

The sodium and chloride in salt play many roles in the horse's body, including maintenance of fluid balance, acid-base balance, normal flow of nerve impulses, and muscular movements. The requirement for salt has not been clearly established because of individual variation in the requirement and the influence of many environmental factors. Environmental temperature is one of the primary determinants of the salt requirement. Considerable amounts of salt are lost in sweat. The average salt requirement is about 50 to 60 grams per day. It can be provided several ways, but the horses should always have free access to salt, in block or granular form. It is a common practice to add 0.5 to 1 percent salt in a ration to supply most of the salt requirement. Free access to additional salt should still be provided. Horses regulate their salt intake and do not consume too much provided they are getting an adequate supply of water. They may overeat salt if they have been deprived of it for extended periods. With adequate water intake, the only result is a loose bowel. Other symptoms of salt toxicity are colic, frequent urination, weakness, and a paralysis of the hind limbs. Deficiency symptoms are depraved appetite, rough hair coat, reduced growth, and reduced milk production.

Calcium and Phosphorus

Calcium and phosphorus are associated in their main function of aiding bone formation and maintenance. They also function in energy metabolism by serving as enzyme cofactors, nerve conduction, muscle contraction, and blood clotting. There are two important points to consider: the amount to feed, and the ratio of calcium to phosphorus content in the ration.

Mature horses need 20 to 60 gm of calcium (45 mg/kg or 2 gm/100 lb body weight) and 15 to 40 gm of phosphorus (35 mg/kg or 1.6 gm/100 lb body weight) per day depending on their size and physiological state. Young, growing horses that are rapidly synthesizing bone and lactating mares require the most calcium and phosphorus. Pregnant mares also need more during the last trimester of pregnancy. Old horses need more calcium because they absorb less.

The calcium–phosphorus ratio (Ca:P) is an extremely important feature of horse rations. The ratio should be not less than 1.1:1, and the average ratio in milk and in growing tissues is 1.6:1. If the horse is fed a ratio less than 1:1 with adequate amounts of phosphorus, calcium is mobilized from bone and replaced with fibrous connective tissue. This is particularly evident in the head. "Big head" or nutritional secondary hyperparathyroidism is the name of the disorder. With a 0.8:1 ratio, it takes about 6 to 12 months for the symptoms to be manifested. If the ratio is 0.6:1, the symptoms occur much faster. The opposite situation is a high Ca:P ratio. Foals can tolerate a 3:1 ratio, and adult horses can tolerate a 6:1 ratio without adverse effects. However, excess calcium interferes with the use of magnesium, manganese, iron, and phosphorus. If excess calcium is fed for prolonged periods, bones may become brittle and dense. Inadequate amounts of calcium and/or phosphorus can lead to rickets, which is improper bone development.

Magnesium

The magnesium requirement is estimated to be 14 mg per kg (636 mg/100 lb) body weight. In areas where grass tetany is common with grazing cattle and sheep, adding 5 percent magnesium oxide to the salt mixture helps protect horses. Symptoms of a deficiency are hyperirritability, tetany, glazed eyes, and eventual collapse. The conditions are usually precipitated by some form of stress.

Iodine

The iodine requirement is about 0.1 parts per million (0.1 mg/kg) in the feed. In iodine-deficient areas, adequate amounts of iodine can be given the horse by feeding iodized salt. Deficiency symptoms include goiter and foals born weak, hairless, or dead. A major problem is iodine toxicity. Oversupplementation of iodine by the use of some seaweeds or other sources can cause the development of goiter and weakness in newborn foals. Some foals may die soon after birth.

Selenium

The requirement for selenium is not more than 0.5 ppm. Deficient animals may develop muscle degeneration—"white muscle disease." Selenium toxicity will occur if the feed contains 5 to 40 ppm of selenium. Toxic horses lose hair from the tail and mane, eventually become blind and paralyzed, and the hoofs may slough off. In certain selenium-toxicity areas in the United States, "alkali disease" may develop in horses.

Potassium

Horses consuming at least 50 percent of their ration as forage or hay should receive adequate amounts of potassium because it is readily absorbed from feeds. The requirement is about 0.6 percent in the ration. Foals on a high-grain diet may need additional potassium.

Other Minerals

Sulfur is not a dietary essential, because adequate amounts of it are obtained by meeting the protein requirement. Requirements for other trace minerals are cobalt, unknown but very low; copper, 5 to 8 ppm; iron, 40 to 50 ppm; manganese, unknown; zinc, less than 50 ppm; and fluorine, unknown. Excessive zinc causes anemia, swelling of the epiphyseal region of long bones, stiffness, and lameness. Lead poisoning is common in horses kept adjacent to smelters and that chew on surfaces painted with a lead-based paint.

FORMULATING AND BALANCING RATIONS

A balanced ration is the amount of a mixture of feeds that provides the horse with its nutrient requirements for one day. A single feed may provide a balanced ration, or it may be necessary to mix several feeds to form a balanced and palatable ration. Rations can be formulated by trial and error, which is time-consuming, or by computer. Details of ration formulation are given in Appendix 5-1.

Rules of Thumb

You can formulate the initial ration in a crude manner by applying rules of thumb. One of the first rules concerns the amount to feed that the horse will consume. Feed consumption is related to body weight and to the use or physiological state of the horse (Table 5-4); for example, lactating mares consume more feed than an idle horse. Table 5-4 is only a guide, because horses vary considerably. Some horses have an excessive appetite, and their feed intake must be restricted; others must be encouraged to eat. The proportion of roughages to concentrate should also be considered (Table 5-4). Heavily exercised horses need a ration with a high-calorie density (Mcal/unit of weight), so they can consume enough feed to meet their energy requirement. If the energy density is too low, the horse develops a "potbelly" because of its large intake of roughage (Tables 5-4 and 5-5). Growing foals need a high-energy ration that contains adequate protein. In fact, all rations should contain minimal energy

TABLE 5-5
Nutrient Concentrations in Diets for Horses and Ponies (90 percent dry matter basis)

	Digestible Energy		Crude Protein (%)	Calcium (%)	Phosphorus (%)	Maximum Ca:P Ratio	Vitamin A Activity	
	Mcal/kg	Mcal/lb					IU/kg	IU/lb
Mature horses and ponies at maintenance	2.0	0.9	7.7	0.27	0.18	6:1	1,450	650
Mares, last 90 days of gestation	2.25	1.0	10.0	0.45	0.30	5:1	3,000	1,400
Lactating mare, first 3 months	2.6	1.2	12.5	0.45	0.30	5:1	2,500	1,150
Lactating mare, 3 months to weaning	2.3	1.1	11.0	0.40	0.25	5:1	2,200	1,000
Creep feed	3.15	1.4	16.0	0.80	0.55	1.5:1		
Foal, 3 months old	2.9	1.35	16.0	0.80	0.55	2:1	1,800	800
Weanling, 6 months old	2.8	1.25	14.5	0.60	0.45	3:1	1,800	800
Yearling, 12 months old	2.6	1.2	12.0	0.50	0.35	3:1	1,800	800
Long yearling, 18 months old	2.3	1.1	10.0	0.40	0.30	3:1	1,800	800
2-year-old, light training	2.6	1.2	9.0	0.40	0.30	4:1	1,800	800
Mature working horses								
light work[a]	2.25	1.0	7.7	0.27	0.18	6:1	1,450	650
moderate work[b]	2.6	1.2	7.7	0.27	0.18	6:1	1,450	650
intense work[c]	2.8	1.25	7.7	0.27	0.18	6:1	1,450	650

Note: Values are rounded to account for differences among Tables 5-1 to 5-8 and for greater practical application.
[a]Examples are horses used in western pleasure, bridle path hack, equitation, and so on.
[b]Examples are ranch work roping, cutting, barrel racing, jumping, and so on.
[c]Examples are race training, polo, and so on.
SOURCE: National Research Council (1978).

content per unit weight. The content varies, depending on the physiological state of the horse. Protein is also an important consideration in the initial stages of formulating rations for lactating mares and for mares in the latter stages of gestation. Therefore, the major concern for idle and working horses is energy. For growing horses it is protein, and for lactating and pregnant mares it is energy and protein.

Rations

The ration should be as simple as possible. Rations for idle and for working horses are relatively simple (Table 5-6). They become more complex for lactating and pregnant mares because more nutrients are needed (Table 5-6). Rations for growing horses are more complex because higher-quality protein is needed (Tables 5-7 and 5-8). The rations in Tables 5-6 to 5-8 are given as guides and should be modified depending on economics and availability of ingredients.

Costs

One of the primary concerns about feed costs or economics of ration formulation is the cost per unit of energy. For example, the most economical hay for energy (comparing alfalfa with oat) can be calculated as follows. The first step is to determine the energy content of alfalfa hay and oat hay. Have the hay analyzed by a laboratory, or assume that it is of average composition and consult Appendix Table 5-6 on feed composition. Unless you are buying large quantities of hay, the published values for the composition of feeds are close

TABLE 5-6
Rations for Nonpregnant, Pregnant, and Lactating Mares (fed with timothy hay)

Ingredients	Nonpregnant or Maintenance (%)	Pregnant (%)	Lactating[a] (%)
Rolled oats	40.0	40.0	40.0
Rolled corn	40.0	40.0	35.0
Wheat bran	10.0	10.0	5.0
Soybean oil meal	5.0	5.0	15.0
Molasses	4.5	3.0	3.0
Defluorinated phosphate	0.5	—	1.5
Dicalcium phosphate	—	2.0	0.5
	100.0	100.0	100.0

[a]Data from Hintz (1977).

TABLE 5-7
Creep Rations for Foals

Example 1		Example 2	
Ingredients	Percent	Ingredients	Percent
Rolled oats	50.0	Rolled oats	23.0
Rolled corn	4.5	Rolled corn	37.4
Soybean oil meal	10.0	Soybean oil meal	33.0
Dried skim milk	5.0	Brewer's yeast	0.5
Dehydrated alfalfa	10.0	Molasses	3.0
Wheat bran	10.0	Dicalcium phosphate	1.0
Molasses	7.0	Ground limestone	1.0
Trace mineralized salt	1.0	Trace mineralized salt	1.0
Dicalcium phosphate	2.0	Vitamins A and D	0.5
Ground limestone	0.5		100.0
Vitamin premix[a]	0.5		
	100.0		

[a] 1×10^6 IU vitamin A, 1×10^5 IU vitamin D, 5×10^3 IU vitamin E, 0.5 gm thiamine, 2.5 gm niacin, 1 gm pantothenic acid, and 2.5 mg vitamin B_{12} per pound (454 gm) of premix.

SOURCE: Example 1, Baker (1971); Example 2, Tyznik (1975).

TABLE 5-8
Rations for Weanlings and Yearlings
(fed with alfalfa-grass hay)

Ingredients	Weanling (%)	Yearling (%)
Rolled oats	40.0	50.0
Rolled corn	12.5	9.5
Rolled barley	10.0	13.5
Soybean oil meal	7.0	6.0
Dried skim milk	10.0	—
Dehydrated alfalfa	10.0	10.0
Molasses	7.0	7.0
Trace mineralized salt	1.0	1.0
Dicalcium phosphate	2.5	3.0
Vitamin premix[a]	0.5	0.5

[a] See Table 5-7 for formula.

SOURCE: Baker (1971).

enough. The average energy content of oat hay is 0.94 Mcal/lb (Appendix Table 5-6, Line 49) or 1,880 Mcal/2,000 lb (1 ton, or 907 kg). Alfalfa hay contains about 1.04 Mcal/lb, or 0.473 Mcal/kg (Appendix Table 5-6, Line 4), or 2,080 Mcal/ton. For this example, figure alfalfa hay to be $82 per ton (907 kg) and oat hay to be $60 per ton (907 kg). Next, one divides the cost by amount of energy by the cost. The alfalfa hay costs 3.94 cents per megacalorie of energy ($82.00/2,080 Mcal), and the oat hay costs 3.19 cents per megacalorie of energy ($60.00/1,880 Mcal). At these prices, the alfalfa hay costs 23 percent more than the oat hay per unit of energy. To make the same calculations for another hay, substitute the cost per ton and divide by the megacalories.

To figure the cheapest source of energy from grains, the procedure is similar. If you consider $10.73/100 lb (or 45.4 kg) for rolled barley, $10.61/100 lb for rolled oats, and $9.70/100 lb for cracked corn, a typical calculation is as follows. Barley contains 1.64 Mcal/lb (Appendix Table 5-6, Line 10) or 164 Mcal per 100 lb (3.6 Mcal/kg), oats contain 1.52 Mcal/lb (Appendix Table 5-5, Line 47) or 152 Mcal/100 lb (3.34 Mcal/kg), and corn contains 1.76 Mcal per pound (Appendix Table 5-6, Line 36) or 176 Mcal/100 lb (3.87 Mcal/kg). Dividing the cost per 100 lb by the number of megacalories per 100 lb or cost per kilogram by number of megacalories per kilogram, we find the cost per unit of energy to be 6.5 cents, 7.0 cents, and 5.5 cents for barley, oats, and corn, respectively. Corn costs less than oats and barley, and barley costs less than oats based on the cost per unit of energy at the above prices. For this example, corn costs about 29 percent less per unit of energy than oats.

Commercial Feeds

When feeding commercial feeds (premixed feeds), you must be able to interpret a feed tag label to determine how much feed should be fed. The information given on a feed bag label must meet certain state requirements. The information you need to know to make a critical evaluation of your feeding program is not given on most labels.

Nutrient Analysis

One requirement is that a basic nutrient analysis must be listed. The basic nutrient analysis includes minimal amounts of crude protein and crude fat and maximum amounts of ash (minerals) and crude fiber. Crude protein (not less than a given percent) merely informs you that a minimal amount of nitrogenous substances is present. It does not give an estimate of the digestibility of the crude protein. However, by looking at the list of feed ingredients, you can determine if the sources of protein are highly digestible or not. Protein quality is not given; you must estimate it by "reading between the lines." Protein quality refers to the balance of amino acids that make up the protein. As tissues or protein compounds are formed in the horse's body, they are made up of specific percentages of each amino acid. For the protein to be formed by the horse's body, all the amino acids must be present at one time. Therefore a low-quality protein in a feed means that the protein has a low percentage of one or more amino acids (improper amino acid balance) in relation to the total amino acid content. That is, one of the protein building blocks required by the horse to build the given protein(s) is limiting the amount of the given protein(s) that the horse's body can build. For example, one of the amino acids that is low in cottonseed meal is lysine.

The crude fat content indicates the minimum amount of fat in the feed. The fat and fatty substances include the fat-soluble vitamins (A, D, E, and K) and fat that is added to keep the "dust content" under control. Fats are concentrated sources of energy and provide about twice the energy that carbohydrates do on a weight basis. The guaranteed maximum amount of crude fiber indicates if the feed is a concentrate feed (low in fiber) or a roughage feed (high in fiber). Most important, no estimate is given of the amount of the fiber that can be digested by the bacteria in the horse's digestive tract. How much of the fiber passes through the horse? The ash content is what is left after a sample of the feed is completely burned. It is an estimate of the mineral content. However, a horseperson needs to know how much of each mineral is present. Is the calcium–phosphorus ratio about 1.5 to 1? Quite frequently on some tags, the nitrogen-free extract (NFE) is also listed. The NFE is an estimate of the more readily available carbohydrate content of the feed. If the TDN (total digestible nutrients) is listed, the energy content of the feed can be calculated. One kg of TDN is equivalent to about 4,410 kcal (1 lb TDN is about 2,000 kcal).

Ingredients Listing

A second state regulation is that all ingredients must be listed. Although the percentage of each ingredient is not given (because the feed company would be giving you their formula), by knowing the ingredients one can estimate protein quality and the digestibility of the feed. You should also visually examine the feed to be sure that the ingredients are of high quality and not moldy or damaged. The list must include any "fillers" that are used. The ingredient list is important if your horse is allergic to a specific ingredient or becomes allergic to the feed because the veterinarian needs to know what the horse has been eating. Vitamin and mineral supplements that are added must be listed.

Other Considerations

A third state requirement is feeding instructions. These instructions are usually quite simple and recommend a given amount of the feed to various classes (pregnant, lactating, and so on) of horses as part of their overall feeding program. Any cautionary measures must also be listed.

A final consideration is the company that produces the feed. It makes no difference if they are large or small, known or unknown, as long as they produce a quality product.

FEEDING MANAGEMENT

Feeding horses is still an "art" that is acquired from experience. A knowledge of the nutrient requirements and the composition and use of the common horse feeds serves as basic information. The individuality of horses and their behavior makes it difficult to feed them to maintain the proper body condition and obtain maximum performance. A certain amount of skill, sound judgment, or experience is necessary for feeding horses well.

Most horses are kept in individual stalls or paddocks. From a management standpoint, this is important if one wants to obtain maximum performance from a horse or maintain an optimum condition. Some horses digest and use nutrients more efficiently than do others. The "hard-keeping" horse is inefficient. Both types must be fed individually, or they will become fat or too thin.

Group Feeding

Group feeding works with some groups of horses, but several management practices should always be used. Adequate feeder space should be available if only a single trough is used. Group feeding is not advisable for mature horses, because most horses are individualistic in their eating habits. Aggressive horses eat more than their share, and timid horses do not get enough. A more satisfactory situation is to provide feeders for each horse, separated by at least 50 feet (15.24 m). This allows a timid horse to eat without undue harassment. Group feeding of young foals usually works without too many difficulties provided adequate feeder space is available. Feeder design is important. The trough should have no sharp edges or projecting points. If the trough is too wide, the horse will pop its knees when eating the feed on the far side of the trough. A narrow trough is preferred. However, overly aggressive eaters cannot be fed from narrow troughs. The grain must be spread out over a larger area so they cannot get a mouthful each time. They are thus forced to take small amounts each time and eat more slowly. Small rocks or other obstacles that cannot be swallowed can be placed in the trough to force the horse to pick the grain from around them. Small salt blocks are excellent for this purpose. Preventing the horse from "bolting" its feed may prevent indigestion, colic, and founder.

Feeding Facilities

Hay can be fed from a manger, hay rack, or net or from the ground. Placing the hay on the ground is discouraged because it becomes contaminated with dirt, feces, and urine. The horse wastes more of the hay by walking on it and scattering it about the stall. Parasite control is more difficult because the horse may ingest the parasite eggs in feces. The hay manger, net, or rack should not be placed so high that the horse must reach up to eat, because hay particles and foreign material fall into the eyes and cause irritation.

Feed troughs and hay mangers and racks should be cleaned regularly. Moldy and spoiled hay or grain can cause colic and other digestive disturbances. Fecal contamination of feeding areas and water troughs should be removed as soon as possible. And dusty feed can lead to heaves or to pulmonary emphysema.

Timid horses or finicky eaters must be encouraged to eat. These horses should be fed individually in an environment where they are not afraid to eat and where they are not disturbed by other horses or by people. Fresh carrots, apples, lettuce leaves, or molasses can be sprinkled on the ration to get the horse to eat it.

Feeding Times

Feed horses at the same time each day. Horses form habits, and they anticipate being fed at the same time each day. Failure to feed regularly results in bad eating habits such as eating too fast or going off feed.

The frequency of feeding is important. Mature, idle horses can be fed once a day, but it is preferable to feed all horses at least twice daily. Hay and grain should each be given in equal parts morning and night. Horses that are being exercised heavily or expected to be in the best physical condition should be fed three times per day. Working horses should be fed 2 hours prior to work and at least 1 hour after hard work. This gives the horse time to digest part of the ration before work and to cool out and relax after work before eating. Some horsepeople prefer to feed one-half of the grain in the morning prior to work, one-fourth of the grain at noon, and one-fourth of the grain and all the hay at night to hard-working horses.

Age and Pregnancy

Age affects the way horses should be fed. Young foals should have the opportunity to eat a mixed concentrate ration within a few days after birth. One method is to place the ration in a box adjacent to the mare's feed trough. The box contains spaced rods above the feed so that the foal can pass its mouth between the rods to eat. This prevents the mare from eating the foal's ration.

If several mares are kept together, a creep area can be placed in the large paddock or pasture. The creep enclosure allows only the foals access to the feed. Foals are creep fed until they are weaned. It is best to wean at least two foals at a time and place them in a box stall. During and following weaning, they should have free access to their ration unless it causes digestive disturbances, in which case it should be limited. Yearlings and 2-year-olds are managed as mature horses, because their feed is restricted but meets their requirements.

There are three management aspects of feeding a pregnant mare. One is the "flushing technique" to increase the conception percentage. Barren, open, and maiden mares are given a ration that allows them to gain condition during the breeding season. The ration is increased about 45 days prior to the breeding season. However, it is important that they do not become obese. This is usually avoided by allowing them to lose some weight prior to starting the "flushing period." At the start of the breeding season, the mare's condition should be so that the ribs can be felt and are just visible. Excessive fat can hinder conception because fat surrounding the reproductive tract may cause obstruction. Pregnant mares have two critical periods for nutrition. Good nutrition is necessary during the 18- to 35-day period after breeding to prevent early embryonic death. Mares on a poor-quality diet have normal fetal growth to Day 25 of pregnancy, but resorption will occur within the next 6 days. If their nutrition regime is increased on Day 18 and extended through Day 35 after breeding, the chances of fetal resorption are quite low. During the last trimester of pregnancy, the mare should be fed adequate amounts of nutrients to allow normal fetal development and a gain in body weight of 5 percent (see p. 187). During lactation, mares must be fed to regain the body condition lost during pregnancy and to meet the demands of milk production. Orphan foals are a special problem and are discussed in Chapter 16.

Care must be taken in feeding old horses. They need high-energy feed that is easy to digest. Good-quality leafy hay should be fed, and you may need to chop it. The grain ration should be crimped to aid digestion. Old horses require more energy to perform work than do younger horses and thus need more feed per unit of body weight. Frequently inspect and care for their teeth and keep their parasite load to a minimum.

Ration Changes

Change rations gradually, over a period of about 1 week. Sudden changes may cause colic, going off feed, loss of condition, and digestive disturbances. One method is to replace 25 percent of the ration with the new ration every 2 days so that it takes 6 days before the horse is eating 100 percent of the new ration. Horses being turned out to pasture should be turned out for 30 minutes the first day and then for increased amounts of time so that after a week they are left out.

Horses returned home from sales or strenuous performance training should be brought down gradually. Their ration and amount of exercise should be reduced gradually, over a 2-week period.

Rations must be closely controlled for horses subjected to strenuous daily exercise when the horse is given a day off or when it gets very tired. Azoturia, or "Monday morning sickness," is caused by feeding the grain portion of the ration on a horse's day of rest. (See Chapter 10 for full discussion of azoturia.)

Overweight and Appetite

Prolonged fasting is not a recommended practice for weight reduction in horses, because hyperlipemia may be induced. This condition is characterized by high concentrations of lipid in the blood. In severe cases, the animal will not start to eat again after food is offered. It may die unless treated properly with glucose and insulin.

One of the most common forms of horse malnutrition is being too fat. Horsepeople may overfeed their horses out of kindness. Often, overfeeding is coupled with lack of exercise. The weight of a horse can be monitored periodically by weighing it on a scale or by estimating its weight by measuring its heart girth with a calibrated tape. "Weigh tapes" are supplied by several feed companies. They are fairly accurate (± 10 percent) and indicate if a horse is losing or gaining weight over a period of time. Thinness in horses is caused by poor-quality feed, inadequate amounts of feed, internal parasites, poor teeth, milk production, pregnancy, and excessive work. These conditions can usually be corrected. If not, the horse may be sick or have a metabolic disorder.

Horses should be fed by weight of the feed, not by volume (see Table 5-1). Nutrient content must also be taken into consideration. Changing from a coffee can of oats to a coffee can of corn increases the energy intake two- to threefold. Therefore an adjustment in volume (actually, weight) must be made to keep from giving the horse excess energy.

Horses that go off their feed may lose considerable condition before they start to meet their nutrient needs. Horses should be exercised at least 30 minutes to 1 hour each day, which stimulates their appetites. Adding molasses to the ration or, in some cases, salt, carrots, or apples may stimulate the horse to start eating. Occasionally an inadequate source of clean, cool water causes the problem.

Other Factors

The feces of horses should be routinely observed. Changes in consistency, odor, color, or composition usually indicate some type of disorder.

Horses kept in close confinement tend to crave unnatural feedstuffs, particularly if they are bored with their surroundings. Those fed pelleted or cubed rations tend to chew on wood and eat hair and dirt. Chewing wood can be alleviated by placing the horse in a stall or paddock where no wood is exposed. Exposed wood can be treated with cresote or similar materials. Excessive consumption of sand or dirt can result in sand colic, an accumulation or impaction of sand in the caecum or colon that prevents the passage of digesta. Horses that graze pastures grown on sandy soil are prone to sand colic. Young foals chew the tails of their dams or other horses. This can be prevented by rubbing some bone oil on all the horses' tails that are exposed to the culprit. Coprophagy (eating feces) is commonly observed in horses. Young foals seem to do it more often than mature horses. It can result in eating the eggs of internal parasites.

Parasite control (see Chapter 10) is an important part of feeding management. Internal parasites decrease digestive efficiency and cause digestive disturbances such as colic and diarrhea. External parasites can annoy horses and cause extra energy expenditure to escape or fight them by running, stomping feet, or continually switching tails.

Feed should be stored in rodent-proof bins. Galvanized garbage cans with lids are excellent to store one to two sacks of grain. Rodents eat the grain out of grain hay and leave the bales as straw. They can be prevented from living and breeding in stored hay with new chemicals that are available.

Exercise (at least 30 minutes per day) is important to maintain muscle tone, stimulate appetite, and help prevent digestive disturbances.

Dental problems can lead to improper mastication and digestive disturbances. Teeth should be inspected at regular intervals (at least yearly) for abnormal wear (hooks) and for broken teeth.

Treats such as carrots, apples, commercial treats, and sugar cubes are given to horses by some horsepeople. Sugar is never recommended. Other treats should be placed in the feed trough, or else the horse will develop bad habits. If fed from the hand, horses look for treats when they are handled, constantly nibbling at one's hand, shirt pocket, or pants pocket. Although horses enjoy them, treats play no role in the nutritional program.

6
Handling

Horses form habits rather quickly, and improper handling leads to undesirable behavior patterns. Once these are established, the horse is not as much fun to work with and may be unsafe to work around. Horses that are trained by competent horsepeople learn to be handled in certain ways. If not handled according to their expectations, they become confused and/or frightened. For full enjoyment, owners should become familiar with accepted and safe procedures for handling horses.

CATCHING AND HALTERING

Catching a horse the proper way prevents it from becoming hard to catch. When approaching the horse, you must make it aware of your presence. If the horse is daydreaming or focusing its attention on something else, it may be startled when it realizes you are about to catch it. Speak to the horse to make it aware of your presence. Approach the horse at an angle, from the side (Figure

FIGURE 6-1
Approaching a horse to catch it.

6-1). Never approach it directly from the front or rear, because if it is startled it may kick or run over someone. A horse cannot see well directly behind and may not recognize what is approaching unless it turns its head. On reaching the horse, allow it to smell a hand and then rub it a couple of times with the back of your hand on the neck or shoulder. Do not try to rub the horse's nose. Once a horse learns this routine, it will let you approach and halter it without any difficulty.

Some horses are hard to catch, a habit that may be classified as a vice and that is hard to cure. They are usually hard to catch because of fear, resentment, or just plain habit. To overcome fear of being caught, the horse must be handled often, and the experience should be pleasant. It is best to handle a horse properly at birth so that it trusts whoever will handle it. The experience can be made pleasant by grooming and talking to it until it relaxes and accepts a person's presence. You can also feed it a few bites of grain from a bucket after it is caught. Most horses will come to you after you catch them a few times and then give them some grain. A horse should never be allowed to eat the grain before being haltered so that it will never learn to play the game of taking a bite and then refusing to be haltered. If this procedure does not work, the horse should be kept confined in a small stall or paddock where it cannot escape from being caught. It should be caught and haltered several times a day until it becomes easy to catch.

Horses should never be chased so that they run. If they run, they are or have become afraid. After they learn the habit of running away, they usually have to be run into a small paddock before they will let someone catch them.

Horses that resent how they are used once they are caught become difficult to catch. Some common causes of resentment are long, hard, confusing training sessions; exhausting rides for poorly conditioned horses; painful wound doctoring; and abuse. Most horses that have been injured and require injections or painful treatments become hard to catch. This can be prevented by catching the horse several times between each treatment session and giving it a pleasant experience by grooming or feeding it. With a varied routine, it does not know what to expect each time it is caught and does not become bored. The horse that resents being caught and tries to kick can be dangerous. Such horses usually watch you approaching them and try to keep their heels directed toward you. As soon as they shift their weight so they can kick, slap them across the croup. Such horses can be taught to face you by hitting them below the hocks with a whip. As soon as they face you, speak to them in a calm voice and approach. In a few sessions, they will always face someone who approaches them.

The horse that is hard to catch only because of habit can be retrained by isolating it and catching it several times a day. After it is caught for a few minutes and given a pleasant experience, it soon forgets about running away. Once the horse discovers that being caught is part of its daily routine and is usually pleasant, it looks forward to being caught and handled.

Do not place the halter on the horse as soon as you rub its neck. This gives the horse an opportunity to escape. Place the lead rope around the horse's neck before slipping the halter on its head so the horse cannot escape.

Keep the halter in good repair so the horse cannot break it easily after being tied up. Do not leave a halter on a horse turned out in the paddock or confined in a stall unless it has a safety release buckle. Horses can easily get the halter hung up on something and have an accident. When a horse scratches its head with a hindfoot, the foot can get caught in the halter and this can lead to injury or death. A safety release buckle can release the halter when sufficient pull is applied. A horse should never be tied up and left unattended. If it decides to pull back and "have a fit," it can break its neck or damage neck muscles. And it may throw itself down and hang itself.

LEADING

Walk beside a horse when it is being led (Figure 6-2). The proper position for the handler's shoulder is between the horse's head and shoulder, with the hand placed about 12 to 18 inches below the lead shank ring. Fold and place the lead rope in the hand opposite to the one with which you are leading the horse. Never coil the lead rope around your hand because an injury will result if the horse suddenly bolts and runs away. If you lead the horse from in front, the horse's feet will strike the back of your feet or legs. Also, if the horse goes too

FIGURE 6-2
Correct position to lead a horse.

far forward, you cannot control it, and it can easily run away or run around
the handler. And you are vulnerable to kicks from a hindleg when the horse
passes by you.

A horse can be led from either side, but it is customary to lead it from the
left side. On a trail, you must sometimes lead from the right side.

Turn the horse to the direction opposite to the side from which you are
leading it. This way, you walk around the horse. If you turn the horse toward
you, it may step on your feet.

If a horse fails to start forward when you cue it to move forward, you can
turn it sharply to one side and lead it a few steps forward. You cannot pull
horses that fail to move forward, because they are too strong. Some horses
will not move forward when you face them. If so, pass a rope around the
horse's buttocks and pull it to encourage forward movement. If a horse rears
up while being led, release the lead rope from the hand adjacent to the halter.
This will prevent the horse from lifting you off the ground or throwing you
down. Immediately move to the side of the horse to avoid its flailing front legs.
From the side, you can pull the horse down. Some horses prance around or
get nervous while being led. They will step on your feet unless you extend your
elbow and place it against the horse's neck.

When you remove the halter, the horse should face you. Do not let it bolt
away once the halter is off. Make sure the horse does not turn and run, kicking
as it passes by.

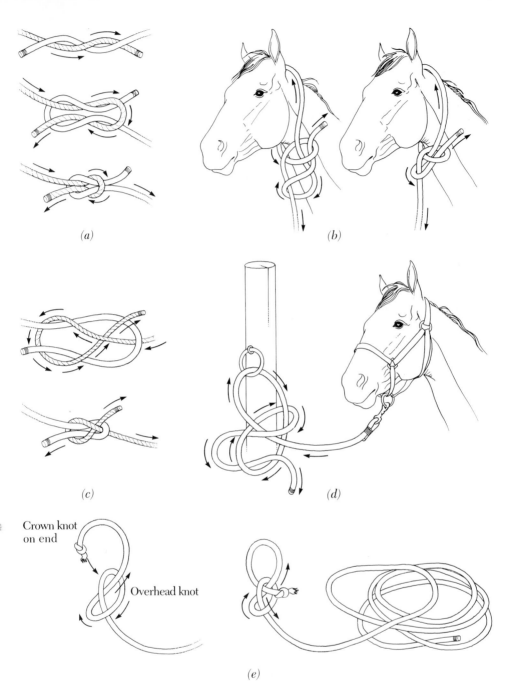

(a)

(b)

(c)

(d)

Crown knot on end

Overhead knot

(e)

FIGURE 6-3

Knots. *(a)* The square knot is useful for tying two ends of the same rope together. *(b)* The bow-line knot will not slip. Every horseperson should be able to tie it. To tie, make a loop over the standing line. Take the free end and run it under and up through the loop. Now bring the end around the standing part and back through the top of the loop. Tighten the knot by pulling the standing part while holding the end loop. *(c)* Use the sheet bend knot for tying together two different ropes of two different diameters; this knot will not slip. *(d)* Use the slip knot to tie a horse to a post or ring. It is easy to tie and can be quickly untied by pulling on the free end of the rope. *(e)* Use the honda to tie a small loop in the end of a rope so that the loop will not slip.

RESTRAINT

There are several methods of restraint, and the use of each depends on specific needs. The safety of the horse and handler should always be considered. Accidents are common during restraint procedures.

Tying

Horses can be tied with the lead rope to hitching rails or posts. The horse should be tied with a quick-release knot so that the horse can be released in an emergency (Figure 6-3). It should be tied about the same height as the withers, and the rope should only allow the nose to barely touch the ground. This prevents the horse from getting a front leg over the tie rope. Cross-tying restricts movement more than a single rope. Two tie ropes are used to cross-tie the horse. The ropes are usually anchored about 6 to 8 feet off the ground and are long enough to allow the horse to stand with its head about level. The ends anchored to the wall or posts are tied with quick-release knots.

The horse may be tied without a halter. Use a bow-line knot on the end placed around the horse's neck so that it will not slip and choke the horse (Figure 6-3).

Release the tie rope before removing the halter. Failure to do so may cause the horse to pull back and break the halter. They may become habitual halter pullers once they do this a few times.

Hobbles

When a horse will not allow someone to hold up its leg while being shod or will not remain still while a leg is being treated for an injury, a variety of hobbles can be used to restrain the horse. Restraining hobbles should be used seldom and should be properly used. They do not absolutely immobilize a horse but are usually more humane than a big battle with the horse. All horses should become accustomed to being hobbled.

First, lift one of the horse's front legs off the ground. If the horse is easy to handle, an assistant may be able to hold the leg. When no assistance is available or if the horse struggles, you can tie up the leg with a strap or soft cotton rope at least ½ inch in diameter (see Figure 6-4). Lay the strap across the pastern in front and then cross it behind. Lift the leg and secure the rope or strap around the horse's forearm. A different technique is used on horses that refuse to let their legs be lifted. Tie a 25- to 30-foot (7.62- to 9.14-m) cotton rope around the horse's neck just in front of the withers with a bowline knot. Wrap a short strap with two rings in the ends around the pastern. The free end of the rope is passed through the rings and then through the loop around the neck. This forms a pulley system to hoist the leg.

(a)

(b)

FIGURE 6-4

(a) Front leg strap: (A) 6.5 inches (16.5 cm). (B) 3.5-inch ring (8.89 cm).
(C) 31.5 inches (80 cm). Strap is 2 inches (5.08 cm) wide. *(b)* Front leg tied up.
(Photos by B. Evans.)

To lift a back leg, use a 25- to 30-foot (7.62- to 9.14-m) cotton rope (¾-inch diameter) and strap with two rings to form a "sideline." If the horse tries to kick or is a confirmed kicker, apply the sideline as in Figure 6-5. Lift the foot until the horse cannot get it closer than 4 to 6 inches to the ground (101.6 to 152.4 mm).

You can use the tail tie method with horses that do not offer as much resistance (Figure 6-6).

Hock hobbles are used to prevent the horse from kicking with the hindfeet. However, they are usually only used on breeding farms to hobble mares during the breeding process (Figure 6-6).

Western hobbles (Figure 6-7) are used to keep a horse from walking very fast. They are used to keep horses from running away from camp on trail rides or if they are ground tied. The hobble consists of two straps that encircle the pasterns and are joined together with a short strap or chain. In an emergency, you can make one by opening up a burlap bag and rolling it across the diagonal, like a bandanna. Wrap it around one pastern and twist and tie it around the other.

FIGURE 6-5

Typing up the hindleg with a pastern strap and a long line. *(a)* Equipment: (A) 30 feet of ¾-inch soft cotton rope (9.14 m of 1.9-cm rope). (B) 8-foot lead rope with snap. (C) Pastern strap. *(b)* A long line is tied around the neck with a bow-line knot, and the free end is passed through one ring of the pastern strap. *(c)* The horse has stepped over the rope with one hindleg. *(d)* The rope is above the hock, and the free end has been passed through the other ring of the pastern strap. *(e)* The pastern strap is placed in position by pulling on the free end of the rope. *(f)* The leg is held up, and the rope is tied off with a quick-release knot. (Photos by B. Evans.)

FIGURE 6-6

Back leg held up with tail tie and pastern strap by pulling rope over the horse's croup. (Photo by B. Evans.)

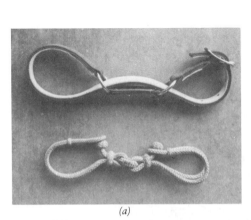

(a)

(b)

FIGURE 6-7

Hobbles. *(a)* Two types of western hobbles. *(b)* Horse hobbled with western hobble. (Photos by B. Evans.)

Twitches

For generations, various types of twitches have been used to restrain horses (Figure 6-8). The simplest twitch is grasping the upper lip of the horse and rolling it upward. The skin on the shoulder area can be grasped and rolled with the hand. An ear twitch is applied by grasping the ear, placing the wrist and forearm against the horse's head, and pulling outward. The ear should never be twisted or bent double and twisted. You should avoid the ear twitch because it makes horses head shy.

More severe twitches include the types that are loops placed around the horse's upper lip and tightened. A chain or rope is used to form the loop, and a handle is used to twist the loop. The twitch may injure the horse's teeth, gums, and inner lip. If the horse jerks its head or rears, the handle of the twitch becomes a lethal club, so you must be very careful when using this type of twitch.

The Wilform twitch is a square frame with a movable bar across the center. A screw ring tightens the bar on the horse's upper lip.

The Kendal twitch is one of the safest lip twitches for the horse and handler (Figure 6-9). Two aluminum bars are slightly flared on one end to form a loop

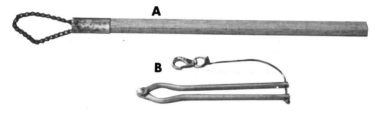

A

B

FIGURE 6-8

Twitches. (A) Chain. (B) Kendal. (Photo by B. Evans.)

FIGURE 6-9

Kendal twitch on upper lip. (Photo by B. Evans.)

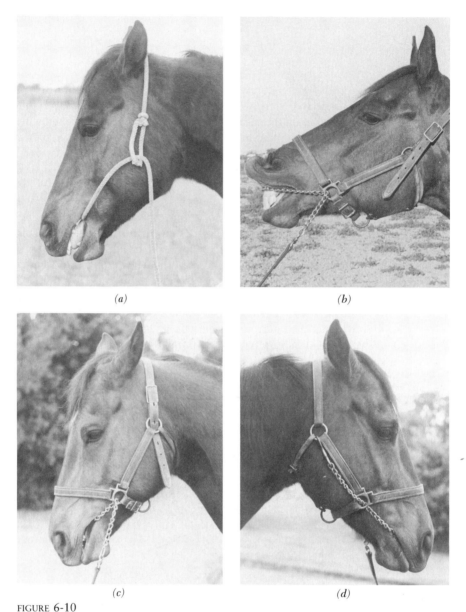

(a)

(b)

(c)

(d)

FIGURE 6-10

War bridles and chains. *(a)* Rope loop over poll and through the mouth. *(b)* Chain across gum. *(c)* Chain in mouth (left side view). *(d)* Chain in mouth (right side view). *(e)* Chain over nose (left side view). *(f)* Chain over nose (right side view). *(g)* Chain over nose and under chin. *(h)* Chain under chin. (Photos by B. Evans.)

and are hinged together like pincers; a cord at the other end of the bars fastens the twitch quickly and safely. The cord is then snapped to the halter. The smooth bars do not injure a horse's mouth, allow some blood circulation, and are not as dangerous to the handler if the twitch is thrown off.

All lip twitches should be snug enough for restraint but not too severe. They should not be left on too long, because they cut off blood circulation.

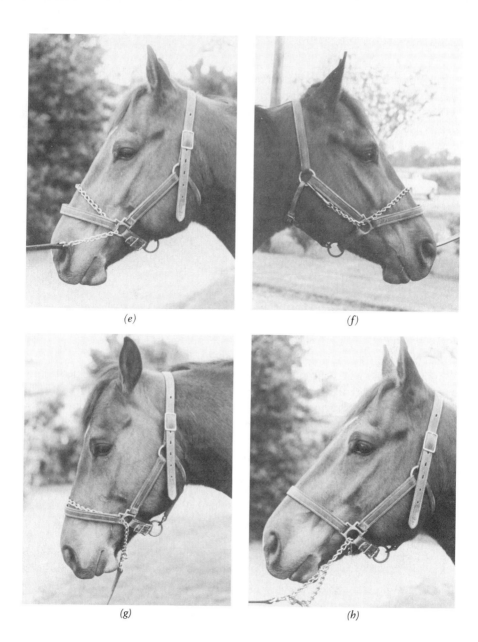

(e)

(f)

(g)

(h)

War Bridles

Horses that are difficult to control while being led can be restrained with a war bridle (Figure 6-10a). The simplest war bridle is formed by tying a honda in the end of a rope and then forming a loop. The loop is placed over the poll and across the gum of the upper incisor teeth. Sharp, quick pulls on the loop are used to correct or restrain the horse. Similar results can be obtained by using a leather halter and a lead shank (stallion lead shank) with a chain and snap on the end. The chain is run through the ring on the left side of the halter, across the gum, through the ring on the right side of the halter, and snapped to the upper ring on the halter (Figure 6-10b). If the chain is passed through the mouth (Figure 6-10c) or over the nose (Figure 6-10d), it will not be as severe. The latter three methods are frequently used to restrain stallions. A less severe use of the chain is to pass it over the nose and snap it under the chin (Figure 6-10e) or just to snap it under the chin (Figure 6-10f).

Chemicals

Many tranquilizers are effective for horses. They should only be used under the direction or supervision of a veterinarian, because horses do not always react as expected. They may go berserk and seriously injure themselves and handlers.

SAFETY

Observe all safety precautions while working with horses. Most accidents happen with horses that are very gentle. The person handling the horse forgets how horses react to a situation, so they receive an injury when the gentle horse reacts. The following are a few safety rules you should always observe.

Actions

Your actions around a horse reflect your ability and confidence in handling a horse. Remain calm and confident at all times. Nervous handlers usually make a horse become nervous and unsafe.

Never tease nor unduly punish a horse. A teased horse may develop dangerous habits for the rest of its life and may lose all respect for its handler.

If you must punish a horse, do so at the instant of its disobedience. If you wait, even for a minute, the horse will not understand why it is being punished. Control your temper at all times, but let the horse know that you are its firm, kind master.

Know the horse's temperament, reactions, and peculiarities. If the horse is unfamiliar to you, ask about the horse before working with it.

Clothes

Wear footgear that protects your feet from being stepped on and from being injured by nails and other small objects around the stables and barnyard. Boots or hardtoed shoes are preferable. Never wear tennis shoes or moccasins. Never go barefoot, because infectious organisms such as tetanus are prevalent around barns, corrals, and fences. Gloves safeguard against cuts, scratches, splinters, and rope burns.

Catching and Releasing

Be careful when approaching a horse to catch it so that the horse does not kick or run over you (see first section in this chapter). It is not safe to leave a halter on a loose horse. When you must do so, check the halter daily or make sure it has a safety release. Some halter materials shrink, so be certain to check the fit. A loose horse may catch a foot in the halter strap or may catch a halter on posts or other objects.

Use judgment when turning a horse loose. It is generally safest to lead the horse completely through the gate or door and then turn the horse around, facing the direction of entry. Make the horse stand quietly while it is being released. Then release the lead strap or remove the halter or bridle. Avoid letting a horse bolt away when released because it may kick as it wheels around or runs by.

Leading

The proper position for leading a horse has been described earlier. When leading a horse, avoid using excessively long lead ropes that may become accidentally entangled. When using lariats or longe lines, watch out for the coils. Never wrap the lead strap, halter shank, or reins around a hand, wrist, or body. When leading, tying, or untying a horse, avoid getting your hands or fingers entangled, and also be careful not to catch a finger in dangerous positions in halter and bridle hardware, including snaps, bits, rings, and loops.

Be extremely cautious when leading a horse through narrow openings, such as a door. Be certain you have firm control and step through first. Step through quickly and get to one side to avoid being crowded.

A horse is larger and stronger than a person. If it resists, do not get in front and try to pull.

Tying

Tie or hold horses when working around them. Always tie a horse in a safe place—with a halter rope, not the bridle reins. Be certain to tie the horse to a strong, secure object so that the rope will not break or come loose if the horse pulls back. Use a quick-release knot (Figure 6-3) so you can release the horse in an emergency.

Never tie the halter rope below the level of the withers, and tie it short enough to prevent the horse from stepping over the halter rope. Tie the horse a safe distance from other horses and from tree limbs or brush where the horse may become entangled.

Handling

Work around a horse from a position as near the shoulder as possible. Also, stay close to the horse, so that if it kicks you do not receive the full impact. Stay out of kicking range whenever possible. When you go to the opposite side of the horse, move away and go around it out of kicking range.

Never stand directly behind a horse to work with its tail. Stand off to the side, near the point of the buttock, facing to the rear. Grasp the tail and draw it toward you.

A good horseperson always lets a horse know what he or she intends to do. When picking up the horse's foot, for example, do not grab the foot hurriedly. This startles the horse and may cause it to kick. Learn the proper way to lift the feet.

A horse should be trained to be workable from both sides, even for mounting and dismounting. Some horses become very irritable and nervous when they are worked on from the right side.

Do not drop grooming tools underfoot while grooming. Place them where they will not trip you and where the horse will not step on them. An accidental slip or stumble can result in injury.

Bridling

When bridling, protect your head from the horse's head. Stand close, just behind and to one side (preferably on the left side) of the horse's head. Use caution when handling the horse's ears.

Maintain control of the horse when bridling by refastening the halter around the neck.

Adjust the bridle to fit the horse before riding it. Three points to check are the placement of the bit, the adjustment of the curb strap, and the adjustment of the throatlatch.

Saddling

Before saddling the horse, check the saddle blanket and all other equipment for foreign objects. Be certain the horse's back and the cinch or girth areas are clean. Foreign objects will make a horse buck or will cause saddle sores.

Bridle reins, stirrup leathers, headstalls, curb straps, and cinch straps should be kept in the best possible condition; the rider's safety depends on these straps. Inspect them frequently, and replace any strap when it begins to show signs of wear (cracking or checking).

When saddling to ride western style, it is safest to keep the off-cinches and stirrup secured over the saddle seat and ease them down when the saddle is on. Do not let them swing wide and hit the horse on the off-knee or belly—this hurts the horse and causes it to shy.

Swing the western saddle into position easily, not suddenly. Dropping the saddle down too quickly or hard may scare the horse. You do not need to, and should not, swing the saddle into position. Lift it and place it into position.

Pull up slowly to tighten the cinch. Check the cinch three times: (1) after saddling, (2) after walking a few steps (untracking), and (3) after mounting and riding a short distance.

When using a double-rigged western saddle, remember to fasten the front cinch first and the rear cinch last when saddling. Unfasten the rear cinch first and the front cinch last when unsaddling. Be certain that the strap connecting the front and back cinches (along the horse's belly) is secure. The back cinch should not be so loose that the horse can get a hindleg caught between the cinch and its belly.

When using English equipment, "run up" the stirrups immediately on dismounting. Dangling stirrups may startle or annoy the horse. When fighting flies, the horse may catch the cheek of a bit or even a hindfoot in a dangling stirrup iron. Dangling stirrups can also catch on doorways and other projections while you are leading the horse. After running up the stirrups, immediately bring the reins forward over the horse's head. In this position, they can be used for leading.

Fasten accessory straps (tie downs, breast collars, martingales, and so on) and adjust them after the saddle is cinched on. Unfasten them first before loosening the cinch.

Mounting

Never mount or dismount a horse in a barn or near fences, trees, or overhanging projections. Sidestepping and rearing mounts have injured riders who failed to take these precautions.

A horse should stand quietly for mounting and dismounting.

Riding

Until you know your horse, ride only in an arena or other enclosed areas. Do not ride in open spaces or unconfined areas until you are familiar with the horse. When the horse is full of energy, exercise it on a longe line or ride it in an enclosed area until it is settled.

Keep the horse under control and maintain a secure seat at all times. Horses are easily frightened by unusual objects and noises that may catch the rider unaware.

Spurs can trip you when you are working on the ground. Take them off when not mounted. Wear neat, well-fitted clothing that will not snag on equipment. Belts, jackets, and front chap straps can become hooked over the saddle horn. And wear boots or shoes with heels to keep your feet from slipping through the stirrups. Wear protective headgear appropriate to the activity in which you are engaged, especially in any form of jumping.

Keep the horse's feet properly trimmed and/or shod. Long toes predispose a horse to stumble and to fall.

There are certain safety precautions to remember when trail riding. First, if you plan to ride alone when trail riding, tell someone where you are going and approximately when you expect to return.

Ride a well-mannered horse. If a horse becomes frightened, remain calm, speak to it quietly, steady it, and give it time to overcome its fear. Then ride or lead the horse past the obstacle.

Courtesy is the best safety on the trail. Do not play practical jokes or indulge in horseplay. Never ride off until all riders in a group are mounted. Ride abreast or stay a full horse's length from the horse in front to avoid the possibility of a rider being kicked, or the rider's horse being kicked. Never rush past riders who are proceeding at a slower gait, because to do so startles both horses and riders, and often causes accidents. Instead, approach slowly, indicate a desire to pass, and proceed cautiously on the left side.

Hold the horse to a walk when going up or down a hill. Allow a horse to pick its way at a walk when riding on rough ground, or in sand, mud, ice, or snow where there is danger of slipping or falling. Use extreme caution at wet spots or boggy places. Walk the horse when approaching and going through underpasses and over bridges.

Select a riding location with care. Choose controlled bridle paths or familiar, safe, open areas. Try to avoid paved or other hard-surfaced roads. Walk the horse when crossing such roads. In heavy traffic, it is safest to dismount and lead across.

If it is necessary to ride on roads or highways, ride on the side required by law. State laws vary as to which side of the road you should ride on. Ride on the shoulders of roads or in ditches, but watch for junk. Riding at night can be a pleasure but is more hazardous than daytime riding. Walk the horse; fast gaits are dangerous. Wear light-colored clothing and carry a flashlight and reflectors when riding at night. Check details of your state's regulations.

When riding on federal or state lands, seek advice from the forest or park officials. Know their rules concerning use of the trails and their fire regulations. Obtain current, accurate maps and information on the area. Become familiar with the terrain and climate. Be certain the horse is in good physical condition and that its hooves and shoes are ready for the trail.

Avoid overhanging limbs. Warn the rider behind you if you encounter one. Watch the rider behind you if you encounter one. Watch the rider ahead so that a limb pushed aside does not snap back and slap your horse in the face.

Learn to handle a rope before carrying one on a horse. Always use caution when working with a rope if the horse is not "rope-broken." Never tie the rope "hard and fast" to a saddle horn while roping from a "green" horse.

EXERCISE

Lack of adequate exercise is one of the most common types of horse mismanagement. Horses need to be exercised for at least 30 minutes each day to maintain some muscle tone. Those that stand in a stall or small paddock and are not exercised become bored. Boredom leads to the common stall vices of weaving, cribbing, chewing wood, and stall walking.

There are several ways of exercising a horse that is not kept in a paddock large enough for free exercise.

Exercise Methods
Hot Walker

A horse can be placed on a "hot walker." A mechanical hot walker (Figure 6-11) keeps a horse moving in a small circle. A properly designed hot walker has several safety features. It must be difficult to pull over, so it must be properly

FIGURE 6-11
Mechanical "hot walker" used to exercise a horse. (Photo by B. Evans.)

anchored to the ground. The arms should be high enough to keep a horse from bumping its head. The drive system should contain a "slip-gear," so the rotation of the arms will stop if a horse falls down. Most of the good hot walkers have two forward and two reverse speeds. For each direction, one speed is for a slow walk and the other is for a fast walk or slow trot.

Horses should not be left on a hot walker too long at one time or they get bored. An exercise period of 30 minutes is usually sufficient. If left on it too long, some horses develop the habit of pulling back to get it to stop.

Treadmills

Treadmills are also used to exercise horses that receive no other exercise or as part of a conditioning program for performance horses (Figure 6-12). Treadmills are constructed so that a continuous belt is driven by an electric motor.

FIGURE 6-12

Horse being exercised on a treadmill. (Courtesy Horsey, Inc., manufacturer of Anamill Treadmill Conditioner, West Lake Rd., Vermillion, Ohio 44089.)

FIGURE 6-13
Horse being exercised on a longe line. (Photo by B. Evans.)

The belt runs over a series of closely spaced steel rollers. Treadmills can be inclined so that the horse is moving up an incline of up to a few degrees (5 to 7°). It is an excellent method of providing exercise and developing muscle tone.

Longeing

Longeing is another method of exercising the horse without riding it (Figure 6-13). To longe a horse, one needs a 20- to 30-foot (6.1- to 9.1-m) longe line and a longe whip. The longe line is made from flat, braided nylon webbing about an inch wide. It has a swivel snap at one end. The longe whip has a stiff or slightly flexible 5-foot (1.5-m) handle and a 4- to 5-foot lash (1.2- to 1.5-m). If one is available, use a longeing cavesion, which has a ring on the nose piece to attach the longe line. If one is not available, attach the longe line to the side ring of an ordinary halter.

To train the horse to longe, it should be trained in a round corral or small corral. The first lesson is to get the horse to walk in a small circle around you. Position the horse next to the wall of the round corral, stepping toward the horse's hindquarters, and tapping it lightly with the whip to encourage it to move forward. As you tap the horse with the whip, give the voice command "Walk." As soon as the horse learns to go in one direction, teach it to go in a small circle in the opposite direction. This can usually be accomplished in one lesson. Starting with the second lesson, allow the horse to go in a larger circle, but within reach of the whip. After it makes a few rounds walking in one direction, teach it to stop. Give the voice command "Whoa," and place the whip in front of the horse. Do not let the horse stop and turn to face you. This teaches them to stop and turn on their forelegs. Later, it is difficult to train them to stop and pivot on their hindlegs. The third lesson is to train the horse to reverse directions. After the horse is stopped by the voice command "Whoa," approach it, give the voice command "Reverse," turn it around to the opposite direction, and give the voice command "Walk." The horse should learn how to reverse in a few repetitions. In later lessons, you can teach it to trot and canter. When cantering, the horse must be using the correct lead. After each good performance, the horse should be rewarded with a good pat on the neck.

Two mistakes are often made while training a horse to longe. One mistake is to let the horse approach you. While longeing, it should never come toward you nor face you. It should be taught to keep its side toward you. Facing you usually results from the second common mistake—pulling on the longe line. By using voice commands and the whip, you never have to pull on the longe line.

Driving

Another form of exercise for the young horse is driving. The horse is worked with driving lines to teach it the basic commands while it is still too young to be ridden. Driving horses is discussed in Chapter 11.

Riding

Riding the horse is the most common method of exercise, because most horsepeople like to ride. The horse should be ridden at least 30 minutes per day, at least 5 to 6 times per week. A common mistake is to ride the horse for several hours on a weekend and not ride it during the week. This is hard on a horse because its muscles are not in physical condition and its bones are not used to the concussion trauma. Horses that are to be used or exercised very hard for long periods need to be conditioned over a period of time.

CONDITIONING

Conditioning is not difficult or stressful for the horse if the handler takes certain precautions. Starting with a horse that is in poor physical condition requires that the conditioning start slowly and increase in intensity as the horse's muscular, respiratory, and cardiovascular systems become conditioned. If the program is too strenuous at the beginning, the horse's muscles become sore and lameness may develop because the bones are not accustomed to the stress of concussion.

Program

It takes about 10 to 12 weeks to condition an average horse for a 25- to 50-mile (40.2- to 80.4-km) competitive trail ride. A typical training schedule that has been used to successfully train several competitive trail ride and endurance horses is as follows:

Week 1: Walk the horse every other day on level ground for 2 hours.

Week 2: Walk the horse every other day on level ground for 3 hours.

Week 3: Walk and trot the horse every other day on level ground for 2 hours. Gradually increase the trotting to 50 percent of the time. However, the trotting periods last for 5 to 20 minutes.

Week 4: Walk and trot the horse every other day on level ground for 3 hours. Gradually increase the trotting until the horse can trot for a full 30-minute period.

Week 5: Walk the horse up hills every other day for 2 hours. Go slowly but make the horse keep moving. During conditioning, never stop for a rest during a climb; stop only when you reach the top.

Week 6: Walk the horse up the steepest hills you can find every other day for 3 hours.

Week 7: Trot the horse up hills every other day for 2 hours. Gradually increase the trotting to 50 percent of the uphill work.

Week 8: Trot the horse up hills every other day for 3 hours. Trot as much as possible.

Week 9: This is the weekend before the ride. Saturday, ride 20 miles (32.2 km); Sunday, ride 15 miles. Ride at a good, steady NATRC pace. (North American Trail Ride Conference). Trot the levels and walk the ups and downs at a good fast walk. Let the horse rest the balance of the week.

Week 10: Go to the ride.

Evaluation of Condition and Stress

When training a horse for long physical performances, the trainer should be able to evaluate pulse and respiration rates and recognize thumps, dehydration, and azoturia.

Evaluation of the pulse *(P)* and respiration *(R)* is necessary to prevent stressing the horse beyond its physical condition and to recognize when the horse becomes conditioned. The methods for determination are discussed in Chapter 10. A well-conditioned horse has a resting pulse of 35 to 40 beats per minute and a respiration rate of 12 to 25 per minute. The $P:R$ ratio should be 3:1 or 2:1. As the horse exercises, both P and R increase but the ratio is maintained in a well-conditioned horse. While working, the pulse rate usually ranges between 75 and 105, and the respiratory rate increases to 24 to 36. At the end of an exercise period, the rates recover rapidly—within 5 to 10 minutes in a well-conditioned horse. Horses in poor condition do not recover for 30 to 45 minutes. When the $P:R$ ratio decreases at elevated levels, it indicates a stress condition of such magnitude as to make further exercise totally inadvisable. For example, at rest a horse may have a pulse rate of 36 and a respiration rate of 12 (thus a 3:1 ratio). After severe exercise, the horse may show a P of 120 and an R of 60 (or a 2:1 ratio). A well-conditioned horse may recover during a 45-minute rest period to a P of 60 and an R of 30, or, again, the 2:1 ratio. Should a horse under the same set of circumstances show a P of 85 and an R of 80 or 90 (a 1:1 ratio), a potentially dangerous level has been reached, and the horse probably should not be allowed to continue to exercise. Many trainers also share the opinion that a horse whose pulse does not drop to 70 or below after a 45-minute rest period is not in good physical condition.

Horses often become dehydrated on long trail rides when a deliberate effort is not made to keep them hydrated. Water is lost from the horse's body by several ways, but during strenuous exercise the primary loss is from profuse sweating. The water loss upsets the water and electrolyte (salts) balance of body tissue. Failure to replenish water and electrolytes leads to disturbance of body functions, weakness, and death.

Dehydration is easily recognized by two methods. One method is to pick up a fold of skin at the point of the shoulder and release it. Normally, the skin flattens out immediately. If the horse is slightly dehydrated, it takes about 5 seconds for the skin to flatten out, and if there is moderate dehydration, it takes about 10 seconds. When the horse is severely dehydrated, the skin does not flatten out, the eyes look sunken in, and no saliva is being produced. Another method is to determine the blood capillary refill time. This is done by opening the upper lip of the horse and pressing the thumb against the gum until the gum turns white (no blood present) under the thumb. On release of the thumb, blood quickly returns to the area. If the horse is dehydrated, the refill time may be 10 seconds, the membranes are dry, and the gum is usually bluish in color, which indicates a lack of oxygen.

Prevent dehydration by making sure the horse has adequate water and that the lost electrolytes are replaced. Prevention starts with a good, balanced diet and free access to salt (sodium chloride) during the training period. During an endurance ride or long training rides, allow the horse to drink at every opportunity to do so. Electrolyte therapy may be required. There are numerous types of electrolyte combinations and methods of administration. Their use for an individual horse should be discussed with a veterinarian who is familiar with the problem.

TRANSPORTATION

All horses should be trained to load into a trailer. Horses that refuse or are difficult to load cause accidents, injuries, and bad experiences. A horse usually refuses to enter a trailer because it fears an unfamiliar trailer or remembers a bad experience while loading. Prior planning prevents poor performance when a horse must be loaded and transported.

Trailers

There are many types of trailers to haul horses. Most are the bumper-hitch type but the gooseneck type is also popular. Single- and two-horse trailers are usually of the bumper-hitch type but are available as goosenecks. Four-horse and larger models are usually goosenecks. The advantages of the gooseneck for larger trailers are weight distribution, ease of handling, radius of turns, and safety. The weight is distributed on the vehicle in front of the rear axle, instead of back on the bumper, where heavy weight tends to pick up the nose of the towing vehicle, which encourages fishtailing.

Rear doors can be hinged or have loading ramps, or have combinations of the two. A hinged door is preferable if the step up is not too high. It is more difficult to train horses to go up a loading ramp. They may step off the edge of it and injure their legs. If the horse will only load by someone leading it into a trailer, an escape door may be advisable.

The wheel wells should be located on the outside of the trailer. Trailers are equipped with electric or hydraulic brakes. The tires should be 6- or 8-ply for added protection.

The partition that separates the horses should be carefully considered. You may need a full-length partition or one that extends just below the hips. If the trailer is not full width, a full partition makes it difficult for the horse to keep its balance, because it cannot spread its legs. This causes some horses to try to "climb the sides" of the trailer on turns. Most horses ride better and are more comfortable with the partial partition. However, a full divider will prevent horses from stepping on each other.

In large trailers, the horses may ride so that one is hauled directly behind another. They may all face forward, or some may face backward, opposite to the direction of travel. Position seems to make no difference to the horse. Occasionally, they are hauled while standing in a diagonal position in the trailer.

Before loading and transporting the horse, inspect the trailer and towing vehicle for safety. The trailer ball should be the proper size and tightly secured to the hitch. The electrical connections should be working, as should all directional and turn signals, and running and stop lights. Manure and urine left too long in the trailer eventually rots out the wooden floors. Clean them out, and check the floor boards for rot. A rubber mat provides a firm foothold and prevents the horse from slipping down. All door latches should be working properly. Butt chains should be present and secure. A properly inflated spare tire, lug wrench, and trailer jack should be in the trailer or towing vehicle. Tires should be properly inflated and checked for abnormal wear. Wheel bearings should be serviced yearly or every 10,000 miles (16,090 km). Safety chains should be in working condition.

Before hauling a horse, the driver should be experienced with towing trailers or should pull the trailer around to get the feel of how it influences the towing vehicle. For the safety of the horse, execute turns more slowly and less sharply than when not pulling a trailer. Avoid sudden stops. Moreover, it takes a longer distance to stop a rig, to start moving, and to reach a desired speed. Backing a trailer is tricky and requires practice. If only one horse is being transported in a two-horse trailer, the horse should be hauled on the driver's side. This way, if the outside wheel drops off the shoulder, there will be less chance of the trailer turning over. For easier towing, each trailer has a recommended height between the tongue and the ground.

Loading

A well-trained horse will go into a trailer without any hesitation but other horses can be loaded only with great difficulty. There are several methods of training horses to load or loading a stubborn horse.

One of the most successful ways of training a green horse to be loaded is by teaching the horse to longe with a whip. While learning to longe, the horse should learn the voice commands "Whoa" and "Walk." The horse should respect the whip. To load the horse, position a tandem trailer so that the right rear door will open adjacent to a smooth fence or wall. Lead the horse to the trailer and allow it to examine it for a few minutes. As soon as it becomes accustomed to the trailer, load the horse as follows (Figure 6-14). Position the horse facing the open trailer and between you and the fence or wall. Hold the lead shank in your left hand, stand at the horse's shoulder, and hold the whip in your right hand. Guide the horse's head with your left hand. Give the horse the command to walk. If it does not step forward, give the command again and tap the horse on the croup with the whip. Constantly tap the horse with the whip until it goes into the trailer. The wall on the horse's off side prevents it from moving or swinging around to the right. If the horse tries to move around to the left, you can apply the butt of the whip firmly to its ribs. If the horse attempts to back away, apply the whip more firmly. If the horse attempts to rear, apply the whip across the withers and back. Each time the horse moves forward, stop tapping with the whip, as a reward. Give the horse the command to walk; if it does not move forward again, apply the whip. As soon as one foot is placed in the trailer, praise the horse but do not allow it to hesitate too

FIGURE 6-14
Training a horse to load into a trailer. (Photo by B. Evans.)

long or it will back out. A little more pressure from the whip usually gets the horse to move all the way in. The butt chain should be secured immediately and the door closed. Praise the horse and allow it to quiet down for a few minutes. Then unload it, making sure it does not bump its head on the top of the trailer. During the first lesson, load and unload the horse at least six times. Repeat the lesson for a couple of days or until the horse walks immediately into the trailer. The first two or three times the horse is hauled should each last only 15 to 30 minutes.

The spoiled or stubborn horse can be very difficult to load, and more drastic procedures may be necessary. One procedure can be to tie two longe lines to the halter, pass them through the ring in the front of the trailer and back alongside the horse (Figure 6-15). Cross the lines behind the horse just above its hocks. One person pulls on each line, and an assistant taps the horse on the croup with a whip. This way, the horse cannot back up. Another method is to place a lip chain on the horse. If it refuses to follow you into the trailer or backs up, give it a jerk on the lead shank. The horse soon learns that it must move forward to get away from the pressure of the lip chain.

Helmets that attach to the halter can be used to protect the top of the horse's head. It may also be advisable to wrap the tail to prevent it from being rubbed out by the butt chain or top of the rear door. The method for wrapping the tail is discussed in the section on grooming.

The front and hindlegs can be protected by shipping boots or by wrapping the legs with track bandages. Be sure to follow correct procedures, or the legs

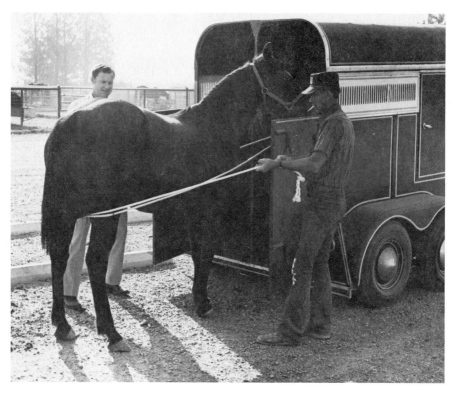

FIGURE 6-15
Loading a stubborn horse into a trailer. (Photo by B. Evans.)

will be injured by lack of blood circulation. First wrap a piece of padding (such as rolled cotton, sheet cotton, or quilts) around the leg. Prepare the track bandage by rolling it up with the ties on the inside of the roll. They will be at the end when unrolled. Place the end of the track bandage about halfway up the cannon and roll it toward the back of the leg. Roll it around the back, toward the front of the leg, and to the back again. Wrap it down the leg so that each wrap overlaps the previous wrap by about 1 inch. Wrap the leg down to where one wrap is taken around the pastern, wrap back up the leg to just below the top of the cotton or quilt, and then back down to the center of the leg. Tie the bandage off on the outside of the leg. It should be tied firmly but not too tightly. If the leg is cushioned by the cotton or quilt and part of the cotton or quilt is left extending beyond the wrap, the circulation should not be cut off.

7
Grooming

Grooming is an essential part of caring for a horse. It is important for the health, hygiene, and appearance of all horses. Daily grooming is the only way to remove dust, dirt, grease, scurf, and dead cells from a horse's skin. Proper grooming stimulates blood circulation and helps maintain muscle tone. Grooming brings out natural oils in the skin and distributes the oil over the hair coat, giving it a glossy sheen. The time spent grooming gives the horse-person and horse time to get to know each other. As the horse becomes accustomed to being groomed, it will relax and become less nervous when handled. While grooming, you can inspect the skin for wounds, cuts, and parasites and inspect the feet for puncture wounds, foot disease, and loose shoes.

BASIC GROOMING

The basic equipment for grooming a horse consists of a hoof pick, rubber curry comb, dandy brush, body brush, mane and tail comb, and a grooming cloth (see Figure 7-1).

Routine grooming usually starts with the feet. Pick the feet up and clean with a hoof pick (see Chapter 8). Remove dirt, manure, and other foreign material

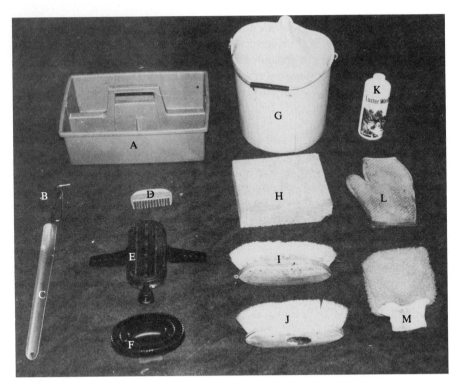

FIGURE 7-1

Basic grooming tools. (A) Storage box. (B) Hoof pick. (C) Sweat scraper. (D) Mane comb.
(E) Scrubber. (F) Rubber curry. (G) Wash bucket. (H) Sponge. (I) Dandy brush.
(J) Body brush. (K) Shampoo. (L) Rubber wash mitt. (M) Rubbing glove for insect repellent.
(Photo by B. Evans.)

from the sole and from the crevices between the frog and base. Clean the feet by working the pick from the heel area toward the toe. If you pull the hoof pick from the toe toward the heel, the hoof pick may become lodged in dirt and if the horse pulls its foot away, the point of the pick may be driven into the sole or frog when the foot strikes the ground. You can remove dirt on the bulbs of the heel with a stiff brush.

Use the rubber curry comb to remove mud and loosen matted hair, dirt, and scurf. Use it in a small circular motion, beginning on the neck and working toward the rear. If you use the rubber curry comb on the head and on the legs below the knees and hocks, do so with caution. It is uncomfortable to the horse when it is used over bony areas. After every few strokes, clean it by tapping it against your heel, the floor, or a post.

The dandy brush has long, stiff bristles and is used to remove deep-seated dirt and scurf. Use it with short, deep strokes in the direction that the hair lies. You can use the dandy brush on the head and below the knees and hocks to remove dirt. However, horses with thin, sensitive skin object to it; you can use a soft body brush or grooming rag on such horses. After brushing the horse with the dandy brush, use the soft body brush to remove surface dirt and dust. This brush is soft enough to be used all over the horse. Brushing in the direction the hairs lie gives the coat a sleek look because it spreads natural oils from the skin.

After brushing, rub the horse with a grooming cloth (usually made of linen, flannel, or terry cloth). Any soft, absorbent cloth is suitable.

When the horse's coat is clean, no white or gray debris appears on your fingers when you run them against the hair.

Then massage the horse with a wisp to help maintain muscle tone. Make a wisp from grass or timothy hay. Lay some hay out in a narrow row on a floor. Twist a handful of hay on one end of the row to form a rope ½ inch thick (13 mm). Twist the hay until a 6- to 7-foot rope is formed (1.83 to 2.13 m). Fold one end of the rope to form two loops. Each loop is about a foot long and is on each side of the long free end. Braid the two loops and the free end. Finally, pass the free end through what remains of the loops. Trim off the bristles, and cut off the excess tail of the free end of the rope. A good wisp lasts 3 to 4 weeks. To wisp the horse, hold the wisp in your right hand and a clean towel in your left hand. Strike the muscles on the neck, shoulder, barrel, hindquarters, and hindlegs. As the wisp hits, the horse's muscle tenses. Then immediately wipe the muscle with the towel in one motion, in the direction of the hair. This wiping causes the muscle to relax. For a good massage, hit each area about 40 to 50 times.

Clean the eyes and nose with damp sponges. A separate sponge should be used to clean the dock.

After the body and legs are cleaned, brush the mane, forelock, and tail with the dandy brush. Some horsepeople prefer to use a mane comb but it tends to break off or pull out the mane and tail hairs. Groom the tail by separating the hairs with the fingers, and then brush each section to remove dirt.

BATHS

Periodically, you may need to give the horse a bath to keep its skin and coat clean. Some horsepeople do not believe horses should be bathed, but many horses are given a bath after each strenuous exercise period. It is true that bathing removes the natural oils from the skin and hair.

Before giving a horse a bath, be sure the horse has enough time to dry before being stabled for the night. The temperature should be at least 50°F (10.6°C) or the horse may catch a cold.

Use a gentle liquid soap or a gentle hair shampoo in warm water. You can hose the horse off to get it wet. Start washing at the neck behind the ears and work toward the horse's tail. Wash its face with a wrung-out sponge, and make sure no water gets in the ears. Always wash under the horse's tail. During hot weather, when horses perspire a lot, a skin irritation can develop under the tail. Clean the mane and tail with a soapy brush, and wash the tail in a bucket of soapy water. Rinse with clear water, making sure that no soap remains. Then rinse with ordinary hair conditioner, which will allow you to remove the tangles easily. Remove excess water from the coat with a sweat scraper to help dry the horse. As the horse dries off, brush it with a soft body brush to keep the hairs lying down. After the horse is dry, brush and rub it vigorously to bring the natural oil up into the coat.

Clean the penis sheath of stallions and geldings periodically to remove dirt, grease, and debris. Part of the debris is smegma, which consists of a fatty secretion and dead cells. The sheath should be cleaned at least once a year. The stallion or gelding must be trained so that you can pull the penis out of its sheath for cleaning. If the horse refuses to allow this, a veterinarian can administer a tranquilizer. Wash the penis and inside of the sheath with a warm, antiseptic soap solution. Make sure that all the folds in the sheath are thoroughly cleaned. The end of the penis around the urethra should be carefully cleaned, because this area tends to accumulate dirt and debris. After the sheath has been thoroughly cleaned, it is important to remove all soap by rinsing with clean clear water. Failure to remove all the soap may result in a skin irritation. If the sheath and penis were very dirty and had a large accumulation of scales, apply a little mineral oil after drying the sheath.

BLANKETING

Blankets are kept on horses for a variety of reasons. Most horses shown at halter or in performance classes are blanketed all the time, to protect their coats. In summer, sun bleaches the coat, leaving it an undesirable color. In winter, blankets keep the horse warm so that it does not grow a heavy coat of hair. Blankets help keep the hair clean. A cooling or summer sheet is placed on some horses such as racehorses while they are being cooled out after exercise, to prevent them from cooling out too fast.

Have blankets fit so that they are cut back behind the withers. This prevents the wither mane from being rubbed off by the edge of the blanket. Also, take care to prevent the tail from being rubbed out by the blanket. If you use a hood to cover the neck and head, it should be lined with linen to prevent rubbing out the mane.

Some horses like to pull their blankets off. You can prevent this by placing a bib on the halter or by placing a cradle on its neck.

TRIMMING

There is a traditional way for preparing each breed of horse for various performance events. The rule book of the American Horse Shows Association describes how horses must be presented for each class. Minor fashionable modifications do occur, and one must be aware of them.

Horses being prepared for sale or to be shown at halter may need a full body clip if they have been at pasture or have been neglected. Clip the horse at least 1 month before being shown, to allow the hairs to even up and the "clipped look" to disappear. Do not clip a horse that you cannot protect from extreme weather.

When preparing a horse to be clipped, allow it to become accustomed to the clippers and the noise they make. Horses are very sensitive about having their ears and heads clipped, so be extremely careful when clipping these areas. Before clipping, give the horse a good bath. Dirt and oil dull the clipper blades. Also, clean, dry hair is fluffy and stands up, allowing a clean, close cut. Cut the hair in the opposite direction of its natural lay. Be careful on the flanks and belly as well as on cowlicks and swirls because the natural lay changes direction. Keep the head of the clippers as flat as possible. A Number 10 blade in an Oster clipper cuts the hair the right length on the body.

There are some problem areas in clipping. On the flank, make one cut from the top of the swirl downward through the center. Then cut the hair toward this center cut. At indentations such as in a muscle or other area, stretch the skin tight to avoid cut marks. Clip the elbow by having someone lift the foreleg, stretch the skin, and move the skin around as you cut. To clip below the knee, flex the foreleg, and take care to clip the grooves around the tendons. Clip the inside of the leg by standing on the opposite side of the horse. Along the mane, hold the mane hairs on the opposite side of the neck and clip downward to get any hair that sticks up. Then clip a line horizontal to the mane. The hairs around the tail head are clipped square across where the tail joins the body by going from one side across to the other.

After the horse has been completely clipped, give it another bath and dry it. Inspect the horse for tracks left by the clippers and remove them. Tracks result from dull blades or blades that get too hot.

Remove the hair from the ears of most performance horses. Clip them with a fine blade. Cut the hair against its natural lay. After removing the inside hair, pass the clippers from the tip of the ear to the base on both inside edges of the ear. Then run the clippers along the edge.

In addition to full body clips, there are various patterns for partial body clips, depending on how the horse is used. Horses can be partially clipped so that they do not get too hot, sweat excessively, and catch colds from taking too long a time to cool out. Working stock horses, trail horses, and hunters may have their bodies clipped and the legs unclipped. The unclipped portion of the legs can be curved downward from front to rear rather than left straight

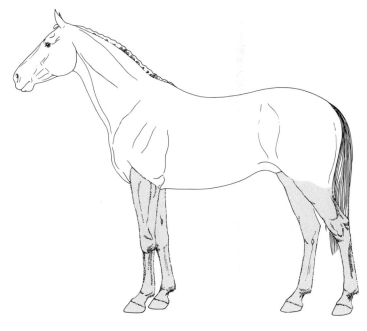

FIGURE 7-2
Body clip for hunter.

across. Leave the legs unclipped to protect them from brush and other obstacles on the trail. For hunters, leave the legs and a saddle patch unclipped (Figure 7-2).

MANES

How long to leave mane hair depends on the type of horse. To roach a mane or cut a bridle path, clip the mane very short. A bridle path is normally about 6 to 8 inches long (152 to 203 mm) starting from the poll but may be only 1½ to 2 inches (38 to 51 mm). For manes that are short rather than full and loose, pull the hair rather than cutting it. To shorten and thin a mane, comb the hair with a mane comb. Grasp the ends of the long hairs with one hand and back comb the remainder of the hairs. Pull the long hairs out. Continue until the mane is thin and has the proper length. If the mane will not lie down or if some of it hangs on both sides, you can train it to lie flat by wetting and braiding it.

The manes of some horses, such as hunters and jumpers, are braided (Figure 7-3). First collect the necessary tools: a pail of water, mane comb, scissors, a large blunt needle (yarn needle), and heavy thread that is the same

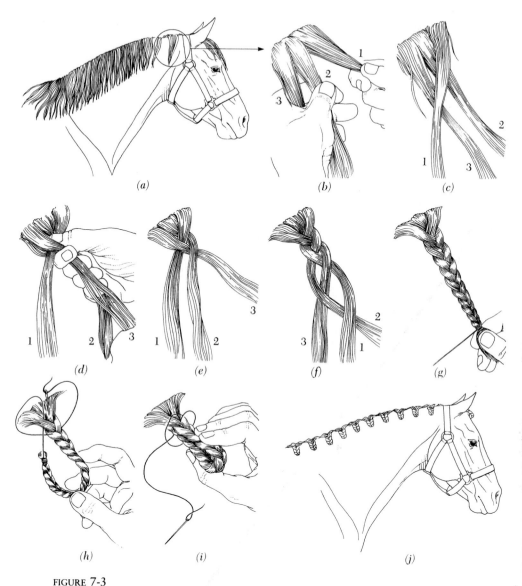

FIGURE 7-3

Braiding the mane. *(a)* Length of mane is divided into braids. *(b, c)* Each length is braided. *(d–h)* Braids are tied and folded under. *(i)* Braided and tied mane. *(j)* Braided mane.

color as the mane. Pull the mane to a length of about 4 inches (102 mm). Divide the length of the neck to allow 15 to 17 braids. Braid a short-necked horse with 17 braids to make the neck appear longer. Braid a long-necked horse with 15 braids to make the neck appear shorter. The braids should be the same size and evenly spaced. Dampen each section with water so that you can hold the hair and make a tighter braid. Starting at the poll, comb each section, divide it into three equal parts, and braid it. To make the braid lie flat against the neck, apply more tension to the hairs underneath than on top. Fold the ends of the braid under and secure with needle and thread. Next bring the braid up almost to the crest of the neck and sew it in place with a few stitches. Cut off the excess thread. The last section to be braided is on the withers where the saddle touches. Brush the remaining hair smooth.

Fine harness horses, ponies, Tennessee Walking Horses, and five-gaited saddle horses display their stable colors as ribbons braided into the forelock and in the upper lock of the mane. Braid the upper lock with a fine braid that is left to fall free down the neck. The rest of the mane is full and loose.

The forelock is prepared different ways depending on the type of horse. You can clip it the same way the mane is roached, leave it loose, or braid it. A colored ribbon can be braided in.

TAILS

The tail may be clipped or long, full, and free flowing. Thoroughbreds, Arabians, and five-gaited horses have full, long tails. Quarter Horse tails are frequently pulled so they are shorter and thinner.

To pull and thin a tail, remove the tangles by combing and brushing. Pull the longer hairs one or two at a time with a quick, jerking motion. Pull the short hairs around the root of the tail to remove the bushy effect. After pulling the short hairs, bandage the tail with damp tail wraps to make the hair lie close to the tail.

Braiding the tail is more difficult than braiding a mane. After removing all the tangles, dampen the tail hair near the croup with water. Then braid the tail to the end of the tail bone (see Figure 7-4). For a neat job, keep each section of hair snug and straight. The braid should be uniform in size. At the end of the tail bone, simply braid out the last strands. Then tuck this end braid under the tail braid and secure it or coil and sew it in place. Sew the initial braid in place with heavy thread and a large, blunt needle.

A "mud tail" is braided to keep the tail clean during rainy weather or to make a hunter with a short tail presentable. Braid the tail as just described, but make one large braid with the hair below the end of the tail bone. Then coil the large braid, if it is not too large, and sew it in position. You can sew it on top of the braid extending down the tail bone. It must be carefully sewn or it will fall down.

After braiding the tail, wrap it in a damp tail wrap to get all the hairs to lie flat (otherwise they get fuzzy). Use extreme caution when wrapping a tail. A tail wrap that is too tight cuts off blood circulation, and the horse will lose its tail. Wrap the tail by starting next to the croup. Each wrap around the tail overlaps about one-half the width of the preceding wrap. Tie off the wrap below the end of the tail bone. Each wrap should be snug enough to hold it

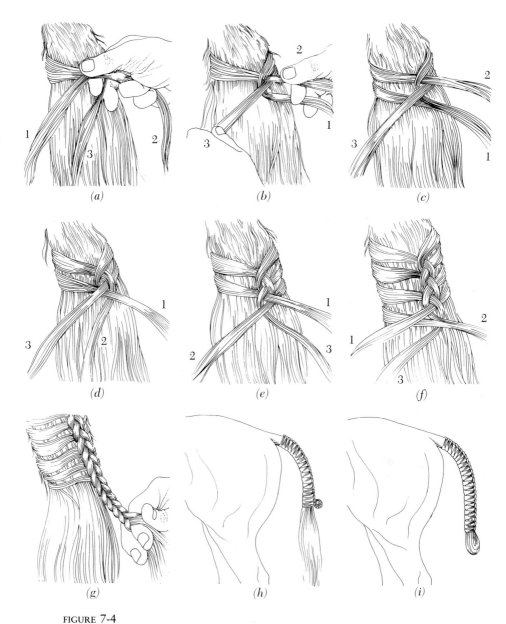

FIGURE 7-4

Braiding the tail. *(a–g)* Tail hairs are braided as shown by diagrams. *(h, i)* Two ways of finishing off the end of the braid.

in place but not tight enough to damage the tail hair roots. To help hold the wrap, a couple of short hairs can be bent over a wrap and then covered with the next wrap. This prevents the tail wrap from sliding downward and becoming loose enough to fall off.

8
Caring for
the Feet

Care for the feet is one of the most important parts of horse care but is often neglected. Neglecting the feet often leads to problems that prevent the horse from working. Diseases such as thrush and scratches develop rapidly in feet that are not cleaned regularly. Hooves that are not trimmed regularly grow too long, and hoof cracks and contracted heels may develop. Other unsoundnesses of feet and legs also can be caused by poor foot care.

HOOF HANDLING

Confined horses should have their feet cleaned and inspected daily. The feet must be raised and held properly, or the horse will be uncomfortable and try to pull away. When you approach the horse to lift one of its feet, make the horse aware of your presence. A calm, deliberate approach gains the horse's confidence and keeps it from becoming frightened or excited. Raise and hold the front feet in position as shown in Figure 8-1. Stand facing the rear of the horse and beside the front leg. Place the hand next to the horse on its shoulder, and move the other hand down the leg so that the horse knows its leg is being handled. Grasping the leg suddenly may startle the horse and make it shy away. Move your hand down to grip the pastern. As you do so, press the shoulder with your other hand so the horse will shift its weight to the opposite leg. When

242

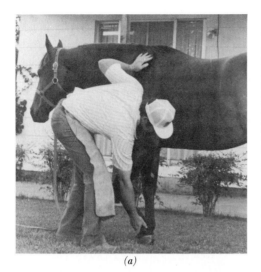

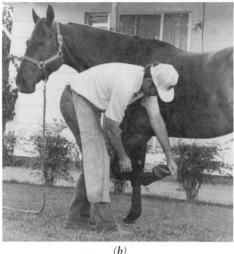

(a)

(b)

(c)

FIGURE 8-1

Lifting and holding forefoot. (a) One hand on the shoulder causes horse to shift weight to opposite leg so the foot can be picked up with other hand. (b) Foot is held up in preparation of restraining it. (c) Proper position for holding foot. The farrier's toes are pointed inward, and the pastern is gripped above the farrier's knees. (Photos by B. Evans.)

the weight is shifted, lift the foot off the ground and place it between your legs just above your knee. To hold the foot securely in a position that is comfortable for both you and the horse, turn your toes in, turn your heels out slightly, flex your knees, bend your back forward. This position forces your knees together to grip the horse's foot with a minimum of effort. While shoeing or trimming the feet, you may need to take a forefoot forward to work on it. This procedure is illustrated in Figure 8-2. First hold the foot up with your hand. Then take one step back toward the horse's head. Turn your body so that you step back under the horse's neck, in line with its front and back legs. Swing the leg and rest the foot between your knees. To work on the opposite side of the foot, turn initially to face the horse, swing the horse's leg forward, and grip it. Instead

(a) *(b)*

(c) *(d)*

FIGURE 8-2

Taking forefoot forward. *(a)* Holding foot by the toe—called the "home position." *(b)* The farrier turns toward the horse and changes hands, holding the foot while taking the foot forward. *(c)* Foot held to work on its medial side. *(d)* Foot held to work on its lateral side. (Photos by B. Evans.)

FIGURE 8-3
Lifting a hindfoot. *(a)* One hand is placed on the hip to shift horse's weight to the opposite leg while other hand grips the pastern and lifts the foot. *(b)* Hindfoot is brought forward. *(c)* The farrier steps under the hindleg. *(d)* Correct position for holding the hindfoot. (Photos by B. Evans.)

of resting the foot on your legs, you can use a foot stand (Figure 8-2). Be careful when using a foot stand to prevent the foot from sliding off and striking your own foot. The foot stand helps prevent excessive back strain while working on the foot. Raise and position a hindfoot as shown in Figure 8-3. While standing and facing toward the rear of the horse and standing forward of the hindleg,

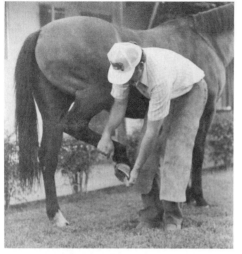

(a)

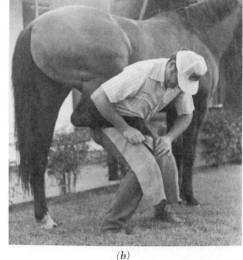

(b)

(c)

FIGURE 8-4

Taking a hindfoot forward. *(a)* From the "home position," the toe is held with one hand. *(b)* The farrier turns in between the horse's body and leg to work on lateral side of foot. He supports the foot on inside of the leg just above the knee. *(c)* To work on medial side, the farrier turns facing the horse and positions the foot. (Photos by B. Evans.)

place the hand next to the horse on its hip, and move the other hand down the hindleg to the fetlock. While gripping the back of the fetlock, use the other hand to push against the horse's hip. As the horse shifts its weight to the other leg, pull the hindleg forward. As the leg is raised up, step forward and under the raised foot. The hand you used to push on the hip now grasps the toe of the foot so that you can correctly position the foot on your outside knee. Hold the foot low to the ground. The hind cannon should be about perpendicular to the ground. To take the hindfoot forward, follow the procedure illustrated in Figure 8-4. By taking the hindleg forward, you can rasp the sides of the hoof wall. Grip the toe, and step backward, and turn so that your back is next to the horse's abdomen. Bring the foot forward and rest it between your knees. To work on the inside part of the hoof wall, step back and turn to face the horse's abdomen. Then place the foot between your knees.

Horses must be trained to have their feet handled. Young foals are trained by getting them accustomed to having their legs groomed. When they no longer pay attention to their legs being brushed, lift each foot off the ground and hold it for a couple of minutes. Someone may have to hold the foal's head while someone else lifts the foot. It is often easier to lift the feet if you back the foal into a corner. After you can lift the feet and hold them in position without the foal struggling and the foal feels comfortable, lightly tap the feet to accustom the foal to the shoeing hammer.

Older horses may develop bad habits that hinder holding and working with their feet. Restless or nervous horses may become excited if they are not handled very often and are taken out of familiar surroundings. Allowing the horse to "settle down" before handling its feet usually makes the horse more cooperative. It may be necessary to let its companion horse stand nearby. If a horse tries to lean on the person holding its foot, it should be disciplined by quickly dropping the foot so the horse must catch its balance or by letting it lean against a sharp object.

Some horses will not allow their feet to be raised or held. There are several methods to restrain their feet (Figures 6-4 and 6-5). A nose twitch can be applied to the horse that continually takes its feet away from a handler (Figure 6-8).

Horses that bite should be held by someone or tied with a short lead shank. If a horse tries to kick or strike the handler, someone should always hold the horse's head. If the horse continues to strike deliberately or kick, it should be disciplined with a whip.

DAILY CLEANING AND INSPECTION

The feet should be cleaned every day. Use a hoof pick (Figure 7-1) to clean the dirt and debris from the bottom of the foot. Run the pick from the heels toward the toe to avoid puncturing or bruising the sole or frog if the horse pulls its foot away and slams it to the ground. Clean the frog area thoroughly, particularly between the frog and bars. Failure to keep the foot clean can lead to thrush, a disease caused by a bacterium that lives in the absence of oxygen. The organism is usually present in manure. When the manure becomes packed in the sole area, the proper conditions are present for the bacteria to grow, multiply, and attack the foot. The signs of thrush are a very foul odor and usually a blackish discharge. If thrush is present, clean the foot with soap and water, using a toothbrush or other similar brush. After the foot is thoroughly dry, apply a germicidal preparation such as Koppertox or a 7 percent iodine solution. Thoroughly clean and treat the foot daily until the disease is cured. It may be necessary to trim the affected part of the frog to remove the diseased tissue.

After cleaning the foot, you should inspect it for puncture wounds, bruises, loose shoes (if shod), abnormal growth, uneven wear, and abnormal hoof conditions. Hoof moisture must be maintained or the hoof wall will become dry and brittle and will start to crack. Moisture is continually being lost from the hoof wall by evaporation. This moisture is replaced by blood and lymph from the coriums of the foot and by environmental water conducted up the hoof wall by osmotic and capillary action. Moisture balance is maintained by the waxy hoof varnish *(stratum tectorium)* on the outside of the hoof wall, which reduces evaporation loss. Usually this varnish wears off of the lower part of the hoof wall. The tubules that form the hoof wall are very dense and flat at the surface to retard moisture loss. Conditions that lead to dry hooves are stabling in dry, dusty, and sandy paddocks; dry bedding; careless rasping of the hoof wall; and application of most commercial hoof dressings. The hoof needs water for moisture, and the best way to help water enter the foot is to stand the horse in a pool of water or clean mud for 20 minutes daily for several days or to pack the feet in wet, clean clay. To stimulate increased blood flow and hence moisture, massage the coronary band for a few minutes each day. Exercise also increases the blood flow to the foot. After the proper amount of moisture is present in the hoof, applying an oil base hoof dressing helps prevent moisture loss.

Abnormal hoof growth can be caused by founder (excessive growth) or by injury to the coronary band (defects in hoof wall). The feet of a foundered horse need special care (see Chapter 2). Defects in the hoof wall as it grows can lead to unsoundness, particularly if the wall splits and the sensitive structures are irritated. Factors that affect hoof growth are age, season, irritation or injury of sensitive structures, front versus hind, neurectomy, and nutrition. As horses get older, the rate of growth decreases. Hoof growth is approximately 15.5 mm (0.6 inch) per month for nursing foals, 11.5 mm (0.5 inch) per month for weanlings, and 8.5 mm (0.34 inch) per month for mature horses. Growth is most rapid during spring and is lowest during summer and winter. Injury to the sensitive structure of the foot increases the growth rate of the hoof wall, as does neurectomy. The hindfoot grows about 12 percent faster than the front foot for young foals, but the difference decreases with age. The level of feed intake is the only nutritional factor demonstrated to significantly affect rate of hoof growth. Adequate intake of a balanced ration encourages maximum hoof growth. Feed supplements are merely a waste of money if a balanced ration is being fed.

Bruises on the sole are indicated by a red or yellowish-red discoloration that is sensitive to tapping or pressure. Severe bruises may need the attention of a veterinarian.

TRIMMING

Most horses that are shod need their feet trimmed every 6 to 8 weeks. Some horses that are kept unshod wear off the hoof wall at a faster rate and may only need their feet trimmed at longer intervals of time.

Failure to properly and regularly trim the feet causes foot problems. The toe grows at a faster rate than the heel, so the toe tends to curl upward as the foot gets too long. This causes the heels to contract. The hoof wall starts to split and crack, and the angle of the pastern is changed, which affects the foot flight arc. A long toe inhibits the foot from breaking over at the toe and may cause forging, overreaching, and scalping. Long toes may cause other limb interferences and faulty gaits.

The tools needed to trim a horse's foot are illustrated in Figures 8-11 and 8-12. The rasp is 14 inches long (0.36 m) and has two sides. One side has larger teeth and is used for rasping down the hoof wall. Use the side with small teeth for the final dressing. Some rasps have regular file teeth on one side to file sharp edges off of horseshoes. Use hoof knives to trim out the dead sole and to trim the frog, and use hoof nippers to cut the hoof wall.

To trim the foot, use the following procedure. First watch the horse move and look for limb interference and faults in way of going. Observe the overall conformation to see if the horse is predisposed to a faulty way of going. Observe the position of the feet when the horse is standing squarely on them.

Examine the feet to determine conformation and balance of the foot. Determine foot balance by sighting across the foot from heel to toe. Both sides of the hoof wall should be equal in length (Figure 8-5). Faults in conformation and in the way of going cause uneven wear on the sides of the hoof wall. Horses that stand toe-wide and wing-inward when traveling wear down the inside wall faster. Toe-narrow and wing-outward horses wear down the outside of the hoof wall faster. If not corrected when the hoof is trimmed, the uneven wear will accentuate the fault in conformation and/or way of going.

If the horse is shod, remove the old shoes (see section on shoeing later in this chapter). Remove the dead sole from the sole area with the hoof knife. Only the dead sole should be removed; stop when the sole is thin enough to

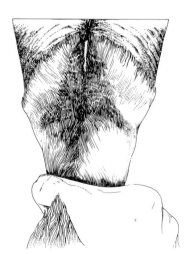

FIGURE 8-5

Foot balance. The opposite heels and opposite quarters should be equal in length. The hoof wall should be flat all the way around.

be flexed by pressure applied with the thumbs. The remaining sole is ¼ to ⅜ inch thick (6.3 to 9.5 mm). Removing too much sole leaves insufficient protection for the underlying sensitive structures. If the sole is extremely hard, use a sole knife to chisel it out. Trim the bars level with the sole to prevent them from being broken or torn, and trim the frog to remove ragged edges and loose pieces. The frog should be trimmed so that it almost touches the ground when the horse stands on a hard surface. Removing too much frog interferes with normal functioning. The frog is sloughed off and replaced about twice a year. Frogs of neglected feet can often be pulled off, because they are in the process of sloughing. The frogs of feet that are well cared for may not slough because they are continually being trimmed. Trim the hoof wall with hoof nippers. Make the first cut at the toe. Be sure to cut the hoof wall flat, not on an angle. Keep the line bisecting the angle of the two handles perpendicular to the hoof wall surface. Continue the cut around the hoof wall to both heels. After trimming the hoof with nippers, check the balance of the hoof wall—both sides should be of equal length. Rasp the hoof wall level with the coarse side of the rasp. Check the balance again. If it is balanced, use the fine side of the rasp to smooth the surface and remove the sharp edge around the hoof wall.

If the foot is now properly shaped and the angle of the hoof wall is proper (50 to 55 degrees), nail the shoe on. The angle of the forefeet should be the same as the angle of the shoulder, and the angle of the hindfeet should be 2 to 3 degrees greater. If the foot is to remain unshod, round off the outer edge of the hoof wall (about one-half the thickness of the wall) to prevent pieces of the wall from chipping off.

Some feet are not shaped normally and must be shaped with the rasp. Any outward distortions ("flares") are removed. The rasped area must be blended into the slope of the undistorted upper part of the wall (Figure 8-6).

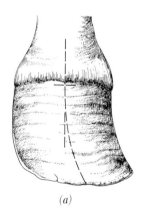
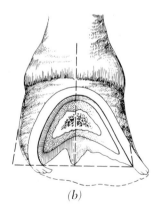
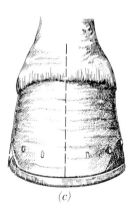

(a)   (b)   (c)

FIGURE 8-6

Principles of shaping the foot. (a) Rough untrimmed foot. (b) Diagram of trimming.
(c) Trimmed and shod. (From Butler, 1974.)

ANATOMY AND PHYSIOLOGY

Anatomy

The external structures of the foot are shown in Figure 8-7. The shape of the hindfoot is slightly different from that of the forefoot. The toe of the hindfoot is more pointed, and the sole is more concave. The internal structures are illustrated in Figures 8-8, 8-9, and 8-10.

Hoof Wall and Laminae

The bulk of the hoof wall is composed of keratinized epithelial cells. These cells are arranged in tubules that run from the coronary band to the ground surface. These tubules are formed from papillae that project from the coronary band (coronary corium). A tubule is formed around each of the papillae. The centers of the tubules are filled by a cementlike substance of loosely packed cells called *intertubular horn*, which holds the tubules together. Tubules toward the outer surface of the hoof wall are flattened and packed closely together, which helps retain hoof moisture. The cells that form the intertubular horn, located next to the outside surface of the hoof wall, contain pigment. The same cells adjacent to the white line do not contain pigment.

The outer surface of the hoof wall is covered by the periople and the stratum tectorium (a varnishlike substance). The periople extends below the coronary band for approximately ¾ inch except at the heels, where it covers

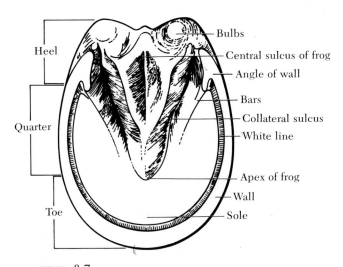

FIGURE 8-7

External structures of normal forefoot. (From Adams, 1974.)

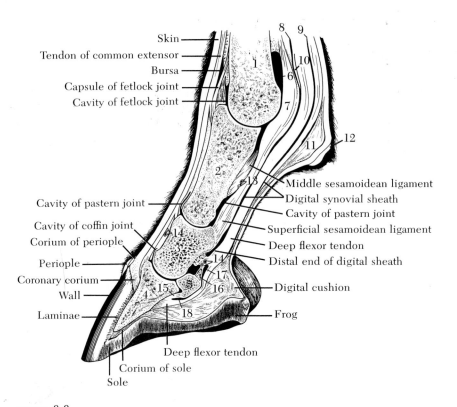

FIGURE 8-8

Sagittal section of digit and distal part of metacarpus of horse. (1) Metacarpal bone. (2) First phalanx. (3) Second phalanx. (4) Third phalanx. (5) Distal sesamoid bone. (6) Volar pouch of capsule of fetlock joint. (7) Intersesamoidean ligament. (8–9) Proximal end of digital synovial sheath. (10) Ring formed by superficial flexor tendon. (11) Fibrous tissue underlying ergot. (12) Ergot. (13, 14) Branches of digital vessels. (15) Distal ligament of distal sesamoid bone. (16) Suspensory ligament of distal sesamoid bone. (17) Proximal end of navicular bursa. (18) Distal end of navicular bursa. The superficial flexor tendon, behind (9), is not shown. (From Getty, 1953.)

the heels and blends in with the frog. The periople protects the junction of the hoof wall with the coronary band. The underside of the periople flakes off, and this material forms the stratum tectorium, which therefore extends from the periople to the bottom of the hoof wall. Friction from dirt, grass, and so on usually wears off this thin layer of cells so that the stratum tectorium is usually not present on the lower part of the hoof wall.

The insensitive laminar layer forms the inner surface of the hoof wall. This layer intermeshes with the sensitive laminae that cover the surface of the third phalanx. The sensitive and insensitive laminae are composed of primary leaves. There are about 600 primary leaves in each hoof. Each primary leaf has about 100 secondary leaves that run parallel to the primary leaves. The sensitive and insensitive laminae intermesh or interlock to hold the hoof wall

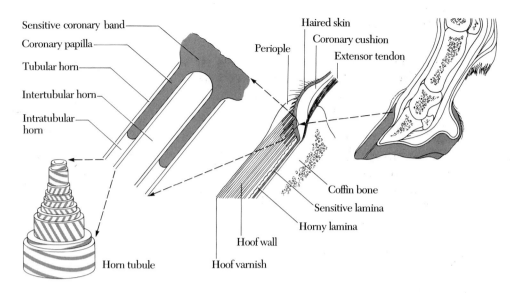

FIGURE 8-9

Magnified diagrammatic section of papillae of the coronary band and horn tubules of the hoof wall. (From Butler, 1974.)

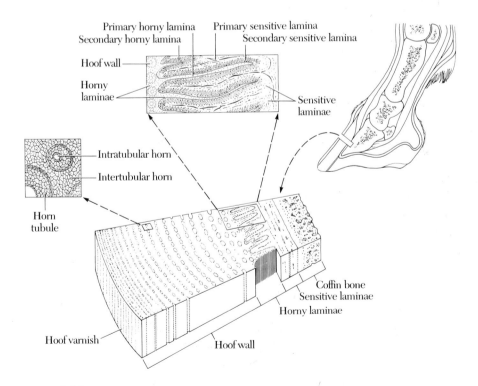

FIGURE 8-10

Diagrammatic representation of the cross section of the interlocking of the sensitive and horny laminae, showing the relative size of the horn tubules of the hoof wall. (From Butler, 1974.)

to the third phalanx. The third phalanx is actually suspended from the inside of the hoof wall. Therefore, the weight of the horse is not exerted on the sole surface but is borne by the hoof wall. The laminar arrangement allows the weight to be distributed over a large surface area.

The white line of the foot is the area where the laminae intermesh. It is composed of sensitive and insensitive laminar leaves and the interlaminar horn produced by papillae at the base of the leaves of the sensitive laminae. This combination of structure gives it a yellowish color. It should not be confused with the white (unpigmented) area of hoof wall tubules adjacent to the white line.

The hoof wall is thickest at the toe and becomes thin at the quarters. At the angle of the wall, the wall is reflected forward to form the bars of the foot. The bars serve as a brace structure to prevent overexpansion of the hoof wall.

Sole

The sensitive sole (sole corium) covers the bottom of the third phalanx. Papillae similar to those of the coronary band form the horn tubules that compose the sole. The sole has self-limiting growth in that flashes of it slough off after it gets about as thick as the hoof wall. The sole varies in thickness from about ⅜ inch (9.5 mm) adjacent to the hoof wall at the toe to about ¼ inch (6.3 mm) at the heel area. The sole is concave at the ground surface. This shape, in conjunction with the continual scaling or sloughing, prevents the sole from bearing weight. The sole is easily bruised, and bruises occur frequently when the sole bears weight.

Frog

The frog is the elastic, wedge-shaped mass that occupies the area between the bars. The bottom surface is marked by a depression called the *central sulcus* or *cleft*. Internally, the frog stay is formed as a result of the wedge shape. The point of the frog toward the toe is referred to as the *apex*.

The horny frog is produced by the papillae that project from the surface of the sensitive frog (frog corium). Horn tubules produced by the sensitive frog are not completely keratinized, so they remain more elastic. Elasticity is enhanced by greasy secretions from the fat glands that pass from the digital cushion into the frog. The moisture content is about 50 percent, which also helps to maintain a pliable condition.

Coriums

Five coriums of modified vascular tissue furnish nutrition to the hoof. The perioplic corium, within the perioplic groove in the coronary band, serves the periople. The coronary corium, responsible for the growth and nutrition of the bulk of the hoof wall, lies within the coronary band. Because the coronary band is responsible for wall growth, injury to the coronary band is quite serious and usually leads to a defect in wall growth and structure. The laminar corium is attached to the top of the third phalanx and bears the sensitive laminae. Consequently, it transports blood and nutrition to the sensitive and insensitive laminae as well as to the white line. The sole corium, on the lower surface of the third phalanx, nourishes the sole. The frog is nourished by the frog corium.

Digital Cushion

The back half of the foot contains the digital or plantar cushion. This fibro-elastic, fatty cushion acts as a shock absorber for the foot.

Bones

Three bones are located in the horse's foot: the second phalanx (or short pastern), the third phalanx (or coffin bone), and the navicular bone. The third phalanx is located mainly to the front of the hoof and slightly to the outer side of the hoof. It is the largest bone in the foot, is quite porous, and resembles a miniature foot in shape. The second phalanx is partly in the hoof and partly above it. The navicular bone is the smallest bone of the foot and increases the articulatory surface and movement of the distal interphalangeal (foot or coffin) joint.

Lateral Cartilages

The lateral cartilages are attached to the wings of the third phalanx (Figure 8-8). They rise above the coronary band and extend approximately one-third of the way around each side of the hoof from the heel.

Physiology

The structures of the foot work together to absorb concussion when the foot strikes the ground. As the foot strikes the ground, the heels are expanded by frog action. The frog is pushed upward but primarily flattens out. Because of its wedge shape, the frog stay forces the digital cushion upward and outward on both sides of the foot. Movement of the digital cushion presses on the lateral cartilages, which in turn move outward. The lateral cartilages compress the blood veins draining the foot. Compression of the blood veins forces blood toward the heart and also causes blood to pool in the foot. The pooled blood then acts as a hydraulic cushion that absorbs concussion. To aid further in concussion absorption, the third phalanx descends slightly and the sole yields slightly. The laminae also help absorb shock. In addition, the navicular bone helps absorb concussion as a result of its placement in the coffin joint. As the weight is transferred from the second phalanx to the navicular bone, the navicular bone yields slightly before the weight is transferred to the third phalanx. The deep flexor tendon supports the navicular bone. The remainder of the concussion is absorbed at the pasterns, knee, and shoulder.

SHOEING

Horses have been shod for centuries for a variety of reasons but usually to protect the hoof wall from excessive wear and from damage, to increase traction, and to correct gaits or prevent limb interference. Shoeing must be done properly if the foot is to function normally. Therefore, it is important to understand the anatomy and physiology of the foot before one learns how to shoe.

Equipment

The basic horseshoeing tools are shown in Figure 8-11. In addition to these tools, the other tools used by farriers are shown in Figure 8-12.

A leather apron is worn for protection against nail cuts and against burns if a forge is used. For safety reasons, strings are used on the apron, tied so they can be quickly released if a horse hooks a nail in the apron and starts kicking. The strings should be long enough to be crossed behind and in front and then tucked under on the sides. Use the hoof or farrier's knife to remove dead sole scales, trim the frog, and cut out corns. Such knives may be purchased with a wide or a narrow blade, with one or two sharp edges, and either left- or right-handed. Initially, learn to use a knife with a single-edged wide blade. If the sole is too hard to cut out with the hoof knife, use a sole knife (sole chisel) to chisel out the sole. To keep the edges of the hoof and sole knives sharp, first clean the sole of all rocks and dirt with a hoof pick. Sharpen knife edges with a tapered, flat oilstone.

Use the blade end of a clinch cutter or buffer to cut or straighten out clinches (bent-over nails) before removing the shoe. After the shoe is removed and the nails cut off, use the pritchel end to take seated nails out of the shoe.

Two types of hammers should be kept as part of the basic equipment: the rounding hammer for shaping shoes on the anvil to fit the horse's feet, and the driving hammer for driving nails and forming and finishing clinches. Use the claw end of the driving hammer to wring (twist) off the nails. Driving

FIGURE 8-11

Basic tools for shoeing. (A) Apron. (B) Hoof pick. (C) Hoof knife. (D) Nippers.
(E) "Pull-offs." (F) Rasp. (G) Driving hammer. (H) Clinch block. (I) Clinch cutter.
(J) Tool box. (K) Rounding hammer. (L) Anvil. (Photos by B. Evans.)

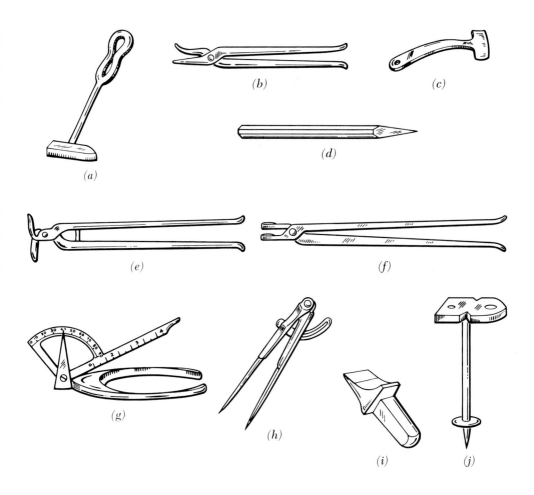

(a) (b) (c) (d) (e) (f) (g) (h) (i) (j)

hammers can be bought in different weights such as 10, 14, or 16 ounces. The 14-ounce hammer is the most commonly used.

Cutting nippers have two sharp edges and are used for removing excess hoof wall. Some farriers prefer a hoof parer, which is similar to the nippers but has a blunt edge and a sharp edge. Farrier's pincers, commonly called the "pullers," are similar in shape to cutting nippers but are used to pull shoes or nails from the foot and may be used to clinch the nails. Usually, a clinch block or an alligator clincher is used to set the clinches. If the pincers are used as a clinch block, they are closed, and the backside of the jaw is held against the nail.

If a forge is not available to shape shoes, use a shoe spreader to spread the heels of the shoe. Use a rasp fitted with a wooden handle to level the bearing surface of the foot and finish the clinches. Use the rough side for rasping the hoof, and the smoother side for final leveling and for finishing the

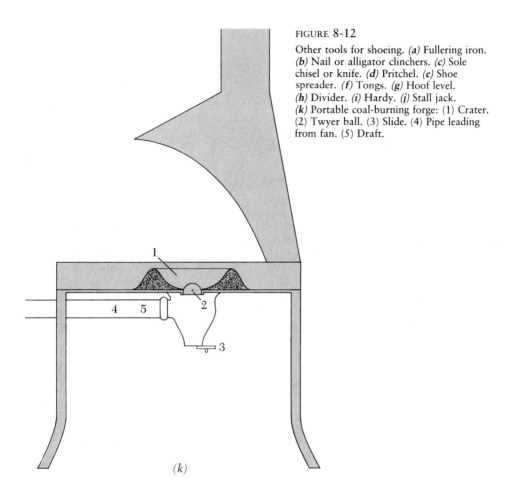

FIGURE 8-12

FIGURE 8-12

Other tools for shoeing. *(a)* Fullering iron. *(b)* Nail or alligator clinchers. *(c)* Sole chisel or knife. *(d)* Pritchel. *(e)* Shoe spreader. *(f)* Tongs. *(g)* Hoof level. *(h)* Divider. *(i)* Hardy. *(j)* Stall jack. *(k)* Portable coal-burning forge: (1) Crater. (2) Twyer ball. (3) Slide. (4) Pipe leading from fan. (5) Draft.

(k)

clinches. In the final leveling process, use a hoof leveler to determine the angle of the wall and to determine if the foot is level. The length of the wall is usually checked with hoof calipers or a divider to make sure both feet are the same and are of the proper length.

Keep the tools just described in a shoeing box so they are handy for work on a horse's feet. The anvil is the farrier's workbench and is used in many ways to prepare the shoe properly. Normally, the anvil weighs 80 to 125 pounds (36.3 to 56.7 kg). The farrier's anvil has a thinner heel and a longer, more tapered horn than the blacksmith's anvil. The horn may have a clipping block to draw out clips. The face or flat area has a hole for the hardy and one or two pritchel holes. The hardy is used for cutting hot metals, such as the heels of the shoe. It is also used to cut blank bars to the proper length to be made into shoes. A pritchel is used to expand the nail holes in a shoe or to remove nails that are broken or cut off even with the web of the shoe. The stall jack is a handy tool used by farriers to shape aluminum plates when shoeing racehorses so that the foot does not have to be put down during the shoeing process.

Most competent farriers are skilled in the use of a forge to make and shape shoes. The coal-burning forge is the most common, and portable ones can be carried in a pickup truck. Some farriers use gas forges. The gas forge keeps the temperature constant and will not burn up a shoe that is left in the forge too long. Coal-burning forges require the use of a fire shovel to add coal and a fire rake to remove clinkers (melting metal particles that bind burned coal into larger masses). Farrier's tongs are used to handle the shoes in the forge and while shaping hot shoes. A fullering iron is necessary to crease a hand-forged shoe or to repair damaged creases during shaping.

In addition to the shoeing tools, some other equipment may be needed. The farrier must carry a halter and lead rope to use on horses when the owner is not present. In some areas where horses are kept together in large numbers and are not handled regularly, a catch rope is necessary. A knee strap 2 feet long (61 cm) is handy to hold up a forefoot of a problem horse. Often, a horse will stand if a twitch is applied to the muzzle. A couple of long, soft ropes may be used to tie up a hindfoot or tie down a vicious horse. If it is necessary to tie up a hindfoot, a 12-inch (30.5-cm) strap with dee-rings in each end will prevent a rope burn around the pastern (see Figure 6-5). Some horses kick, thus requiring the use of a hoof hook to pick up a hindfoot.

If a person shoes horses all day, a 14-inch-high (35.6-cm) foot stand can prevent excessive back strain. However, such stands are not safe to use around nervous horses and teach young horses to lean on the farrier. A farrier who can properly handle a horse's foot seldom uses a foot stand.

A 5-inch (12.7-cm) vise is useful, particularly when filing the heels or turning calks on a shoe. A crease nail puller is used to grasp a nail head seated in the crease and enables the farrier to pull out the nail. Nail cutters are used to cut off the ends of the nail to the same length before they are clinched. This is important if the horse is to be shown.

Shoes

The shoe selected depends primarily on the intended use for the horse. Other factors influencing the selection are types of terrain, need for traction, way of going, and conformation. A typical shoe and its parts are shown in Figure 8-13. Shoes may be custom-made by a farrier but most shoes are keg (machine-made) shoes (Figure 8-14). A variety of kinds and sizes of keg shoes are used. Machine-made "hot shoes" are made with long branches so they can be cut off at the proper length or so that heel calks can be made. A forge is required to shape them. Cowboy shoes are designed like machine-made "hot shoes" but are fitted without a forge. Because they are difficult to shape, they are not used very often. This type of cowboy shoe is not to be confused with the keg shoe,

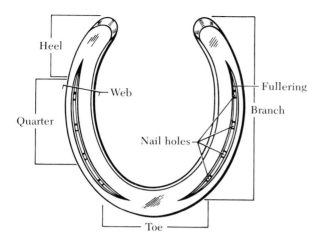

FIGURE 8-13

Ground surface view of a horseshoe.

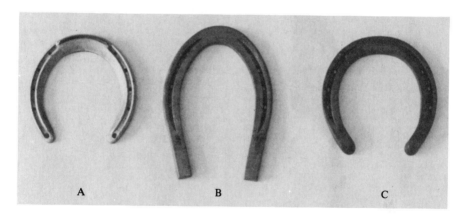

FIGURE 8-14

Basic types of shoes. (A) Aluminum racing plate. (B) Shoe for hot shoeing.
(C) Cowboy shoe for cold shoeing. (Photo by B. Evans.)

which is the same as a cowboy, bronco, or saddle horse shoe. Ready-made shoes are popular because they are shaped and ready to be used without much alteration for a good fit. They can be fitted without a forge. Racehorses are shod with aluminum or steel plates. Barrel racing and polo horses are shod with polo shoes. Calks and toe grabs can be added to the shoes to increase traction and help prevent limb interference. Shoes are made in front or hind patterns and shapes. Mules and ponies require different patterns and shapes.

The size of the shoe needed for a foot is determined by the size of the hoof, the position of the nail holes, and the length of the shoe heels. The last nail hole on a front shoe should not be behind the widest part of the foot. Nails

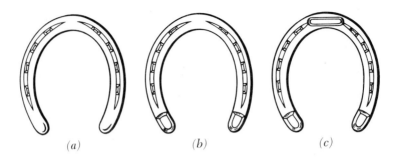

FIGURE 8-15

Keg or cowboy shoes. (*a*) Plain. (*b*) Heeled, with heel calks.
(*c*) Heeled and toed, with heel and toe calks.

placed too far back inhibit expansion of the heels. In the hindfoot, the last nail may be a little further back. The branches must be long enough to support the entire hoof wall. If they are too long on the forefeet, a hindfoot may overreach and pull the front shoe. If too short, the heels of the horse will grow out beyond the heels of the shoe, and the heels of the shoe will rest on the sole, producing corns.

The most commonly used shoes are keg shoes. Keg shoes are presized and often fitted "cold." There are several types of keg shoes. The saddle horse, western, or cowboy shoe (Figure 8-15) is used on most pleasure horses and on working cow horses. The cowboy shoe is available in several sizes (Table 8-1) to cover the range of light horse hoof shapes. Standard Diamond brand shoe sizes range from the No. 000, which is $4^7/_{16}$ inches long (11.3 cm) and $3^{11}/_{16}$ inches wide (9.4 cm), to the No. 3, which is $6^3/_{16}$ inches long (15.7) cm and $5^9/_{16}$ inches wide (14.1 cm). The No. 1 shoe fits the average horse. Another size system is used by the Multi-Products Company. In this system, the No. 3 is equivalent to the standard No. 00, the No. 6 is equivalent to the standard No. 1, and the No. 9 is equivalent to the standard No. 3. The No. 4 is slightly smaller than the standard No. 0, and the No. 5 is slightly larger than the standard No. 0. The No. 7 is slightly smaller, and the No. 8 is slightly larger than the standard No. 2. These so-called half-sizes almost completely eliminate the necessity to shape standard shoes to obtain a good fit. Pony shoes come in two sizes (Table 8-2). A No. 2 pony size is equivalent to the standard No. 000.

Cowboy shoes are made in several weights for each size (Table 8-1). They are classified as light, extra-light, and extra-extra-light. The heavier the shoe, the less speed and agility and the more leg fatigue. Therefore, most pleasure horses wear extra-light or extra-extra-light shoes. Keg shoes that are fitted "cold" are extra-light; the extra-extra-light are fitted "hot." The heels of the extra-extra-light keg shoes must be turned inward.

Light and extra-light cowboy shoes are also classified as plain, heeled, or toed and heeled (Figure 8-15). The heeled shoes have heel calks, and the heel

TABLE 8-1
Specifications for Diamond Brand Cowboy Shoes

Size No.	Length (inches)	Width (inches)	Approx. Weight (ounces)	Approx. No. in 50-lb Carton
Light, Plain				
000	$4^{7}/_{16}$	$3^{11}/_{16}$	$7\frac{1}{2}$	107
00	$4\frac{3}{4}$	$4^{9}/_{32}$	9	90
0	$4^{31}/_{32}$	$4\frac{5}{8}$	$10\frac{3}{4}$	78
1	$5^{5}/_{16}$	$4^{13}/_{16}$	12	66
2	$5^{21}/_{32}$	$5\frac{1}{8}$	$14\frac{1}{4}$	53
3	$6^{3}/_{16}$	$5^{9}/_{16}$	$18\frac{1}{4}$	44
Light, Heeled Only				
000	$4^{7}/_{16}$	$3^{11}/_{16}$	$7\frac{3}{4}$	103
00	$4\frac{3}{4}$	$4^{9}/_{32}$	$9\frac{3}{4}$	82
0	$4^{31}/_{32}$	$4\frac{5}{8}$	11	73
1	$5^{5}/_{16}$	$4^{13}/_{16}$	$12\frac{1}{2}$	62
2	$5^{21}/_{32}$	$5\frac{1}{8}$	15	53
3	$6^{3}/_{16}$	$5^{9}/_{16}$	19	42
4	$6^{21}/_{32}$	6	$22\frac{1}{2}$	35
Light, Toed and Heeled				
00	$4\frac{3}{4}$	$4^{9}/_{32}$	$9\frac{3}{4}$	79
0	$4^{31}/_{32}$	$4\frac{5}{8}$	$11\frac{1}{2}$	71
1	$5^{5}/_{16}$	$4^{13}/_{16}$	$13\frac{3}{4}$	61
2	$5^{21}/_{32}$	$5\frac{1}{8}$	16	49
3	$6^{3}/_{16}$	$5^{9}/_{16}$	$19\frac{1}{2}$	42
4	$6^{21}/_{32}$	6	$22\frac{3}{4}$	34
Extra-Light, Plain or Heeled				
S00	$4\frac{1}{2}$	$4\frac{1}{8}$	$6\frac{1}{4}$	128
S0	$4\frac{3}{4}$	$4^{9}/_{32}$	7	114
S1	5	$4^{21}/_{32}$	$8\frac{1}{4}$	97
S2	$5\frac{3}{8}$	$4^{13}/_{16}$	9	89
Extra-Extra-Light (Hot Shoe)				
0	$5^{9}/_{16}$	$4^{9}/_{32}$	8	100
1	$5^{13}/_{16}$	$4\frac{5}{8}$	9	82
2	$6^{1}/_{16}$	$4^{25}/_{32}$	11	79

TABLE 8-2
Specifications for Diamond Brand Pony and Mule Shoes

Size No.	Length (inches)	Width (inches)	Approx. Weight (ounces)	Approx. No. in 50-lb Carton
Pony Shoe				
0	$3\frac{3}{4}$	$3^{5}/_{16}$	$5\frac{3}{4}$	139
1	$4^{1}/_{16}$	$3^{7}/_{16}$	$6\frac{1}{4}$	128
Plain Mule Shoe				
2	$5\frac{1}{8}$	4	$10\frac{1}{2}$	75
3	$5\frac{1}{2}$	$4\frac{1}{4}$	$12\frac{3}{4}$	62
4	6	$4\frac{1}{2}$	$15\frac{3}{4}$	49
Heeled Mule Shoe				
2	$5\frac{1}{8}$	4	$11\frac{1}{4}$	70
3	$5\frac{1}{2}$	$4\frac{1}{4}$	$13\frac{1}{2}$	58
4	6	$4\frac{1}{2}$	$16\frac{1}{2}$	46

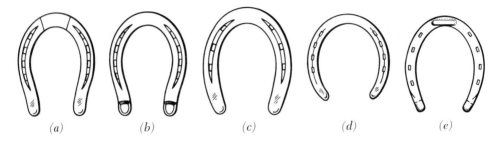

FIGURE 8-16

Shapes of pony and mule shoes (Diamond Brand). *(a)* Plain mule shoes. *(b)* Heeled mule shoes. *(c)* Plain pony shoes. *(d)* Plain. *(e)* Block heels toed.

and toed shoes have heel and toe calks. Calks increase traction on slick, soft, or rough terrain. Extra-extra-light shoes' heels are long enough so that heel calks can be formed without cutting off the heels. To form the calks, heat the shoe with a forge and bend the end of the heel back against the heel. If toe calks, or "toe-grabs," are required, they must be welded to the toe.

Mule shoes are a variation of keg shoes and are similar to cowboy shoes. However, their sizes (Table 8-2) are slightly different, and they are shaped to fit a mule's feet (Figure 8-16).

Plates are a type of keg shoe that is used on racehorses. Steel plates are very seldom used on running horses because aluminum is lighter. The sizes and weights of steel plates are given in Table 8-3. They are available in two sizes—light or heavy—and can be obtained with a variety of combinations for

TABLE 8-2
Sizes and Weights of Phoenix Brand Steel
Racing Plates

Type	Size	Weight (ounces)
Plain, light	00	3¾
	0	4
	1	4¼
	2	4½
	3	4¾
	4	5
Plain, heavy	00	4
	0	4¾
	1	5
	2	5¼
	3	5½
	4	5¾
Hind, toed, and block heels	00	5¼
	0	5½
	1	5¾
	2	6
	3	6¼
	4	6½

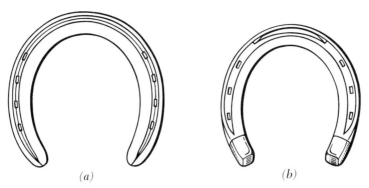

FIGURE 8-17
Shape of steel racing (running) plates. *(a)* Plain. *(b)* Block heels toed.

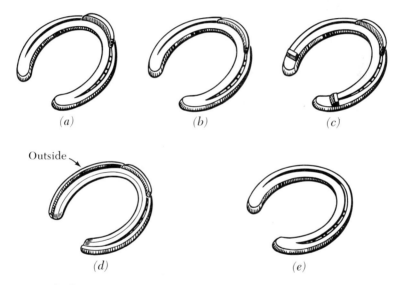

FIGURE 8-18
Aluminum racing plates for front feet. *(a)* Regular toe. *(b)* Low toe. *(c)* Jar calk.
(d) Level grip or outside rim. *(e)* Queens plate—flat (no calks).

toe and heel calks (Figure 8-17). Aluminum racing plates (Figures 8-18 and 8-19) weigh 2 to 3 ounces (0.7 to 1 mg). The front plates can be obtained plain, which means that there are no calks on the plate. A regular toe has a toe calk. If the toe calk is lower than the regular height, it is a low toe plate. Heel calks on the front plates are referred to as "jar calks." Hind aluminum racing plates are also available with various toe and heel calk combinations. The heel calk on a hind plate is referred to as a "block" or "sticker," depending on its shape. The sticker, or "mud-calk," is set across the heel, whereas the

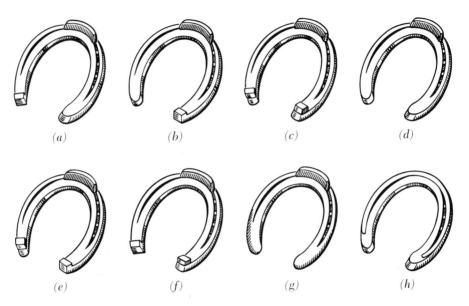

FIGURE 8-19

Aluminum racing plates for hind feet. *(a)* Right sticker or mud calk. *(b)* Left sticker or mud calk. *(c)* Block heel. *(d)* Plain heel. *(e)* Left block and sticker. *(f)* Right block and sticker. *(g)* Level grip or outside rim. *(h)* Queens plate—flat therapeutic (no calks).

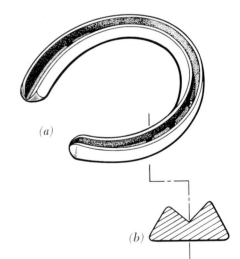

FIGURE 8-20

(a) Polo shoe. *(b)* Cross section of polo shoe, showing that the inside rim is higher than the outside rim.

block is set lengthwise. Blocks cause a horse's hindfeet to break over faster, thus preventing the track surface from burning the ergot area. Stickers increase the traction on muddy tracks.

Polo shoes (Figure 8-20) differ from cowboy shoes in that the inside rim on the web is raised above the outside rim. This shape increases traction,

TABLE 8-4
Specifications of Phoenix Brand Polo Shoes

Size	Weight (ounces)	Width (inches)	Height (inches)
0	6½	4¼	4⅝
1	7	4⅝	5
2	7¾	4¾	5½
3	8½	5¼	5¾
4	9¼	5⅜	6¼

prevents sliding, and enables the foot to roll over faster. Polo horses and western barrel-racing horses make many turns at high speeds. The polo shoe allows the horse to pivot on the shoe and maintain a toe grip regardless of where the foot breaks over. The sizes and weights of polo shoes are given in Table 8-4.

Farriers modify shoes with the aid of a forge. Toe and/or heel clips (Figure 8-21) can be drawn on shoes to help hold the shoe in position. They are essential for horses that have weak hoof walls and whose hooves do not hold nails very well. Gaited and walking horses need toe and heel clips to help hold the extra-heavy shoes. Horses that must stop and turn suddenly are difficult to keep shod unless toe and heel clips are added to prevent the hoof wall from sliding off the shoe and tearing out the nail holes. They can also support the hoof wall in the areas of cracks.

Borium can be applied to horseshoes with an oxyacetylene torch to improve the grip and life of the shoe. The roughened surface increases grip on ice, pavement, and dry grass. Borium (a metal alloy) is harder than any substance except diamonds, so it doubles or even triples the life of the shoe. Application of borium is particularly important before long trail rides over rough terrains; borium should also be applied to the shoes of pack horses and mules.

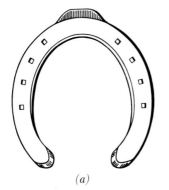 

(a) (b)

FIGURE 8-21
(a) Toe clips. (b) Heel clips.

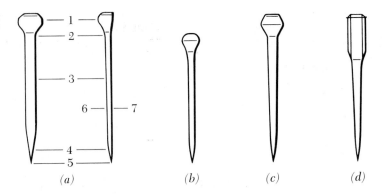

FIGURE 8-22

Horseshoe nails. *(a)* Parts: (1) Head. (2) Neck. (3) Shank. (4) Bevel.
(5) Point. (6) Inner face. (7) Outer face. *(b)* City shape. *(c)* Regular shape.
(d) Frosthead shape.

Horseshoe Nails

Horseshoe nails (Figure 8-22) are made so that one side of the shank is flat and
the other side is concave, with a bevel near the point. The head of the nail is
tapered and roughened on the same side as the bevel. Horseshoe nails differ
according to size and head shape. The most commonly used sizes are 4, 5, and
6, but sizes range from 2½ to 12 (small to large). After being driven, a regular
head does not fit completely into the fullering. The city head fits into the fullering and is used for racing plates. Nails with regular heads are used when the
shoes are reset and when the nail holes are slightly larger. Sharp, beveled heads,
called *frostheads*, are used to increase traction on icy surfaces. Nails with coun-
tersunk heads can be obtained but are seldom used except for handmade shoes.
Use a pritchel to open the nail holes in a shoe before driving the nails (Figure
8-12).

Shoeing Procedure

The initial step is to remove the old shoes (Figure 8-23). First, cut the clinches
by placing the sharp edge of the clinch cutter under the clinch (bent-over nail)
and striking the cutter with the driving hammer. After the clinches are cut or
straightened out, remove the shoe with "pull-offs." Insert them under one heel
of the shoe and rock them forward and toward the center of the toe. Repeat
the procedure for the opposite side. Then gradually work the pull-offs toward
the toe until the shoe is removed.

    Examine the shoes for abnormal wear, and give the horse the same pre-
liminary examination as for trimming the feet. As previously described, trim

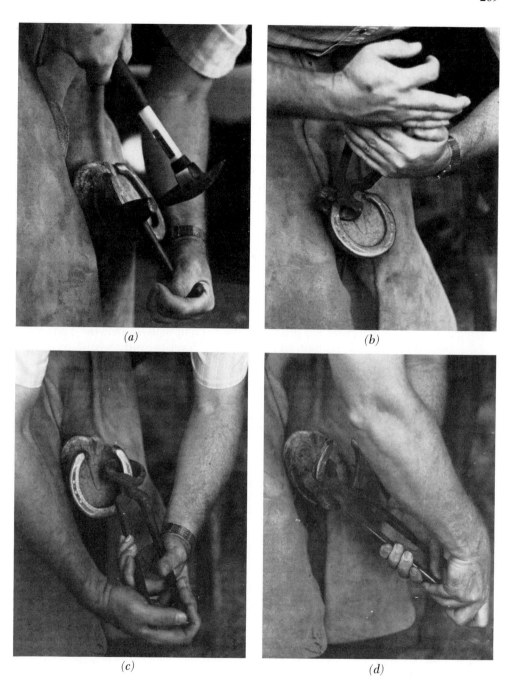

(a)

(b)

(c)

(d)

FIGURE 8-23

Cutting clinches and removing old shoe. (a) Cutting clinch. (b) Inserting pull-offs
under one heel. (c) Inserting pull-offs under opposite heel and rotating pull-offs toward toe.
(d) Pulling shoe off. (Photos by B. Evans.)

the feet and select the appropriate shoe. Then shape shoes to fit the horse's feet. Using a forge to heat the shoe so it can be shaped more easily is called "hot shoeing." In cold shoeing, the shoe is shaped without heat. The only advantage of a forge is to make shaping easier. The shoe must be flat after it is shaped so the surface touches the hoof wall all the way around and does not rock on the foot. Each high spot causes excess pressure, which in turn cracks the hoof wall. The shoe fits the foot if the outside edge of the hoof wall is flush with the outer edge of the shoe except at the heels. Shoes should be about $\frac{1}{16}$ to $\frac{1}{8}$ inch (1.6 to 3.2 mm) wider at the heels to allow for normal heel expansion when the horse moves. The inside edge of the shoe should not press on the sole. Heels must be long enough to cover the foot buttress but not long enough to be stepped on by the other feet. Nail holes should be in a position so that the nails can be properly placed and directed.

Nailing the shoe on the foot is the next step (Figure 8-24). Most shoes have four nail holes in each branch, but three nails are usually enough. Four nails are used if the horse is used heavily, worked at speed, or ridden in rough country. After positioning the shoe on the foot, drive the first nail. Any nail can be driven first, but it is usually easier if one of the two center nails in either side is driven first. The point of the nail should exit the hoof wall about $\frac{3}{4}$ inch (19.1 mm) above the ground surface. Hold nails so that the bevel side faces the center of the foot and the flat side faces the outside surface of the hoof wall. Point the nail toward the point where it should exit, and drive it with the driving hammer. Light taps with the hammer drive the nail deeper; a sharp blow causes the nail to exit. Therefore, hit the nail head with a sharp blow just before it is $\frac{3}{4}$ inch (19.1 mm) into the hoof wall. Immediately after the point exits and the nail is driven completely in, bend it over and wring it

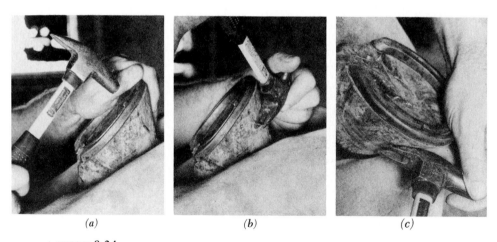

(a)   (b)   (c)

FIGURE 8-24

Nailing on shoe. (a) Start nail in one of the center nail holes; it should exit about ¼ inch from ground surface of the hoof wall. (b) Bend nail over. (c) Insert nail between jaws of the hammer and wring off. (Photos by B. Evans.)

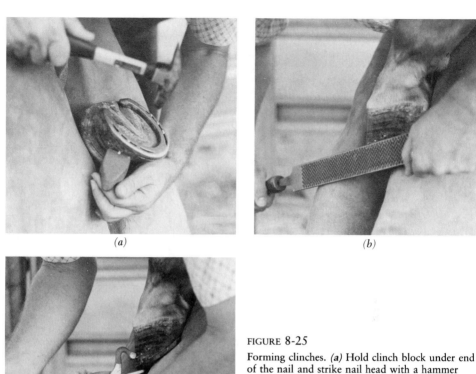

(a)

(b)

(c)

FIGURE 8-25

Forming clinches. *(a)* Hold clinch block under end of the nail and strike nail head with a hammer to bend the nail and form clinch. *(b)* Rasp hoof wall lightly under the bent nail to clear the hoof wall. *(c)* Bend clinch over with clincher (or a hammer). (Photos by B. Evans.)

off with the claws of the hammer. Drive in the remaining nails and cut them off. Remove and reposition nails that are driven in too far or not far enough. Take care not to drive a nail into the sensitive tissue (such an accident is called "quicking"). If this does occur, remove the nail and treat the wound with an iodine solution. The horse should have been immunized against tetanus; if it has not, give it a tetanus antitoxin injection for protection.

Form the clinches by holding the clinch block under the nail and hitting the nail head a couple of times with the hammer (Figure 8-25). If the nail stub places too much pressure on the hoof wall, the hoof wall will tear or crack. The stub should project about $1/16$ inch (1.6 mm) from the wall before forming the clinch. Cut clinches to their final length with pull-offs or a pair of nail cutters. If necessary, rasp them to the proper length. After forming the clinches, rasp the hoof wall lightly immediately under the nails to remove the burr of horn that appears under the nails. Then bend the clinches flat against the hoof wall by hitting them with a hammer or by using the clinchers.

Evaluation

After the farrier has completed the job, it should be evaluated. Several points should be considered (Figure 8-26). Examine the feet first while they are on the ground. The feet should be the same size. The angle of the hoof wall should be the same as the pasterns and the shoulder. The toe should not be "dubbed off," and the branches should extend about $^{1}/_{16}$ inch (1.6 mm) out from the heels to allow for expansion. Shoes should be the right length. The foot should be in balance with the leg and be properly shaped. Clinches should be about ¾ inch (19.1 mm) above the hoof wall ground surface, and they should be evenly

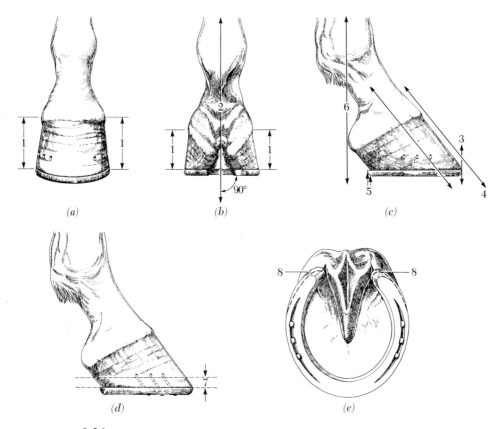

FIGURE 8-26

Properly shod foot. *(a)* Front view: 1's are equal length. *(b)* Back view: 1's are equal length: 2 is at right angles to ground surface. *(c)* At 3 shoe is even with edge of hoof wall. Heel of shoe extends up to ¼ inch (6.4 mm) beyond heel of foot at 5. Hoof wall has proper angle at 4 and is continuous with slope of pastern. Line 6 bisects leg, touches bulb of heel, and is perpendicular to the ground surface. *(d)* Nails exit (7) ¾ to 1 inch (19.1 to 25.4 mm) above shoe and are in a straight line at (7). Drive nails parallel to tubules in hoof wall. *(e)* Shoe extends beyond heels and extends about $^{1}/_{16}$ inch (1.6 mm) from hoof wall at 8.

spaced. The hoof wall should not have been excessively rasped except for necessary corrective shaping of the foot. The entire surface of the web should be in contact with the surface of the hoof wall; that is, there should be no space between the shoe and hoof wall.

Then lift the foot off the ground and examine it. The nails should have been driven home, and the frog should have been trimmed but not excessively.

Finally, observe the horse moving to make sure it is not lame and has no defects of gait caused by shoeing.

CORRECTIVE TRIMMING AND SHOEING

Corrective trimming and shoeing are done to change the way the foot travels or lands to avoid limb interference.

Foals

Many faulty positions of stance or incorrect ways of going in young foals can be helped or corrected by proper trimming of the feet. Corrective trimming for young foals is based on three principles.

First, adequate exercise is important if the foal is to be straight-legged when it matures. Support muscles increase in size and strength as the body size increases. However, when foals are kept in close confinement, the muscles required for shoulder propulsion and rotation are not used as much as they should be to stimulate normal development. Consequently, these muscles shrink and weaken as the foal grows, allowing the legs to become base-wide. Then the feet become toed-out and the inside hoof wall is worn off because the foot rolls over the inside wall. To prevent this sequence of events, foals should be kept where they have room to run and play, or they should be vigorously exercised for a few minutes every day.

Second, pressure across the growth plate or epiphysis of the leg bones influences the rate of bone growth. Unevenly distributed pressure across the epiphyseal plate causes improper bone growth. The objective of corrective trimming for young foals is to evenly distribute pressure across the epiphysis and, when necessary, to redistribute pressure to stimulate or depress bone growth to strengthen the leg. In the toe-wide stance, excess pressure is occurring on the lateral (toward the outside) side of the epiphysis, and less pressure is occurring on the medial (toward the center of the horse) side. Excess pressure decreases bone growth and less pressure increases it so that the medial side of the bone grows at a faster rate than the lateral side. The condition worsens if the feet are not correctively trimmed.

Third, corrective trimming must begin when the foal is a few days old and the hoof wall is hard enough to trim. All corrections must be accomplished

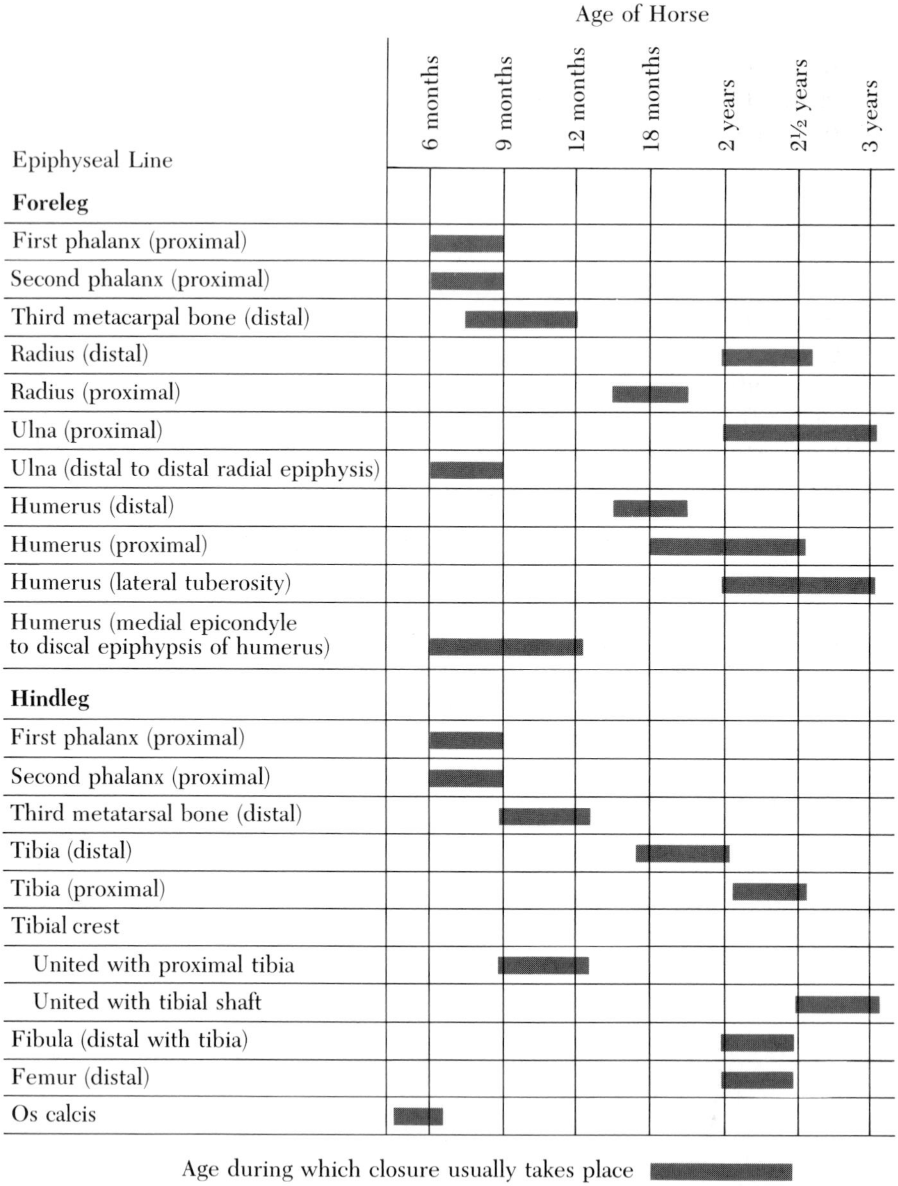

Age of Horse

Epiphyseal Line	6 months	9 months	12 months	18 months	2 years	2½ years	3 years
Foreleg							
First phalanx (proximal)	▰						
Second phalanx (proximal)	▰						
Third metacarpal bone (distal)		▰					
Radius (distal)						▰	
Radius (proximal)				▰			
Ulna (proximal)						▰	
Ulna (distal to distal radial epiphysis)	▰						
Humerus (distal)				▰			
Humerus (proximal)					▰		
Humerus (lateral tuberosity)						▰	
Humerus (medial epicondyle to discal epiphypsis of humerus)		▰					
Hindleg							
First phalanx (proximal)	▰						
Second phalanx (proximal)	▰						
Third metatarsal bone (distal)			▰				
Tibia (distal)					▰		
Tibia (proximal)						▰	
Tibial crest							
United with proximal tibia			▰				
United with tibial shaft							▰
Fibula (distal with tibia)					▰		
Femur (distal)					▰		
Os calcis	▰						

Age during which closure usually takes place ▰▰▰▰▰

FIGURE 8-27

Closure of epiphyseal plates. Age of epiphyseal closure is based on Quarter Horse and Thoroughbred breeds, both male and female. The age of epiphyseal closure given here is based on radiographic determination, which will be earlier than actual closure determined histologically. (From Adams, 1974.)

before the epiphyseal plate closes for a particular bone. Radiographic data indicate that the epiphyses of the first and second phalanges close at approximately 9 months (Figure 8-27). The epiphysis at the distal end of the cannon bone closes at 9 to 12 months. Most of the remaining bones' epiphyses close at 1 to 2 years.

Mature Horses

The hoof wall grows at the rate of ¼ to ½ inch (6.4 to 12.7 mm) per month. Therefore, the hooves should be trimmed or shod every 6 to 8 weeks. They need to be trimmed more frequently if they are growing abnormally fast, such as after founder, or if they show uneven wear. Horses that wear their hoof walls unevenly usually have crooked feet or legs. Uneven wear of the hoof wall accentuates faulty conformation and way of going. Keeping the hoof wall level does not correct the fault in mature horses, but it may prevent leg interference. Quite often corrective trimming only requires rasping down one side of the hoof wall until it is level with the opposite wall.

The outside wall wears down at a faster rate than the inside wall of base-narrow and toe-in or toe-out horses. Consequently, the inside wall must be rasped level with the outside wall. Remove additional hoof wall if the wall is too long after it is leveled. Horses that are base-wide and toe-out or toe-in wear the inside wall at a faster rate. If the horse only toes out, the inside wall wears faster; and if it toes in, the outside wall wears faster.

The toe of the hoof wall grows faster than does the heel. A long toe decreases the expansion of the wall when the hoof strikes the ground. Because the contraction is greater than the expansion, the walls contract inward. Contracted heels put excess, painful pressure on the third phalanx. Failure to trim long hooves leads to contracted heels, faulty gaits, and limb interference. Long toes and low heels inhibit the foot rolling over during forward motion and may cause forging, overreaching, and scalping.

The basic principle of corrective shoeing is to trim the foot level and then to apply corrective shoes to improve the faulty gait. Drastic changes in the flight of the foot and leg can put excess strain or pressure on other bones and structures of the foot and leg. The new stresses can cause pathological changes, so use the least severe corrective measure.

Several horses may share the same defect in way of going, but each horse may require different corrective measures. The initial step is a visual inspection of the horse's way of going. Note any "breakover" and "landing" areas of the hoof wall. In the base-narrow, toe-in fault in foreleg conformation, the foot breaks over and lands on the outside wall, and if the horse is unshod the outside wall will be worn down. The inside wall must be lowered to level the foot, and then several corrective measures can be used to force the foot to break over the center of the hoof (see Figure 8-28). One of the simplest and mildest methods of correction is to put on a square-toed shoe. Other measures,

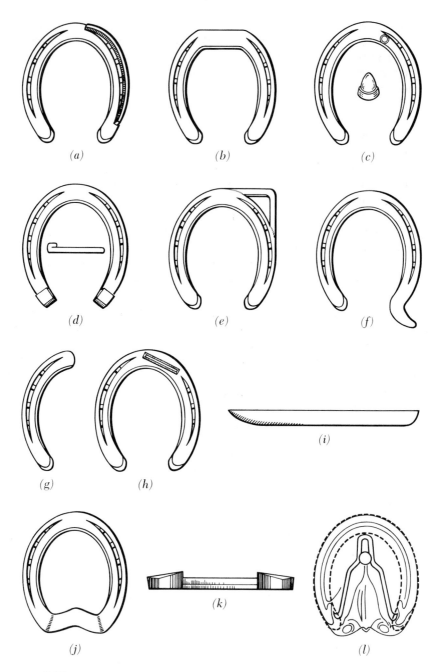

FIGURE 8-28

Corrective and therapeutic shoes. *(a)* Half-rim shoe. *(b)* Square toe. *(c)* Calk at first outside nail hole. *(d)* Heel calks. *(e)* Lateral toe extension. *(f)* Shoe with trailer. *(g)* Half shoe. *(h)* Bar across breakover point. *(i)* Rolled toe. *(j)* Bar shoe. *(k)* Slippered heels. *(l)* Chadwick spring.

such as an outside half rim (outside rim) shoe, calks, a calk at the first outside nailhole, or a lateral toe extension, encourage the foot to break over the toe. If the horse is base-narrow but toes out, the foot will break over the outside of the toe, wing inward, and land on the outside wall. Successful correction methods to encourage the foot to break over the center of the toe include outside half rim shoes, square-toed shoes, outside toe extensions, and half shoes on outside wall.

Horses with a base-wide, toe-out conformation tend to break over and land on the inside aspect of the toes. Correct the fault by modifying the shoe to make it difficult for the foot to break over the inside of the toe but easier to break over the center of the toe. Several modifications, such as heel calks, square toes, or bars across the breakover points, force the foot to break over the center of the toe. Base-wide horses that have their toes pointing the opposite way (that is, toe-in) land on the inside wall. To improve their way of going, the inside branch of the shoe must be raised.

Cow-hocked horses tend to break over the inside toe and thus strain the inside of the hock. This fault in conformation is particularly straining when the horse makes a sliding stop. The main objective in correcting the fault is to force the hindfoot to break over the center of the toe by braking the inside of the hoof when it lands and by rotating the toe inward. A lateral extension toe or a short inside trailer usually accomplishes the objective.

The stance of a sickle-hocked or camped-under horse can be improved by shortening the hind toes if they are too long. This correction causes the hindlegs to move backward. The opposite effect can be accomplished for the camped-behind horse by lowering its hind heels if they are too long.

Other less obvious faults in conformation may lead to faulty gaits. Forging and overreaching result when the hindfoot breaks over faster than the forefoot in such a way that the toe of the hindfoot strikes the sole of the heels of the forefoot. Horses that have short backs and long legs or short forelegs and long hindlegs or that stand under in front or behind tend to forge and overreach. The corrective measure is to speed up the breaking over of the forefoot and to retard the breaking over of the hindfoot and/or increase the height of the forefoot flight. A variety of corrective measures can be used. The fault can be corrected in some horses by rolling the toes (and shoes) of the forefeet and leaving the toes of the hindfeet slightly longer than usual or by placing heel calks on the hind shoes. In extreme cases, a ½-inch (12.7-mm) bar can be welded across the heel of the front shoes and across the toe of the hind shoes.

When the toe of the forefoot strikes the hairline of the hindfoot ("scalping"), rolling the toe enables the forefoot to break over faster and may prevent contact. The same corrective principle and objectives apply to horses that scalp and/or hit their shins as to horses that forge and overreach.

Pacers that toe-out in front and toe-in behind tend to cross-fire. To prevent the hindfoot from hitting the opposite forefoot, all feet must be shod to encourage breaking over the center of the toe. Corrective measures include those discussed earlier for correcting the front toed-out and hind toed-in conditions.

Interference usually results when a horse is base-wide or base-narrow and has a toed-out conformation in the forelegs or is cow-hocked in the hindlegs. Correction of these conditions has been previously discussed and applies also to interference.

A farrier often encounters horses with pathological conditions that require therapeutic shoeing. The most common conditions are contracted heels, ring bone, side bone, navicular diseases, and toe or quarter cracks.

There are several ways, or combinations of ways, to reestablish proper foot function in a horse with contracted heels. Frog pressure can be increased with a bar shoe so that the bar maintains constant pressure on the frog. Other types of shoes include shoes with slippered (beveled) heels or a Chadwick spring to force the wall outward. The expansion process should be gradual, but additional expansion aids can be used; for example, cutting a horizontal groove below the coronary band, cutting several vertical grooves in the quarter area, or thinning the wall at the quarters with a rasp.

Horses with ring bone have impaired or no action of the pastern and/or coffin joints. The objective of correction is to enable the foot to break over more easily so that the required action of the joint or joints is reduced. The simplest corrective measure is to roll the toe. When side bone is the problem, the action of the coffin joint needs to be decreased, so rolling the toe is helpful. To restore the ability of the foot to expand and relieve pain if the horse is lame, thin the quarters of the wall.

The objective of corrective shoeing for a horse with navicular disease is to make it easier for the foot to break over and to reduce the anticoncussive activity of the deep flexor tendon against the navicular bone. Raising the heels and rolling the toe let the foot break over faster and reduce concussion to the navicular bone. To further reduce trauma to the area, use a bar shoe or pads. This type of correction contracts the heels because frog pressure is reduced. The tendency for the heels to contract can be reduced by slippering the heels of the shoe so that they slope to the outside. This causes the wall to slide outward as the foot strikes the ground. The expansion of the wall can be further aided by thinning the quarters with a rasp.

When shoeing a horse with flat feet, trim the sole slightly but do not trim the frog. Prevent the sole from dropping further by making sure that the shoe covers the entire wall and white line and covers only a small part of the outside edge of the sole. Pads are usually necessary to prevent sole bruising.

Toe, quarter, and heel cracks are quite frequent. Each crack requires individual analysis with regard to the corrective treatment and shoeing. The general principle is to use a toe clip on each side of the crack to prevent wall expansion and to lower the wall under the crack so the wall will not bear weight. Plastics may be used to seal the crack and prevent expansion and contraction of the crack during foot action.

9
Recognizing Illness and Giving First Aid

There are several components to a successful health care program for the horse. One is prevention of injury and sickness. This is accomplished by parasite and disease control, immunizations, teeth and foot care, safety, exercise, keeping accurate records, and other measures. When an injury or a disease occurs, the horseperson must know proper procedures for immediate care of the horse and where to obtain assistance. Most important, the horseperson must recognize symptoms that indicate a horse needs medical attention.

RECOGNIZING ILLNESS

Horses must be observed at regular intervals—at least daily—to determine if they need special attention. Certain vital signs should be checked for horses that appear to be sick: temperature, heart rate, respiration, gut sounds, mucosal color, and skin pliability. In addition to the vital functions, feces, urine, appetite, depression, nasal discharge, coughing, swelling, and behavior should be monitored.

Other signs can indicate illness. Rapid weight and appetite losses indicate a sudden change in health and are usually the first signs of illness. The horse may become depressed and isolate itself from other horses.

279

Respiration

The normal respiration rate for a horse is 8 to 16 respirations per minute. To determine the respiration rate, watch the nostrils or the flanks. Nostrils flare and contract with each breath, and the abdomen rises and falls with each breath. Any type of distress increases the respiration rate. If the respiration rate exceeds the heart rate, the horse has a serious problem.

Temperature

Normal body temperature for a horse is 100°F or 38°C. It does vary from 99.5°F (37.5°C) to 101.5°F (38.6°C) for normal horses. It also varies during the day for the same horse. Internal and environmental factors can increase body temperature. Exercise, excitement, and hot weather raise temperature, as will pain and illness. A fever is classified as mild (102°F or 38.9°C), moderate (104°F or 40°C), or high (106°F or 41°C). If it exceeds 106°F (41°C), the chances of recovery are low, and the horse needs immediate veterinary care. A low temperature may be observed in horses that are in shock. The increased body temperature is also classified as continuous (constant), remittent (marked increases and decreases), intermittent (normal periods between marked ups and downs), and recurrent (long periods of increased temperature followed by long periods of normal). When you describe a horse's condition to a veterinarian, he or she will want to know the temperature and its classification.

To take the temperature, use a veterinary thermometer. It contains a loop on one end to which an 8-inch (203-mm) string with an alligator clip or clothespin on the other end can be attached. Shake the mercury level down to below 95°F (35°C) before inserting the thermometer. Coat the thermometer with a small amount of lubricant such as petroleum jelly for easy insertion. Raise the horse's tail and insert the thermometer bulb end first about 75 percent of its length into the anus with a rotating motion. Attach the alligator clip or clothespin to a few tail hairs to prevent the thermometer from being lost in the rectum or from being broken if it falls out. After at least 2 minutes, remove and read the thermometer. Clean it in cool, soapy water and dip it in alcohol before storing or using it again.

Heart Rate

Heart rate or pulse varies with the age of a horse. Normal mature horses have a pulse of 28 to 40 beats per minute. Newborn foals have 80 to 120; foals, 60 to 80; and yearlings, 40 to 60. To determine an accurate pulse, the horse should be calm, rested, and relaxed. Many things increase heart rate. Exercise, fright, excitement, and hot weather increase the rate. Illness causes pulse rates of 80 to 120 for a prolonged period of time.

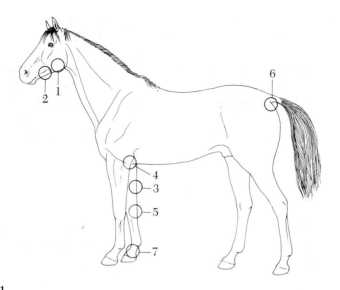

FIGURE 9-1

Points at which the horse's pulse can be taken. (1) The back edge of the lower jaw (the cheek), four inches below the eye (facial artery). (2) The inner surface of the groove under the lower jaw (external maxillary artery). (3) The inside of the foreleg (median artery). (4) Inside the left elbow, up and forward, against the chest wall (heart). (5) Behind the carpus, or knee (digital artery). (6) Under the tail, close to the body (medial coccygeal artery). (7) Medial or lateral pastern (digital artery).

Pulse rates per minute are determined by counting the pulse for 30 seconds and multiplying the value by 2. It may be necessary to determine 15-second rates, but in this case several determinations should be used, because pulse rates can vary widely over short time periods.

To determine the pulse, press your fingers against an artery. Each throb is a pulse or heartbeat and is the increased pressure in the artery caused by the heart contracting and forcing blood into the arteries. There are several locations where the pulse can be felt (see Figure 9-1), but practice is usually necessary to feel it easily.

Skin and Mucous Membranes

Skin pliability is a test for dehydration. Normally, the skin is pliable and elastic. If you pick a fold of skin up on the neck and release it, it quickly returns to its original position. But if the horse is dehydrated the skin returns slowly and tends to stay in a fold. Other signs of dehydration are loss of moisture on mucous membranes and dull, sunken eyes.

The color of the mucous membranes and the rate of capillary refill (rate of return of blood to the area) give an indication of the quantity and condition of the circulating blood. These signs help determine the presence of anemia, colic, congestion, and shock. Several mucous membranes can be checked: the gums, the inside lips of a mare's vulva, nostrils, and conjunctiva should be pink. A fire-engine red color usually denotes illness. Anemia causes a pale color, and lack of circulation causes a bluish-purple color.

Determine capillary refill time by pressing your thumb on the horse's gum and then releasing it. It should take about 2 seconds for the blood and normal color to return to the area. Longer capillary refill times indicate dehydration or a circulatory problem.

Excreta

Normal urine from a horse is yellow to brown in color. The cloudy color is caused by calcium carbonate crystals. The longer the time period between urinations, the cloudier the urine. It is somewhat viscous because mucus and cellular debris are present. Normally an average-size horse produces about 5.5 quarts (or 3.5 liters) each day.

Reduced urine volume may mean the horse is dehydrated. Other abnormalities that should be checked for are difficulty in urinating (straining), free blood present in urine, or coffee-colored urine. These conditions should be called to the attention of a veterinarian.

Examine the feces for consistency, and determine the frequency of defecation. Most horses defecate about every 2 hours, but if excited they defecate very frequently. Failure to defecate usually indicates a disorder such as impaction, colic, or severe dehydration. Normal feces are tan or yellow to dark green in color. The fecal balls should be well formed and not too hard or soft. Loose stools may mean a disorder, and diarrhea can lead to severe dehydration in a short period of time (2 to 3 hours in a severe case). Severe diarrhea is very serious in a young foal.

Gut sounds can help you diagnose a sick horse. The sounds can be heard by placing an ear on the flank area or using a stethoscope in the same area. The sounds are caused by the normal contracting and relaxing movements of the digestive tract during the digestion process. It requires considerable practice to gain experience in distinguishing excessive from normal gut sounds. Listen to several normal horses as a basis for comparison. Absence of gut sounds is usually more critical than excessive sounds.

RECOGNIZING LAMENESS

Lameness can be located by using several diagnostic aids. Because lameness is caused by pain, the site of the pain must be located by visual inspection, manipulation, pressure, and checking for swelling, rapid pulse, and heat.

Several symptoms of lameness should be taken into account during a visual inspection. If the pain is severe, the horse may hop around on three legs and may refuse to place weight on the affected leg when standing still. If the pain is not very severe, only a disturbance of the gait will be observed. The horse may step short with the affected leg rather than take a full-length step.

Two of the most useful signs are a bobbing of the head or hip. With a front leg lameness, the horse will bob its head upward when the lame leg strikes the ground during the trot. Have the horse trot around you on a level surface in a 20- to 30-foot radius (6.1 to 9.1 m) and watch the head movement. A sound horse will carry its head steady without any up and down motion when it is trotting.

With a hindleg lameness, the horse extends its hip upward as the lame leg strikes the ground during the trot. To observe this, stand behind the horse and have it trot away from you.

Often, a lame leg can be located by comparing pulse strength. An inflammation makes the pulse in the affected leg stronger than in the unaffected leg. Make the comparison several times to take into account shifts in body weight.

Manipulating the legs and their joints may evoke a response to the pain caused by the pressure or tension placed on the damaged area. The knee and hock joints and the joints located below them can easily be manipulated. Pressure is a useful tool during the manipulation because it aggravates the injury. Starting at the knee or hock, use your hands to circle the leg and place pressure on all the parts of the leg. As you feel the leg, check for swelling, heat, and abnormalities. Compare each leg with the opposite leg. Examine the knee by flexing and probing it with your fingers. Thoroughly examine the back of the leg for tendon problems. When examining the hindleg, increase pressure slowly to discourage the horse from kicking. While you are applying pressure, the horse's behavior may indicate pain by withdrawing the affected part, putting the ears back, jerking the head, and grunting.

Flex tests place pressure on the hoof and fetlock, knee, elbow, and shoulder in the foreleg and the hoof and fetlock, stifle, and hip joints in the hindleg. The flex tests produce pain from an injured joint that will be noticeable when the horse trots away from you immediately afterward. To test the hoof and fetlock joints in the foreleg, grasp the toe of the foot with one hand and the front of the knee with the other hand. Flex the hoof toward the knee for about 30 seconds. For a knee flex test, fold the front leg in a "V" and apply upward pressure on the cannon bone for about 30 seconds. Flex the elbow by facing the horse, grasping the leg behind the knee, and pulling the leg upward until the horse's forearm is at least parallel to the ground. For a shoulder flex test, lift the forefoot up about knee high, grasp the knee, and rotate the leg toward the horse's belly. For a hock flex test, face the hindleg, grasp the back of the cannon and fetlock, and draw the leg up toward the stifle. Hold this position for about 1 minute. Test the hind fetlock and foot joints by grasping the toe and rotating the foot upward while pushing down on the point of the hock with the other hand. Flex the stifle by facing the side of the hindleg, grasping the leg in the hock, and pulling the leg upward and back. The hip flex is difficult: Pick up the hindleg and extend it toward the horse's elbow without bending the hock joint too much.

VETERINARY ASSISTANCE

It is important for horsepeople to be acquainted with a veterinarian and use his or her services when they are needed. The ability to obtain the services of a veterinarian during an emergency or when their aid is indicated pays dividends because properly treated horses recover faster. If you suspect illness but cannot diagnose it, call a vet immediately. The veterinarian may use other diagnostic aids, including nerve blocks, x-ray, Faradism, hoof testers, and wedge tests. Many illnesses that require a veterinarian also require first aid before he or she arrives, and every horseperson should be skilled in giving first aid.

FIRST AID

Confined horses are accident prone because of their behavior patterns, negligence in stabling, and subjection to stresses and strains beyond their capabilities. Horses protect themselves by running away from adversaries or frightening situations. In so doing, they frequently try to run through or over obstacles in their path. Their means of immediate defense is to kick other horses, stall walls, people, and so on. Horses are inquisitive and may paw at objects or may stick their legs through fences, holes in the stall wall, or pieces of equipment. As a result, they frequently cut themselves or receive puncture wounds. The frequency of injury increases when they are kept in surroundings with protruding nails, sharp edges, barbwire, miscellaneous junk, and other dangerous objects. And many common unsoundnesses result from injuries from stresses and strains beyond a horse's capacity. Horses forced to run too far and too often, jump too many or too high fences, and stop too quickly damage tendons, ligaments, muscles, and bones.

First aid is the emergency treatment you give a horse until the services of a qualified veterinarian can be obtained. In many instances, the horse must be transported to a veterinary hospital. The following sections discuss first aid procedures commonly employed to prevent further injury to the horse and to reduce pain.

Be prepared for emergencies by keeping a first aid kit readily available. First aid kits should be kept at all stables and carried in every horse trailer or van. Kits that fit on the back of a saddle should be carried on trail rides. First aid kits are commercially available but they can be made up by buying the necessary supplies. A commercially available kit from Springbrook Forge (Figure 9-2) contains a shoe puller and spreader combination (also used for cutting

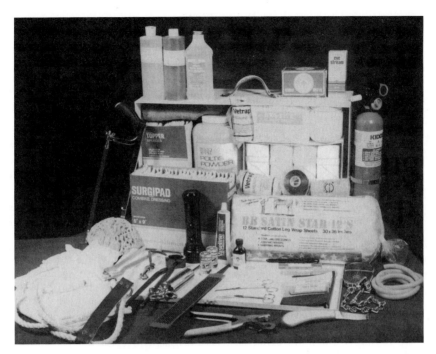

FIGURE 9-2
Equine first aid kit. (Courtesy Springbrook Forge.)

nails and wire); a wide Frost hoof knife; hack saw; rasp; handmade rope halter, braided and spliced; a lead; aluminum humane twitch; straight forceps; bandage scissors; 12 cotton leg wraps (30 inches × 36 inches, or 76.2 × 91.4 cm); latex tubing tourniquet; 12 3-inch (76.2-cm) kling gauze rolls; a roll of 3-inch (76.2-cm) sterile cotton; 25 5-inch × 9-inch (12.7-cm × 22.9-cm) surgical pad dressings; toppers; 50 4-inch × 4-inch (10.2-cm × 10.2-cm) surgical sponges; 4 track bandages with tie tapes; 1-inch (2.54-cm) porous adhesive tape; a 5-inch × 5½-inch (127-mm × 139.7-mm) natural sponge; a 4-inch (10.2-cm) elastic bandage; 1 pint (0.47 liter) of surgical soap; a tube of vaseline; a self-tying canvas poultice boot; a large container of poultice powder; 1 pint (0.47 liter) of alcohol; 1 pint (0.47 liter) of iodine; soft nail brush; 5 yards (4.6 m) of 4-inch (10.2-cm) Vetrap bandaging tape (two rolls); stream spray eye wash; 1 pint (0.47 liter) of soluble salts; veterinary thermometer with ring top and clip case; reusable containers for medicines, pills, and salves; ¾-inch (19-mm) polyethylene plastic electrical tape with dispenser; flashlight with batteries; vinyl examination gloves; plastic aprons; revolving leather punch; steel storage box (26½ inches × 18½ inches × 10½ inches, or 67.3 cm × 47.0 cm × 26.7 cm); first aid instruction booklet; and a Kidde triclass fire extinguisher with a clip attachment outside of the kit (with a limited 2-year warranty).

Cuts and Tears

Wounds require immediate first aid treatment. The seriousness of a wound depends on its location, depth, type of cut or tear, amount of tissue damage, and type of tissue affected. All serious wounds should be treated by a veterinarian. Proper wound treatment before the veterinarian arrives enhances veterinary services and the healing process.

Emergency Treatment

First, bleeding must be controlled. Blood from a severed artery spurts and is bright red. Venous blood is dull red and flows, rather than spurts. Direct pressure controls bleeding for almost all injuries. Very seldom, if ever, is a tourniquet needed. Pack gauze into large or deep wounds and apply pressure until the bleeding stops. Apply direct pressure on a wound by firmly pressing a sterile pad over the wound and holding it in place by hand or by a tight bandage. As soon as the bleeding stops, a tight bandage should be removed. Cuts in the coronary band are serious and need immediate treatment by a veterinarian. The bleeding can be controlled by pressure.

Bleeding or loss of blood seems to panic most horsepeople. A 1,000-pound (454-kg) horse has about 50 quarts of blood and can lose about 10 percent of its blood before the loss becomes critical. Rapid blood loss is critical; a horse can tolerate more slow blood loss. If you find an injured animal that has been bleeding for some time, evaluate it for shock. If there has been excessive blood loss, the inside of the mouth will feel cold to the touch and the gums will be pale. Other signs of shock include an inability to stand and profuse sweating. Do not move the horse, but cover it with blankets.

The second step is to clean the wound with warm water and a mild soap. Remove all dirt and foreign matter. The best way is to use a squeeze bottle or a hose with running water to clean the wound. If these methods are unavailable, use gauze. Rubbing and swabbing wounds cause further tissue damage, so just press gauze on dirt to remove it. Do not use cotton, because pieces of it will remain in the wound. Peroxide, irritating disinfectants, or other cleaning agents cause further tissue damage and slow healing. The hair should be clipped or shaved from the edges of the wound, but this can be delayed until the veterinarian arrives.

The third step is to immobilize the wound to prevent further damage. Hold the horse at halter or place it in a box stall. Keep it from chewing the wound or the bandages. A neck cradle prevents chewing of most wounds. Cayenne pepper or hot-pepper sauce applied to a bandage discourages the horse from chewing it.

If the wound must be sutured, the veterinarian should do so within 12 to 24 hours for best results. The quicker the veterinarian arrives, the better the chances of successful suturing.

TABLE 9-1
First Aid Treatments

Wounds and Injuries	Medication and Treatment Recommended	Medication Not to Use
Minor scrapes or abrasions	Vaseline, Corona, or Topazone spray	—
Minor skin scratches	Furacin solution or Topazone	—
Rope burn		
Superficial	Furacin solution or powder, or Topazone	—
Deep[a]	Furacin or Dermafur salve and bandage	Astringents or caustics
Small skin wound, left open	Furacin solution or powder or Topazone	Astringents or caustics
Large skin wound, to be sutured[a]	Furacin or Dermafur under bandage	Astringents or caustics
Large skin wound, left open[a]	Furacin powder or Topazone spray	Astringents or caustics
Jagged skin wound, to be sutured[a]	Furacin or Dermafur salve and bandage Furacin solution or powder if not able to bandage	Astringents or caustics
Jagged, deep muscle wound, to be sutured[a]	Furacin or Dermafur salve and bandage Furacin solution or powder if not able to bandage	Astringents or caustics
Deep open wound, over 48 hrs old[a]	Furacin powder or Topazone	Astringents or caustics
Open wound		
Healed to skin level	Furacin powder or Topazone	Astringents or caustics
With proud flesh[a]	Tissue should be trimmed flat; caustic powder	Tissue stimulants
Puncture wound		
In soft tissue[a]	Furacin solution or powder	Astringents or caustics
In hoof[a]	Kopertox or tincture of iodine or Betadine	Caustics
Minor wound to hoof coronary band	Kopertox or Topazone	Caustics
Minor sprain or swollen leg	Cold water or ice and supportive bandage; aspirin twice daily in grain (50–60 g/1,000 lb, or 50–60 g/454 kg)	Heat applications, such as liniments, sweats, leg paints, or blisters

[a]Wound should be under the care of a veterinarian.
SOURCE: Naviaux (1974).

The fourth step is to prevent infection. The procedure should be under the supervision of a veterinarian. Systemic antibiotics may be necessary, or a nonirritating wound dressing may be used (Table 9-1). Clean and treat minor skin wounds with nitrofurazone.

The fifth step is to bandage the wound to protect it from dirt or other irritation. The bandage may help decrease movement of the wound and allow a faster healing process. However, bandaging is dangerous unless done properly. An improperly applied bandage aggravates the original damage or creates serious new problems. Bandaging is a basic skill that requires practice, and all horsepeople should be familiar with the process (see the next section).

The final step is to prevent tetanus by immunization. If the horse has not been immunized, tetanus antitoxin should be administered. A tetanus toxoid booster may be given if it has been 8 to 12 months since the last toxoid booster or since the two tetanus toxoid vaccinations.

Bandaging

When bandaging a leg, apply the medication and cover it with a nonstick surgical pad. Telfa or Adaptic squares are excellent. Hold the pad in place with a primary wrap such as kling gauze. Apply padding over the primary wrap. There are many different commercial padding materials. Rolled cotton, quilted leg wraps, and baby diapers are excellent. The final wrap is used to hold up the padding. This wrap can be an adhesive and elastic-type bandage or a non-elastic, nonadhesive bandage. Vetrap and Ace bandages are elastic, and the traditional track bandages are nonelastic.

The proper way to wrap an injured foreleg below the knee or hock is shown in Figure 9-3. Before starting the leg wrap, lay out all materials within easy reach. Apply medication to the nonsurgical sterile pad used to cover the wound. Hold the pad in position by wrapping the leg with kling gauze, which is very elastic and does not restrict blood circulation. Stretch the kling gauze just enough to hold the pad in place. Start the wrap about 3 to 4 inches (7.6 to 10.2 cm) above the wound and extend it 3 to 4 inches below the wound. Each wrap overlaps the previous wrap about one-half the width of the gauze. Do not tie the primary wrap off; leave the end free. Apply the padding to the leg, placing the edge of the padding on the side of the leg to start the wrap and also finishing on the side. Starting on the front of the cannon bone or on the tendons can cause injury. Padding should be ½ to 1 inch thick (1.27 to 2.54 cm). After the padding is on, apply the final wrap by starting just below (1 inch) the upper edge of the padding and wrapping downward to within 1 inch of the lower edge. Wrap the leg upward again and then finish by taking a couple of wraps downward. When making the wraps, keep the pressure uniform and use an even overlap.

Wounds on the forearms of horses are difficult to bandage (Figure 9-4). The bandage is easy to apply but it easily slips off because the forearm tapers. Apply two or three long strips of adhesive tape vertically to the horse's leg so that the ends extend below the knee. Put the Telfa pad with medication over the wound and wrap it with kling gauze. Sheet cotton can be wrapped around the leg. Apply an elastic wrap plied over the cotton and incorporate the ends of the tape into the elastic wrap. Pressure from the bandage holds the tape to the horse's leg, which helps hold up the bandage.

Bandages over the knee and hock joints are more difficult to apply. The same principles as the simple leg wrap apply; however, be very careful not to

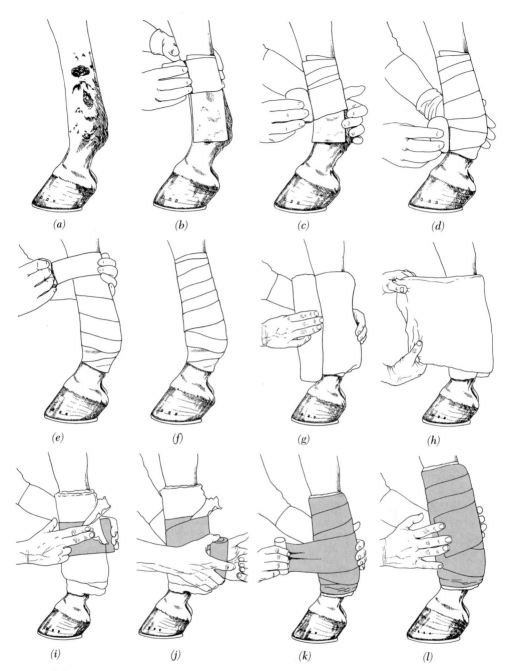

FIGURE 9-3

Bandaging a leg below the knee or hock. *(a)* Injured leg ready for routine dressing. *(b)* Applying medicated covering and primary wrap together (note hand position). *(c)* Progressive turns of the primary wrap spiraling downward. *(d)* Turning back up the leg, well below the injury. *(e)* Upper limit of the primary wrap, well above the injury. *(f)* Finished primary wrap. *(g)* Applying padding to relieve tension. *(h)* End of padding carefully smoothed for finishing off. *(i)* Securing end of the self-adherent stretch bandage with a second turn. *(j)* Successive turns of final spiraling downward. *(k)* All the way up and back to the center, unrolling final turn for application without tension. *(l)* Pressing final wrap so that it adheres to itself. (From *Equus Magazine.* Copyright © 1978. Gaithersburg, MD 20760.)

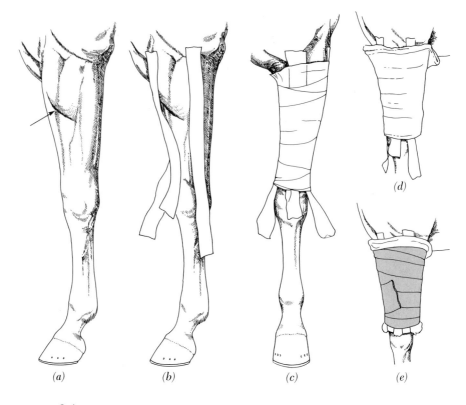

FIGURE 9-4

Bandaging the foreleg. *(a)* Wound is ready for dressing. *(b)* Apply adhesive tape. *(c)* Apply medicated pad and wrap with primary wrapping. *(d)* Wrap cotton around leg. *(e)* Apply elastic wrap. (From *Equus Magazine,* Copyright © 1978. Gaithersburg, MD 20760.)

place pressure where bony prominences are situated just under the skin. Otherwise, the skin will be damaged.

The proper way to wrap the knee is shown in Figure 9-5 and to wrap the hock in Figure 9-6. Start the knee wrap by placing the medicated surgical pad on the wound and taking two or three wraps above the knee with kling gauze. Continue wrapping in a figure-eight pattern until about a dozen wraps around the knee are completed. Then apply a final adhesive elastic wrap. Start the final bandage by spraying tincture of benzoin on the hair above the knee to aid in sticking the adhesive wrap to the skin. At 6 to 8 inches (15.2 to 20.3 cm) above the primary wrap, start the elastic final wrap and take two wraps around the leg. Use moderate tension as you start wrapping downward. As you continue wrapping down the knee, do not cover the bony prominences on the inside and at the back of the knee. Continue wrapping until you have

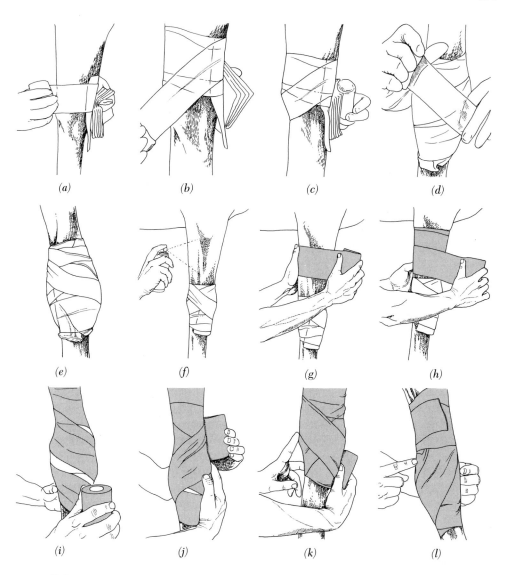

(a) *(b)* *(c)* *(d)*

(e) *(f)* *(g)* *(h)*

(i) *(j)* *(k)* *(l)*

FIGURE 9-5

Bandaging the knee. The steps are shown in succession; see text explanation. (From *Equus Magazine*. Copyright © 1978. Gaithersburg, MD 20760.)

applied four layers of final bandage. Apply the last two wraps without tension so the adhesive will stick and not come loose.

When bandaging the hock, do not cover the point of the hock and the bony protrusion on the inside of the leg right above the hock. Start the primary wrap above the hock and take a couple of wraps. Start wrapping downward

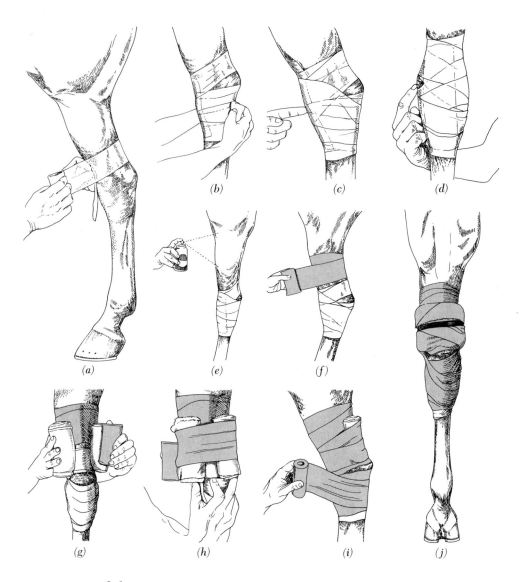

FIGURE 9-6

Bandaging the hock. The steps are shown in succession; see text explanation. (From *Equus Magazine*. Copyright © 1978. Gaithersburg, MD 20760.)

by dropping below the point of the hock. About two layers of primary wrap are sufficient. After spraying the area above the primary wrap with tincture of benzoin, start the final bandage about 6 to 8 inches (15.2 to 20.3 cm) above the primary wrap. Spiral the wrap downward, and do not use tension over the Achilles tendon. After about three wraps, place two soft rolls of gauze or toweling on each side of the Achilles tendon so that they fill up the holes just above the point of the hock. The rolls should avoid tension being placed on the Achilles tendon. Continue by covering the rolls, spiraling downward and then back up to the top. A couple of wraps with electrician's tape will hold the loose end of the bandage.

Three other types of bandages are used. The shipping bandage protects a horse from leg injury during transport and provides support to prevent excess strain and "stocking up." Exercise bandages are used for support while the horse is being worked. Stable bandages are also used for support.

Apply the exercise bandage in almost the same way as the bandage for an injury to the leg above the fetlock and below the knee or hock. The medicated pad and primary wrap are not used. Because less padding is necessary, the proper tension applied to the wrap is critical to prevent injury. Extend a stable bandage down just below the fetlock. Start a shipping bandage below the knee or hock and cover the rest of the lower leg, including the bulbs of the heel. Before using stable, exercise, or shipping bandages, rub a leg brace, consisting of 50 percent veterinary absorbine and 50 percent alcohol, on the legs to stimulate circulation.

Healing

As cuts and tears heal, proud flesh may develop. Proud flesh is granulation tissue formed to fill in the wound space. The excess granulation tissue bulges above the skin surface and inhibits the growth of the edges of skin, so the wound heals slowly. Proud flesh can be removed by surgery or by caustic chemicals, such as silver nitrate or copper sulfate; caustic chemicals must be used with caution.

Punctures, Burns, and Abrasions

Puncture wounds are caused by sharp protruding objects and are usually deep and penetrating. They should be attended by a veterinarian because of the danger of tetanus. If necessary, treat them by stopping the bleeding and then irrigating with hydrogen peroxide. Normally, the next treatment step is to clip the hair around the area and to make sure the wound can drain. Routine care during healing consists of daily cleansing and irrigation with hydrogen peroxide. Nail punctures in the foot need veterinary attention. The veterinarian may enlarge the hole to provide drainage and fill the hole with iodine. Bandage the foot to keep it clean.

Burns are treated with tannic acid jelly. If the burn must be cleaned before the veterinarian arrives, use gauze soaked in saline solution: 1 teaspoon (4.9 ml) table salt per pint (0.47 liter) of water.

Abrasions can be treated with a coating of vaseline if only the hair has been rubbed off. If some bleeding and oozing of serum occurs, wash the abrasion with an antiseptic solution, and apply nitrofurazone. Treat minor rope burns as abrasions, but deep rope burns need attention by a veterinarian.

Wound Treatment—"Don'ts"

There are some "don'ts" for wound treatment as discussed by J. L. Naviaux (1974):

1. *Don't* apply any astringent or caustic, such as gentian violet, tincture of iodine, copper sulfate, or alum, to any fresh wound more severe than a superficial scrape, abrasion, or scratch.

2. *Don't* apply *any* medicine to a wound that will need to be sutured unless it is recommended by your veterinarian to be used as a first aid until he or she arrives. Washing it with a mild soap and water is safe. Peroxide can irritate the tissues and should not be used in a wound to be sutured.

3. *Don't* apply a salve, such as vaseline, corona, or nitrofurazone, to any *open* deep wound, because salve accumulates dirt and debris.

4. *Don't* wash a wound frequently with water, peroxide, or *any* liquid as this slows healing and stimulates growth of proud flesh.

5. *Don't* wrap a leg without good padding under the bandage, because an unpadded bandage can cut into the leg and/or restrict blood circulation. This is especially true for young foals.

6. *Don't* leave a stretch bandage on for more than 24 hours. Daily removal is necessary to ensure proper circulation.

7. *Don't* use sugar or salt water in the eyes. Fresh water or boric acid rinses are safe.

Other Emergency Situations

In addition to wounds, there are other situations requiring first aid.

Electric Shock

A horse found dead in the pasture may have been killed by electric shocks or lightning. The teeth are usually tightly clenched, and there is rectal bleeding. The mouth may be full of fresh feed. If still alive, the horse may be in shock, unconscious, or dazed. A weak pulse and convulsions are typical, and veterinary treatment is indicated. Survivors frequently have impaired vision, bad coordination, or paralysis. Full recovery may take months.

Eye Injuries

Failure to treat eye injuries can lead to permanent damage and blindness. Wash the eye with clear water in a squeeze bottle or syringe to remove dirt and debris. Foreign bodies that will not wash out but are not imbedded in the eye can be removed with a wet, cotton-tipped applicator. If the eyeball has been penetrated, keep the horse in a clean, dust-free stall until a veterinarian arrives. Any eye inflammation needs veterinary diagnosis and treatment.

Bruises

Bruises are common injuries, because horses frequently kick each other. As soon as a bruise occurs, treat it with cold packs or cold water from a hose. If blood vessels are damaged, a hematoma ("blood blister") will form under the skin. After a day or two, the hematoma can be treated with hot packs and massaged with witch hazel. After a couple of weeks, a veterinarian may drain it. It usually takes 1 to 2 months for the hematoma to disappear. Large hematomas need the care of a veterinarian.

Sprains and Strains

Sprains and strains result when a muscle has been pulled or twisted. There may be a partial rupture of the ligaments and tendons. These injuries need immediate medical attention by a veterinarian. Rest the horse and restrict its movement until it is treated.

Saddle Sores

Saddle sores are frequently encountered when horsepeople use poorly designed and poorly fitted saddles or when insufficient padding is used under the saddle. Minor saddle sores are swollen, tender areas. First aid is rest and cold water or ice packs to reduce swelling. If the area is raw, zinc oxide salve or gentian violet may be helpful. Continued use of the horse under similar conditions can lead to large open wounds, requiring a veterinarian's help. They can be treated with nitrofurazone. If a saddle sore develops on a trail ride or camping trip, cut a hole in a sponge saddle pad to prevent continued damage to the tissue.

Heat Stroke

In hot, humid weather, horses may suffer heat stroke when being exercised. The symptoms are rapid breathing, weakness, dry skin, and no sweating. First aid consists of immediately lowering the body temperature. Place the horse in a cool place and spray with cold water. In extreme cases, use ice packs and place ice in the rectum. It may be necessary for a veterinarian to restore lost body fluids.

Severe Chilling

Horses are very tolerant of extreme cold, but young foals can become chilled. They should be placed under heat lamps and can be given a warm water enema.

Snake Bites

In some areas, snakes frequently bite horses on the head, neck, and other body parts. The only first aid treatment is to make sure that the swelling does not obstruct the airways. It may be necessary to pass a rubber hose down the trachea. All snake bites should be attended by a veterinarian but almost all horses survive even if left untreated.

Fractures

All fractures are serious, but many can be repaired with sophisticated surgical techniques and equipment. It is important than the fracture not be made worse than it is. Do not move the horse and call a veterinarian to make a temporary splint before moving the horse to a veterinary hospital. Successful treatment usually depends on immediate and continued immobilization during healing. Fractures in the pastern area always have a poor prognosis even though the area can be immobilized, because large calcium deposits usually develop and cause lameness. Fractures in heavily muscled areas are very difficult or impossible to immobilize. Euthanasia is often recommended for horses suffering such fractures.

If a veterinarian cannot arrive in a few minutes, immobilize the leg fracture with a pillow. Wrap the pillow around the leg as tightly as possible and wrap it tightly with bandages. Place two sticks along the leg and wrap them tightly to give extra stiffness. If the splints are properly applied, the horse can then be moved to a veterinary hospital.

Lameness

First aid for the acutely lame horse with injured and inflamed tissues must be given before the veterinarian arrives. Applying cold is an accepted means of first aid, because it reduces pain, swelling, hemorrhage, and inflammation.

Cold constricts the blood vessels, which in turn reduces blood flow to the area. Use cold treatment as soon as possible and only during the first 24 to 48 hours after an injury. There are several ways to apply cold, such as running water from a hose, ice packs, or standing the horse in a cool stream or a bucket of cold water. Apply cold for 20 to 30 minutes and then allow an hour before starting another treatment. If the treatments are too long or too often, the opposite effect is obtained—pain, swelling, and inflammation increase.

Heat is used to increase the blood supply to an area by dilating blood vessels and increasing their permeability. If infection is present, do not use heat because it allows the infection to enter the bloodstream and spread to other parts of the body. Heat can be applied with a heat lamp, warm water, a hot water bottle, liniments, and poultices. Heat is usually applied on the advice of a veterinarian.

Massage also increases blood flow and reduces tissue swelling. Increased blood flow pulls excess fluid in the tissue into the bloodstream, which carries it away. Soaking the injury in an epsom salts solution (2 cups epsom salts per gallon of water) draws the excess tissue fluid out of the tissue and reduces swelling. Poultices work on a similar principle and effectively draw fluid out of puncture wounds.

Older injuries require different treatment. A common treatment is the use of a counterirritant. Its purpose is to convert an old chronic injury into an active form to begin an active healing process. There are several types of counterirritants that vary in strength. Braces (such as Absorbine) are mild forms and are usually used after a workout. They produce mild heat. Tighteners are a little stronger than braces but are still mild counterirritants. These drug mixtures reduce fluid filling in a joint or tendon sheath and are used after hard workouts to reduce swelling of the lower legs. Tighteners are applied with a massage—which may be the most beneficial part of the treatment. Liniments are mixtures of drugs that increase blood flow to an area by producing mild heat. There are many types of liniments, and the advice of a veterinarian is helpful in choosing one. Sweats cause an accumulation of moisture on the skin. Wrapping the area with plastic after applying the sweat increases its effectiveness. Sweats should never be used without the advice of a veterinarian. Blisters and leg paints produce a profound counterirritation. They are used for bowed tendons and chronic bone unsoundnesses. Their effectiveness is debated among veterinarians because the effect may be limited to the skin. Firing is another means of producing counterirritation. Firing involves piercing various parts of the horse's leg with a red-hot iron or needles. It may be used on splints, bowed tendons, osselets, ring bone, and sesamoiditis. Some people feel that the rest following the treatment is the most beneficial aspect of many treatments.

Drugs are also used for therapy. Some antiinflammatory drugs are useful while others only serve to mask a lameness. Masking a lameness usually leads to more serious consequences that could have been avoided by resting the horse.

DENTAL CARE

Proper dental care for the horse is a necessary part of the grooming and health care program. Bad teeth or other dental problems result in poor mastication (chewing), which may cause colic. Some horses with painful dental problems may refuse to eat and will rapidly lose body weight. You can recognize problems in the early stages of development, because the horse may tend to wallow the feed around in its mouth before it swallows and a lot of feed may fall out of its mouth. There may also be clumps of hay in the horse's mouth a couple of hours after it has eaten.

The teeth of young horses grow about twice as fast as mature horses until the horse is about 5 years old. Therefore, the teeth of young horses should be examined every 6 months. The teeth of mature horses should be examined at least yearly. Examine the teeth by opening the mouth and pulling the tongue out to one side. With a flashlight, examine the incisors and the molars on that side. Then pull the tongue to the other side of the mouth to see the remaining molars. Some horses resist the procedure; use a mouth speculum (Figure 9-7) to examine their teeth (but avoid being hit if the horse swings its head).

One of the most common dental problems is "hooks" or sharp edges on the molars that develop because the upper and lower molars do not meet evenly (Figure 9-8). The upper molars are set slightly wider than the lower molars. The horse chews in a circular grinding motion, which causes the inside edges of the upper molars and outer edges of the lower molars to wear off at a faster rate than the rest of the top surface. Sharp edges develop on the outside edge of the upper and the inside edge of the lower molars. Hooks can cause mouth sores by striking the opposite gums and can cut the cheek or tongue. A horse with a sore mouth eats poorly. Remove the sharp edges with a special rasp called a *float* (Figure 9-7). Floating is an easily learned process. The supplies needed are floats, mouth speculum, flashlight, and a bucket of warm, soapy water. When floating teeth, take care not to hit the gums or the back of the jaw. This hurts the horse, which will flinch and pull away. Floats of various shapes are made for upper and lower jaws and have a fine and coarse side. The fine side is used for the final dressing. Use warm, soapy water to keep the floats warm and clean.

Wolf teeth are small teeth located in front of the premolars. There may be one to four wolf teeth in the horse's mouth. Normally only the upper wolf teeth erupt. If not removed, they can cause dental problems. A bit hitting the wolf teeth can be very painful to the horse. A horse with sore wolf teeth may carry its head to one side to escape pressure from the bit. If a wolf tooth is broken off, a chronic sore may develop, and it is almost impossible to use a bit until the problem is corrected. The lower wolf teeth usually do not erupt, but the bulge they make can bruise and become sore.

Wolf teeth are poorly developed and can be removed easily. However, those broken off below the gum margin can be difficult to pull out. The upper wolf teeth are lost at 5 to 6 months of age, but all young horses' mouths

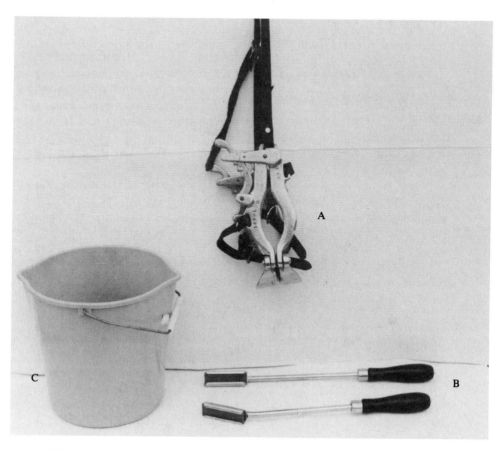

FIGURE 9-7

Equipment needed to float the teeth. (A) Mouth speculum. (B) Floats. (C) Bucket of warm, soapy water. A flashlight is also helpful for examining teeth. (Photo by B. Evans.)

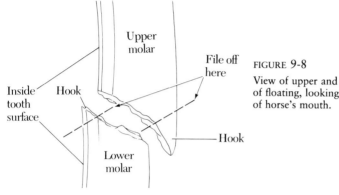

Upper molar

File off here

Inside tooth surface

Hook

Lower molar

Hook

FIGURE 9-8

View of upper and lower molars in need of floating, looking at left jaw toward rear of horse's mouth.

should be examined for wolf teeth before training. Older horses that sling their heads or carry them off to one side when wearing a bridle should be examined for wolf teeth or a broken wolf tooth.

The canine teeth of older horses may grow too long and strike the opposite gum, causing sores. A veterinarian can clip them to correct the problem. Sharp edges that develop can be dressed with a float.

The tooth opposite a tooth lost by accident or surgery can hinder chewing. Because it is not worn off by an opposed tooth, it may become too long. Rasp the tooth off at about 6- to 12-month intervals to keep it from getting too long. A too-long tooth can be clipped off.

Temporary teeth that fail to fall out may cling to the gum when they are being replaced by permanent teeth. These so-called caps interfere with the horse's ability to chew. Until the horse is 5 years old, the molars must be checked every 6 months to make sure the caps fall out. If they do not, pull them off.

Molars may wear unevenly so that they do not meet and cannot grind properly. Affected teeth can be floated. Usually the correction is made in two to three steps at 6-month intervals unless only a minor correction is required.

10
Knowing About Parasites and Diseases

Two major concerns of horse owners are to keep the parasite infestation of their horses to a minimum and to prevent their horses from getting common communicable diseases that can be prevented by immunization and a disease prevention program.

PARASITES

Parasites are small organisms that live on or in a host organism. They derive their food from the host organism. Parasites can be beneficial (such as symbiotic cecal bacteria) or detrimental to the horse. Our primary management concern is to control harmful parasites, which can lower efficiency, performance, and digestion and can even kill a horse.

The symptoms of parasitism may develop slowly and may not be recognized by the person who works with the horse every day. Typical symptoms are weakness, unthrifty appearance, emaciation, "potbelly" (large, distended abdomen), tucked up flanks, rough hair coat, and slow growth. The horse may lose its desire to perform and seem lazy. Some symptoms of acute parasitism are colic, diarrhea, cough, and lameness. These symptoms are discussed for each parasite.

Two types of parasites infest horses—internal and external. Internal parasites spend part of their life cycle in their host, whereas external parasites live on and/or derive their food from their host.

Internal Parasites

Many internal parasites affect horses, mules, and donkeys. From a management standpoint, they are generally classified as roundworms and insects with an internal stage of development. There are six groups of roundworms: Ascarids; small strongyles; large strongyles (bloodworms); threadworms; pinworms; and stomach worms. Insects that have an internal stage of development in the horse are stomach bots and cattle grubs. To minimize parasitism, the life cycle must be broken. An understanding of the life cycles can help you make management decisions.

Ascarids

Ascarids, commonly called *large roundworms,* are the largest parasites that live in horses. They grow 12 to 15 inches long (30.5 to 38.1 cm) and are about the diameter of a lead pencil. Ascarid infestation is primarily a problem in young horses, because older horses develop an immunity to them. The Ascarid life cycle (Figure 10-1) starts with mature female worms laying eggs in the digestive tract of the horse. The eggs pass out of the horse in the feces. After the eggs are outside the horse for about 2 weeks, they develop to an infective stage. The eggs are swallowed by the horse as it eats grass, feed or feces, and drinks water. Once inside the horse, the eggs hatch into larvae. These larvae burrow into the wall of the small intestine and then enter the bloodstream, which carries them to the liver and heart. They migrate through the liver and heart to the veins that drain these organs and thus travel to the lungs. They are coughed up from the lungs and swallowed. They reenter the digestive tract and mate, and the female produces eggs.

Ascarids damage the liver, heart, and lungs. In the bloodstream, they can block small capillaries and other blood vessels. Heavy infestations of mature worms can block the small intestine, causing colic. The intestine may rupture, causing peritonitis and death.

Ascarid eggs are very resistant to environmental conditions and can live for years before being eaten by a horse. Therefore, paddocks, stalls, and pastures easily become very contaminated with eggs.

Managerial control of Ascarids requires clean stalls for foaling, preventing fecal contamination of feed and water troughs, frequent manure removal from paddocks, proper manure disposal, and frequent treatments of the mature worms with drugs. The primary objective is to keep the environment relatively free from egg contamination. This is critical, because once the infective egg is

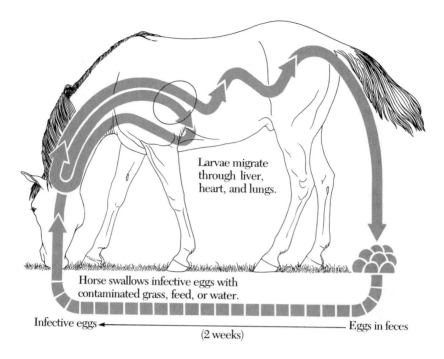

Larvae migrate
through liver,
heart, and lungs.

Horse swallows infective eggs with
contaminated grass, feed, or water.

Infective eggs ⟵   Eggs in feces

(2 weeks)

FIGURE 10-1
Cycle of infection by large roundworms.

swallowed the larvae travel their migratory route before they return to the
digestive tract and are killed by drugs (worm medicines). Worming the foal
at 2-month intervals beginning at 2 months of age keeps the number of mature
worm females and the number of eggs passed in the feces to a minimum.

Strongyles

Of the many parasites affecting horses, the most serious threat to the horse's
health is the strongyle. Approximately 54 species of strongyles (either large or
small) infest horses. Three species of large strongyles do the most harm to
horses. The small strongyles are much less pathogenic. Severe infestations of
both types are common in horses grazed on permanent pasture. Symptoms of
severe infestation are grouped under the condition known as *strongylosis* and
include low appetite, anemia, emaciation, rough hair coat, sunken eyes, digestive disturbances, and "tucked up" appearance. Sometimes the horses have
posterior incoordination.

Their life cycle (Figure 10-2) has some unique, important features. The
eggs are laid by adult worms in the intestine. They pass out in the feces and

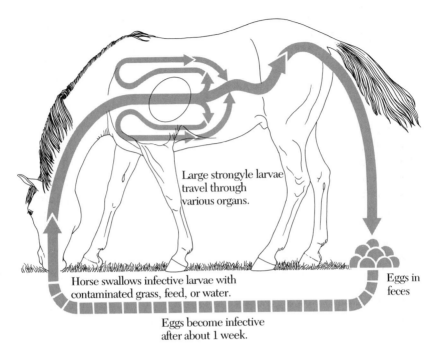

Large strongyle larvae travel through various organs.

Horse swallows infective larvae with contaminated grass, feed, or water.

Eggs in feces

Eggs become infective after about 1 week.

FIGURE 10-2
Cycle of infection by large strongyles or bloodworms.

hatch into larvae in 1 to 2 days. Larvae feed on manure and continue to develop. After about a week, the larvae become infective. The infective larvae move up on the blades of grass, and the horse eats them. Once inside the horse, they migrate to various organs and tissues where they do extensive damage. After the young adult worms return to the small intestine, they attach themselves to the gut wall, feed on the gut lining, reach maturity, and lay eggs. The life cycle of the small strongyle is different from that of the large strongyle in that they do not migrate from the small intestine.

Strongyles cause arterial damage and the development of small blood clots in the arteries. After the artery becomes damaged, aneurisms develop and burst. The horse bleeds internally and dies. Blood clots may break loose and become lodged in smaller arteries, blocking the blood supply to a particular organ or tissue. Frequently, the blood supply to parts of the digestive tract is blocked, leading to colic and necrosis of the affected part. When the blood supply to muscle groups is partially blocked, the horse may become lame.

Strongyle control is based on the life cycle. The larvae become infective after about 1 week outside the horse, so manure must be removed at least weekly and preferably twice weekly. Outside, infective strongyle larvae are not as resistant to the environment as Ascarid eggs and usually cannot live beyond 1 year, so pasture rotation or rest for a year helps control them. Grazing the pasture with cattle or sheep, which are not harmed by them, also helps remove the larvae. Routine worming with appropriate drugs must be used to keep larval numbers to a minimum and to minimize damage to the intestinal mucosa.

Pinworms

Pinworms do not cause much damage to the horse and their life cycle is simple (Figure 10-3). The mature worms live in the colon of the large intestine. One quite small species, *P. vivipara*, completes its entire cycle in the host. The species Oxyuris equi is much larger—up to 3 inches long (7.6 cm). The egg of *O. equi* are laid in the large colon and pass out in the feces, or the female worm may crawl out the anus and deposit the eggs on the skin around the anus. This causes an itching sensation, so the horse rubs its tail on fences, stall walls, posts, and so on. The eggs are quite sticky and yellow in color. They stick to stall walls, feed mangers, buckets, and other equipment. Therefore, the environment becomes contaminated with eggs and increases the opportunities for reinfestation through water and feed. Pinworms are controlled by manure removal and medication. To avoid egg contamination, cloths or sponges used to clean the perianal region should not be used to clean the mouth and nostrils of the horse.

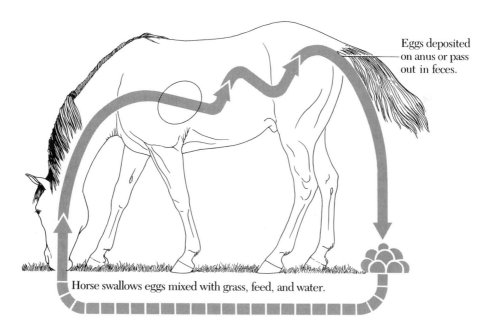

Eggs deposited on anus or pass out in feces.

Horse swallows eggs mixed with grass, feed, and water.

FIGURE 10-3
Cycle of infection by pinworms.

Threadworms

Foals are particularily susceptible to threadworms. The infestation can become severe in a short time, because the life cycle can be completed within 2 weeks. Infestation occurs from the dam's milk as early as 4 days after foaling and from penetration of the skin by larvae in the bedding. Control requires daily removal of feces from the stall and keeping mares relatively free of parasites.

Stomach Worms

Stomach worms attach themselves to the stomach wall or they may be free. Their life cycle (Figure 10-4) must be taken into account for management practices. The eggs are laid in the stomach and pass out in the feces. The maggots of flies that breed in horse manure eat the eggs. The worm egg hatches

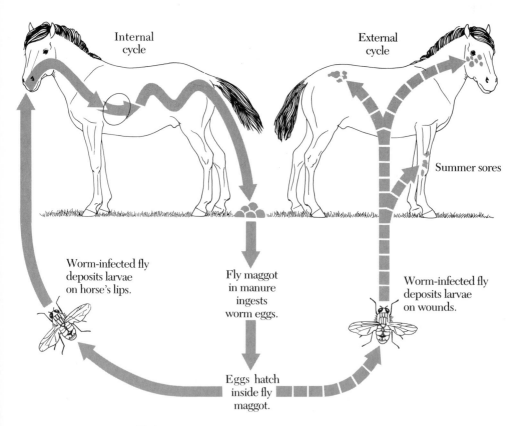

Internal
cycle

External
cycle

Summer sores

Worm-infected fly
deposits larvae
on horse's lips.

Fly maggot
in manure
ingests
worm eggs.

Worm-infected fly
deposits larvae
on wounds.

Eggs hatch
inside fly
maggot.

FIGURE 10-4

Cycle of infection by stomach worms as internal and external parasites.

and stays inside the fly maggot until an adult fly is formed. The larva leaves the fly when it is feeding around the liquid secretions of the horse's lips. The larvae are swallowed and go to the stomach, where they mature and lay eggs. They can cause stomach inflammations and colic. Controlling stomach worms requires manure removal, fly control, and worm medication.

Stomach worms are responsible for a condition known as "summer sores." When the fly is feeding around the edges of a wound, the larvae leave it and enter the wound but are unable to complete their life cycle. Therefore, they remain as larvae for 6 to 24 months. The sore becomes a granulomatous mass and is very difficult to heal. Chronic sores that fit this description should be treated by a veterinarian.

Stomach Bots

Three species of bots, or botflies, affect horses (see Figure 10-5). Parts of their life cycle are similar and others are different. The adult botfly resembles a bee. They do not feed and are an energy-consuming stage of the life cycle. Botflies are active from the first part of the summer until the first hard frost in winter.

Common botflies, *Gasterophilus intestinalis*, start their life cycle by laying eggs on the hairs of the horse's legs, chest, belly, and neck. Eggs on the hair are yellow and can be easily seen. About 1 to 2 weeks after they are laid, the eggs are capable of hatching. They hatch only when they are exposed to moisture, warmth, and friction. These conditions are met when the horse licks the area. When they are licked, they immediately hatch and enter the mouth, burrow into the tongue, and are eventually swallowed. Inside the stomach, they grow to maturity attached to the stomach lining. On maturity, they detach themselves from the stomach and pass through the digestive tract and out the anus. During the movement, they change to a pupal stage. Once outside, they burrow into the ground. Adult flies develop in the pupal casts and emerge during warm weather. Therefore the cycle takes 1 year.

The throat or chin botfly, *G. nasalis*, has a different life cycle from the common botfly in that the eggs are laid around the chin and throat. Their eggs incubate in 4 to 6 days and hatch without any stimulation. The larvae then crawl inside the mouth and enter the gum tissue around the teeth. They leave the mouth and enter the stomach and first part of the small intestine. The remainder of the cycle is similar to the common botfly.

The third botfly is the nose botfly, *G. hemorrhoidalis*. It lays black eggs in the area of the nostrils and lips. Eggs hatch in 2 to 4 days without stimulation and burrow into the inner lip membranes. After 5 to 6 weeks, they enter the stomach.

Bots cause stomach damage and can block the entrance into the small intestine, causing colic and rupture of the stomach. Most horses have a few bots, which will not cause problems. To control or prevent heavy infestations,

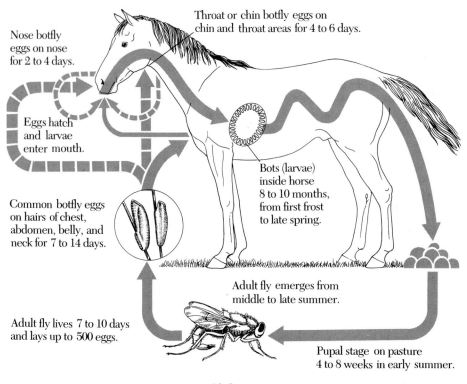

Nose botfly eggs on nose for 2 to 4 days.

Throat or chin botfly eggs on chin and throat areas for 4 to 6 days.

Eggs hatch and larvae enter mouth.

Common botfly eggs on hairs of chest, abdomen, belly, and neck for 7 to 14 days.

Bots (larvae) inside horse 8 to 10 months, from first frost to late spring.

Adult fly emerges from middle to late summer.

Adult fly lives 7 to 10 days and lays up to 500 eggs.

Pupal stage on pasture 4 to 8 weeks in early summer.

FIGURE 10-5
Life cycle of horse botflies.

the life cycle must be broken by killing the flies, egg removal, and worming. Botflies may be killed by sprays but it is difficult to control them this way. Once the eggs are laid, they can be removed from the hairs by scraping them with a knife or sandpaper. For effective control, they should be removed at least twice a week. If there are large numbers of common bot eggs on the horse, they can be induced to hatch by washing them with warm water (110° to 120°F, or 43.3 to 48.9°C) containing an insecticide. Washing the eggs once a week will hatch and immediately kill them. However, the empty yellow eggs will remain on the hairs. The horse should be treated with an antihelmintic medication at least twice yearly and preferably four times a year. At the end of the botfly season, after the first hard frost, worm the horse to remove internal bots. The frost should kill the botflies, and no new larvae should enter the horse. The second time to worm the horse is in the spring, to kill any larvae that were in the tissues at the previous treatment and to prevent any live pupae from passing out in the feces. For more effective control, worm the horse two more times, during the middle of the winter and summer.

Cattle Grubs

The larvae of heel flies, which are cattle parasites, can also cause problems in horses. The flies lay their eggs on the lower extremities. Larvae hatch and enter the horse through the skin. After a long migratory period, they emerge on the back of cattle by cutting a hole in the skin. The horse is the wrong host, so they cannot complete their life cycle and thus appear as hard knots on the neck, back, and withers. If the grub is mashed, a sore will form that takes a long time to heal. Sometimes it is best to have a veterinarian remove the grub before it becomes a sore.

Controlling Internal Parasites

To control internal parasites, feed horses from mangers and troughs rather than from the ground. Prevent fecal contamination of feed and water. Water should be fresh and not from stagnant water holes.

Remove manure from stalls and paddocks at least weekly. If possible twice weekly or daily removal is preferred. Scatter manure in pastures with a harrow to dry out the feces and expose eggs to the environment.

If pastures are used, temporary pastures that are plowed up or rotated yearly are preferred to permanent pastures. The more space that is available per horse, the less chance it will have to eat contaminated forage. Treat all horses with an antihelmintic before putting them at pasture.

Keep the general area clean and free from materials that encourage fly populations. Clean stalls daily to remove feces. This helps control flies. Compost the material removed from the stalls and all manure. The heat generated during composting kills the eggs and larvae.

Regularly deworm all horses. No single drug has been found to control all types of worms consistently and safely (Table 10-1). Usually a combination of drugs is necessary. New drugs are continually being produced, and you must be aware of their use and efficiency in killing internal parasites. There are several ways to worm horses. A veterinarian can pass a stomach tube and administer a combination of drugs. This method is effective because the dose can be given quickly so that the concentration is sufficient to kill the worms. Paste deworming is popular because of its effectiveness and ease of administration and because there is less chance of injury. The individual horse owner can obtain and administer the paste. Feed additives can be used, but palatability can be a problem. Some horses refuse to eat their feed containing the medicine.

For large horse farms, it may not be economical to routinely worm horses at predetermined intervals. Fecal examinations can be made and the horses treated when the average egg count indicates the need. Trained personnel are necessary to determine fecal egg counts, but the individual horse owner can check for strongyle eggs. The procedure is to put a fresh fecal ball in a quart

TABLE 10-1
Worming Chart

Name	Manufacturer	Chemical	Percent Effectiveness on Large Bloodworms	Percent Effectiveness on Small Bloodworms	Percent Effectiveness on Roundworms	Percent Effectiveness on Pinworms	Percent Effectiveness on Bots	Percent Effectiveness on Stomach Worms	Percent Effectiveness on Threadworms
Shell® Horse Wormer	Shell	Dichlorvos	90–100	85–100	99–100	99–100	91–100	90–100	0
Equigard®	Shell	Dichlorvos	90–100	85–100	99–100	99–100	91–100	90–100	0
Equigel®	Shell	Dichlorvos	90–100	85–100	99–100	99–100	91–100	90–100	0
CUTTER® Paste	Cutter	Febantel	95–100	95–99	100	100	0	0	% not available
Banminth	Pfizer	Pyrantel Tartrate	98	90–99	94	81	0	0	% not available
Strongid T	Pfizer	Pyrantel Tartrate	98	90–99	94	81	0	0	% not available
Wonder Wormer	Farnam	Piperazine	40–60	90–100	90–100	70–80	0	0	0
Pheno Sweet	Farnam	Phenothiazine	50–75	85–95	0	0	0	0	0
Equivet TZ	Farnam	Thiabendazole	Effective, but % not available	Effective, but % not available	Effective, but % not available	Effective, but % not available	% not available	0	90–100
Top Form	Merck	Thiabendazole	90–100	90–100	10–30	90–100	0	0	90–100
Equizole	Merck	Thiabendazole and Piperazine	90–100	90–100	95–100	90–100	0	0	0

Name	Manufacturer	Chemical	Percent Effectiveness on Large Bloodworms	Percent Effectiveness on Small Bloodworms	Percent Effectiveness on Roundworms	Percent Effectiveness on Pinworms	Percent Effectiveness on Bots	Percent Effectiveness on Stomach Worms	Percent Effectiveness on Threadworms
Foal Wormer	Farnam	Piperazine phosphate	40–60	90–100	90–100	70–80	0	0	0
Equivet 14	Farnam	Thiabendazole and Trichlorfan	98	99	99	99	98	90–100	90–100
Alfalfa Pellet	Farnam	Piperazine phosphate monohydrate	40–60	90–100	90–100	70–80	0	0	0
Parvex plus	Upjohn	Phenothiazine, piperazine, carbondisulfide	50–100	90–100	95–100	50–75	70–80	60–90	0
Telmin	Pittman Moore	Mebendazole	90–100	80–90	95–100	95–100	0	0	% not available
Dyrex TF	Fort Dodge	Trichlorfon, piperazine, phenothiazine	90–100	90–100	90–100	90–100	90–100	90–100	0
Dyrex Granule	Fort Dodge	Trichlorfon in organic phosphate	90–100	90–100	90–100	90–100	90–100	90–100	0
Dyrex Cap-tab	Fort Dodge	Trichlofon in organic phosphate	90–100	90–100	90–100	90–100	90–100	90–100	0

[a]The manufacturer presently seeks clearance for use of CUTTER® Paste Wormer in pregnant mares on the basis of data already submitted. When the original application was made, these data were not available.

SOURCE: *Horse Lovers National* (October 1978).

(continued)

TABLE 10-1 (continued)

Name	Sales	Administration	Pregnant Mares	Use on Foals	Recommended Frequency of Use	Resistance	Dosage	Comments
Shell® Horse Wormer	Over the counter	In feed	Safe	Not recommended for sucklings and weanlings	4–6 weeks	None	7.8 gm/ 100 lb (45.4 kg)	Has no taste, but horse can smell it. If frozen, dichlorvos is released slowly from plastic encasement and does not smell as strong. Must withhold water, or bots on stomach ceiling won't be touched.
Equigard®	Through vets	In feed	Safe	Not recommended for sucklings and weanlings	4–6 weeks	None	7.8 gm/ 100 lb (45.4 kg)	Same as above.
Equigel®	Through vets	Paste syringe	Safe	Not recommended for sucklings and weanlings	4–6 weeks	None	Not available	When injected into mouth, it starts to break down and kills migrating bots in lips.
CUTTER® Paste	Through vets as Rintal Paste, over the counter as CUTTER® Paste	Paste syringe or mix in feed	ª	Safe	Every 8 weeks	Not available	6 gm/ 1,000 lb (454 kg)	Outstanding features: broad spectrum and low-volume dosage.
Banminth	Over the counter	In feed	Safe	Safe	Every 6 weeks	Not available	Not available	Outstanding palatability.
Strongid T	Through vets	Tube, syringe, or in feed	Safe	Safe	Every 6 weeks	Not available	3 gm/ 100 lb (45.4 kg)	Can be given with carbon disulfide to kill bots.
Wonder Wormer	Over the counter	In feed	OK until foaling	From 3 months of age	8–10 weeks	Not available	1 oz/ 250 lb (113.5 kg)	For small horses and ponies.
Pheno Sweet	Over the counter	Flavored granules sprinkled on feed	OK until last 4 weeks	Not for use on very young foals	2 times yearly	May develop	5.7 gm/ 100 lb (45.4 kg)	Also used on cattle; animal may be sensitive to sunlight after administration.

Name	Sales	Administration	Pregnant Mares	Use on Foals	Recommended Frequency of Use	Resistance	Dosage	Comments
Equivet TZ	Over the counter	Paste syringe	Data unavailable	Data unavailable	6-8 weeks	May develop	2 gm/ 100 lb (45.4 kg)	
Top Form	Over the counter	Pellet feed	Safe	OK for all ages	6–8 weeks	Not available	5 oz/ 1,000 lb (454 kg)	
Equizole	Through vets	Tube or in feed	Use only on vet's advice	Use only on vet's advice	6–8 weeks	Not available	Not available	
Foal Wormer	Over the counter	Sugar granule in feed	Not meant for adult horses	Not for use on unweaned foals without vet's advice; OK for weanlings	6–8 weeks	Not available	1.85 gm/ 250 lb (113.5 kg)	
Equivet 14	Over the counter	In feed	Not to be used in last month	Not for use on foals under 4 months old	6–8 weeks	Not available	3.8 gm/ 100 lb (45.4 kg)	
Alfalfa Pellet	Over the counter	Apple base pellet feed	Safe	Not for use on foals under 3 months old	8–10 weeks	Not available	50 mg/ 1 lb (454 gm)	Requires a vet's followup.
Parvex plus	Through vets	Tube	Use only on vet's advice	Use only on vet's advice	Vet's advice	Not available	Vet's advice	
Telmin	Through vets	Tube or syringe	Use only on vet's advice	Use only on vet's advice	Vet's advice	Not available	Not available	
Dyrex TF	Through vets	Tube	Can use until 30 days prior to foaling	Not for use on foals under 3 months old	Vet's advice	Not available	Vet's advice	
Dyrex Granule	Through vets	In feed	Can use until 30 days prior to foaling	Not for use on foals under 3 months old	Every 30 days	May develop	3.6 gm/ 100 lb (45.4 kg)	
Dyrex Cap-tab	Through vets	Paste syringe	Can use until 30 days prior to foaling	Not for use on foals under 3 months old	Every 30 days	May develop	2.8 gm/ 100 lb (45.4 kg)	

313

jar. Place the jar, with a loose lid on top, in a dark room at room temperature for a week. Drops of condensed moisture should develop on the sides of the jar. If, on daily examination, the walls are dry, add 10 to 20 drops of water to the jar. At the end of a week, strongyle larvae will be visible with a hand lens in the moisture drops if the horse has the parasites. A heavy strongyle infestation will be evident because of the large numbers of the larvae, giving the inside of the jar a frosted appearance. With training by the local veterinarian, egg counts can be made by a horse owner. The sugar flotation test is easy to use. Weigh a piece of feces (about one-quarter of a fecal ball) and place it in 20 cc of a solution of equal parts sugar and water. Stir the solution until the feces is in suspension. Strain the mixture through a piece of gauze or a tea strainer. Let the fluid portion sit for 15 minutes, and then put a few drops on a microscope slide for 15 minutes. Count eggs with a microscope and a 100 or 200× lens with a 10× objective lens. Counts of over 1,000 per gram of feces indicate that large numbers are being passed daily. The count should be around 50 or less under a good parasite control program.

External Parasites

In addition to internal parasites, many external ones also annoy or harm horses.

Ticks

Three kinds of ticks commonly infest horses. The winter tick is widely distributed in the northern and western states. Pacific Coast ticks are found primarily in coastal areas. Spinose ear ticks are found in the arid and semiarid regions of the West and Southwest. Horses grazed on pasture normally acquire heavy tick infestations. Frequently, the ticks go unnoticed on a horse with a long winter hair coat. The first visible symptoms are rubbing, itching, or biting of the affected areas. Ear ticks may cause drooping ears, head shaking, moist exudate, or excessive wax secretion from the ear.

Tick removal is important because heavy infestations of winter ticks can cause weakness, emaciation, loss of appetite, anemia, and a general state of poor health. Pacific Coast ticks transmit many diseases, such as Rocky Mountain spotted fever and Colorado tick fever. It only takes a few of these ticks to cause paralysis. Ticks transmit piroplasmosis and African horse fever.

Many insecticides are available for tick control (Table 10-2). The entire horse can be sprayed with either emulsifiable concentrates or wettable powders. An excitable horse can be sponged off. In 2 to 4 weeks, repeat the process. Single ticks can be removed by swabbing them with cotton soaked in alcohol or chloroform. Because ticks breathe through spiracles or holes on their abdomens, they are suffocated by the alcohol or chloroform.

Areas where horses are confined that become infested with ticks should be sprayed with insecticide. Surfaces such as the underparts of feed troughs, corral posts, stall walls, barn walls, and tree trunks should be carefully sprayed. Areas where horses congregate and stand, such as under shade trees or loafing sheds, should also be thoroughly sprayed.

TABLE 10-2
Parasite Control for Horses (nonfood use)

Chemical Formulation and Concentration			Application Method	Amount	Remarks

- Horn Flies, Stable Flies (Treat during warm months when abundant.)
- Lice (Treat during fall and early winter, with repeated application in 10 to 14 days if necessary.)

Chemical Formulation and Concentration			Application Method	Amount	Remarks
Ciodrin®[a]	Aerosol	2%	Mist spray	Direct application	Apply to areas where flies congregate (head, neck, withers). Repeat as necessary.
Couma-phos (Co-Ral®)	WP[b]	25%	Hand wash or spray[c]	½ lb/25 gal water, or 226 gm/94.6 liters (0.06%)	Do not use in conjunction with internal medications, especially phenothiazine, natural or synthetic pyrethroids and their synergists, or with organic phosphates. Repeat as necessary. Apply to areas where flies congregate (hand, neck, chest, withers).
	EC	11.6%		½ qt/25 gal water, or 226 gm/94.6 liters (0.06%)	
Dichlorvos (Vapona®)	EC	20–24% (2 lb/gal)	Hand wash or mist spray	½ oz/1 qt water, or 14 gm/.75 liter (0.5%), ½ oz/animal	Apply to forehead, neck, wither, or chest areas where flies congregate. Repeat as necessary.
Lindane[d]	WP[d]	25%	Hand wash or mist spray	4 oz/25 gal water, or 113 gm/94.6 liter (0.03%)	Do not treat overheated or sick animals.
Malathion	WP[b]	25%	Hand wash or spray	1 lb/25 gal water, or 454 gm/94.6 liters (0.12%)	Do not treat overheated or sick animals.

[a]May be sold in combination with or without 0.25% dichlorvos (Vapona®) and/or pyrethrins + synergist.

[b]Wettable-powder formulations are preferred over emulsifiable-concentrate solutions because of certain solvents contained in EC formulations, which even upon dilution with water may cause hair damage and skin irritation to some horses.

[c]Hand washing or sponging may be more desirable than spray applications because some horses become excitable and difficult to control when sprayed.

[d]Restricted material. A permit is required to purchase, apply, and possess this material.

SOURCE: Bushnell and others (1976).   (continued)

Methoxy-chlor	EC	5%	Hand wash or spray	1 pt/1 gal water, or .47 liter/ 3.78 liters (0.7%)	Do not treat overheated or sick animals
Pyrethrins + Synergist (Piperonyl butoxide)	EC	0.1% + 1%	Hand wash or spray	Full strength	Do not treat overheated or sick animals.

● Face Flies[e] *(Treat during warm months when flies are abundant.)*

Stirofos (Rabon®)	Smear	1%	Wipe on	Direct application	Apply to areas where flies congregate (face area; under eyes, atop nostrils in line to forehead). Repeat as necessary.
Couma-phos (Co-Ral®)	Dust	1%	Dust bags	Add 2 to 3 lb dust/bag (.9 to 1.3kg). Recharge as necessary. Do not allow more than 1 oz (28.4 gm) /head-day use.	Coumaphos dust may be used in dust bags made from narrow-gauge burlap sacks and grommeted at open end for hanging from a chain or board; placement depends on number of horses, size, and habits. In some instances, dust transferred to homemade bags hung like punching bags may prove more effective. Also pour dust into sack and apply by hand.

● Screwworms *(Treat when present.)*

Ciodrin®[a]	Aerosol	2%	Thoroughly mist spray wounds and sur- rounding area with not more than 2 oz (56.8 gm)	Direct application	Repeat once weekly until wound is healed.
Couma-phos (Co-Ral®)	Dust	5%	Thoroughly dust wounds and surround-ing area	Direct application	
Lindane[d]	EQ 335 smear		As directed on label	Direct application	
Ronnel (Korlon®)	Spray	2.5%	Spray, dust, or smear	Spray, smear or dust into and around wounds. Use as directed.	
	Dust bottle	5%			
	Smear	5%			

- *Spinose Ear Tick (Treat when abundant. May be present all seasons, particularly in spring and fall.)*

Ciodrin®[a]	Aerosol	2%	Mist spray or dust into ears as deeply as possible.	Repeat as necessary.
Couma-phos (Co-Ral®)	Dust	5%		
Ronnel (Korlon®)	Smear	5%	Smear into and around ears. Use as directed.	Repeat as necessary.

- *Ticks (Treat when abundant.)*

Couma-phos (Co-Ral®)	WP[b]	25%	Hand wash or spray	1 lb/25 gal water, or 454 gm/ 94.6 liter (0.12%)	See instructions given for horn flies
	EC	11.6%		1 qt/25 gal water, or .75 liter/ 94.6 liters (0.12%)	
Lindane[d]	WP[d]	See instructions given for horn flies.			
Malathion	WP[b]				

- *Botfly Eggs (on body, lips, and jaws of animals)*

Couma-phos (Co-Ral®)	WP	See instructions given for horn flies. Hand wash using warm water.	Wipe body hair areas, lips, or jaw regions where botfly eggs are observed. Refer to remarks given for horn flies.
Lindane[d]	WP		
Malathion	WP		

[e]These flies are extremely annoying to horses, congregating mainly on muzzles and eyes. They closely resemble the common house fly, which is more numerous on horses having access to pasture, open rangeland, or when in confined conditions (pens) when these are adjacent to pastured cattle. Animal sprays do not give satisfactory control of the face fly, and because the fly concentrates on the face, frequent use of face smears or daily dust-bag use by horses offers more promise of satisfactory control.

Note: Formulas to Calculate Insecticide Dosages:

1. To prepare a spray of certain percentage strength from a wettable powder (WP), use

$$\frac{\text{gal spray needed} \times \text{\% spray recommended} \times 8.3^*}{\text{\% of WP}} = \text{lb WP to add}$$

Example: 100 gal of 0.25% coumaphos is needed from a 25% WP package

$$\frac{100 \times 0.25\% \times 8.3}{25} = 8.3 \text{ lb to add}$$

(continued)

TABLE 10-2 *(continued)*

2. To prepare a spray of certain percentage strength from an emulsifiable concentrate (EC) use

$$\frac{\text{gal spray needed} \times \% \text{ spray recommended} \times 8.3^*}{\text{lb active ingredient/gal in EC} \times 100} = \text{gal EC to add}$$

Example: 100 gal of 0.06% coumaphos is needed from a 11.6% EC can (1 lb active ingredient/gal);

$$\frac{100 \times 0.06 \times 8.3}{1 \times 100} = 0.49 \text{ gal (2 qt) to add}$$

Various emulsifiable concentrates usually contain the following lb active ingredient/gal:

Less than 20 = 1 lb
21–25 = 2 lb
40–45 = 4 lb
50–57 = 5 lb
58–67 = 6 lb
72–80 = 8 lb

(There are exceptions to this, so be sure to read the insecticide label.)

*8.3 = density in lb/gal for water. If mixing insecticide in #2 fuel oil for backrubber, use 7.3.

Lice

Two kinds of lice infest horses: the sucking louse and the biting louse. Their life cycles are similar, but the sucking louse probably does more damage. Sucking lice can live only a few days after removal from their host. Their eggs—commonly called *nits*—are laid on the hair and are firmly attached to it. After 10 to 20 days, they hatch. The newly hatched nits reach sexual maturity and lay eggs when they are about 11 to 12 days old. Therefore, their life cycle is 20 to 30 days. Infestation usually shows up in the winter, when the animals have long hair. They spread quite quickly by direct contact and through hair left in grooming equipment. The biting louse lives by feeding on the external layers of skin, hair, and exudates. Sucking lice actually penetrate the skin and suck blood. The itching sensation causes the animal to rub, bite, and kick. If they are present in large numbers, the horse's coat becomes rough and patches of hair are rubbed off.

Lice can be seen by pulling a tuft of hair and holding it up to the sunlight. Treatment consists of applying an insecticide, which will also kill mites and ticks (Table 10-2). One treatment usually suffices, but it is best to give another treatment about 10 to 14 days later to get the ones that hatch out after the first treatment.

Mites

Mange or scabies of horses is a specific skin disease caused by small mites that live on or in the skin. There are three types of mites—sarcoptic, psoroptic, and chorioptic.

The sarcoptic mites penetrate the upper layer of the skin and form burrows in which to mate and lay eggs. This causes irritation and itching. Each female forms a separate burrow and lays 10 to 25 eggs. The eggs hatch in 3 to 10 days. The young mites begin laying eggs in new burrows when they are 10 to 12 days old. The average life cycle is 14 to 15 days. The first symptoms are intense localized itching. Later on, blisters and nodules form, break, and discharge serum. The first lesions usually form on head, neck, shoulders, flanks, and abdomen but spread to the whole animal.

Psoroptic mites do not form burrows but live in colonies on the skin surface. The female lays about 24 eggs that hatch in 4 to 7 days. They reach maturity and lay eggs 1 to 12 days later. They have a 2- to 3-week cycle. Psoroptic mites spread faster than the sarcoptic variety. They usually start in the mane and tail. They puncture the skin for food and secrete a poisonous substance into the wound. This causes itching, inflammation, formation of vesicles, and exudation of serum.

Chorioptic mites are usually found below the hocks and knees. They live on the surface and cause symptoms similar to psoroptic mites. The horse tries to relieve the irritation by pawing, kicking, and biting its lower limbs.

Treatment for mites requires a thorough weekly application of an effective insecticide (Table 10-2). Dusts are not effective for the sarcoptic mites, because the insecticide must penetrate into the burrows.

Chiggers

Chiggers, commonly called *redbugs,* may also infest horses. Massive infestations cause severe dermatitis. They can be controlled by application of a detergent wash containing an insecticide (see Table 10-2).

Gnats

Several species of blood-sucking gnats that feed on horses are widely distributed throughout the United States. Some black fly species are attracted to the horse's ears, where they suck blood and cause irritations and tissue damage. Lesions resulting from the bites make the ears very sore. Other species attack the head, neck, and belly. When they are feeding, horses toss their heads, swish their tails, twitch their skin, and restlessly move about their paddock. Other species of gnats are small and difficult to see. Their bites are painless but later cause itching. The horse loses hair where it rubs and scratches.

Gnats are difficult to control so sprays, wipe-ons, or smears must be used daily (Table 10-2). It may be necessary to have the horse wear an ear net if the ears get very sore and there are many gnats.

Mosquitos

Mosquitos should be controlled because they are a vector for transmitting "sleeping sickness" viruses. Their biting annoys horses, and skin reactions may develop. Large populations of mosquitos are usually found close to their water breeding sources. However, wind can carry them several miles. To control them, eliminate standing water or treat the water areas with insecticides. Look for standing water in old containers, water troughs, plugged rain gutters, low spots in corrals and pastures, ditches, irrigated pastures, creeks, tree holes, and poorly covered septic tanks and drains.

Flies

Several species of flies annoy horses and must be controlled. House flies transmit stomach worms. They feed and lay their eggs in waste matter such as manure, bedding, wet feed, and decomposed plant material. Stable flies are vicious biters and suck blood. They also transmit stomach worms. Their eggs are laid in horse manure and piles of decomposed plant matter. Face flies are pests to horses grazing with cattle. They irritate the mouth, eyes, and nose by sucking mucous secretions. Their control is difficult because repellants and contact insecticides do not last long enough. Fly masks can help keep face flies from the eyes and general face region. Screwworm flies lay their eggs in fresh wounds. The larvae develop in the wound and feed on the tissue; large sores develop as a result. In areas where screwworm flies are a problem, inspect horses frequently for minor cuts and abrasions. The navel cords of newborn foals are prime targets, and blowflies lay their eggs on the vulva after the mare has foaled.

Fly control on horse ranches is based on reducing natural fly attractants, eliminating breeding areas, and chemical (Table 10-2) or biological control. Eliminate attractants and breeding grounds by manure and water management and general farm sanitation. Remove manure from stalls and paddocks daily or at least twice weekly. It should be composted, taken to a sanitary landfill, or spread very thin on a field. Avoid accumulations of manure, bedding, and/ or feed with water. No part of the water supply system should leak, and water should not accumulate around water troughs. General farm sanitation requires prompt removal of all animal and vegetable matter. Remove garbage, spoiled feed, afterbirths, and so on, as quickly as possible.

There are several ways to control flies chemically. The choice of the appropriate insecticide and method of application depends on the fly population and physical facilities. Persistent insecticides can be applied to surfaces in the stable area. Large populations can be reduced quickly with quick-knockdown insecticides. Baits can be used around areas that tend to attract flies. Space sprays or fogs can be used in enclosed areas to quickly kill large numbers if no residual effect is necessary.

Finally, biological control by the use of fly predators such as parasitic wasps is effective, economical, and labor saving. A major benefit of biological control is that it avoids the adverse side effects of chemical controls. Many of the chemicals are toxic nerve poisons that accumulate in fat depots of horse and human bodies; for example, DDT and organophosphates.

DISEASES

The ability to recognize diseases or disease symptoms is important because the symptoms indicate the need for veterinary services. A knowledge of the common diseases, their means of transmission, and ways of prevention enables the horse owner to mount an effective disease control program. (See Table 10-3 for a summary of infectious diseases.)

Common Diseases
Anthrax

Anthrax is an acute infectious disease caused by bacteria—*Bacillus anthracis*. There are two routes of infection—ingestion or biting insects. Horses eat anthrax spores during droughts when the pastures are overgrazed or after a flood that spreads the spores. Spores resist environmental conditions and live for several years. The ingested spores cause colic, enteritis, and septicemia. Horses also eat the feces of infected animals or contaminated feed and feed products. Biting insects transmit the disease, and the symptoms resulting from insect bites include fever, hot and painful edematous swelling of the throat, lower chest, abdomen, prepuce, and mammary glands. The symptoms last 48 to 96 hours. Treatment is usually not successful or necessary, because mortality is very high. In areas where anthrax is a problem, it can be prevented by annual vaccinations. If a horse dies from anthrax, bury the carcass at least 6 feet deep. Flesh eaters can spread the disease, so dispose of the carcass immediately.

Equine Encephalomyelitis

The common name for equine encephalomyelitis is "sleeping sickness." It is a brain disease that results in degeneration of the brain's vascular system. The disease is caused by several different viruses, each located within a given geographical area. The three most common virus types are Eastern (EEE), Western (WEE), and Venezuelan (VEE). Viral transmission is primarily by mosquitos.

TABLE 10-3
Summary of Infectious Diseases

Disease	Cause	Primary Characteristics	Prevention	Treatment
Influenza	Myxovirus	Fever of 101–105°F (38.3–40.5°C), watery nasal discharge	Vaccinate; isolate new animals	Good nursing care
Strangles	Streptococcus equi	Fever of 103–105°F (39.4–40.5°C); enlarged lymph nodes, nasal discharge—watery at first, then purulent	Isolate new animals; vaccinate all animals with killed vaccine	Antibiotics
Pneumonia	Virus or bacteria	Fever of 102–105°F (38.9–40.5°C); chest pains, lung congestion, difficulty in breathing	Good management; avoid chills and stress, provide proper ventilation	Antibiotics when cause is bacterial
Rhinopneumonitis	Herpes virus	Fever of 102–106°F (38.9–41.1°C); coughing, clear nasal discharge; abortion primarily in last third of gestation		No treatment
Viral arteritis	Herpes virus	Fever of 102–106°F (38.9–41.1°C); abortion	No vaccine available	Good nursing care
Equine encephalomyelitis	Virus	Brain lesions, drowsiness, fever; lower lip drops; difficulty in walking	Vaccination	Good nursing care
Equine infectious anemia	Virus transmitted in blood of infected horses	Fever of 104–108°F (40–42.2°C); weakness, jaundice, edema of ventral abdomen	Use of Coggins test to detect carriers	No treatment
Tetanus	Clostridium tetani	Muscular rigidity; prolapse of third eyelid	Tetanus toxoid or antitoxin	Mortality rate is high
Anthrax	Bacillus anthracis	High fever, edema about throat, lower neck, and chest; lasts approximately 48–96 hours	Vaccines—use only in an epidemic area	Antibiotics, antianthrax serum
Piroplasmosis	Protozoans	Fever, depression, anemia, tearing eyes and swollen eyelids	Tick control	No treatment

Wild birds serve as the principal reservoir for the virus, even though it is also carried by wild rodents and other animals. People and horses get the disease but do not transmit EEE and WEE to other horses or people. VEE can be transmitted from human to human or horse to horse by moisture in the breath.

Early symptoms include high fever for 1 to 2 days, restlessness, and nervousness. Brain lesions form, and the horse becomes drowsy, holds its head low, has droopy ears, and may walk around aimlessly. This is followed by paralysis and possibly death within 2 to 4 days after the first symptoms are noticed. Mortality for WEE is about 20 to 30 percent. Horses that recover are seldom useful because of brain damage. EEE and VEE have mortality rates of 70 to 90 percent.

All three types of sleeping sickness can be prevented by yearly vaccinations. Vaccine is available to the horse owner at tack shops and animal care centers.

Equine Infectious Anemia

Equine infectious anemia (EIA) is a viral disease that has received considerable attention by horsepeople ever since a means of detecting carrier animals was developed. The disease affects less than 1 percent of the horse population. It takes three forms—subacute (chronic), acute, and inapparent (in carriers). The three forms differ widely in their impact on the horse.

The temperature of an acutely infected animal rises to 105 to 108°F (40 to 42.2°C). The horse appears depressed, becomes weak, and loses appetite and weight. Fluid (edema) may collect under the skin, in the legs, chest, and abdomen. Membranes lining the natural body openings may be congested and yellow in color. There is a marked drop in the number of red blood cells (anemia). The horse usually dies in a few days.

Subacute forms are less severe, so the horse survives. The chronic form is sometimes classified as subacute. Such horses may display acute symptoms at any time or periodically.

An inapparent carrier usually shows no symptoms of EIA except that it may be unthrifty. Such horses react positively to the test for EIA. The virus can be transmitted by some of these horses by various means. About 85 percent of positive reactors are inapparent carriers.

The disease is primarily transmitted from acute and subacute cases to other horses by biting insects (primarily horse flies) and contaminated syringe needles. Therefore insect control and use of sterile or disposable needles is important to prevent spread of the disease. There is no immunization for protection against EIA.

The current recognized test for EIA is the agar-gel immunodiffusion (AGID) test developed by Dr. LeeRoy Coggins. The test is not for the EIA virus but for antibodies developed to fight the disease. A horse that reacts positively to the test is classified as a carrier of EIA. The test is very accurate and demonstrates that the horse has been infected with EIA virus. Many states have regulations concerning positive reactors and require tests before horses are shipped into the state.

Influenza

Two specific type viruses—myxovirus A/Equi and myxovirus A/Equi 2—are responsible for most cases of influenza (flu). About 48 hours after exposure to the virus, symptoms develop, including fever (101 to 106°F, or 38.3 to 41.1°C), cough, runny nose, depression, and loss of appetite. Flu is quite common where large numbers of horses are congregated for shows, rodeos, and sales. The virus is spread by water droplets moving through the air. A coughing horse can spread the virus as far as 35 feet (10.7 m). It settles on feed and in water troughs and is ingested. It is inhaled in water droplets. Flu is not fatal and only causes temporary incapacity. However, if the horse is not rested and is exposed to bad weather, serious secondary complications may arise.

Horses bought at sales should have their temperature taken a couple of times per day for about a week after they are brought home. A rapid rise in temperature is the first sign. The horse can be immediately treated and isolated to prevent the flu from spreading. All horses should be vaccinated against influenza and booster shots given annually. Horses that are constantly being shown or exposed to large groups should be given booster shots at 6-month intervals.

Piroplasmosis

Piroplasmosis was discovered in the United States in 1961. This is a tick-borne disease caused by either of two protozoans that invade the red blood cells by the following process. The tick attaches itself to the horse, and the protozoan leaves the tick and enters the bloodstream and the red blood cells, multiplies, and destroys the cells. The released protozoa enter other red blood cells and multiply. Symptoms include fever, depression, anemia, thirst, tearing eyes, and swollen eyelids. The urine is yellow to reddish in color. Mortality rate is 10 to 15 percent. Prevention consists of tick control and sterilizing all needles and medical instruments.

Pneumonia

The all-inclusive term *pneumonia* refers to any lung inflammation. Pneumonia can be caused by a bacterium, a virus, or a combination of the two. The symptoms are fever (102 to 105°F, or 38.8 to 40.5°C), nasal discharge, loss of appetite, difficulty in breathing, chest pains, and lung congestion. Prompt treatment by a veterinarian is usually successful. However, in some circumstances treatment is useless. Foals that have combined immunodeficiency disease (CID) (see Chapter 16) die after they contract adenoviral pneumonia.

Pneumonia can develop in a horse as a result of liquid being passed down a stomach tube inadvertently placed in the trachea. Therefore stomach tubes should be passed only by trained personnel.

Rabies

Rabies is common among horses in some areas. The presence of a rabid animal on the premises and fresh wounds on the muzzle, face, or lower limbs usually indicate that exposure has occurred. Quarantine the horse and closely observe it for 3 to 6 months. Exposed animals should be immediately treated by a veterinarian. Initial signs of rabies include personality change, lack of appetite caused by inability to swallow, and incoordination. Some horses with rabies do not become vicious. Others are very vicious and must be destroyed. In problem areas, horses should be vaccinated against rabies.

Ringworm

Several kinds of fungi cause ringworm. They are spread by direct or indirect contact with contaminated stalls, feed, mangers, gates, and grooming equipment. Some types can be spread to other animals, including people. Each type of ringworm requires a certain type of treatment, so a veterinarian should take skin scrapings, identify the type, and then prescribe appropriate treatment. A diagnosis of ringworm can usually be made from its appearance. Small, reddish, round lesions appear, covered with small scales. The hair breaks off just above the skin level. Older lesions may heal in the center, but the edges are quite active.

Rhinopneumonitis

Rhinopneumonitis is a viral infection passed from horse to horse and similar to flu. It affects horses of all ages. The respiratory symptoms are usually quite mild. Body temperature may increase 1 to 2°F (0.6 to 1.1°C), nasal discharge is clear and thick, and the cough is shallow. In pregnant mares, it causes abortions or stillborn foals. Abortions normally occur during the last trimester of pregnancy and may occur as long as 4 months after the respiratory symptoms. Occasionally, rhinopneumonitis may attack the central nervous system, causing ataxia (inability to walk or stand).

No vaccine is recommended. Vaccinating horses with live or attenuated live virus may cause problems. If the disease is present, a planned exposure program may be employed. Seek and follow recommendations by the local veterinarian.

Strangles

Strangles is also called *distemper* or *shipping fever*. It is a highly communicable disease caused by a bacterial infection. The bacteria *Streptococcus equi* causes the lymph glands in the jaw and throat to become infected and swollen. The incubation period is about 3 weeks. Symptoms are fever of 103 to 104°F (39.4 to 40°C), nasal discharge, cough, swollen lymph glands, and difficulty in swallowing. The lymph glands may abscess and burst. Severe cases may result in abscesses forming in other organs of the body. Treatment consists of penicillin treatment by a veterinarian and good husbandry (clean water and feed; clean, dry stall; and quiet surroundings).

Once strangles bacteria are present in an area, they persist for years. Horses are exposed and develop the disease, so it is a common problem of stables that have a transient trade or of horse traders. To avoid strangles, all horses should be isolated when they are first brought to the premises. Quarantine sick horses so that no feed buckets, equipment, or personnel handle the sick horse and then other horses unless first disinfected. After the horse recovers, thoroughly disinfect the area and all its contents. If the stall floor is dirt, the disinfectant must penetrate at least 6 inches (15.2 cm).

Horses can be vaccinated against strangles but the side effects are severe. The best procedure is to avoid contact with horses that have the disease. After recovering, horses develop varying degrees of immunity to it. If they have it again, the severity is greatly reduced, and many horses develop a lifetime immunity.

Tetanus (Lockjaw)

Tetanus is a serious disease caused by a wound infection by a bacterium called *Clostridium tetani*. The bacteria are unique in that they grow in the absence of oxygen and light. Tetanus bacteria spores are present in most soils. These spores are dormant and resist heat, cold, and dryness for several years. When the spores are carried into a wound on the end of a nail or in dirt and when the right conditions exist, they change back to a normal bacterial cell. The bacteria produce toxin that affects the nervous system. The initial symptoms, which occur about 7 to 10 days after the infection, are stiffness in all four legs, stumbling when walking, and cocked-out ears. The third eyelid may cross the inside aspect of the eye, or it may flip across the eye when the horse is startled. In 2 to 3 days, the horse is unable to walk and stands like a "sawhorse." The ears are cocked out and the tail raised and held out behind. Muscle spasms and twitches occur. All muscles seem to contract. If the horse falls down, it will be unable to rise.

Treatment should be by a veterinarian, but few horses recover. Prevention is by two means. All horses should be immunized against tetanus with tetanus toxoid and given booster shots at recommended intervals. If the horse has not been protected by immunization and sustains a wound, tetanus antitoxin gives temporary protection for 2 to 3 weeks. Tetanus antitoxin should be given within 24 hours after an injury occurs. Young foals can be immunized by tetanus antitoxin after birth or by boosting the mare's immune system with tetanus toxoid about a month before foaling. Sufficient antibodies are produced and secreted in the milk to protect the foal.

Viral Arteritis

Viral arteritis is an upper respiratory disease that resembles rhinopneumonitis and influenza. Symptoms include a fever (102 to 106°F, or 38.8 to 41.1°C) and nasal discharge. The symptoms are more severe than for influenza and rhinopneumonitis but mortality is usually very low. The virus also causes abortions in about 50 percent of affected pregnant mares.

The disease is highly communicable for 8 to 10 days from an infected horse and is spread by inhalation of respiratory droplets from infected horses or by injection of contaminated material. After exposure, horses acquire a natural immunity.

Warts

Warts frequently appear around the lips and muzzles of horses as well as the underparts such as the inside of the thighs and on the prepuce. One wart or masses of them may be present. They usually cause no harm unless they are irritated by tack. They will spontaneously disappear in 3 to 4 months if they are not treated. They are transmitted to humans and other horses because they are caused by a virus. A vaccine is available for treatment.

Metabolic and Other Diseases
Azoturia and "Tying Up" Syndrome

Azoturia and "tying up" syndrome are two metabolic disorders involving various muscles of the horse. Azoturia is frequently called "Monday morning disease" and is technically known as *paralytic myoglobinuria*. There is controversy as to whether or not they are similar but distinct diseases or different manifestations or varying degrees of the same disease. The more serious is azoturia. The circumstances under which it occurs are as follows. The horse is being exercised regularly, and typically it is being fed a ration high in grain. The horse rests for 1 or 2 days and eats its full ration during that period. The horse returns to work, and shortly thereafter symptoms of the disease appear: signs of muscle stiffness, tremors, and pain. If exercise continues, the severity increases. In advanced stages, the horse is unable to move and may fall down. Anxiety causes profuse sweating. The urine becomes reddish-brown to almost black. The change in urine color is caused by the release of myoglobin (red pigment in muscle tissue) when the muscle fibers break down. All movement should be stopped and a veterinarian called.

Symptoms of "tying up" are less severe but similar to azoturia. The symptoms occur after the horse has been exercised for a while. Muscle pain and stiffness are noticeable, and the stride becomes short and stilted. Most horses are reluctant to move. These horses usually have a history of being fed a high-grain diet.

To prevent azoturia, reduce the amount of grain or do not feed grain on rest days. Horses that "tie up" should receive less grain if they have been receiving large amounts. Other preventative measures include vitamin E and selenium therapy.

Colic

The term *colic* is used to describe a number of painful digestive disturbances with many different causes. Improper feeding practices such as abrupt changes in the horse's feed or overfeeding commonly cause colic. One of the overwhelming causes is parasites. The bloodworm, *Strongylus vulgaris,* migrates through the circulatory system, causing arterial damage and blood clots. When blood clots block the flow of blood to part of the intestines, irreparable damage occurs. Roundworms may block the intestine and prevent the movement of digesta through the tract. Twists may occur in the long digestive tract, which causes part of the tract to necrose. Therefore, the extensive folding and length of the digestive tract make the horse vulnerable to colic. Infections with bacteria and viruses may cause enteritis and colic.

The symptoms of colic are essentially the same regardless of the cause. Heart rate and body temperature increase. Sweating may occur and may become very profuse as the pain increases. The horse will be restless, paw the ground, try to roll, frequently get up and down, bite its abdomen, kick at its abdomen, refuse to eat, may or may not drink water, show a change in feces, or fail to defecate for several hours. The pain is caused by the digestive tract becoming stretched by gas or digesta accumulation, inflammation resulting from blockage of blood supply or blood leaking into the peritoneal cavity, or muscle spasms in the gut wall.

Because colic is the number one killer of horses, it is a serious problem. Prompt action is essential to successful treatment. Professional assistance is necessary whenever abdominal pain is moderate to severe; abdominal distension is present; temperature, pulse, and respiration have increased; the color of the mucous membranes darkens, or no feces have been passed for 24 hours. Treatment may be simple such as a mild analgesic and walking the horse around or it may be complicated and involve surgical correction.

The incidence of colic can be reduced by good feeding management practices and by a well-planned and -executed internal parasite control program.

Heaves

The common names for pulmonary emphysema are "heaves" and "broken wind." It can be caused by an allergic reaction to milk, by chronic cough in dusty surroundings, or by eating dusty hay. The condition is characterized by a chronic cough, difficulty in forcing air out of the lungs, possible nasal discharge, unthriftiness, and lack of stamina. Heaves cannot be cured but relief from the symptoms can be obtained by proper management precautions. One of the most obvious visual signs is the marked jerking of the flanks to force the air out of the lungs. Extensive use of these muscles causes the formation of a "heave" line in the flank area. Air must be forced out of the lungs because the alveoli, small air sacs in the lung, have lost their elasticity. Because expiration is passive, the air will not move out of the lung after the elasticity is lost. Lack of stamina results from a decreased ability to get oxygen to the tissues.

To prevent heaves or to manage the horse with heaves, never feed moldy or dusty feed. Keep bedding as free as possible of dust. Avoid keeping the horse in confinement conditions where dust is a problem. Horses should be kept at pasture if possible. Severe cases render the horse useless for anything except breeding purposes. Horses with mild cases can be used for pleasure if they are not strenuously exercised.

Melanomas

Melanomas are benign growths that occur on the tail and anus and on the head of gray or white horses. They are usually rare in horses 6 to 7 years old, but about 80 percent of gray horses over 15 years old have them. They start out about the size of BB shot and gradually increase in size. There is no treatment except surgical removal. New ones may appear at a later time. They can become malignant and invade vital organs, causing death. The major problem is interference with tack, particularly on the head.

Disease Prevention

Disease prevention is an important part of managing a horse farm or keeping a pleasure horse healthy. The age of the horse affects the efficiency of the program. Young foals are not capable of developing antibodies until they are about 3 months of age, when passive immunity acquired from the mare begins to decline. Therefore, they should not be immunized until they are 3 months of age. All horses should be immunized against tetanus, influenza, and sleeping sickness (EEE, WEE, and VEE). It may be advisable to immunize against strangles, anthrax, and rabies depending on the circumstances.

Each farm or stable should have an isolation area where incoming horses can be kept for 2 to 3 weeks for observation. The incubation periods for most diseases are less than 3 weeks. Horses that become sick should also be isolated to prevent the spread of communicable diseases. The tools and equipment used to take care of a sick horse should not be used for other horses until they have been decontaminated. They can be washed in disinfectants but a veterinarian should choose the disinfectant, depending on the disease. The stall area should be decontaminated. After all dirt and debris are removed, a thorough scrubbing with water and soap followed by a thorough application of moist steam under pressure is an excellent method because it kills all disease organisms except some spores. Steam cleaners can be rented from local tool rental agencies. If such cleaners are unavailable, the stall can be sprayed with a disinfectant such as one of the phenolic germicides or hypochlorites. Each disinfectant has limitations as to which bacteria, fungi, spores, and viruses that it will kill, so a veterinarian should be consulted for the exact product to use.

Bedding and manure that is removed should be burned. Aborted fetuses should be removed and buried.

Rats and mice must be controlled because they are a source of contamination. All grain and feed bins should be rodent-proof. Flies and other insects serve as vectors of disease and must be controlled.

Records of all vaccinations, treatments, health care, wormings, hoof care, and dental care should be maintained for all horses. Review them periodically to ensure that each horse is receiving appropriate care.

III
USING AND
ENJOYING
THE HORSE

11
Riding
and Driving

Riding or driving a horse is a popular means of recreation. Horses are ridden several different ways, such as western, forward, sidesaddle, or saddle seat. Each method of riding was devised to meet a particular purpose or obtain a particular type of performance. Driving horses is also becoming increasingly popular. A large recreational industry has developed, including riding stables (both private and public), breeding farms, training facilities, tack manufacturers, clothing companies, transportation, and so on.

To ride or drive a horse successfully and safely, one must become familiar with the necessary equipment for various methods of riding and driving, which are discussed in this chapter.

RIDING EQUIPMENT

Western

Saddles

The western saddle was developed because horsepeople who spent long hours riding needed a comfortable saddle for the horse and for themselves and a strong saddle that could hold a roped steer. The basic western saddle is shown in Figure 11-1. In selecting a western saddle, one of the main considerations is the seat, which should be flat, not built up in front. A seat that is built up in

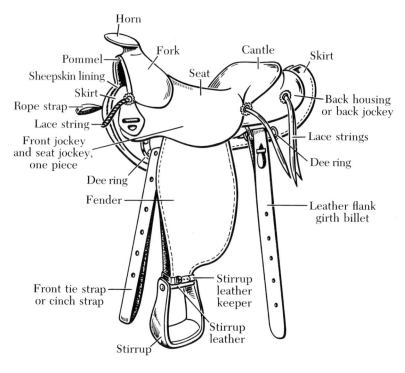

FIGURE 11-1

Basic western saddle.

front prevents the rider from sitting forward over the horse's center of gravity. The rider then sits too far back on the loins of the horse, which fatigues the horse too quickly. The placement of the stirrup leathers determines the position (forward or backward) of the legs and feet and thus determines balance when riding. Many saddles are made with the stirrups too far forward, which puts the rider's legs also too far forward.

The tree is the basic unit of the saddle. Saddle trees (Figure 11-2) are made of fiberglass, aluminum, or wood covered with rawhide. The way a saddle fits on a horse's back is determined by the width and height of the gullet and the flare of the bars. Gullet height and width are important in determining what type of horse the saddle tree will fit. A low, wide gullet will probably rub against the withers of a narrow-chested, high-withered horse, whereas a narrow or regular-width gullet will not fit down on a wide-backed horse. The bars of a saddle tree must have the proper angle and "twist" to fit a horse. The bars must be long enough so that they do not irritate the back over the loin. Longer bars give more bearing space on the horse's back and distribute the rider's weight more evenly. These factors often determine the difference between a low-cost saddle that causes saddle sores on the shoulders and back, and a

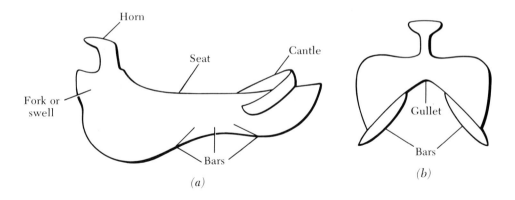

FIGURE 11-2

(a) Side view and *(b)* front view of a western saddle tree, around which the rest of the western saddle is built. (From Hyland, 1971.)

good, usable saddle. The size of the tree determines the size of the finished saddle. The average size of a tree is 15 or 15½ inches (38.1 or 39.4 cm) measured from where the seat joins the fork to the inside of the cantle after the saddle is completed. The saddle is rigged as centerfire, ⅝, ¾, ⅞, or full, depending on the position of the front cinch ring (Figure 11-3). The centerfire-rigged saddle is rigged with a single cinch. Cowboys found that such saddles tended to tip and slip forward when they roped cattle so the front cinch was moved forward and a flank cinch was added (Figure 11-1). Today, most saddles are rigged double—full or ¾. Western saddles weigh approximately 40 pounds (18.5 kg) but may be lighter or heavier. The saddles used by single-steer ropers take a great deal of strain and weigh 43 to 45 pounds (19.4 to 20.4 kg).

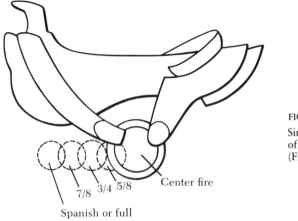

FIGURE 11-3

Single rigging position of a western saddle. (From Hyland, 1971.)

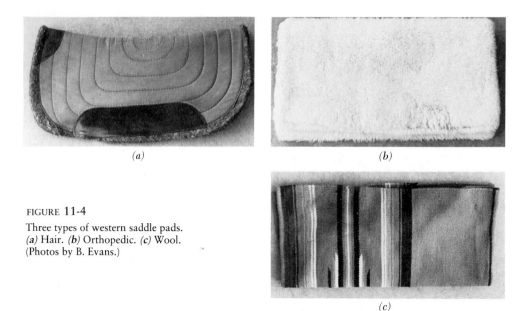

FIGURE 11-4

Three types of western saddle pads.
(a) Hair. *(b)* Orthopedic. *(c)* Wool.
(Photos by B. Evans.)

Pads

Western saddles require a pad between the saddle and the horse's back. These pads are made of a variety of materials (Figure 11-4). Pads made of synthetic material can be washed and kept clean quite easily. Pads made from the material used to make orthopedic hospital pads have become very popular. The "hair pad" has been a standby over the years. It does cake with dirt when it is wet with sweat, but the dirt can be brushed out. Failure to clean hair pads results in a bruised or irritated area on the horse's back. Wool blankets are used on horses that are not worked very hard and are ridden for short periods.

Bridles

Western riders use a bridle composed of a set of reins, a bit, and a headstall (Figure 11-5). The type of rein varies with the use of the horse. Braided rawhide reins with a romal are quite popular with many riders. Rope horses are usually ridden with a single closed rein. The split reins are an old standby used on ranch, pleasure, and show horses. Many variations of headstalls are used. The simplest headstall consists of a leather strap that goes over the poll and is split so that it encircles one ear. Other types of headstalls have a browband and/or throatlatch to keep the headstall in the proper position on the horse's head.

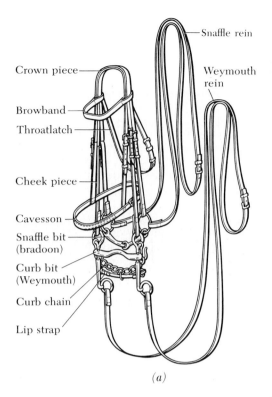

Crown piece

Browband

Throatlatch

Cheek piece

Cavesson

Snaffle bit
(bradoon)

Curb bit
(Weymouth)

Curb chain

Lip strap

Snaffle rein

Weymouth
rein

(a)

FIGURE **11-5**

Three types of bridles. *(a)* Use Weymouth bridle,
composed of snaffle and curb bits with double
reins, in showing three- and five-gaited horses.
(b) Use pelham bridle, a single-bitted, double-
reined bridle, on hunters, polo ponies, and
pleasure horses. *(c)* Use one-ear or split-ear bridle
on working stock horses (horses that work
cattle); the type shown has a roping rein and a
curb bit.

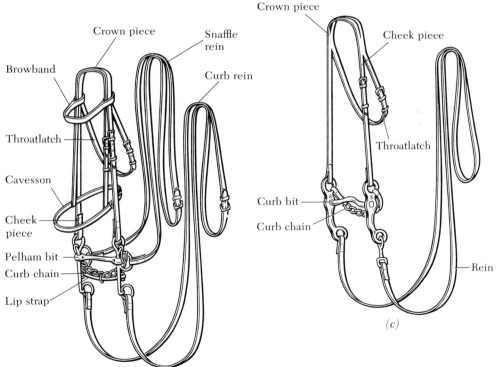

Crown piece

Browband

Throatlatch

Cavesson

Cheek
piece

Pelham bit

Curb chain

Lip strap

Snaffle
rein

Curb rein

(b)

Crown piece

Cheek piece

Throatlatch

Curb bit

Curb chain

Rein

(c)

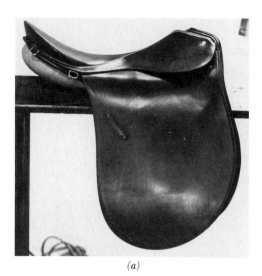

FIGURE 11-6

Basic types of forward saddles. *(a)* Dressage saddle.
(b) General-purpose forward seat saddle.
(c) Show saddle. (Photos by B. Evans.)

(a)

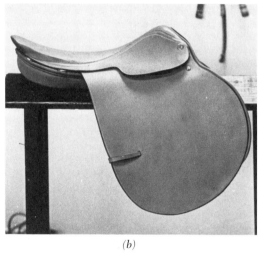

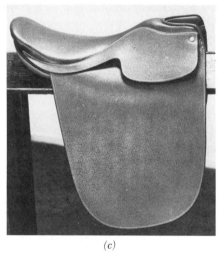

(b)   *(c)*

Forward

Saddles

Forward saddles (Figures 11-6 and 11-7) are made in three basic designs. The "flat saddle" is the general-purpose saddle for all-around riding and for dressage. The seat of the flat saddle is a little deeper than that of the show saddle. In recent years, many forward seat riders are using the "forward seat" or "jumping" saddle. The flaps are cut well forward and may incorporate padded knee rolls to help the rider when going over jumps. The forward seat saddle may be cut back. The third type, the show saddle, has a flat seat that causes

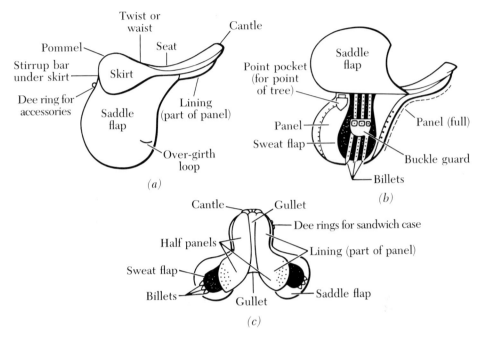

FIGURE 11-7
Forward (hunt seat) saddle. *(a)* Side view. *(b)* Side view, flap raised. *(c)* Saddle flap.

the rider to sit back toward the cantle. The flaps are straight down, and usually the pommel is cut back. Show saddles are used in saddle seat classes and for saddle-bred and gaited horses that are ridden with long stirrups. These saddles are measured from the front end of the pommel (at the screw) to the top of the cantle. The average rider takes a 20- to 22-inch (or 50.8- to 55.9-cm) saddle (Figure 11-8).

The girths that hold the forward-type saddle on the horse are made of several types of material. Sturdy, washable cotton webbing or folded leather are the most popular materials. A tubular, sheep-wool girth cover can be slipped over the girth to prevent chafing.

If the girth is not sufficient to hold the saddle in position, a breastplate is attached to the sides of the pommel and to the girth. Breastplates are frequently used on hunters. A plump-bellied pony may require a crupper under the tail to prevent the saddle from slipping forward.

Stirrup leathers for forward-type saddles are sold separately. To keep the stirrups adjusted to the same length, the holes must be punched at exactly the same length on both stirrups. Leathers should be switched from one side of the saddle to the other quite regularly because mounting tends to stretch them.

Stirrup irons must be approximately 1 inch (2.54 cm) wider than the rider's foot. Many novice riders use safety stirrups (Figure 11-9) because they auto-

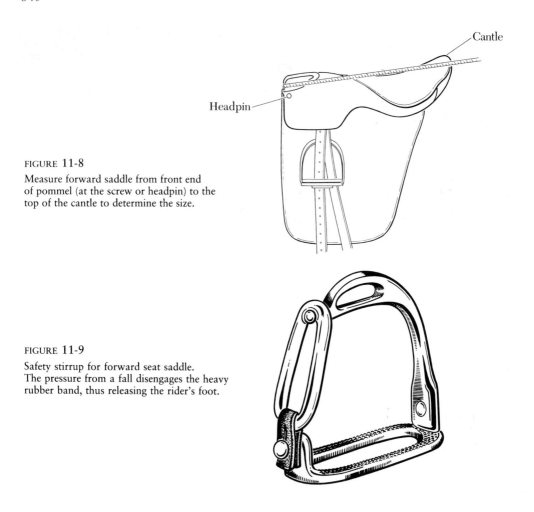

Cantle

Headpin

FIGURE **11-8**

Measure forward saddle from front end
of pommel (at the screw or headpin) to the
top of the cantle to determine the size.

FIGURE **11-9**

Safety stirrup for forward seat saddle.
The pressure from a fall disengages the heavy
rubber band, thus releasing the rider's foot.

matically open if the rider falls. Rubber stirrup treads are quite popular but
are dangerous to use because the rider's foot does not slip out of the stirrup
as easily when an accident occurs.

Pads

Saddle pads or "nunnals" are not necessary for English-type saddles but are
used by many horsepeople, particularly during ordinary riding. The pad is
attached to the saddle by running the billets through two straps. Pads are
usually made of sponge rubber, synthetic fibers, sheepskin, or felt (Figure
11-10). They absorb sweat (thus protecting the saddle) and protect the
horse's back and withers.

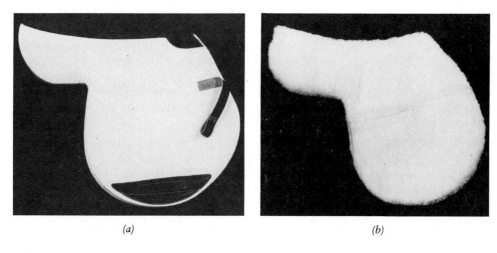

(a)   *(b)*

FIGURE 11-10
Forward saddle pads. *(a)* Felt. *(b)* Orthopedic. (Photos by B. Evans.)

Bridles

English bridles (Figure 11-5) come in a variety of types, depending on the use of the horse and type of bit it needs. The so-called pelham bridle is composed of a double set of reins, the pelham bit, cavesson, headstall, browband, and throatlatch. It is commonly used on hunters, polo ponies, and pleasure horses. The Weymouth bridle is used to develop flexion while maintaining the head carriage induced by riding the horse into the bradoon (small ring snaffle). Consequently, it is used on pleasure, three-gaited, and five-gaited horses.

The cavesson (noseband) can be used with a bridle to prevent the horse from evading bit action by opening its mouth (Figure 11-5). The effectiveness of the noseband can be increased by dropping the chin portion forward and under the mouthpiece of the bit and moving it to cross the chin groove.

Endurance and Competitive Trail Riding
Saddles

On endurance or competitive trail rides, most riders use western and forward seat saddles. However, special saddles have been developed to meet the needs of these performance events (Figure 11-11). Modifications of the McClellan saddle have been popular (Figure 11-12). Most are equipped with a crupper and/or breastcollar to prevent the saddle from slipping backward or forward on the horse's back. The cinch must be positioned far enough behind the horse's

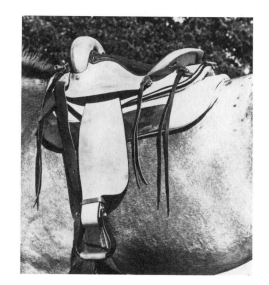

FIGURE 11-11

Endurance saddle designed by Sharon Saare
for American Saddlery of Chattanooga, Tennessee.
(Courtesy S. Saare.)

FIGURE 11-12

McClellan saddle. (Courtesy R. A. Woodworth.)

elbow to prevent galling the horse behind the front legs, so western saddles are
rigged with a three-quarter rigging. The saddle horn of western saddles is in
the rider's way and prevents one from riding a horse properly up steep inclines,
so it is usually removed.

Pads

Endurance and competitive trail riders use typical western and forward saddle pads. Good-quality pads and blankets are very important to prevent back sores. The padding must be thick and heavy enough so that it will not wrinkle. When actually competing, most riders have enough saddle pads so that they can start out with a fresh, dry pad after each veterinary check. They are transported by the rider's crew.

Bridles and Bits

Almost all types of headgear are used by trail riders. Although it is not advisable, some people ride with only a halter. The bridle and bit should be simple but should allow easy control of the horse and let it drink easily out of shallow streams.

Other Equipment

Riding long distances requires equipment for the horse's and rider's safety and comfort. Water canteens are usually carried in warm climates. If water is not available for the horse, the crew may have to carry water to the rest stops. A small pouch attached to the saddle is handy to carry a flashlight, chapstick, sun lotion, salt tablets, gloves, sunglasses, and a snack. A water sponge, hoof pick, and a rubber boot (in case of a lost shoe) are carried for the horse.

Bits and Bitting

Bitting a horse has traditionally been done by trial and error and probably will continue to be. Each person has a different sense of touch for a given horse's mouth with a particular bit. You may have to use a different bit on the horse to obtain maximum performance. Riders with "light hands" may use a harsh bit with a punishing effect but beginning riders or riders with "heavy hands" need to use a bit that does not punish a horse. Similarly, each horse may require a different bit for maximum performance because of differences of jaw width, tongue thickness, head length, and amount of pressure applied by the bit to specific areas.

A bit must fit the horse's mouth so that the horse is comfortable. Bit selection also depends on the horse's training and temperament and on the kinds of riding the owner does. The wrong bit can decrease control or delay training. Proper selection requires an understanding of the types and uses of bits.

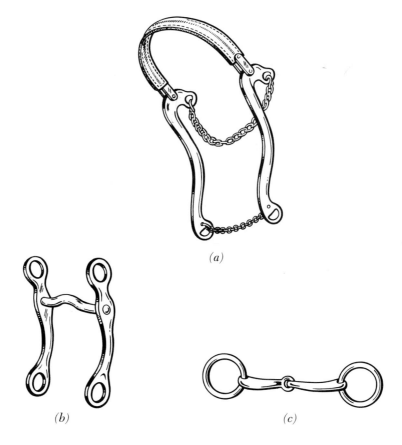

FIGURE 11-13
Three basic types of bits. *(a)* Hackamore. *(b)* Curb. *(c)* Snaffle.

Types

Bits are classified as hackamore, curb, and snaffle (Figure 11-13). English and western riders are acquainted with the many types and uses of snaffle bits. Western riders use curb bits that have many variations, whereas English riders use kimblewicks (commonly called "kimberwickes"), curbs, and pelhams. The snaffle bit comes in a variety of forms (Figure 11-14) and is used for many purposes. The common feature of all snaffle bits is that the mouthpiece is jointed or straight, with rings for the reins at both ends. The rings on the ends of the mouthpiece may be of four types: round, egg butt, racing dee, or full check. The snaffle is a mild bit, and the severity is regulated by the circumference of the mouthpiece; that is, the smaller the circumference, the more severe the

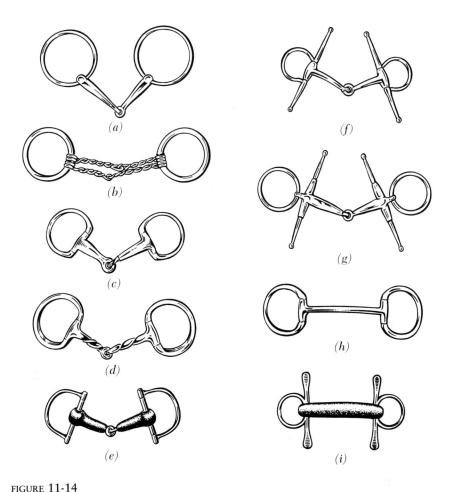

FIGURE 11-14

Types of snaffle bits. *(a)* Jointed mouth. *(b)* Flat ring, Y or W twisted wire snaffle. *(c)* Jointed mouth, German egg butt snaffle. *(d)* Jointed mouth, twisted egg butt snaffle. *(e)* Rubber-covered racing dee snaffle. *(f)* Egg butt full-cheek snaffle. *(g)* Australian loose-ring or Fulmer snaffle. *(h)* Metal mullen mouth egg butt snaffle. *(i)* Soft rubber mullen mouth full-spoon cheek snaffle. (From DuPont, 1973.)

bit. Twisted mouthpieces also increase the severity of the bit and, if used improperly, can ruin the bars of a horse's mouth. The pull of the rein exerts a direct pull on the horse's mouth equal to the force exerted by the rider. Therefore, lowering a snaffle in a horse's mouth from the correct position of one or two wrinkles at the corners of the mouth increases its severity. The round rings must be large enough not to pull through the mouth. Quite often the lips are pinched by the rings. To prevent this, the racing dee and egg butt snaffles have

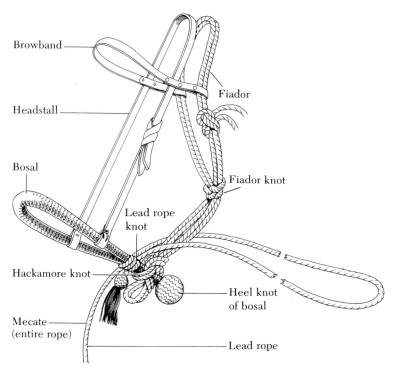

Browband

Headstall

Bosal

Fiador

Fiador knot

Lead rope
knot

Hackamore knot

Heel knot
of bosal

Mecate
(entire rope)

Lead rope

FIGURE 11-15
A hackamore (jaquima) with a fiador.

metal casing around the joints. The full cheek snaffle has two small bars inside the rings that prevent the rings from sliding into the mouth and that serve as a steering aid for the young horse.

The hackamore bit (Figure 11-13) is often confused with a hackamore (Figure 11-15). Hackamore bits do not have mouthpieces. They are used on horses that will not accept a bit or that have mouth injuries that would be irritated by a mouthpiece. Many hackamore bits have long shanks, and thus a great deal of pressure can be applied to the nose and chin groove without much pull on the reins.

A true hackamore (Figure 11-15), derived from the Spanish word *jaquima,* is used to start colts in training. The hackamore is used to apply a direct pressure on the nose and chin with the bosal. Bosals can be made of any of several materials such as rope, horsehair, or leather, but a braided rawhide bosal with a rawhide core is preferred. The bosal must fit closely around the horse's nose and low near the soft cartilage. The bosal is held in place with a headstall and a fiador. The fiador (Figure 11-15), which is a small double rope attached to the heel knot, passes over the poll and serves as a throatlatch. The mecate (Figure 11-15) is a soft, large-diameter rope that is easy to grip and is long enough to form a lead rope after the continuous rein is adjusted so that it reaches the saddle horn. Attachment of the mecate to the bosal forms the lead rope knot. The size of the bosal is adjusted by the hackamore knot (Figure 11-16) so that a slight lift to the heel knot causes the cheeks of the bosal to apply pressure to the chin. The heel knot should be heavy enough so that chin pressure is relieved as soon as rein pressure is decreased. Both reins of a hackamore are used simultaneously in the early stages of training so that the horse learns to respond to a direct rein and a bearing rein. The direct rein applies lateral pressure to the horse's head, and the bearing rein applies pressure to the side of the neck. Later, only a bearing rein is used.

The kimblewick bit (Figure 11-17) is similar to a snaffle bit except that there are loops or slots at the top of the rings for attachment of the bridle and a curb chain. The kimblewick uses single reins that exert light pressure on the mouthpiece. Increasing the pull on the reins causes the dee rings to act as short shanks and apply curb chain pressure. The curb pressure is not severe because of the short lever action of the dee rings and because the rider does not have much control over the amount of curb action.

To gain greater control than that provided by snaffle and kimblewick bits, English-style riders use the double-reined pelham bit (Figure 11-17). One set of reins attaches to rings that pull on the mouthpiece, and the other set is attached to rein loops on the ends of shanks. Pulling on both reins simultaneously or on the snaffle rein gives snaffle action. Pulling only on the curb rein activates the curb action, which also applies pressure to the curb chain groove and on the poll. The pelham bit is popular because it is easy for a rider to learn to use.

The double bridle allows the English-style rider to use snaffle and curb action either simultaneously or separately. The two bits, called a *Weymouth set* (Figure 11-18), are composed of a bradoon (small ring snaffle) and a curb. They are used primarily on dressage and gaited horses and sometimes on hunters.

Curb bits for western riders come in numerous shapes and combinations of mouthpieces and shanks. The basic curb bit is shown in Figure 11-19. The distance between the center of the mouthpiece and the center of the bridle loop should be 1⅞ to 2 inches (4.8 to 5.1 cm) for proper bit action. The distance between the center of the mouthpiece and the center of the rein loop (shank length) determines the leverage or amount of pull from the reins that is exerted on the bit. The shanks of most curbs are 5 to 5½ inches long (12.7 to 14 cm). Mouthpieces are 4½ to 5 inches long (11.4 to 12.7 cm). Solid mouthpieces usually have a rise (port) in the middle that relieves part of the tongue pressure and applies pressure to the roof of the mouth. The "halfbreed" has a high port (1½ inches maximum, or 3.8 cm) with a roller in it. The curb chain must be kept fairly tight to prevent the port from injuring the roof of the mouth. Most horses play with the roller with their tongues and, in so doing, keep their

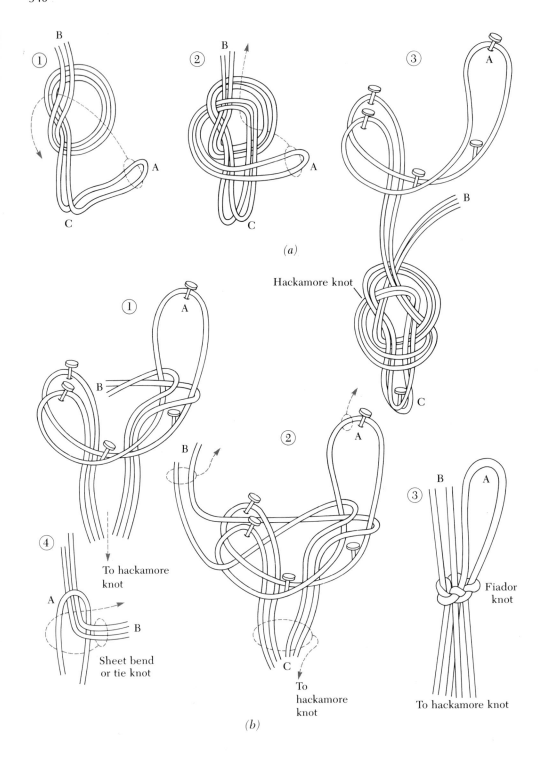

① B

② B

③ A

C

A

C

A

B

Hackamore knot

C

(a)

① A

B

② A

B

Hackamore knot

④

To hackamore
knot

A

B

Sheet bend
or tie knot

C

To
hackamore
knot

③ B A

Fiador
knot

To hackamore knot

(b)

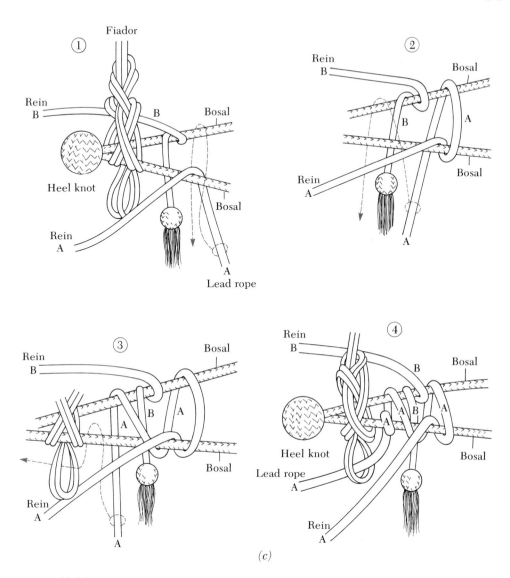

FIGURE 11-16
Steps in preparing (a) jaquima (hackamore), (b) fiador, and (c) lead rope knots for a jaquima.

mouths moist. Quite often, rollers and the ports of bits are covered with copper, which increases the flow of saliva to keep a horse's mouth moist. The spade bit (Figure 11-19) is very severe (3½-inch port, or 7.8 cm) and must be used with a tight curb chain and very light rein pull. The spade bit also has a roller (cricket) and springs.

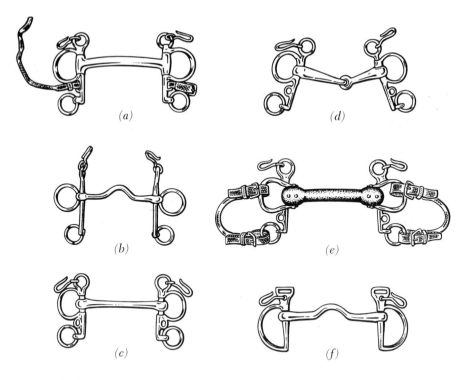

(a)

(d)

(b)

(e)

(c)

(f)

<small>FIGURE</small> 11-17

Pelham and kimblewick (kimberwicke) bits. *(a)* Mullen-mouth egg-butt steel pelham with lip starp. *(b)* Loose-ring, fixed port mouth steel pelham. *(c)* Mullen-mouth, sliding cheek, Tom Thumb pelham. *(d)* Jointed mouth steel pelham. *(e)* Hard nylon Tom Thumb pelham with converter straps. *(f)* Steel kimblewick. (From DuPont, 1973.)

Bit Action

A bit and bridle can exert pressure on one location or on a combination of several different locations on a horse's head. The shape, weight, and size of the mouthpiece and the thickness of the horse's tongue determine the amount of pressure exerted on the tongue. A common mistake is to use a thick mouthpiece on a thick-tongued horse so that the animal cannot comfortably close its mouth. When the mouth is closed, too much pressure is applied on the tongue by the mouthpiece so that the horse wants to keep its mouth open. The curb chain applies pressure to the curb chain groove. Proper adjustment for different types of bits is necessary for proper action. A "halfbreed" bit with a roller or a spade bit requires a tighter curb chain than does a curb bit. This limits the rise of the port, avoiding possible injury to the roof of the mouth. Curb chains that are too loose pinch the corners of the mouth when the reins are pulled tight and

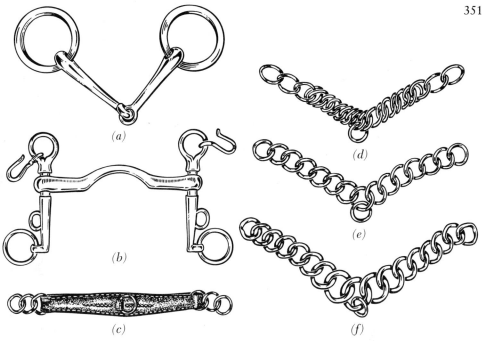

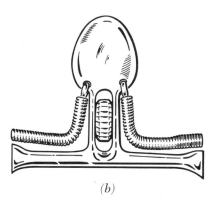

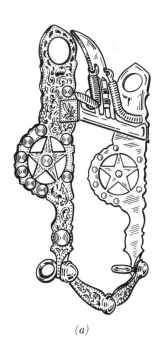

FIGURE 11-18

Weymouth set. *(a)* Flat ring bradoon. *(b)* Sliding cheek curb. *(c)* Leather curb strap. *(d)* Double-link curb chain. *(e)* Single-link curb chain. *(f)* Wide, single flat-link curb chain. (From DuPont, 1973.)

FIGURE 11-19

Santa Barbara *(a)* silver mounted spade bit and *(b)* spoon spade mouthpiece.

do not give the desired pressure on the chin groove. A general rule is that you should be able to insert two fingers between the curb chain and the chin groove.

The bars carry the weight of a snaffle bit primarily when the bit is passive, unless the horse properly flexes its head at the poll to relieve the pressure. The curb bit presses on the bars when the curb chain becomes tight. Because of the fulcrum action of the mouthpiece, most curb bits exert some pressure on the poll right behind the ears when the reins are pulled. The pressure results from movement of the curb strap loop forward and downward. Bridle adjustment varies the amount of pressure applied by a curb bit to the poll (a loose bridle applies less pressure). The lips receive pressure at the corners of the mouth and over the bars, which they cover until heavy pressure is applied to a curb or snaffle bit. The amount of pressure applied to the roof of the mouth is controlled by the type and movement of the curb bit port. As previously discussed, the curb chain adjustment limits the rise of the port. Certain types of bits, such as the hackamore bit (Figure 11-17), apply pressure to the nose.

The balance or collection of a horse can be controlled to a certain extent by the bit it carries. With curb bits, the ratio of the distance from the curb strap loop to the mouthpiece and from the mouthpiece to the rein loop (Figure 11-20) controls the horse's balance. A horse that is heavy on the forehand (that is, its center of balance is too far forward) will pick up its head, bring in its nose, and flex at the poll so that the center of gravity is shifted backward when the ratio is increased (for example, increased from 1:1 to 1:3). In contrast, a horse with a sensitive mouth performs much better for a heavy-handed rider when the ratio is decreased and less pressure is applied for a given pull on the reins.

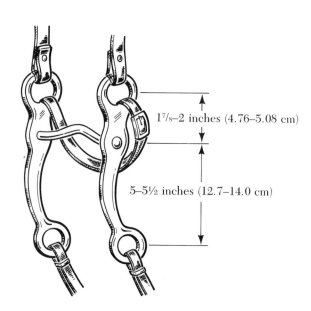

1⅞–2 inches (4.76–5.08 cm)

5–5½ inches (12.7–14.0 cm)

FIGURE 11-20
Curb bit.

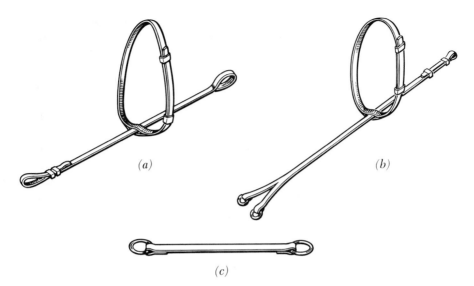

FIGURE 11-21
Martingales. *(a)* Standing. *(b)* Running. *(c)* Irish.

Martingales

A standing martingale keeps a horse from carrying its head too high (Figure 11-21). It attaches to the noseband, is supported by a neck strap, and is connected to the girth. A running martingale (Figure 11-21) works on the mouth to keep the head low. The reins are run through the rings on the end of the straps, which change the direction of the rein pull. It is used to get the proper headset during training and to correct a horse that throws its head upward. The Irish martingale (Figure 11-21) is a leather strap with rings at each end. It links the reins together in front of the horse's neck to prevent them from being flipped over onto the same side of the neck during a stumble or other sudden, unexpected movement of the horse's head. It is merely a safety device and does not affect a horse's performance.

Care, Storage, and Cleaning
Saddles

Saddles and other tack are expensive and require adequate care if they are to have a long and useful life. A few minutes of care after each ride can keep tack in excellent condition. Soft, pliable leather is comfortable for the horse as well as for the rider. Cleaning leather that is exposed to sweat, moisture, and dirt with saddle soap prolongs its usefulness. Cleaning after use also allows an opportunity to inspect the saddle and tack for damage and excessive wear.

Every 3 to 6 months, the saddle and all tack should be stripped, thoroughly cleaned with saddle soap, and preserved with neat's-foot oil. Commercial preparations combine neat's-foot oil with other compounds to form an excellent preservative and waterproofing compound. During these two to four cleanings each year, give all tack components a thorough safety inspection. Clean the sheepskin of western saddles with a brush. Dirty, lumpy sheepskin causes saddle sores. Remove excessive dirt from tack by soaking in warm water and castile soap. After the dirt is soft, remove it with a stiff brush. Allow the leather to dry partially, but wipe it with neat's-foot oil while it is still damp. Usually two or three coats of neat's-foot oil are needed to restore it to a usable condition. You can rejuvenate old, dried, cracked leather by applying warm neat's-foot oil at least once a day over a period of several days.

The saddle and other tack should be stored in a dry room because moisture and mildew cause rapid deterioration. To keep saddles and other tack clean, the room should be free from dust and a dust cover placed on the saddle. The room should be free of rodents and insects. Rodents chew on leather to obtain salt and can ruin a great deal of tack in a short time. Store saddles on saddle racks to prevent the tree from being damaged and to keep the sheepskin clean. Place the cinch straps and cinch across the seat, or hang them from the saddle horn. Store English saddles on a rack with the irons slid up the stirrup leathers next to the saddle flap. Place the girth over the seat so that the saddle is ready to be placed on a horse.

Clean saddle pads and blankets with a brush after drying. Dirty blankets frequently cause saddle sores. The new synthetic pads are machine washable and are easy to keep clean.

Bridles

Clean the leather parts of the bridle in the same manner as was described for a saddle. When the bridle is thoroughly cleaned, remove the metal components, if possible, and clean them with warm water and soap. Clean sterling silver parts with any of a variety of commercially available compounds.

It is important to hang the bridle over a round rack that is approximately 4 inches (101.6 mm) in diameter and similar to a coffee can, to keep the headstall in the proper shape. Bridles and reins that are hung over nails tend to crack and develop weak spots.

Bits

You can clean bits with a damp cloth. If you remove a bit from a sick horse, boil it in water before using it again. This helps prevent the spread of disease.

Vaseline prevents bits that are not made of stainless steel from rusting. Other types of oils leave a bad taste in the horse's mouth even after they are wiped off.

RIDING

Saddling

Before the saddle is placed on the horse, make sure the horse's back and cinch areas are clean and all the hairs are lying in their natural direction. Check the saddle blanket and all other equipment for foreign objects.

Forward

With the forward saddle or hunt seat (Figure 11-6), lay the girth (leather strap that goes under the belly) across the seat and pull the stirrups up high before placing the saddle on the horse's back. A saddle pad is not necessary, but if you use one, connect it to the saddle before saddling up. Place the saddle and pad gently on the horse's back from the near (left) side and just behind the shoulder blades. Then move the saddle back approximately 1½ inches (38.1 mm) so that the hairs are straightened. Never move the saddle forward because if you do the hairs become ruffled and will cause saddle sores. Next pass the girth—attached to the off (right) side billets (Figure 11-7)—under the horse's belly. Pull the girth up and buckle it to the near (left) side billets. Before tightening the girth enough to ride, smooth out all skin wrinkled under it and lead the horse around briefly. Then tighten the girth enough to keep the saddle in position during the ride. Adjust the stirrup leathers on a hunt seat saddle for proper length, and pull the stirrups up high again until you are ready to mount. The stirrup length should be such that the stirrup iron reaches the armpit when you extend your arm with the fingertips touching the stirrup buckles. Some people adjust (shorten or lengthen) their stirrups to a position in which the iron hits just below the ankle bone. A riding instructor may suggest changing the standard stirrup length to improve the rider's position.

Western

To saddle a horse for western riding, place the saddle pad on the horse's back in front of the withers. Then prepare the western saddle (Figure 11-1) for saddling by placing the cinches and right stirrup across the seat. Grip the saddle by the pommel and cantle and place it gently on the saddle pad and horse's back just behind the shoulder blades. Move the saddle and pad backward to a position where the bars will not rub the shoulder blades. After lowering the cinches and right stirrup, cinch the saddle up, but not too tight. If the saddle has front and back cinches, tighten the front cinch first to prevent the horse from bucking the saddle off. After walking the horse around, tighten the cinches.

Before using a saddle on a horse, determine if the saddle fits the horse or, in some instances, if the horse fits the saddle. To find out if the saddle fits, the following steps have been suggested. Place the saddle, without any padding, far forward on the back, and slide it back into the place it is usually fastened. Before doing up the girth, run a hand under the saddle. If at any point it becomes difficult or impossible to slide the hand between the horse and saddle without lifting up the saddle with the other hand, that part is putting excessive pressure on the horse. Now fasten the girth just tight enough to take the slack out of it, but not tight enough to mount. Stand to one side and look down the gullet (the "backbone bridge") from the front (withers) end. You should be able to insert two fingers between the withers and the gullet at the top; there should also be space on either side of the withers. The padding of the bars should rest ¾ inch (19 mm) below the side edge of the withers. Again standing to the side, look through the gullet from the back (loin) end. There should be ½ inch (12.7 mm) of clearance between backbone and bars on either side, and the tree must not touch the spine. Face the saddle from the side, and grab the pommel with one hand and the cantle with the other. Rock the saddle, front to back, exerting moderate pressure. If either end rocks up more than 1 inch, the tree is too hollow for that back; in other words, that saddle belongs on a horse that is more swaybacked. If there is no motion at all as you rock, the saddle is too flat and the horse's back too hollow. The saddle is resting only on its ends, not putting any pressure at all on the center of the back. Now look under the back of the saddle again, this time at the padding on the bars. The center of the padding should be resting on the back muscles. If the inner edges of the bars are the only parts touching the horse, the saddle is too flat and broad and the horse will wind up with pressure injuries along either side of its spine. If the outer edges touch, the saddle is too narrow—or else the horse is too fat.

With the saddle still in the position of normal use, look at where it bears on the withers and shoulder blade region. If the bars contact the shoulder blade, the saddle is placed too far forward. Rotate the shoulder blade by extending the horse's foreleg. While the edge of the scapula may rotate *under* the saddle, it should not make contact.

Now, with an assistant, tighten up the girth and get into the saddle (or have a friend of your approximate build mount up). Repeat all the tests: there should still be daylight through the gullet, you should be able to tell where most of the pressure from the bars is bearing, and the saddle should not rock with the rider's weight in riding position. Put a hand between the saddle and the shoulder blade; as the foreleg is extended, there should not be severe pinching as the hand is run under any part of the saddle as you ride. If it pinches the hand, it is pinching the horse.

Several signs of an improperly fitting saddle may appear after a horse has been ridden. The horse shows pain when the saddle is placed on its back and may sink its back down sharply. The horse may also do this when it is mounted. Some horses will not stand still while saddled and will move with a very stiff gait. Both reactions indicate pain from pressure on the back. You can locate tender spots by rubbing a hand over the back—the horse will flinch. After a saddle is removed, dry spots in the middle of a large sweaty area indicate that pressure has prevented the sweat glands from functioning in that area. When the pressure is removed, the area fills with fluid (edema). If you see dry areas, replace the saddle and loosen the cinch slowly during the next 2 hours. Failure to replace the saddle results in dry, crusty skin and hair falling out. Frequently, the hair is white when it grows back. Any swelling, broken hairs, or bald spots indicate scalding and/or abrasion. Scalding is an irritation or burning of the skin caused by a saddle or piece of tack chafing a sweaty hide. Abrasion curls or breaks off the hair. The spontaneous appearance of wrinkles in an area indicates that the skin is losing elasticity because of excess pressure on that area by the saddle.

Bridling

After saddling the horse, bridle it. When bridling a horse, hold the crown (Figure 11-22) in your right hand. As your right hand pulls the bridle upward, your left hand guides the bit into the horse's mouth. If you bang the teeth with the bit during a few bridlings, the horse usually becomes head shy and hard to bridle. If the horse does not open its mouth for the bit, insert your thumb and forefinger between the bars, and the horse will usually open its mouth. After inserting the bit in the mouth, adjust the bridle so that there are one or two wrinkles at the corners of the mouth. Adjust the throatlatch strap, if present, so that you can insert three or four fingers between it and the horse's jaws. The noseband (English bridle) and curb strap must also be properly adjusted for the horse (see the section on bitting). When removing the bridle, place your left hand on the horse's face to keep its nose down and tucked in, so the bit will not hit the teeth when you remove it.

Follow several safety precautions when bridling the horse. Protect your head from the horse's head. Stand in close, just behind and to one side (preferably on the left side) of the horse's head. Handle the horse's ears carefully. Maintain control of the horse when bridling by refastening the halter around its neck. Adjust the bridle to fit the horse before riding it. Check the placement of the bit, the adjustment of the curb strap, and the adjustment of the throatlatch.

FIGURE 11-22
Bridling a horse with a curb bit. Place bit in mouth by holding it
with fingers and pressing on bar of mouth with the thumb.
(Photo by B. Evans.)

Western Riding

Mounting

A western horse is customarily mounted from the near (left) side. However, all
horses should be accustomed to being mounted from the off (right) side because
some conditions preclude mounting from the near side. Grip the reins properly
in the left hand, which also grips the mane in front of the saddle. When mounting, face the rear of the horse and place your left foot securely in the left stirrup.
With your right hand, grip the saddle horn, and swing your right leg over the
saddle. The right leg swing should be close to the saddle, your left toe should

not hit the horse, and your left knee should not exert excess pressure on the horse's side. Then place the right foot in the right stirrup. On cold, frosty mornings when horses feel good and may buck, or when mounting horses that may buck, pull the horse's head toward you with your left hand to prevent the horse from bucking until it is mounted. The procedure for dismounting is opposite from the procedure for mounting.

Before riding the horse, adjust the stirrups for length. Each person needs a different length stirrup to maintain the proper position, but a general rule is to adjust the stirrup so that it strikes 1 inch (25.4 mm) above the ankle bone. If the rider stands up in the saddle, there should be a 4-inch (101.6-mm) clearance between the crotch and saddle.

Never mount or dismount a horse in a barn or near fences, trees, or overhanging projections. Side-stepping and rearing mounts can injure riders who fail to take these precautions.

A horse should stand quietly for mounting and dismounting. To be certain of this, you may have to control its head through the reins.

Position

After learning to mount and dismount, the rider's next consideration is position. You must have good position to control the horse and to be able to use the riding aids. A rider with good position is graceful and relaxed when the horse is in motion. Each rider usually modifies the basic position slightly, and the positions that are popular in the show ring are different in different parts of the country. The American Association of Health, Physical Education and Recreation (Division for Girls' and Women's Sports and the Division of Men's Athletics) lists four elements of good position: it (1) allows unity of horse and rider, (2) is nonabusive to the horse, (3) ensures security for the rider, and (4) permits the effective use of aids or controls. Proper riding position is shown in Figure 11-23. Flex your ankles, and turn your toes outward approximately 30 degrees. When your ankles are flexed so that your heels are down, your ankles can act as shock absorbers. Drop the weight of your body into your heels. The inside area of the ball of the foot is slightly lower than the outside area. For a good position that allows you to be balanced, rest the ball of your foot on the stirrup treads, and make sure the inside of the ball of your foot is lower than the outside (Figure 11-24). This foot position, with your toes turned out slightly, keeps the inner side of your calf in close contact with the saddle fender for good frictional grip. To aid in shock absorption, flex your ankles so that your heels drop downward. When your body weight is carried by your heels, the stability of your lower leg increases. Complete the positioning of your lower leg by keeping the inside of your calf in contact with the saddle so that your leg, from the knee down, allows the stirrup to hang vertically. You can thus use your lower leg efficiently and effectively to signal the horse.

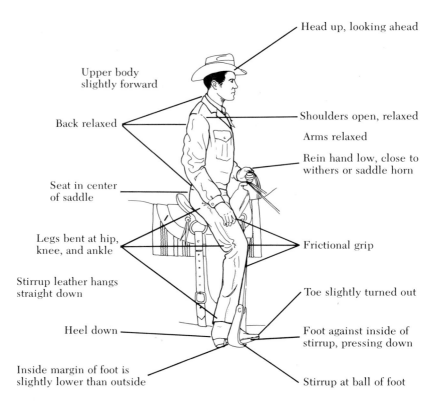

Upper body
slightly forward

Head up, looking ahead

Back relaxed

Shoulders open, relaxed

Arms relaxed

Rein hand low, close to
withers or saddle horn

Seat in center
of saddle

Legs bent at hip,
knee, and ankle

Frictional grip

Stirrup leather hangs
straight down

Toe slightly turned out

Heel down

Foot against inside of
stirrup, pressing down

Inside margin of foot is
slightly lower than outside

Stirrup at ball of foot

FIGURE 11-23
Basic position for western riding. (From Shannon, 1970.)

One of the most common mistakes made by beginning riders is keeping the ankle stiff. The concussion, rather than being absorbed by the ankle, is transferred upward, and the rider bounces in the saddle. The bouncing irritates and tires both the horse and the rider. Flex your knees to help absorb concussion and to slightly grip the saddle. Keep the inner surfaces of your thighs flat against the saddle to complete the leg-gripping action.

To keep your balance on a horse, relax your hip joints and keep them mobile. This allows the upper part of your body to adjust for changes in speed and direction. Maintain three-point contact between the rider's seat and saddle (that is, at your two seat bones and your crotch).

Keep your back straight except for a slight arch in the small of your back. Tilt your upper body slightly forward to maintain balance and to align with the horse's center of gravity. Keep your shoulders open and even. Hold your head erect, and look straight forward.

Allow your left arm to drop naturally beside your body. Your left hand

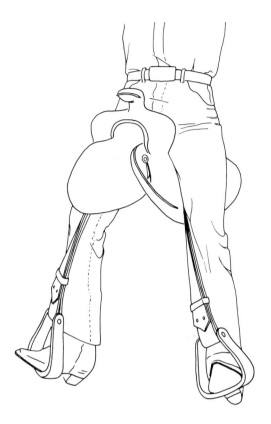

FIGURE 11-24

Frictional grip: thighs, knees, and calves held in contact with horse by weight flowing into heels. The rider's weight is supported by the crotch and seat bones. Hold the inner surface of your thigh flat against the saddle, and position your knee and lower thigh against the fender of the saddle and barrel of the horse. The inner surface of your upper calf is in contact with the barrel of the horse; the lower part of your leg does not grip the knee. Use the lower leg to signal the horse to indicate the movements you desire. Position the ball of the foot in the stirrup in such a way that the inside margin is lower than the outside margin, and position the foot against the inner curve of the stirrup. (From Shannon, 1970.)

holds the reins (Figure 11-24) just above the saddle horn. Wrist and finger motion and carriage are important to obtain proper response from the horse. Hold your wrist straight but relaxed. To feel the horse's mouth, keep your fingers relaxed and "soft." A rider's lack of self-confidence or nervousness is quickly transmitted to the horse via the rider's hands. The style of holding the reins varies from locality to locality and under different conditions. Some styles are shown in Figure 11-25.

Carry your right arm in a relaxed position, resting on your thigh or holding the romal (Figure 11-26).

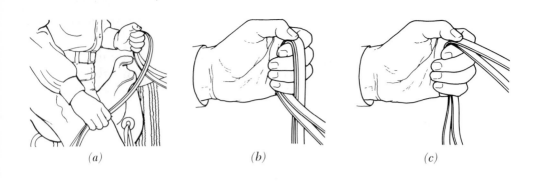

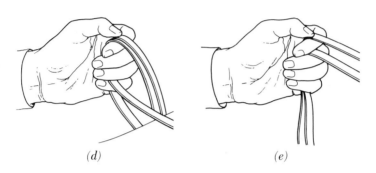

FIGURE 11-25

Proper way to hold reins during western riding. *(a)* Free hand holding split ends of reins or romal. *(b)* Reins in one hand. *(c)* Reins passing down through the thumb and index finger. *(d)* Reins separated by the little finger. *(e)* Reins separated by the index finger. (From Shannon, 1970.)

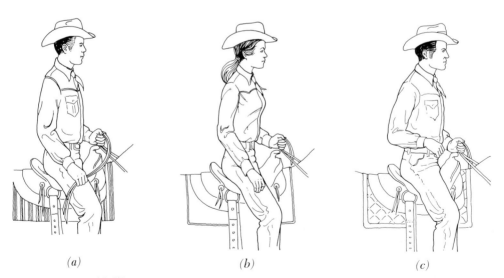

FIGURE 11-26

Position for free hand during western riding. *(a)* Holding split reins or romal (juniors). *(b)* On thigh. *(c)* At belt. (From Shannon, 1970.)

As the speed of the horse increases, increase your frictional grip of the thighs, knees, and calves and tilt your upper body forward to maintain balance. If your upper body gets behind the center of gravity of the horse and you do not shift more weight into your lower leg, you lose balance, control of the horse, and effective, efficient use of the riding aids. Most riders also feel that, with speed, it is easier to carry the free forearm parallel to the belt.

Forward Riding

Mounting

The sequence of mounting the horse for forward riding is shown in Figure 11-27. Address the reins (pick them up and adjust them) and hold them with the bight (loop) on the right side. Stand facing the rear of the horse and just forward of its shoulder, and place your left hand on its withers. Use your right hand to position your left foot in the stirrup. After positioning the stirrup, grip the off side of the cantle with your right hand. To gain momentum to swing up, take one or two hops on the right foot. Swing your right leg over the horse's hips, and ease into the seat. Position your right foot in the stirrup. In general, dismounting is a reversal of mounting. After swinging your right leg over the horse's hips, step down or slide down, depending on your size and the horse's size. Immediately on dismounting, "run up" the stirrups. Dangling stirrups may startle or annoy the horse. Horses can catch a cheek of the bit or even catch a hindfoot in a dangling stirrup iron when fighting flies. The dangling stirrup can also catch on doorways and other projections while you are leading the horse. After running up the stirrups, immediately bring the reins forward over the horse's head. In this position, they can be used for leading.

Position

The basic position (Figure 11-27) has been described as follows by George H. Morris, former United States Equestrian Team member. Place the ball of your foot in the middle of the stirrup with your heel down and just behind the girth. When you turn the toe out (to a maximum of 15 degrees), your calf and inner knee bones establish equal contact with the horse. Your thighs grip the saddle with the same pressure as your calves and knees. Sit forward in the saddle as close to the pommel as possible and as far from the cantle as possible. Carry your upper body erect but relaxed. An imaginary plumb line should be able to fall from your ear, to your shoulder, down through your hip bone, to the back of your heel. Carry your head so that you are looking directly ahead. Do not carry your shoulders forward or backward. Position your hands so that there

FIGURE 11-27

English equitation—mounting, correct seat, and dismounting. (a) Prepare to lead. (b) Stand at horse's head. (c) Mount, first step. (d) Mount, second step. (e) Position right foot. (f) Address the reins. (g) Position hands. (h) Basic position in saddle. (i) Ready to ride. (j) Prepare to dismount. (k) Dismount. (l) Step down. (From Bradley and Hardwicke, 1971.)

(c)

(d)

(g)

(h)

(k)

(l)

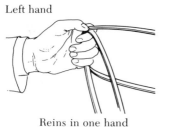

Left hand

Reins in one hand

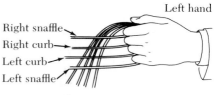

Left hand

Right snaffle
Right curb
Left curb
Left snaffle

Reins in one hand

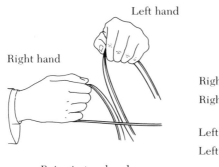

Left hand

Right hand

Reins in two hands

(a)

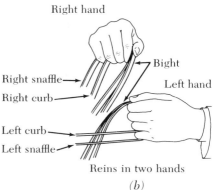

Right hand

Bight

Right snaffle
Right curb

Left hand

Left curb
Left snaffle

Reins in two hands

(b)

FIGURE 11-28

Methods of holding the reins for forward-style riding. *(a)* Single-rein bridle. *(b)* Double-rein bridle.

is a direct line between your elbow and the horse's mouth. Your hands should be flexible, held above and slightly in front of the withers. Your thumbs should be 2 or 3 inches (50.8 to 76.2 mm) apart and as straight as possible. The correct position of the reins is illustrated in Figure 11-28.

When the horse is in motion, your body is farther forward so that your center of gravity is also shifted forward. This allows the rider to follow the motion of the horse. The faster the motion, the further forward your body is shifted.

Saddle Seat

Mounting

The sequence of mounting the horse was discussed in the previous section on forward riding (Figure 11-27).

Position

Your seat and hips should be underneath, in a balanced position. Each leg should be in the middle of the saddle skirt with the stirrup leather under each knee (Figure 11-29). The lower half to two-thirds of the calf of your legs should be slightly out from the horse's side. Each stirrup should be at a right angle to the horse and flat across the ball of your foot. Stirrup length is correct when the stirrups strike your ankles (Figure 11-29). Keep your heels slightly down, a little lower than the toes. Turn your toes out about 15 degrees. You should be able to look down and see your toes. A line perpendicular to the ground can be drawn through the middle of your head, through your shoulder, side, and ankle. Your arms should hang in a natural position from your shoulders. Do not tuck them in too tight. Your elbows should not come back past the sides. Hold your shoulders square with your head, and hold your neck erect. Your hands hold the reins as in Figure 11-28. Hand position depends on the height of the horse's head carriage. Keep your hands low if the horse carries its head low. When the horse trots or canters, your basic position changes a little so that you can follow the motion of the horse. You should post at the trot, and you should post on the correct diagonal. When the horse trots to the left, rise as the right foreleg goes up. When traveling to the right, post on the left diagonal. As the horse's foreleg moves downward, move downward. The purpose of posting on the correct diagonal is to allow the horse to keep its balance.

Some common errors made while riding saddle seat are made with the hands and/or legs. Holding the reins too long forces the rider to use excess arm motion to control the horse. If the stirrups are too long, the rider's legs are forced too far forward and the seat moves to the back of the saddle. Riding too far forward in the saddle forces the legs behind the center of balance and tips the shoulders forward.

Sidesaddle

Mounting

Pick up the reins with your right hand, which also holds the riding crop. While holding the reins and crop, grasp the pommel of the saddle. Stand facing the front of the horse with your right shoulder next to the horse (Figure 11-30). An assistant slightly stoops in front of you. Place your left foot in the assistant's hands and your left hand on the assistant's shoulder. Flex your left leg and spring up sideways onto the flat of the saddle. Lift the right knee over the horn.

To dismount, place the reins in your right hand, and remove your left foot from the stirrup. Bring your right leg up over the pommel and to the left so that you are sitting facing out left. With your right hand, grip the top of the pommel as you dismount lightly and land facing the horse's head.

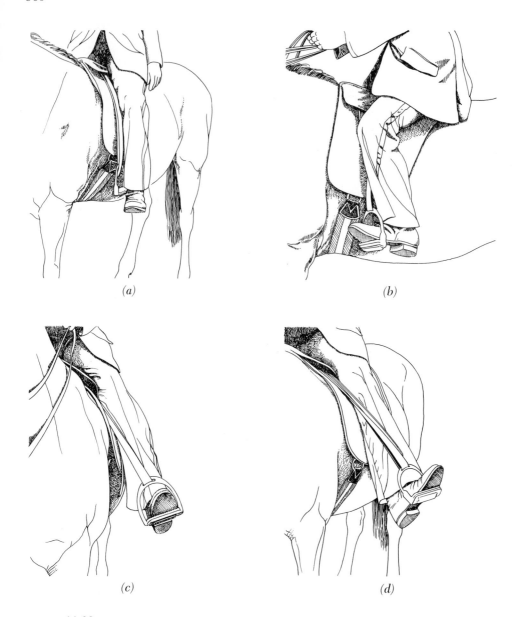

(a) *(b)*

(c) *(d)*

FIGURE 11-29

Correct and incorrect leg positions for saddle seat equitation. *(a)* With your leg hanging free, the stirrup should be ankle height. *(b)* With your foot in the stirrup, your knee should rest on the stirrup leather. Your calves should be just slightly out from the horse's side. *(c)* Correct position. Swing your heel down and out to achieve the proper contact, with the inside thigh and knee on the saddle. *(d)* Incorrect position of foot—toe is turned out, and ankle bent in an unnatural position.

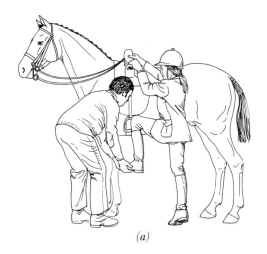

FIGURE 11-30

Mounting the sidesaddle. *(a)* Mount with assistant to give a leg up. *(b)* Mount from ground or with mounting block. (From Floyd, 1979a.)

Position

Your left leg should hang perpendicularly from the knee. Depress your left heel, and hold your foot parallel to the ground (Figure 11-31). Sit erect in the middle of the saddle with your back slightly hollowed and perpendicular to the horse's back. Hold your arms naturally by your sides with your hands positioned a few inches apart (Figure 11-32). Hold your hands as low as possible, just above your knees.

Post when riding at the trot. To keep from bouncing off the saddle, hold your left foot parallel to the horse, which keeps your knee close against the saddle and prevents your leg from moving around.

FIGURE 11-31

Proper leg position for riding a sidesaddle.

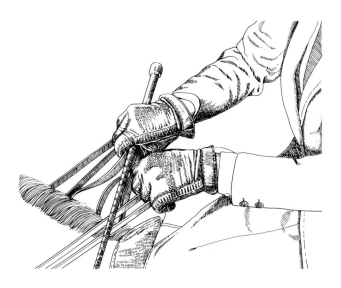

FIGURE 11-32

Proper hand position and way to hold reins while riding side saddle.

DRIVING

There has been a rapid growth of interest in driving horses, which is an enjoyable way to exercise a horse. Many types of harnesses and hitches can be used. A complete description of them is beyond the scope of this book, but we will mention some of them.

Harnessing

A popular and common harness is the single fine harness. The basics for "putting to" in fine harness are as follows. When driving a heavy vehicle, the horse should wear a collar; a light vehicle may be driven with a breastcollar. Collars come in two types: (1) the Kay or closed collar is usually used with a fine harness and (2) the open collar opens at the top and is closed with a strap (Figure 11-33). A collar must fit the horse so that it lies comfortably and flat on the shoulders with enough room at the bottom to allow a hand to pass

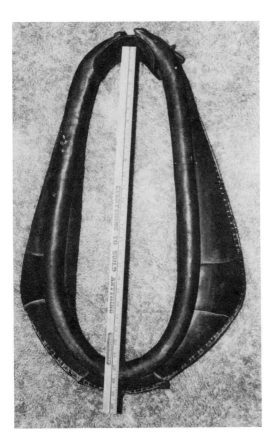

FIGURE 11-33
Way to measure a horse collar.
The collar is of the open type.
(Photo by B. Evans.)

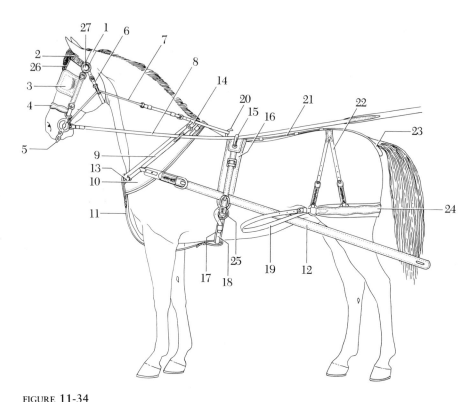

FIGURE 11-34

Single surrey harness. (1) Headpiece. (2) Browband. (3) Winker. (4) Noseband. (5) Liverpool bit. (6) Throatlatch. (7) Bearing rein. (8) Rein. (9) Hame. (10) Collar. (11) False martingale. (12) Trace. (13) Bottom hame strap. (14) Top hame strap. (15) Terret. (16) Saddle. (17) Girth. (18) Bellyband. (19) Breeching strap. (20) Saddle hook. (21) Back strap. (22) Hip or loin straps. (23) Crupper. (24) Breeching. (25) Tug. (26) Winker stay. (27) Rosette.

freely between the collar and neck. Morgan horses usually wear a 21- or 22-inch collar (533.4 or 558.8 mm). Determine the size by measuring from the inside, top to bottom (Figure 11-33). Place a Kay collar on the horse by slipping it over the horse's head. So the head will pass through it, turn it upside down. You may need to stretch open the sides. Then rotate and place it in the proper position. Fasten the hames in their grooves in the collar, with the leather strap on top and a kidney link on the bottom. The harness' saddle back band, bellyband, girth, crupper, loin strap, and breeching are in one unit that is placed on the horse's back (Figure 11-34). Place the saddle in the center of the back, and hang the breeching down around the horse's hindquarters so it rests about a foot below the dock. Place the tail through the crupper. Then move the saddle forward until the back strap is tight and the crupper is under the root of the tail. Take care that no tail hairs are between the crupper and the tail, because they will irritate the horse. The saddle should be resting so that the elbows do not rub on the girth. Tighten the girth and buckle it. Attach a martingale to the bottom of the collar, pass it between the forelegs, and fasten it to the girth. Pass the reins through the saddle pad terrets and the hame terrets and leave them hanging until the horse is bridled. When you place the bridle on the horse, it should be tight enough so that the bit makes one to two wrinkles in the corner of the mouth. This prevents the winkers (blinders) from bulging out when the reins are pulled. Adjust the winkers so that the eyes are in the middle of them. Connect the reins to the bit. The two common driving bits that are used are the snaffle and the Liverpool (Figure 11-35). Attach the overcheck to the saddle ring.

After harnessing the horse, hitch it to the vehicle. Keep the horse still by cross tying it or by having an attendant hold its head. Always draw the vehicle to the horse. With the shafts held high, draw the vehicle into position and place the shafts in the tugs. Uncoil and attach the traces to the vehicle. Next, fasten the breeching to the vehicle so that if the vehicle moves forward about 6 inches (15.2 cm), the breeching will take hold and keep it from striking the horse's hindlegs. Then pass the bellyband over the traces and fasten it around the shafts. Tighten it, but not as tightly as the girth. The shafts should be free enough to move slightly up and down.

To unhitch the horse, reverse the procedure. Be sure to move the vehicle away from the horse so that the horse does not step on the shafts. Also, take care not to drop the shafts and strike the horse's legs or feet.

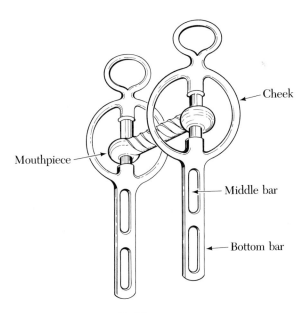

FIGURE 11-35
Liverpool bit for fine harness horse.

FIGURE 11-36
Correct way to hold harness reins and whip.

Driving

The whip (driver) mounts the vehicle from the off (right) side and sits on that side. As you prepare to mount, hold the reins in your right hand and transfer them to your left hand immediately on entering the vehicle. Hold the reins so that the near-side rein lies on your index finger and the off-side rein goes between your middle and ring fingers (Figure 11-36). Keep equal tension on the reins. To turn the horse, apply pressure to one of the reins by moving your wrist. Some horses require the use of both hands when turning. Carry a whip in the right hand, holding it at the balance point. Most whips are about 6 to 7½ feet long (1.8 to 2.3 m).

12
Showing
and Racing

Of the 8.5 to 10 million horses in the United States, 80 percent are used for nonprofessional purposes. The amateur owners and their enjoyment of the horse are the foundation on which our pleasure horse industry is built. At no previous time in our country's history have so many people used the horse for recreational purposes. Interest in racing, rodeo, horse shows, endurance rides, competitive trail rides, cutting horses, polo, combined training, gymkhana, and fox hunting has continued to increase during the past few years. In the following sections, the most popular ways that horsepeople enjoy their horses are described.

HORSE SHOWS

In New York City in 1917 the Association of American Horse Shows, Inc., was founded. In 1933, the name of the organization was changed to the American Horse Show Association, and was changed again shortly thereafter to its present name, American Horse Shows Association. The AHSA has grown steadily since its inception, but its growth has been phenomenal since 1960.

In addition to sanctioning various types of horse shows, the AHSA keeps all records, licenses judges and stewards, handles disciplinary matters, awards

375

annual prizes, and compiles the annual rule book of the horse show world. There are two types of membership in AHSA: group and individual. Group memberships are of the following classes: regular show, local show, combined training event, dressage competition, honorary show, and affiliated. Individual membership consists of senior, junior, and affiliated. Specific rules govern the selection of each type of membership.

The AHSA is a member of the Fédération Equestre Internationale, the world governing body of equestrian sports. The Fédération Equestre Internationale is the certifying agency for equestrian athletes for the Olympic Games and other international competitions.

Horses are shown in many divisions under the auspices of the AHSA: Appaloosa, Arabian, combined training, dressage, equitation, gymkhana, hackney and harness pony, hunter, hunter and jumper pony, jumper, junior hunter and jumper, Morgan, Paint, Palomino, Parade, Pinto, Pony of the Americas, roadster, Saddle Horse, Shetland Pony, Tennessee Walking Horse, Welsh Pony, and western. Each breed has specific classes, and there are open shows where the horse can be of any breed or nonregistered. Some of the most popular classes are the equitation, hunter, jumper, western, gaited, roadster, and halter.

Halter Classes

Most halter classes are judged on type, conformation, quality, substance, and soundness. Some breeds are also judged on disposition, manners, and action. Each breed organization has specific guidelines that are used by the judge. There are classes for certain ages and sex such as weanlings, yearlings, fillies, colts and stallions (Figure 12-1), mares, and geldings. Horses are also shown in breeding classes, such as produce of dam (two animals from the same dam), get of sire (three animals by same sires, as in Figure 12-2), and sire and get (sire and two of get).

Equitation Division

The equitation division is divided into three classes: hunt, saddle, and stock seat. Only the rider is judged in equitation classes. Therefore, any horse suitable for a particular style of riding and capable of performing the class routine is acceptable.

In hunt seat equitation, the rider uses a forward saddle with knee rolls (Figure 12-3). At horse shows, the class may or may not be over obstacles. If the class is over obstacles, their number, design, and height depend on the class of the entries. In nonjumping classes, the entries are required to perform at the walk, trot, canter, and hand gallop and to execute individual figures. Competitors in the saddle seat division ride a park saddle (Figure 12-4). They

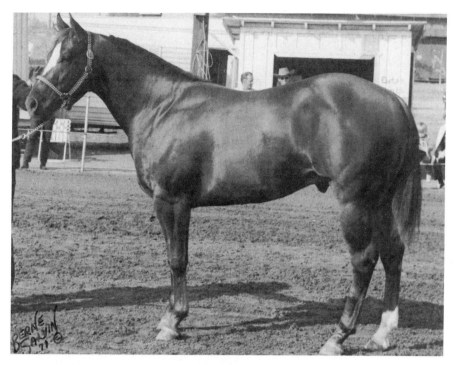

FIGURE **12-1**

Jimmie the Kid, 1971 Champion 2-Year-Old Stallion of the Pacific Coast Quarter Horse Association being shown at halter at Norco, California. He is at stud at Taylor Ranch, Davis, California. (Photo by Salvin, courtesy K. Taylor.)

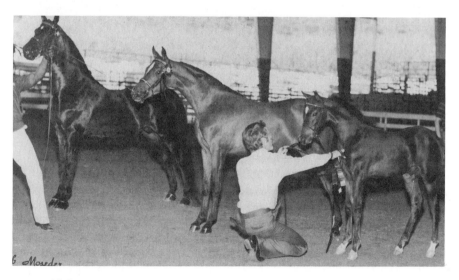

FIGURE **12-2**

Get-of-Sire Class at Dixie Cup Morgan Classic. Sire is UVM Promise, and the get are TVM Majesty and TVM Emperor. (Photo by Madeder, courtesy Tennessee Valley Morgan Horse Farm.)

FIGURE 12-3

Hunt seat equitation. Toni Dirschel riding Mr. Jutona, an AQHA gelding. (Courtesy T. Dirschel.)

FIGURE 12-4

Ruth Anne Lewis riding Shamrock's Lady Luck in a saddle seat equitation class. (Photo by Donaldson, courtesy S. Lewis.)

FIGURE 12-5

Stock seat equitation. Marie Graver riding College Vixen. (Photo by Fallow, courtesy M. Graver.)

are required to demonstrate the walk, trot, canter, and individual tests. Individual tests may include a figure eight at trot or canter, a serpentine at trot or canter, a change of leads, and other specified performances. Stock seat equitation is quite different from hunt and saddle seat equitation in that the rider uses a western stock saddle (Figure 12-5). In stock seat equitation competition, the horse and rider are scored as a team. The rider is scored on seat, hands, performance of horse, appointments of horse and rider, and suitability of horse to rider.

Western Division

The western division is composed of stock, trail, and pleasure horse classes. Horses may be of any breed or combination of breeds as long as they are at least 14.1 hands, serviceably sound, and of stock horse type. Riders must be dressed as for the stock seat equitation class and carry a lariat or reata. A rain slicker may be required in trail and pleasure classes.

FIGURE 12-6
Bobby Ingersoll riding Royal Cutter, 1977–1978 winner of $10,000 Stock Horse Classic at San Francisco Cow Palace, in a stock horse class. (Photo by McNabb, courtesy B. Ingersoll.)

Stock Horse

The stock horse is judged on its reining work, conformation, manners, and appointments. Some classes also require the horse to work cattle (Figure 12-6). A good AHSA stock horse has the following characteristics.

1. The horse should have a good manner.

2. The horse should be shifty, smooth, and have its feet under it at all times. (The hindfeet should be well under the animal when it stops.)

3. The horse should have a soft mouth and should respond to a light rein, especially when turning.

4. The horse's head should be kept in its natural position.

5. The horse must remain under the rider's control but still work at a reasonable speed.

Horses are faulted for swishing tail, exaggerated opening of the mouth, hard mouth, nervous throwing of head, lugging on bridle, and anticipation of maneuvers—particularly stopping.

The AHSA has suggested that, when stock horses are not worked on cattle, the working pattern in Figure 12-7 be followed.

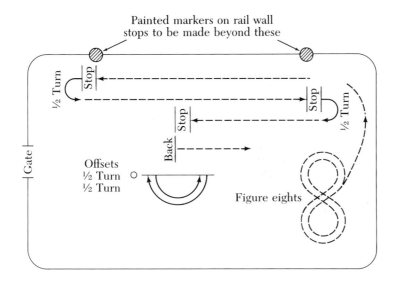

FIGURE 12-7

Stock horse dry work pattern as suggested by American Horse Shows Association.

Jaquima

The jaquima class (Figure 12-8) is similar to the stock horse class except that the horses are up to 5 years of age and have never been shown in a bridle other than a snaffle bit. The rider uses both hands on the mecate while showing the horse. Jaquima horses do not work as fast as stock horses and they are not required to work cattle in most shows.

Trail

Trail horses (Figure 12-9) are shown at a walk, trot, and lope both ways in the ring. They must work on a loose rein and without undue restraint. They are also required to work through and over obstacles that might be found on the trail. These tests include crossing bridges, mounting and dismounting, carrying objects from one part of the arena to another, backing through obstacles, and riding through water or over other types of obstacles. Trail horses are always required to allow the rider to open a gate, pass through it, and close it while maintaining control. Performance and manners are scored 60 percent; appointments, equipment, and neatness are scored 20 percent; and conformation is scored 20 percent.

Pleasure

The western pleasure horse class is popular. In this class, the horses are required to be shown at walk, trot, and lope. They must work both ways in the ring on a reasonably loose rein and be able to extend the gaits. The horses are also

FIGURE 12-8

Burnt Girl, being ridden by Bobby Ingersoll in a jaquima class. (Photo by Fallow, courtesy B. Ingersoll.)

FIGURE 12-9

April, owned by J. Sprafka and being shown by Maggi McHugh in a trail horse class. (Photo by Fallow, courtesy M. McHugh.)

required to back in a straight line. Finalists in a class may be required to do further ring work. They are judged on performance (60 percent), conformation (30 percent), and appointments (10 percent).

Jumper Division

The jumper division of AHSA is a popular division. Jumping horses may be of any breed, height, or sex. In this division, unsoundness does not penalize an entry unless it severely affects the horse's performance. Each horse is classified as preliminary, intermediate, or open, depending on the amount of money or points the horse has won. An amateur-owner jumper is any horse that is ridden by an amateur owner or an amateur member of the owner's immediate family. The horses are required to go over a course that should allow the exhibitor a fair opportunity to demonstrate the horse's capabilities and training. Each course must be carefully designed to fit the capabilities of the horses to be exhibited. Ideally, the faults should be evenly distributed over the course after the first or second jump, which is a single jump. A reasonable percentage of the horses should be able to negotiate the courses without faults, and, if no horse has a "clean round" (no faults), the course is too severe a test. Variation of obstacles on the course is important; however, they must be fair, "jumpable" tests of a horse's ability. The jumper is scored on a mathematical basis according to the number of penalty faults. Disobedience, falls, knockdowns, touches, and time penalties are considered penalty faults. Refusals, run-outs, loss of gait, and circling are disobediences. The horse is said to refuse when it stops in front of an obstacle to be jumped, whether or not the horse knocks it down or displaces it. When a horse evades or passes an obstacle to be jumped or jumps the obstacle outside its limiting markers, the horse is penalized for a run-out. Gait is lost when it loses after crossing the starting line. A penalty is also given if the horse crosses its original track between two consecutive obstacles anywhere on the course. A horse is penalized if it falls on the course. Knockdown faults are given when any part of an obstacle that establishes the height of the obstacle is lowered, whereas touching any part of the obstacle is considered a touch fault. The horse with the best score wins the class. If a tie for first place exists, a jump-off is held over the original course but a different sequence of obstacles. At least 50 percent of the obstacles can be raised 6 inches (15.2 cm) for each clean round.

Hunter Division

The hunter division of the AHSA is open to all horses of either sex (Figure 12-10). The division is usually divided into breeding, conformation, working, or miscellaneous classes. In each class, a green hunter is a horse in its first or second year of showing, whereas a regular hunter is a horse of any age that is not restricted by the number of years shown previously in any division. Breed-

FIGURE 12-10

Fran Walter going over a jump in a hunter class. (Photo by C. Goldberg, courtesy F. Walter.)

ing classes are judged on conformation, quality, substance, and suitability to become or, in the case of sires and dams, apparent ability to produce and beget hunters. Conformation classes are judged on conformation and performance. Depending on the class, conformation may count toward 25 to 60 percent of the total score. Working hunters are judged on performance and soundness. Conformation and working hunters are exhibited over a course with obstacles that are between 3 feet 6 inches (1.07 m) and 4 feet 6 inches (1.37 m). The fences usually simulate obstacles found in the hunting field, such as natural post and rail, stone wall, white board fence or gate, chicken coop, aiken (parallel bars with brush in the middle), and hedge. Hunters should keep an even hunting pace and have good manners and style. They are also scored on faults and way of moving over the course, as well as soundness. Faults include falls, touches, knockdowns, and disobediences.

Saddle Horse Division

Classes for three-gaited and five-gaited horses are offered within the saddle horse division (Figure 12-11). The horses must be registered with the American Saddle Breeders Association. Three-gaited horses are required to be shown at the walk, trot, and canter. Five-gaited horses must also be shown at a slow gait (slow pace or foxtrot) and the rack. Horses are worked both ways in the ring at all gaits. Finalists are usually requested to do further work as directed by the judge. Gaited horses are shown in any bridle that suits the horse and with a flat, English-type saddle. The exhibitor wears informal dress during morning and afternoon classes and dark-colored riding habit with accessories during

FIGURE 12-11

Having a Good Time, American Saddlebred three-gaited pleasure horse of the year in 1977, being shown by Sue Camins. (Photo by Inman, courtesy S. Camins.)

the evening classes. For formal wear, women usually wear tuxedo-style jackets with matching jodhpurs, whereas men usually wear dark suits. Both wear saddle derbies or silk top hats.

Saddle horses are also shown in breeding classes and are judged on conformation, finish, soundness, and natural action. In the fine harness section, they are shown with an appropriate vehicle, usually a small buggy with four wire wheels and no top. The horses are shown at a walk and at the park trot.

Roadster Division

Horses shown in the roadster division of AHSA may be of any breed (Figure 12-12). The only qualifications for the horse shown in this division are attractive appearance, good conformation, and good manners that make the horse a safe risk in the ring. The horse must also be serviceably sound. Horses are shown as bike roadsters or road wagon roadsters. Some horses can be shown in either class but the road wagon roadster is larger than the bike roadster. Roadsters are shown at a walk, slow jog trot, fast road gait (also a trot), and full speed trot. When working in the show ring, the roadster has animation

FIGURE 12-12
Cedar Creek Beydala, bred and owned by Cedar Creek Farm, being shown in a park harness class. (Photo by Moseder, courtesy R. M. Burger.)

and brilliance. When asked for speed, this horse maintains form and shows full speed. Their gaits must be even and smooth at all times, particularly in turns. The exhibitor in the bike class wears his stable colors and a cap and jacket that match. In the road class, the exhibitor wears a business suit and hat.

Three-Day Events

The three-day competition demands extensive training and devotion to horsemanship (Figure 12-13). Originally a trial for cavalry patrol mounts, it became a recognized form of equestrian competition in 1912, when it was included in the Olympic Games. It is now defined as "the most complete combined competition, demanding of the rider considerable experience in all branches of equitation and a precise knowledge of the horse's ability, and of his horse a degree of general competence resulting from intelligent and rational training."

On the first day of the three-day event, the dressage test is given—a series of movements that are designed to show the horse's fitness, suppleness, and sensitive obedience. Each movement is marked by judges. The dressage test was

FIGURE 12-13
Catherine Mill taking a water jump on Basque Bay during a three-day event. (Courtesy C. Mill.)

originally designed to show the horse's capability on the parade ground. The horse must perform in close formation maneuvers and must perform various movements involved with reviewing troops.

The speed endurance and cross-country test is held on the second day. It tests the speed, endurance, and jumping ability of the true cross-country horse. At the same time, it demonstrates the rider's knowledge of pace and the coordination of effort between horse and rider. The test is composed of four phases.

Phase 1: The first phase is the roads and tracks, in which the horse must cover a given distance over country roads and tracks at a brisk trot.

Phase 2: At the finish of the course (Phase 1), the horse must immediately start the steeplechase. In this phase, the horse must negotiate a course of fences at a speed of at least 26 miles (41.8 km) per hour.

Phase 3: The third phase is another roads and track course. After Phase 3, the horse is given a 10-minute rest and is examined by a veterinarian for soundness.

Phase 4: The last phase is a cross-country obstacle course.

The test given on the third day is stadium jumping. Its object is to prove that, on the day after a severe test of endurance, the horse has retained the suppleness, energy, and obedience necessary to continue in service.

Gymkhana

Gymkhana events or games on horseback are held by many riding clubs, 4-H Clubs, and other interested horsepeople. They are very popular with children but are also enjoyed by many adults because they are competitive and require harmony between rider and horse. The rules of gymkhana events may be established by the contestants, or they may be closely governed by breed or horse show organizations. The standard gymkhana events are keyhole race, figure-eight stake race, cloverleaf barrel race, quadrangle stake competition, pole bending, scurry competition, figure-eight relay race, rescue competition, and speed barrel race. In the gymkhana riding competition, the horses are shown on the rail after they have completed a cloverleaf barrel competition pattern.

Barrel Racing

Barrel racing is a standard gymkhana event. The rules vary depending on the organization that sanctions the gymkhana. The standard positioning of the three barrels and the distance between them are illustrated in Figure 12-14. The event is sometimes called "cloverleaf barrel racing" because the race pattern resembles a cloverleaf. The rider's horse has a running start when it crosses

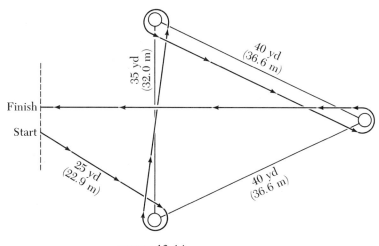

FIGURE 12-14

Cloverleaf barrel racing pattern.

the starting line. Knocking over a barrel or failing to follow the prescribed course results in a time penalty or disqualification. The horse and rider with the fastest time win the race.

Pole Bending

Pole bending is a timed event in which the horse must travel a pattern around six poles (Figure 12-15). The 6-foot-high poles (1.83 m) are set on top of the ground, and if a horse knocks down a pole it is disqualified. There are two types of patterns that can be used. In one pattern (Figure 12-16), the six poles are placed 20 feet apart (6.1 m) in a straight line. The timing line is perpendicular to the line of poles and located even with Pole 1. The horse crosses the

FIGURE 12-15

Jackie Holley riding Shotgun around a pole to win the single-stake championship, California Gymkhana Association. (Courtesy C. Holley.)

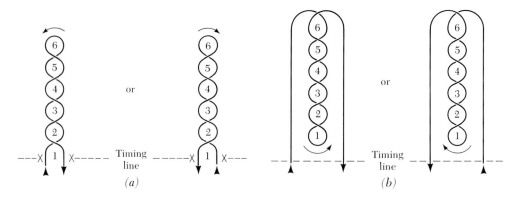

FIGURE 12-16
Pole-bending patterns. *(a)* Poles are 20 feet (6.1 m) apart. *(b)* Poles are 21 feet (6.4 m) apart.

timing line, bends between the poles, circles Pole 6, returns through the course by bending between the poles, and crosses the timing line. In the other pattern (Figure 12-16), the poles are 21 feet apart (6.4 m) and the timing line is 21 feet from Pole 1. The horse crosses the timing line, moves parallel to the line of poles, makes a 180° turn around Pole 6, passes between Poles 6 and 5, bends through the poles to Pole 1, circles Pole 1, bends through the poles to Pole 6, makes a 180° turn around Pole 6, and moves parallel to the line of poles to cross the finish line.

Keyhole Racing

The keyhole race is run over a course that is laid out with a limed keyhole on the ground. The throat of the keyhole (4 feet wide and 10 feet long, or 1.2 m wide and 3.05 m long) is perpendicular to and faces the timing line. The center of the keyhole (20 feet in diameter, or 6.1 m) is 100 feet (30.5 m) from the timing line. The horse crosses the timing line, enters the throat of the keyhole, goes to the center of the keyhole, makes a 180° turn in either direction, and returns across the timing line. The horse is disqualified if it steps on or over the limed keyhole at any point or if it fails to turn around in the center of the keyhole.

Figure-Eight Stake Racing

Young riders find the stake race a challenge during their early stages of learning to ride. The rider has to run a figure-eight pattern around two upright markers that are 125 feet apart, or 38.1 m (Figure 12-17). If an upright marker is knocked down, the entry is disqualified.

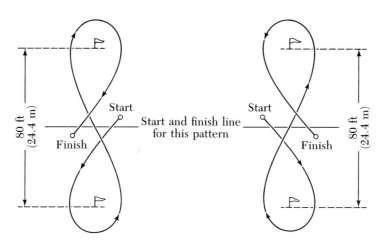

FIGURE 12-17
Figure-eight stake race pattern.

Other events, such as quadrangle stake, scurry, figure-eight relay, rescue, speed barrel, and potato race competitions, are also exciting, and a successful competitor requires a well-schooled horse.

DRESSAGE

Dressage (pronounced drə-säzh′) is a system of skilled horsemanship that draws its foundations from the earliest link between rider and animal. Dressage teaches a horse to be obedient, willing, supple, and responsive (Figure 12-18). It is a complex activity that nonetheless emphasizes fundamentals, combining the best of sport and art. The principles of dressage first appeared in a book written by the Greek statesman and general Xenophon in 400 B.C. The philosophy he expressed is still current in today's dressage riding: "Anything forced and misunderstood can never be beautiful. If a dancer were forced to dance by whip and spikes, he would be no more beautiful than a horse trained under similar conditions."

Similarly, the American Horse Shows Association states:

The object of dressage is the harmonious development of the physique and ability of the horse. As a result it makes the horse calm, supple, loose, and flexible, but also confident, attentive and keen, thus achieving perfect understanding with his rider.

These qualities are revealed by: a) The freedom and regularity of the paces; b) The harmony, lightness, and ease of the movements; c) The lightness of the forehand and the engagement of the hind quarters, originating in a lively impulsion; d) The acceptance of the bridle, with submissiveness throughout and without any tenseness or resistance.

The horse thus gives the impression of doing of his own accord what is required of him. Confident and attentive he submits generously to the control of his rider, remaining

FIGURE 12-18
Hilda Gurney on Keen,
1976 Olympic Gold Medal
winner in dressage competition.
(Courtesy USDF.)

absolutely straight in any movement on a straight line and bending accordingly when moving on curved lines.

His walk is regular, free and unconstrained. His trot is free, supple, regular, sustained, and active. His canter is united, light and cadenced. His quarters are never inactive nor sluggish. They respond to the slightest indication of the rider and thereby give life and spirit to all the rest of his body.

By virtue of a lively impulsion and the suppleness of his joints, free from the paralyzing effects of resistance, the horse obeys willingly and without hesitation and responds to the various aids calmly and with precision, displaying a natural and harmonious balance both physically and mentally.

In all his work, even at the halt, the horse must be "on the bit." A horse is said to be "on the bit" when the hocks are correctly placed, the neck is more or less raised and arched according to the stage of training and the extension or collection of the pace, and he accepts the bridle with a light and soft contact and submissiveness throughout. The head should remain in a steady position, as a rule slightly in front of the vertical, with a supple poll as the highest point of the neck, and no resistance should be offered to the rider.

Dressage competitions are held in 20- × 40-meter or 20- × 60-meter arenas (65.6 × 131.2 feet or 65.6 × 196.8 feet). The smaller arena may be used for tests through second level. Depending on the size, the arenas are laid out according to Figure 12-19. Letters identify specific parts or areas of the

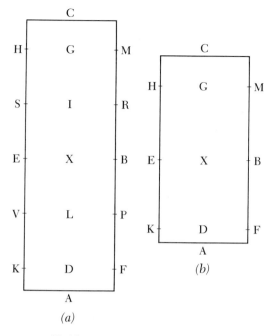

FIGURE 12-19
Dressage arenas. *(a)* 20 × 60 meters (65.6 × 196.8 ft).
(b) 20 × 40 meters (65.6 × 131.2 ft).

arena. Horses and riders are tested at several levels depending on their abilities: training, first, second, third, fourth, International Level B (Prix St. Georges and Intermediaire 1),and International Level A (Intermediaire 2 and Grand Prix). A typical test for third level is given in Table 12-1.

United States Dressage Foundation

The United States Dressage Foundation (USDF) is a nonprofit educational organization founded in 1973 by a group of dedicated dressage enthusiasts from across the country.

USDF seeks to promote and encourage a high standard in dressage, primarily educating its members and the general public. This goal is advanced by USDF's national publicity program and its growing network of 57 member organizations situated in all areas of the United States and divided into seven national regions. There are more than 7,000 USDF members and over 140 USDF-recognized dressage competitions. These competitions vary from one-day, one-judge competitions with 50 riders to shows lasting several days with several judges and 500 to 600 riders.

TABLE 12-1
Typical Third-Level Test

Purpose: The third-level tests are of medium difficulty and are designed to determine that the horse has acquired an increased amount of suppleness, impulsion, and balance, so as to be light in hand and without resistance, enabling the rider to collect and extend its gaits.

Conditions:
1. To be ridden in a plain snaffle with a noseband or in a simple double bridle.
2. Arena size: 20 m × 60 m (65.6 × 196.8 feet).
3. Time allowed: 9½ minutes.

Exercise	Letter in Arena	Test	Directive Ideas	Points	Coefficient	Total	Remarks
1	A X	Enter, collected trot Halt, salute Proceed, collected trot	Entry (straightness) Halt (immobility) Transition from halt				
2	C M–X–K K	Track to right Change rein, medium trot Collected trot	Extension Balance Transitions				
3	F–B B B–M	Shoulder in Circle left 8 m diameter (26.2 ft) Shoulder in	Flexion and balance Bearing of horse Regularity of trot				
4	H–X–F F	Change rein, medium trot Collected trot	Extension and balance Transitions				
5	K–E E E–H	Shoulder in Circle right 8 m diameter (26.2 ft) Shoulder in	Flexion and balance Bearing of horse Regularity of trot				
6	M–X–K K	Change rein, extended trot (sitting) Collected trot	Extensions Balance Transitions				
7	A	Halt Rein back 3 or 4 steps Proceed at medium walk	Halt Rein back Transitions				

No.		Test	Directive ideas						
8	F–X–H H	Change rein, extended walk Medium walk	Extensions and balance Transitions						
9	R	Half-turn on haunches right Proceed at medium walk	Regularity of half-turn Walk		2				
10	S	Half-turn on haunches left Proceed immediately at collected canter right lead	Regularity of half-turn Walk Canter depart						
11	M–F F	Medium canter Collected canter	Extensions and balance Transitions						
12	E E	Circle right 10 m diameter (32.8 ft) Simple change of lead	Regularity of circle Flexion Change of leg						
13	C M–X–K	Circle right 10 m diameter (32.8 ft) Retaining left lead Changing rein, no change of lead	Regularity of circle and counterlead Balance and flexion						
14	F–M M	Medium canter Collected canter	Extension and balance Transitions						
15	H–K K	Extended canter Collected canter	Extension and balance Transitions						
16	B B	Circle left 10 m diameter (32.8 ft) Simple change of lead	Regularity of circle Bearing and flexion Change of lead						
17	C H–X–F	Circle left 20 m diameter (65.6 ft) Retaining the right lead Change rein, no change of lead	Regularity of circle and counterlead Balance and flexion						
18	K–H H	Extended canter Collected canter	Extension and balance Transitions						

(continued)

SOURCE: American Horse Shows Association (1977).

395

TABLE 12-1 (continued)

Exercise	Letter in Arena	Test	Directive Ideas	Points	Coefficient	Total	Remarks
19	C M–X–H K	Collected trot Change rein, extended trot (sitting) Collected trot	Extension Balance Transitions				
20	A L	Down centerline Halt (6 seconds) Proceed, collected trot	Straightness Halt (immobility)				
21	G	Halt, salute Leave arena, freewalk on a loose rein	Halt (immobility) Relaxation in free walk				

Collective Marks:

		Gaits (freedom and regularity)			2		
		Impulsion (desire to move forward, elasticity of steps, relaxation of back, and engagement of hindquarters)			2		
		Submission (attention and confidence; harmony and lightness and ease of movements; acceptance of bit)			2		
		Position, seat of rider, correct use of aids			2		

Subtotal _____

Errors (–)

Time Faults (–)

Total Points _____

Time: _____ minutes _____ seconds _____

Scoring: 10 Excellent; 9 Very good; 8 Good; 7 Fairly good; 6 Satisfactory; 5 Sufficient; 4 Insufficient; 3 Fairly bad; 2 Bad; 1 Very bad; 0 Not performed or fall of horse or rider.

Penalties: Time: ½ point for each commenced second overtime.

Errors: 1st error, 2 points; 2nd error, 4 points; 3rd error, elimination; leaving arena, elimination.

Rider Award Program

The USDF has a horse and a rider award program. The general requirements are:

1. Riders must be members of USDF (individual or participating) at the time their scores were received.

2. Scores are cumulative and need not be earned in one year. Any number of horses may be used.

3. Only scores from USDF-recognized competitions are counted. Riders are personally responsible for notifying USDF of scores, using official rider award report forms available from USDF.

4. To assure accuracy in recording, scores should be reported no later than October 1 of the award year in which they were earned. (The USDF award year is from October 1 to September 30 of the following calendar year.)

5. If two or more judges are scoring one ride, the average of their scores counts as one.

6. Only scores earned riding AHSA tests will be counted.

The Keystone Rider Award is given to riders with four scores of 60 percent or better at the training level, earned at four different competitions under four different judges. On completing the requirements, the rider receives a certificate and is eligible to purchase a patch to be worn on a warmup jacket or other apparel.

The Bronze Medal Rider Award is given to riders with two scores of 60 percent or better, from two different judges for two different rides, at the first, second, and third levels.

The Silver Medal Rider Award is given to riders with two scores of 60 percent or better, from two different judges for two different rides, at the fourth level and Prix St. Georges.

The Gold Medal Rider Award is awarded to riders completing two scores of 60 percent or better, from two different judges for two different rides, at Intermediaire and Grand Prix.

Horse of the Year Awards Program

The requirements for the USDF horse of the year awards are:

1. Scores are recorded from results submitted directly to USDF by USDF-recognized competitions.

2. Only scores earned from currently approved AHSA tests ridden in open classes at USDF-recognized competitions count toward awards.

3. Scores are recorded only for USDF-registered horses owned by participating USDF members. Membership number and horse registration number should be included on entry blanks. If the horse is owned by more than one person, or by a stable, at least one of the owners must be a participating member.

4. No scores are recorded for the horse until it is registered.

To qualify, the horse must have a minimum of five scores earned at four competitions under four judges, with a final median average of at least 55 percent. The median is the middle score when all scores are ranked from the highest to the lowest.

Awards are awarded as follows: 25 at the first level, 25 at the second level, 10 at the third level, 10 at the fourth level, 10 at International Level B (Prix St. Georges and Intermediaire 1), 5 at International Level A (Intermediaire 2 and Grand Prix).

RACING

Racing horses has been a hobby for many centuries. Its history traces from the domestication of the horse to the early Olympiads and finally to modern racing. Racing attracts one of the largest numbers of spectators of any sport. The racing industry is continuing to grow and currently involves several types of horses: Thoroughbreds, Quarter Horses, Standardbreds, Arabians, trotting ponies, Appaloosas, and mules.

Types of Races

A horse can enter several types of races, depending on its ability. The types of races to be run each day are determined by the racing secretary of the racetrack, and the horses that meet the conditions of the race may be entered by the trainers. The types of races and the conditions of each race are published every few days in the "condition book."

A stakes race is a top-quality race in which the owners pay nominating fees, entry fees, and starting fees. In addition to these fees, the racetrack may add money to the purse to guarantee a certain value.

In handicap races, the racing secretary takes entries, evaluates the horse's past performance, and assigns weights to be carried so that all horses have an equal chance of winning. The horses with better records or potential carry more weight than those with poorer records. Their purse value is determined by nomination and starting fees and by added money. Fees are not paid when the race is an invitational handicap race.

In allowance races, the eligibility and the weight a horse carries are based on the number of races or amount of money the horse has won. A basic weight is assigned for the race, and horses with poorer records are given weight allowances. An allowance can be an invitational race.

In claiming races, all horses entered are for sale at the price they are entered. They may be claimed or purchased by any owner who has started a horse at that race meeting. The claiming race is a method of classifying horses in order to produce races involving competition of equal quality.

Maiden races are for horses that have never won a race recognized by the sanctioning organization for the breed.

Derby races are for 3-year-olds.

Supervision of Racing

To supervise and maintain the integrity of racing, each state has a racing commission. The commissioners' duties are to implement state rules and regulations of racing and to protect the public interest. They limit the number of racing days for each racetrack or meet, grant racing franchises to tracks, approve purse schedules, approve appointment of officials at each track, and supervise licensing of all racetrack personnel.

To protect the public, the state racing commission is also responsible for the drug testing program. Each winning horse has samples of saliva and urine tested for illegal use of drugs. The tests are made by a member laboratory of the Association of Official Racing Chemists.

At each racetrack, a number of personnel are required to supervise the racing. Three racing stewards preside over the racing at each track and supervise all other racing officials at that track. They watch each race for rule infractions. If an objection is lodged, they must determine if a foul occurred, its cause, who was responsible, and the extent of responsibility, and they take appropriate action. If a foul occurred, they may or may not disqualify the horse involved and penalize the jockey. The stewards are assisted by patrol judges, who observe the race from elevated platforms located along the racecourse. They report rule violations they observe to the stewards. Three placing judges help the stewards determine the order of finish of each race. Videotapes are made of most races to help the stewards, patrol judges, and placing judges make decisions. The videotapes are available the following day for viewing by the trainers and jockeys to analyze a horse's performance.

The racing secretary is responsible for formulating the race program for each day. He or she must determine the types of races to be run and take the entries from the trainers. The racing secretary approves stalls for racing stables, creates a purse and race program that will attract horses (condition book), and creates racing cards that will attract the public's interest. The identity of each horse is verified by its lip tattoo number and markings by the official track identifier as the horse enters the paddock area, to prevent "ringers" from being run in the race. A paddock judge is in charge of the paddock area, where the horses are saddled prior to each race. The paddock judge controls who enters the paddock area, verifies racing equipment, supervises the saddling, gives the command for the jockeys to mount, and starts the post parade. The jockey's room is under the supervision of the clerk of scales. The clerk also is responsible for seeing that each jockey rides with the proper equipment, is weighed before the race to ensure that the horse carries the prescribed weight, and is weighed at the end of the race to ensure that the horse carried the weight to the finish line. A number of employees in the jockey's room assist the clerk: assistant clerk of scales, color man (cares for and stores racing colors for each stable), number cloth and equipment custodian, porters, guards, and valets. The track veterinarian is responsible for preventing injured and disabled horses from racing so that all the horses racing are physically capable of attaining their performance capability. The starter is responsible for starting each race and ensuring a fair start. The clocker is responsible for timing the daily workouts and publishing the times for the public.

Parimutuel Betting

Parimutuel wagering means that members of the public are betting against each other instead of against some type of organization. The racing association is responsible for holding the bets and paying the winners.

There are three basic types of bets and several types of special bets. The three basic bets are win, place, and show. A win bet is for a horse to win the race. A place bet is for a horse to finish the race in first or second position. A show bet is for a horse to finish first, second, or third. Special types of bets are the daily double, quinella, perfecta, exacta, and combination. A combination bet means that a horse is bet to finish first, second, or third. There the three basic bets are combined in one bet. If the horse wins the race, the bettor collects the payoff for win, place, and show. If the horse finishes in second place, the bettor collects the place and show bets. When the daily double is bet, the bettor must select the winners of the first and second races in order to win. A quinella wager is to select the first two horses that finish a race regardless of which of the two horses wins the race. The exacta and perfecta wagers are a variation of the quinella in that the player must select the winner and the second-place horse in the exact order of finish.

The odds and the eventual payoffs are determined by the amount of money wagered on the entries. The approximate pay to win for the odds are given in Table 12-2. These are figured by the totalisator computer. The payoff is computed according to the following scheme. The payoff for the bets on the winning horse are calculated by taking the total amount of money wagered for horses to win and deducting the percent of money held out for state taxes, racetrack operator's commission, purses paid to owners of the horses, and breeders' and stallions' awards. In California, the percentage withheld from the sum of money wagered is 15.75 percent. After this figure is determined,

TABLE 12-2
Approximate Pay to Win (for $2) for the Odds

Odds	Pays	Odds	Pays	Odds	Pays
1–5	$2.40	5–2	$ 7.00	15	$ 32.00
2–5	2.80	3	8.00	18	38.00
1–2	3.00	7–2	9.00	20	42.00
3–5	3.20	4	10.00	25	52.00
4–5	3.60	9–2	11.00	30	62.00
Even	4.00	5	12.00	40	82.00
6–5	4.40	6	14.00	50	102.00
7–5	4.80	7	16.00	60	122.00
3–2	5.00	8	18.00	70	142.00
8–5	5.20	9	20.00	80	162.00
9–5	5.60	10	22.00	90	182.00
2	6.00	12	26.00	99	200.00

SOURCE: *American Racing Manual* (1978).

the amount of money bet on the winning horse is subtracted and the remainder divided by the amount of money bet on the winning horse. For example, if $150,000 is wagered on horses to win, the amount remaining after the 15.75 percent is withheld is $126,375. If the winning horse was a "long shot" and only $20,000 was bet for it to win, the next calculation will be to deduct the amount wagered on the winning horse from the win pool ($126,375 – $20,000 = $106,375). The $106,375 is then divided by $20,000 to determine the payoff for each winning dollar bet ($106,375 / $20,000 = $5.32). All payoffs are calculated to the lowest $.05; therefore, the $5.32 becomes $5.30. The $.02 goes into the "breakage pool" that is kept by the state and the track. The $5.30 is the odds for each dollar bet; therefore, a $2 wager (minimum wager) returns $12.60. The $12.60 includes the $10.60 odds plus the $2.00 wagered.

The payoffs for place and show bets are calculated by dividing the total money bet to place into two equal pools (for the two horses) and dividing the total money bet to show into three equal pools (for the three horses finishing first, second, and third). Each pool is divided among the winning bets as out-lined for the win bet.

Some states allow off-track betting. In these states, the bettor need not attend the races to place wagers. Each state governs off-track betting in that state if it is allowed. Illegal betting takes place by bettors placing their bets with "bookmakers." However, in some states like Nevada, bookmakers are licensed and the bets placed with a licensed bookmaker are legal.

In order to properly bet and handicap horses, one must be able to read the daily racing program and past performances. The results of the races for the previous day are also published. The information contained in the daily program is given in Figure 12-20. If more than 12 horses are entered in a race,

Distance of Race

Arrow points to where race begins, as distances vary with each day's program

Purse and Conditions for which horses are running.

Track Record for Distance

Program No. also same on tote for betting

Entry—Explained on page No. 12

Position in Starting Gate

Jockey

Weight to be carried by jockey as stated in conditions of race and assigned by Racing Secretary. Note after race jockey weighs in with saddle.

Owner

Morning Line or probable odds as suggested by handicapper. Public sets actual odds by Pari-Mutuel wagering.

Chestnut Gelding 6 years old. For further explanation turn to page 5

Horse's Sire and Dam

Name of horse

Trainer

Apprentice allowance see page 15

Owners colors. These are important to watch as colors, easier to see than numbers, will enable you to see exactly where your horse is throughout the race.

FIGURE 12-20

Daily racing program. (Courtesy Golden Gate Fields.)

the horses carrying numbers greater than 11 are considered to be field horses. By betting on the Number 12 to win, place, or show, the bettor wins if any of the field horses finish first, second, or third, respectively. When two or more horses are entered by one trainer, they are coupled together in the wagering as one entry. A bet on one horse in the entry is a bet for all the horses entered by the trainer.

In addition to the daily racing program, the *Daily Racing Form* publishes the daily program for all the recognized tracks in America. In the program, the past performances of each horse for each race at all the recognized tracks in America are given, as well as the conditions of the race. How to read the past performances is shown in Appendix Table 12-1.

The past performances of horses raced in America are published in the *Daily Racing Form* in the official charts for each horse entered in each race at every racetrack in America. The chart describes a race. It offers a complete, accurate, easy-to-understand word picture of the running from the time the horse is lined up at the gate to the moment the result and prices are official.

Harness Racing

All harness racing in the United States and in the Maritime Provinces of Canada is conducted under the guidance of the U.S. Trotting Association (USTA). This association was organized in 1938–1939 by a group of representative horse-people from all sections of the country who previously held membership in one of the three sponsoring or governing bodies. This move welded all Standardbred racing interests for the first time in the history of the sport and gave great impetus to the sport because it centralized authority and made administrative procedure more efficient and effective than ever before.

Before this time, various sections of the country managed their version of the sport much as they pleased. Three outstanding groups preceded the USTA organization: United Trotting Association, American Trotting Association, and National Trotting Association. Until they were combined, trainers and owners often never knew for sure just which organization they were racing under and therefore were always in doubt as to what racing rules were to be followed.

Harness racing on the modern parimutuel basis first was staged at Roosevelt Raceway, located at Westbury, Long Island, near New York City, in 1940, and has mushroomed into one of the nation's top spectator sports (Table 12-3). With night raceways, modern presentations, and the introduction of the starting gate, the sport changed rapidly from America's smaller towns and an essentially rural atmosphere to an attraction for metropolitan sports fans.

Major raceways now are located in or near such cities as Albany, Boston, Buffalo, Chicago, Cincinnati, Cleveland, Columbus, Dayton, Detroit, Lexington, Los Angeles, Miami, New York, Philadelphia, Pittsburgh, Portland, San Francisco, Syracuse, Toledo, Washington, Wilmington, and many others. In 1977, 491 racetracks had Standardbred racing (Table 12-3).

TABLE 12-3
The Growth of Harness Racing

Year	Purses	Horses Racing	Colts Registered	U.S. Trotting Association Members	Track Members
1950	$ 11,527,711.94	10,281	4,386	8,411	546
1960	31,581,922.54	17,702	6,794	15,593	454
1970	89,280,537.00	35,465	11,981	32,725	429
1977	179,095,788.00	46,888	13,929	40,861	491

SOURCE: United States Trotting Association.

Attendance at parimutuel plants has skyrocketed since 1943, and the millions who have watched racing at fairs and nonbetting meets have pushed the spectator figure up among the top five sports.

From a total of only a few thousand dollars in 1940, the wagering had passed the billion-dollar mark in 1963 with $1,067,130,007 in parimutuels. In 1977, $6,060,622,126 was bet.

The rise of harness racing at the raceways is reflected in purses paid to horsepeople and the consequent boom in the breeding industry and horses in action. Purses have reached $190 million, and over 47,000 trotters and pacers see action during a year.

The rise in wagering meetings has made harness a year-round sport with racing in California and Florida during the winter months and at northern tracks from mid-February to mid-December.

Like every other sport, harness racing has its "big league." In this case, it is the Grand Circuit. Organized back in 1873 with four member tracks "mainly for the purpose of setting up a definite itinerary for horse owners," it not only has grown to include the best of the sulky world in racing competition but also has developed into a gigantic circuit for America's top trotters and pacers. Each track sponsoring a Grand Circuit meeting is considered a member and has representation on the board of stewards. This board formulates policies, awards dates, elects officers, and conducts general affairs of the organization.

Harness racing's counterpart to the famed Kentucky Derby remains the Hambletonian, sponsored by the Hambletonian Society and first raced at Syracuse, New York, but the nation's 3-year-old trotters now have four lucrative targets each year. The Hambletonian has been raced at Syracuse; Lexington, Kentucky; Yonkers, New York; and Goshen, New York, where it was a fixture for many years before being moved to Du Quoin, Illinois, in 1957. Conducted on the two-in-three heat plan, the Hambletonian is the goal of all trainers and owners.

Oldest of the "Big Four" features is the Kentucky Futurity, a feature of the Lexington Trots meeting at Lexington each autumn and usually the grand finale for strictly 3-year-old trotting competition.

Competitive newcomers are the Yonkers Futurity, raced on the half-mile Yonkers Raceway oval in Yonkers and the Dexter Cup on the Roosevelt Raceway half-mile track in Westbury, New York. Speedy Scot in 1963 was the first horse to win all four events.

The 3-year-old pacers were given their first major purse at which to aim when the Little Brown Jug Classic was inaugurated at Delaware, Ohio, in 1946. Held annually since, the "Jug" now has companion attractions in the William H. Cane Futurity at Yonkers Raceway and the Messenger Stake at Roosevelt Raceway, both metropolitan New York ovals.

All the "Triple Crown" pacing events are conducted over half-mile tracks with the heat plan in effect for the Little Brown Jug and single dashes deciding the Cane and the Messenger. Two of the races were named after famous horses. Little Brown Jug was a famous world champion pacer of the nineteenth century, while Messenger was a stallion imported to this country in the late eighteenth century (almost all harness horses trace back to him).

Quarter Horse Racing

Quarter Horse racing had its beginning in colonial America. The races were usually match races between two horses and the horses ran down village streets or across small areas of cleared ground. The first recorded races were held in Enrico County, Virginia, in 1674. These types of sprint races continued to be popular until the mid-1700s, when races for long distances became popular. As the frontier moved westward, the popularity of sprint races also moved westward into the Midwest, Southwest, and West. The first organized Quarter Horse racing was held at Hacienda Moltacqua in Tucson, Arizona. Hacienda Moltacqua also had Thoroughbred, trotting, and steeplechase races but Quarter Horse racing popularity resulted in the construction of Rillito Park, which was designed exclusively for Quarter Horse racing. During the 1900s to 1945, other racetracks were built for Quarter Horse racing and this gave birth to our modern-day Quarter Horse racing.

Today, Quarter Horse racing is conducted at over 102 racetracks (Figure 12-21). Parimutuel betting is allowed at 79 of these racetracks. The total parimutuel handle at these tracks has increased rapidly during the last 20 years (Figure 12-22). The number of races recognized for Quarter Horses is continuing to increase (Figure 12-23) as the total purse that is paid to the owners of the horses also is increasing (Figure 12-24). The All-American Futurity, which is held at Ruidiso Downs in New Mexico on Labor Day, has the largest purse distribution of any horse race that is held each year. The winning horse collects over $330,000.

Quarter Horses race at distances of 220, 250, 300, 350, 400, 440, 500, 660, 770, and 870 yards (201.2, 228.6, 274.3, 320, 365.8, 402.3, 457.2, 603.5, 704.1, and 795.5 m). In addition to establishing world records for racing a given distance, Quarter Horses become recognized for their speed

FIGURE 12-21

Location of North American Quarter Horse racetracks.
(From *Quarter Running Horse Chart Book*, 1978.)

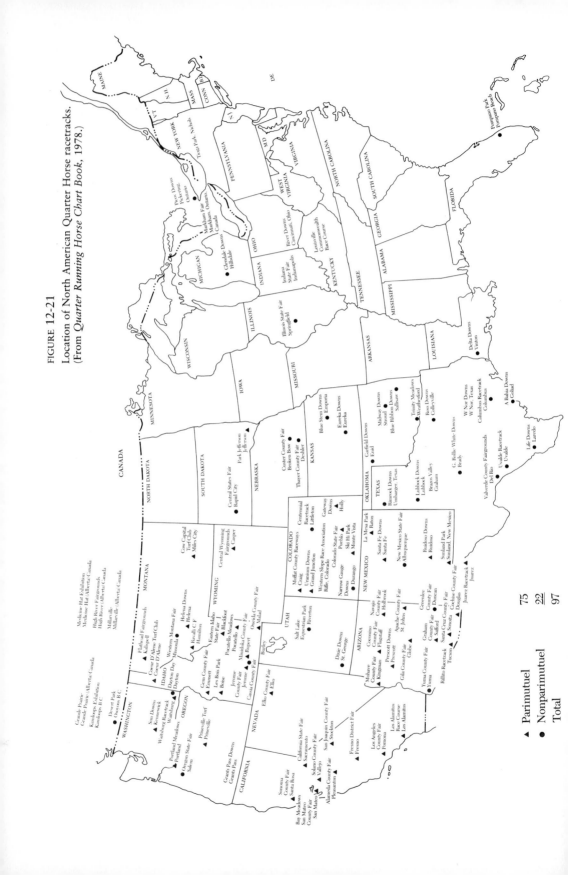

▲ Parimutuel   75
● Nonparimutuel   22
   Total   97

Units of $1 million

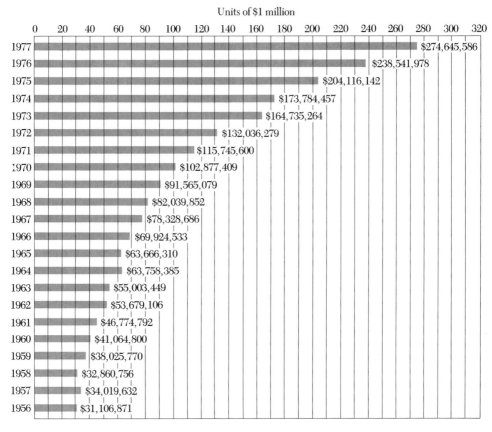

FIGURE 12-22

Total parimutuel handle on races for Quarter Horses, 1956–1977.
(From *Quarter Running Horse Chart Book, 1978.*)

index and for the total amount of money they win. Speed index ratings are based on an average time for horses running given distances at the track on which they race. Based on a formula, each fraction of a second is equal to one speed index point. The basis for calculation is explained in the official handbook of the American Quarter Horse Association. The leading money-earning horse since 1949 has been EASY DATE (Appendix Table 12-2). EASY DATE is by EASY JET, which is one of the leading sires of money earners since 1949 (Appendix Table 12-3). LEO is the leading maternal grandsire of Register of Merit qualifiers for racing (Appendix Table 12-4). Dams of racing Quarter Horses attain fame by producing outstanding horses. Several mares have had outstanding produce records by producing 10 or more horses that earned their Register of Merit in racing (Appendix Table 12-5).

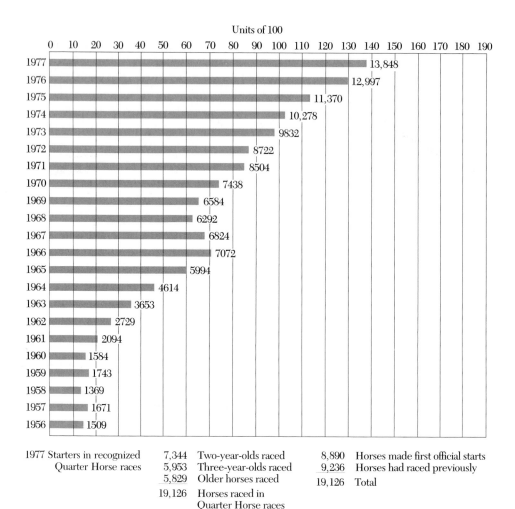

Year	Value
1977	13,848
1976	12,997
1975	11,370
1974	10,278
1973	9832
1972	8722
1971	8504
1970	7438
1969	6584
1968	6292
1967	6824
1966	7072
1965	5994
1964	4614
1963	3653
1962	2729
1961	2094
1960	1584
1959	1743
1958	1369
1957	1671
1956	1509

1977 Starters in recognized Quarter Horse races			
7,344	Two-year-olds raced	8,890	Horses made first official starts
5,953	Three-year-olds raced	9,236	Horses had raced previously
5,829	Older horses raced	19,126	Total
19,126	Horses raced in Quarter Horse races		

FIGURE 12-23

Number of recognized races for Quarter Horses, 1956–1977.
(From *Quarter Running Horse Chart Book,* 1978.)

At the conclusion of each racing year, the American Quarter Horse Association Racing Committee votes to determine the champion horses for the various classifications of racing horses. Since SHUE FLY and WOVEN WEB in the 1940s and GO MAN GO in the 1950s, DASH FOR CASH (Figure 12-25) is the only horse to be named World Champion Running Horse for two consecutive years (Appendix Table 12-6). SHUE FLY, WOVEN WEB, and GO MAN GO were each voted the title for three consecutive years.

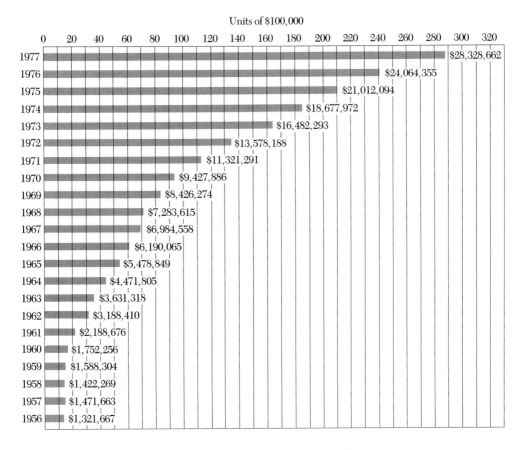

FIGURE 12-24

Total purse distribution in recognized races for Quarter Horses, 1956–1977. (From *Quarter Running Horse Chart Book*, 1978.)

Thoroughbred Racing

Thoroughbred racing traces its foundation to the establishment of the Jockey Club in Newmarket, England, in about 1750. Prior to then, there was no standardized set of rules under which races were run. However, racing was popular in England many years prior to 1750, and it developed into a sport during the reigns of the Tudor and Stuart monarchs.

The first racetrack in America was established at Hempstead Plains on Long Island. From the area of New York, racing spread to the other colonies. The first Jockey Club was founded in 1734, prior to the establishment of the Jockey Club in Newmarket. Over the years, two major racing centers have

FIGURE 12-25
Dash for Cash winning the 1976 Champion of Champions Race. (Courtesy Pacific Quarter Horse Racing Association.)

developed. California has the largest gross distribution in stakes and purses (Appendix Table 12-7). New York has almost the same distribution. However, more races are run in Pennsylvania than in any other state. The gross distribution of purses and stakes and number of racing days has continued to increase each year until the more than 69,406 races with a distribution of over $414 million are run each year (Appendix Table 12-8).

As a result of the sport of racing, many horses and horsepeople have become famous. The leading money-winning horses each year (Appendix Table 12-9) and for their lifetime (Appendix Table 12-10) attract considerable attention, particularly when they win the Triple Crown of Racing—Kentucky Derby, Preakness, and Belmont. In recent years, Affirmed (1978), Seattle Slew (1977), and Secretariat (1973) have accomplished this feat. Each year, the best horses of each type of race and for age are elected (Appendix Table 12-11). Of particular interest to breeders are the leading sires for the racehorses (Appendix Table 12-12). During the 1960s, Bold Ruler was the leading sire, and his get, such as Secretariat, are now becoming leading sires. The owners of racehorses are interested in obtaining the best jockey to ride their horses. The leading jockeys each year have been limited to a few successful jockeys (Appendix Table 12-13). Willie Shoemaker, Eddie Arcaro, and Johnny Longden have dominated the yearly results. Owners are also concerned about trainers for their horses. In recent years, Charlie Whittingham and Laz Barrera have been the leading trainers in terms of total dollars won (Appendix Table 12-14).

13
Rodeo, Trail, and Sports Events

RODEO

Rodeo, a series of contests in cowboy sport, has a colorful history tracing back to the Great Plains area and to the trail drivers. Because the trail drivers seldom were able to go into town for recreation, they made a sport out of their daily skills. Each outfit developed its own riding and roping champions. History did not record the first rodeo when several outfits got together to match their champions against each other. The prize money—the cowboys' own bets—was held in a hat. At Cheyenne, Wyoming, in 1872, townsfolk gathered to watch some Texans try to ride the wild stock. In 1883, at Pecos, Texas, they penned the longhorns on the courthouse lawn and roped them down the main street. This was the first public cowboy contest for prizes. The first rodeo held for competition in front of paying spectators was held at Prescott, Arizona, in 1888. The Cheyenne Frontier Days rodeo started in 1897 and has continued since then. The Pendleton Oregon Roundup started in 1910. Two other famous rodeos are the California Rodeo at Salinas and the Calgary Alberta Stampede, which started in 1911 and 1912, respectively. Since these early days, the sport of rodeo has continued to flourish, organized by several organizations: the National High School Rodeo Association, National Intercollegiate Rodeo Association, Girls' Rodeo Association, Professional Rodeo Cowboys Association, and various state amateur rodeo associations. Today, the contestants compete for over $100,000 to $200,000 at the top rodeos and draw over 14 million spectators. Rodeo has become "big business."

TABLE 13-1
Annual Statistics, Association Rodeo, 1953–1978

Year	States with Rodeos	No. of Rodeos	No. of Performances	Members	Permit Holders	Total Purse	Entry Fees	Total Prize Money
1953	35	578	1,779	3,001	—	$1,486,805	$1,006,051	$2,492,856
1954	34	550	1,721	3,284	—	1,485,488	1,240,702	2,726,190
1955	36	542	1,751	3,184	—	1,549,769	1,280,215	2,829,984
1956	33	519	1,699	3,329	—	1,516,552	1,345,973	2,862,525
1957	33	458	1,475	2,958	1,263	1,467,032	1,328,163	2,795,195
1958	35	475	1,535	2,809	2,131	1,450,109	1,341,998	2,792,107
1959	38	493	1,566	2,781	4,256	1,491,045	1,701,700	3,192,745
1960	35	509	1,583	2,820	3,709	1,578,915	1,508,005	3,086,920
1961	37	542	1,667	2,795	2,795	1,551,528	1,461,715	3,013,243
1962	37	540	1,587	2,670	2,837	1,548,728	1,531,374	3,080,102
1963	41	582	1,738	2,829	3,294	1,672,836	1,838,411	3,511,247
1964	44	591	1,795	3,105	3,361	1,690,379	1,975,091	3,665,469
1965	42	542	1,666	3,205	3,291	1,661,276	1,907,084	3,568,360
1966	44	524	1,602	3,007	2,344	1,660,727	1,791,817	3,452,544
1967	42	537	1,673	3,039	2,362	1,778,657	1,871,098	3,649,755
1968	41	521	1,592	3,159	2,379	1,688,636	1,996,993	3,685,629
1969	40	533	1,615	3,346	2,569	1,659,635	2,090,710	3,850,345
1970	42	547	1,653	3,446	2,821	1,811,080	2,303,941	4,155,021
1971	42	539	1,621	3,153	2,296	1,845,722	2,181,021	4,026,743
1972	42	567	1,691	3,144	2,707	1,944,417	2,430,693	4,375,110
1973	41	600	1,755	3,519	3,384	2,011,913	2,971,054	4,982,967
1974	40	590	1,816	3,583	4,304	2,162,395	3,224,489	5,386,884
1975	39	594	1,817	3,651	5,084	2,432,580	4,000,000	6,432,580
1976	36	586	1,974	4,027	4,025	2,383,867	4,306,233	6,690,100
1977	40	579	1,763	4,153	3,784	2,524,705	4,442,205	6,966,910
1978	39	620	1,890	4,820	4,363	2,817,745	5,238,289	8,056,034
						$46,972,542	$58,315,025	$105,286,566

SOURCE: *Official Pro Rodeo Media Guide* (1978).

Rodeo Associations

Professional Rodeo Cowboys Association

The initial organization of rodeo took place at Salinas, California, in 1926. The Rodeo Association of America (RAA) was formed to standardize the events and rules and to name actual world champions based on the total prize money won. The first organization of the rodeo cowboys (contestants) was the Cowboys Turtle Association. The CTA was formed in late 1936 because of the payoff at the Boston Garden Rodeo. After waging a successful strike for more purse money, the CTA grew rapidly. In 1945, the CTA reorganized to form the Rodeo Cowboys Association. Later the name was changed to the Professional Rodeo Cowboys Association (PRCA). The PRCA organized a circuit system in 1975 to give professional cowboys identity on the local level. There are ten circuits laid out on a national scale according to the number of rodeos in given areas.

Since 1953, professional rodeo has continued to grow (Table 13-1). Today, there are over 9,000 contestants, who compete for over $8,000,000 in total prize money. The all-around champion may win over $100,000 per year in prize money (Table 13-2). PRCA world champions may win $20,000 to $60,000 depending on the event. The world championship for each event is determined by the cowboy who wins the most money during the season. The top 15 leaders in the PRCA championship standings for each are eligible to compete at the National Finals Rodeo for the world championship title for each event. The all-around championship is awarded to the contestant winning the most money by competing in two or more events (Table 13-3).

TABLE 13-2
World Records Set by Members of the Professional Rodeo Cowboys Association (PRCA)

Record	PRCA Member	
Most All-Around Championships	Larry Mahan, Tom Ferguson	(6)
	Jim Shoulders	(5)
Most Consecutive All-Arounds	Tom Ferguson	(6)
	Larry Mahan	(5)
Most World Championships	Jim Shoulders	(16)
	Dean Oliver	(11)
Most Total Money Won (Career)	Dean Oliver (1952–1979)	$543,172
	Tom Ferguson (1972–1979)	$535,874
Most Money Won in One Year	Tom Ferguson (1978)	$103,733
	Tom Ferguson (1976)	$ 96,913
Most Money Won at One Rodeo	Tom Ferguson (1979, Houston, Tex.)	$16,445
	Larry Ferguson (1978, Houston, Tex.)	$13,083
Most Money Won at National Finals Rodeo	Tom Ferguson (1978)	$20,000
	Dave Brock (1978)	$15,500

(continued)

TABLE 13-2 *(continued)*

Record	PRCA Member	
Most Money Won in Single Event for One Year	Bruce Ford (1979, bareback riding)	$80,260
	Roy Cooper (1978, calf roping)	$67,153
Most Saddle Bronc Titles	Casey Tibbs	(6)
	Pete Knight	(4)
Most Bareback Titles	Joe Alexander	(5)
	Eddy Akridge, Jim Shoulders	(4)
Most Bull-Riding Titles	Jim Shoulders	(7)
	Harry Tompkins, Smokey Snyder, Don Gay	(5)
Most Calf-Roping Titles	Dean Oliver	(8)
	Toots Mansfield	(7)
Most Steer-Wrestling Titles	Homer Pettigrew	(6)
	Everett Bowman, Jim Bynum	(4)
Most Team Roping Titles	Jim Rodriguez Jr.	(4)
	Asbury Schell, Leo Camarillo	(3)
Most Steer-Roping Titles	Everett Shaw	(6)
	Shoat Webster	(4)
Youngest World Champion	Jim Rodriguez Jr. (1959, team roping)	18
	Bill Kornell (1963, bull riding)	19
	Guy Allen (1977, steer roping)	19
Oldest World Champion	Ike Rude (1953, steer roping)	59
	Joe Glenn (1967, team roping)	53
Highest Scored Ride	Denny Flynn (1979, bull riding) on Steiner's Red Lightning at Palestine, Ill.	98 points
	Don Gay (1977, bull riding) on RSC's Oscar at San Francisco, Calif.	97 points
	Doug Yold (1979) on Transport at Meadow Lake, Sask.	95 points
Highest Score in Saddle Bronc	Joe Marvel (1976) on Franklin's Transport at Calgary, Alta.	92 points
	John Mulock (1978) on Barnes' Crystal Springs at Indianapolis, Ind.	90 points
Highest Score in Bareback	Joe Alexander (1974) on Beutler Bros. and Cervi's Marlboro at Cheyenne, Wyo.	93 points
	Jim Dix (1970) on Kesler's Moonshine at Medicine Hat, Alta.	89 points
	Joe Alexander (1976) on Kesler's Shady Spot at Calgary, Alta.	89 points
Fastest Times on Record	(No official times are kept because arena conditions and sizes vary.)	
Calf Roping	Bill Reeder (1978, Assinoboia, Sask.)	5.7 sec.
	Jerry Jetton (1979, Watonga, Okla.)	7.3 sec.
	Paul Tierney (1979, Forsyth, Mont.)	7.3 sec.
Steer Wrestling	Oral Zumwalt (1930s, without barriers)	2.2 sec.
	Jim Bynum (1955, Marietta, Okla., with barrier)	2.4 sec.
	Gene Melton (1976, Pecatonica, Ill.)	2.4 sec.
	Carl Deaton (1976, Tulsa, Okla.)	2.4 sec.
Team Roping	Reg Camarillo and Jerold Camarillo (1976, Salt Lake City)	4.9 sec.

TABLE 13-3
All-Around Rodeo Champions, 1929–1979

Year	Champion	Year	Champion
1929	Earl Thode, Belvedere, S. Dak.	1955	Casey Tibbs, Fort Pierre, S. Dak.
1930	Clay Carr, Visalia, Calif.	1956	Jim Shoulders, Henryetta, Okla.
1931	Johnie Schneider, Livermore, Calif.	1957	Jim Shoulders, Henryetta, Okla.
1932	Don Nesbitt, Snowflake, Ariz.	1958	Jim Shoulders, Henryetta, Okla.
1933	Clay Carr, Visalia, Calif.	1959	Jim Shoulders, Henryetta, Okla.
1934	Leonard Ward, Talent, Ore.	1960	Harry Tompkins, Dublin, Tex.
1935	Everett Bowman, Hillside, Ariz.	1961	Benny Reynolds, Melrose, Mont.
1936	John Bowman, Oakdale, Calif.	1962	Tom Nesmith, Bethel, Okla.
1937	Everett Bowman, Hillside, Ariz.	1963	Dean Oliver, Boise, Idaho
1938	Burel Mulkey, Salmon, Idaho	1964	Dean Oliver, Boise, Idaho
1939	Paul Carney, Galeton, Colo.	1965	Dean Oliver, Boise, Idaho
1940	Fritz Truan, Long Beach, Calif.	1966	Larry Mahan, Brooks, Ore.
1941	Homer Pettigrew, Grady, N. Mex.	1967	Larry Mahan, Brooks, Ore.
1942	Gerald Roberts, Strong City, Kans.	1968	Larry Mahan, Brooks, Ore.
1943	Louis Brooks, Pittsburg, Okla.	1969	Larry Mahan, Frisco, Tex.
1944	Louis Brooks, Pittsburg, Okla.	1970	Larry Mahan, Frisco, Tex.
1945	Bill Linderman, Red Lodge, Mont.	1971	Phil Lyne, George West, Tex.
1946	Gene Rambo, Shandon, Calif.	1972	Phil Lyne, George West, Tex.
1947	Todd Whatley, Hugo, Okla.	1973	Larry Mahan, Frisco, Tex.
1948	Gerald Roberts, Strong City, Kans.	1974	Larry Mahan, Frisco, Tex.
1949	Jim Shoulders, Henryetta, Okla.	1975	Leo Camarillo, Donald, Ore.,
1950	Bill Linderman, Red Lodge, Mont.		and Tom Ferguson, Miami, Okla.
1951	Casey Tibbs, Fort Pierre, S. Dak.	1976	Tom Ferguson, Miami, Okla.
1952	Harry Tompkins, Dublin, Tex.	1977	Tom Ferguson, Miami, Okla.
1953	Bill Linderman, Red Lodge, Mont.	1978	Tom Ferguson, Miami, Okla.
1954	Buck Rutherford, Lenapah, Okla.	1979	Tom Ferguson, Miami, Okla.

National Intercollegiate Rodeo Association

The National Intercollegiate Rodeo Association (NIRA) was founded in 1948 and incorporated in 1949. Its primary purpose is to provide leadership in organizing and promoting college rodeos on a nationwide scale by providing rules and procedures. The NIRA is divided into 10 regions in which 165 junior colleges, 4-year colleges, and universities are members. During the 10-year period of 1968 to 1978, the membership increased from approximately 1,450 individual members to 3,100. All individual members who compete in NIRA-approved rodeos must maintain a grade-point average of C or better.

The end of each rodeo year is climaxed by the College National Finals Rodeo (see Table 13-4). Each region is represented by its two top teams and two best individuals in the six men's events and four women's events. Bareback

TABLE 13-4
National Intercollegiate Championship Rodeo Records, 1949–1978

Year	Location	Team	All-Around Champion
1949	Cow Palace, San Francisco, Calif.	Sul Ross	Harley May, Sul Ross
1950	Cow Palace, San Francisco, Calif.	Sul Ross	Harley May, Sul Ross
1951	Will Rogers Coliseum, Fort Worth, Texas	Sul Ross	Dick Barrett, Okla. A&M
	(Rodeo in 1951 did not have national representation, so points are based on season total.)		
1952	Rose Festival, Portland, Ore.	Sul Ross	Dick Barrett, Okla. A&M
1953	Abilene, Texas (1st year All-Around Girl)	Hardin-Simmons Univ.	Tex Martin, Sul Ross Libby Prude, Sul Ross
1954	Finals Rodeo not held—points based on season total	Coloado A&M	Howard Harris, Univ. of Idaho
1955	Lake Charles, La.	Texas Tech	Ira Akers, Sam Houston
1956	Colorado Springs, Colo.	Sam Houston	Ira Akers, Sam Houston Kathlyn Younger, Colo. A&M
1957	Colorado Springs, Colo.	McNeese State	Clyde May, McNeese
1958	Colorado Springs, Colo.	McNeese State	Jack Burkholder, Texas A&I
1959	Klamath, Ore.	McNeese State	Jack Roddy, Cal Poly, SLO Pat Dunigan, NMSU
1960	Clayton, N.M.	Cal Poly, SLO	Ed Workman, Lubbock CC Karen Magnum, Sam Houston
1961	Sacramento, Calif. (Girls' Team)	Univ. of Wyoming Sam Houston	Ed Workman, Lubbock CC Sue Burgraff, Montana State
1962	Littleton, Colo. (Girls' Team)	Sul Ross Sul Ross	Ed Workman, Lubbock CC Donna Jean Saul, Sul Ross
1963	Littleton, Colo. (Girls' Team)	Casper College Colorado State Univ.	Shawn Davis, W. Montana C. Leota Hielscher, Colo. State
1964	Douglas, Wyo. (Girls' Team)	Casper College Colorado State Univ.	Pink Peterson, Casper Maris Mass, So. Colo. State
1965	Laramie, Wyo. (Girls' Team)	Casper College Sam Houston	Pink Peterson, Casper Becky Bergren, Sam Houston

Year	Site	Team Champion	Individual Champion
1966	Vermillion, S.D.	Casper College	Dave Hart, Idaho State Univ.
	(Girls' Team)	Arizona State Univ.	Carol O'Rourke, Montana St.
1967	St. George, Utah	Tarleton State	E. C. Ekker, Univ. of Utah
	(Girls' Team)	Eastern N.M.U.	Barbara Socolofsky, K.S.U.
1968	Sacramento, Calif.	Sam Houston	Phil Lyne, SWTSU
	(Girls' Team)	Sam Houston	Donna Kinkead, ENMU
1969	Deadwood, S.D.	Eastern N.M.U.	Phil Lyne, Sam Houston
	(Girls' Team)	Tarleton State	Nancy Robinson, Cal Poly
1970	Bozeman, Mont.	Cal Poly, SLO	Tom Miller, BHSC
	(Girls' Team)	Tarleton State	Linda Sultemeir, ENMU
1971	Bozeman, Mont.	Cal Poly, SLO	Tom Miller, BHSC
	(Girls' Team)	Tarleton State	Jan Wagner, MSU
1972	Bozeman, Mont.	Montana State Univ.	Dave Brock, SCSC
	(Girls' Team)	Eastern N.M.U.	Linda Munns, Utah State Univ.
1973	Bozeman, Mont.	Cal Poly, SLO	Dave Brock, SCSC
	(Girls' Team)	Univ. of Arizona	Lou Ann Herstead, Univ. of Wyoming
1974	Bozeman, Mont.	Eastern N.M.U.	Dudley Little, Cal. State-Fullerton
	(Girls' Team)	Sam Houston	Jimmie Gibbs, Sam Houston
1975	Bozeman, Mont.	Montana State Univ.	Skip Emmett, Univ. of Tennessee
	(Girls' Team)	New Mexico State Univ.	Jennifer Haynes, NMSU
1976	Bozeman, Mont.	Southeastern Oklahoma Univ.	James Ward, SEOKS
	(Girls' Team)	New Mexico State Univ.	Janet Crowson, NMSU
1977	Bozeman, Mont.	Southeastern Oklahoma Univ.	Tony Coleman, Univ. of Tennessee
	(Girls' Team)	Utah State Univ.	Shelly Muller, U. Wisconsin-RF
1978	Bozeman, Mont.	Southeastern Oklahoma Univ.	Hank Franzen, Casper College
	(Girls' Team)	Central Arizona College	Barrie Beach, CAC

417

SOURCE: National Intercollegiate Rodeo Association.

riding, calf roping, saddle bronc riding, team roping, steer wrestling, and bull riding comprise the men's events; the women compete in barrel racing, break-away roping, goat tying, and team roping. The champion teams and champion all-around individuals are given in Table 13-3.

National High School Rodeo Association

The National High School Rodeo Association sanctions rodeos for high school students in several states and one Canadian province. Claude Millins, the father of high school rodeo, has said, "It is the inherent desire of every American youth to play cowboy. They want to be good cowboys, for good cowboys made America. It is our responsibility to provide them the chance to participate in a good, clean American sport."

Girls' Rodeo Association

The Girls' Rodeo Association (GRA) started in 1947 during a meeting of the contestants at an all-women's rodeo at Pecos, Texas. It was incorporated in 1948 with rules similar to the PRCA. The original membership was 76, and they approved 36 rodeos for the first year. They compete in events comparable to men's events just described. They compete in one specialized event known as the "cloverleaf barrel race." The popularity of barrel racing is evident, because it is a standard event at all PRCA-sanctioned rodeos. The GRA membership exceeds 500 members, who compete for $500,000 in prize money each year.

Standard Rodeo Events

Standard rodeo events are classified as timed or riding events. The timed events are calf roping, team roping, and steer wrestling. The riding events are bareback riding, bull riding, and saddle bronc riding.

Steer Wrestling

Steer wrestling is one of the fastest events in rodeo. Winning times frequently do not exceed 4 or 5 seconds. Originally, steer wrestling was called "bull dogging" and was developed as a skill by Bill Pickett, a black cowboy. It is a team event in which two cowboys mounted on horses participate—"the dog-ger" and "the hazer." The event starts with a horned steer weighing 450 to 700 pounds (204 to 317.5 kg) in a chute, with the hazer positioned on the right side and the dogger on the left side of the steer. Because this is a timed event, the dogger starts from behind a barrier. If the dogger breaks the barrier before it is released when the steer crosses a designated score line, the dogger receives a 10-second penalty. The hazer is responsible for keeping the steer running straight so the dogger can move into position to jump from his running horse. As the dogger moves into position, he reaches out and grabs the steer's right horn in the crook of his right elbow. Simultaneously, his left hand pushes down on the left horn while his horse veers off to the left. As the dogger leaves his horse, his heels touch the ground ahead of his body and at a 45-degree angle to the path the steer is traveling. The steer's head is tipped toward the center of a left-hand turn. As the steer slows down, the dogger's left hand grabs the steer's upturned nose and twists the neck. The steer falls on its side with all four feet in the air. Time is signaled by a field judge.

Calf Roping

The skill of roping calves was developed on the early ranches while branding or doctoring sick animals. A skill that is still required of cowboys, it has developed into a highly competitive rodeo event (Figure 13-1). To start the event, a mounted cowboy is positioned on one side of a chute containing a 200- to 400-pound calf (90.7 to 181.4 kg). A rope barrier is stretched across in front

FIGURE 13-1

Anita Southard roping a calf at the 1978 state finals of the California High School Rodeo Association. (Courtesy Flying U Rodeo Company.)

of the cowboy and horse. When the calf is released and crosses a score line placed a measured distance in front of the chute, the barrier is released. The roper starts his horse so that it reaches the barrier as it is released. If the barrier is broken, a 10-second penalty is given. The calf is roped and the horse stops. As the horse is stopping, the roper dismounts and races to the calf. He must throw the calf on its side and tie three legs with a piggin string (a small rope about 6 feet long, or 1.8 m). While the calf is being thrown and tied, the horse must keep the rope tight to make the roper's job easier. On completing the tie, the time is over. The calf must remain tied for at least 6 seconds. Winning times usually range between 10 and 15 seconds, depending on arena conditions. The horse plays an important role in a roper's ability to quickly rope and tie a calf.

Over the years, calf roping has been dominated at times by Dean Oliver, winner of eight world championships; Toots Mansfield, winner of seven world championships; and Don McLaughlin, winner of five world championships. Top contestants may win $40,000 to $60,000 per year roping calves.

Team Roping

The art of heading and heeling steers developed out of the need to handle and restrain large cattle on the open range. The event consists of one cowboy, a header, roping a steer around the horns and his partner, a heeler, roping the hindlegs (Figure 13-2). There are two styles of team roping. In dally roping,

FIGURE 13-2
Team roping at 1977 Ventura Fair Rodeo. (Courtesy Flying U Rodeo Company.)

the roper must take a turn around the saddle horn with his rope to hold the steer. In hard and fast roping, the ropes are tied to the saddle horn. Today, most team ropings are dally style. Both ropers start from a roping box as the steer is released from a chute. If he breaks the barrier before the steer crosses a score line, a 10-second penalty is given. There are three legal head catches: around both horns, around the neck, and around one horn and the neck. If the heeler catches only one hindfoot, a 5-second penalty is given.

A variation of team roping is team tying. After the steer is roped, the header must dismount and tie a short rope around the steer's hindlegs.

Bareback Riding

Bareback riding was never a part of daily ranch routine but developed from the daredevil desires of one cowboy to outdo the next (Figure 13-3). This is the newest of the three standard riding events, started in the middle 1920s. The horse is ridden with a rigging consisting of a 10-inch-wide leather pad with a

FIGURE 13-3
Bareback riding. (Photo by Foxie, courtesy Flying U Rodeo Company.)

handhold that is cinched to the horse's back. The rider mounts the horse in a chute, and the chute gate is opened when the rider is ready. On the first jump out of the chute, the rider must spur the horse over the break of the shoulders as the horse's front feet strike the ground. Failure to do so disqualifies the rider. If successful, he is then judged on his ability to ride the horse by staying in command, maintaining proper position, and spurring properly. During the ride, the rider holds on with one hand and waves the other hand in the air. If his free hand touches the horse, the rider is disqualified. Horses are scored on how they kick, how hard they jump, how difficult they are to ride, and in how many directions they go. The ride is scored by two judges on a 25-point system for both horse and rider for a maximum of 100 points. The horse must be ridden for 8 seconds for a qualified ride.

Good bucking horses are being bred by the Flying U Rodeo Company for their athletic ability, stamina, agility, size, and personality. High Tide, 1976 World Champion Bareback Bucking Horse, is a grandson of Man o' War. Southern Pride, Sling Shot, and Joker are his full brothers, and all have been chosen for the National Finals Rodeo.

Saddle Bronc Riding

The sport of riding bucking horses with a saddle started as a form of entertainment on the great cattle drives. Therefore, saddle bronc riding is one of the classical events of rodeo (Figure 13-4). The broncs are ridden with an association saddle that meets certain specifications according to PRCA rules. The horse wears a halter with a 6-foot-long, 1½-inch-thick rein (1.8 m long, 38.1 mm thick). The rider holds the rein and uses it to maintain balance. As the bronc leaves the chute, the rider must keep his spurs over the horse's shoulders until the first jump is completed. The rider then spurs from front to back in rhythm with the horse as it bucks. The spurring stroke must be smooth and in a long arc. The score is based on how difficult the horse is to ride, the amount of control exercised by the rider, and how he spurs. Riders are scored from 1 to 25 points and the horse is given 1 to 25 points by each of two judges.

Bull Riding

Bull riding is one of the most dangerous events in rodeo (Figure 13-5). Very seldom can the rider avoid being butted, kicked, stepped on, or hit by a bull when he gets bucked off or jumps off. Bulls are ridden by placing a rope with a handhold around the bull's chest. When the rider places his gloved hand in the handhold, another cowboy pulls the slack out of the rope. When the rope is tight enough, the free end is laid across the palm and wrapped behind the hand and across the palm. The rider clenches it with all his strength. A bell is attached to the rope to serve as a weight to pull the rope free after the rider

FIGURE 13-4
Bill Ward riding the saddle bronc Sea Lion. (Photo by Helfrich, courtesy Les Winget.)

FIGURE 13-5
John Bland riding Flying High. (Courtesy Flying U Rodeo Company.)

releases it. The only rules are that the bull must be ridden 8 seconds and that the free hand cannot touch the bull. They are scored by two judges who give the rider 1 to 25 points and the bull 1 to 25 points. Judges look for the bull's ability to jump and kick high and to change directions suddenly. Riders are scored on their ability to stay in control of the ride, remain reasonably erect, and use their legs to spur the bull.

Cloverleaf Barrel Racing

Girls' barrel racing became a part of rodeo in 1948. It consists of a race against time to make a cloverleaf pattern around three barrels (Figure 13-6). The standard pattern is to cross a score line that is 20 yards (18.3 m) from the nearest barrel. The rider races to the first barrel, turns around it, races to the second barrel 30 yards away (27.4 m), turns around it, and races to the third barrel 35 yards away (32 m). After turning around it, the rider races back across the score line. There is a 5-second penalty for each barrel knocked over.

FIGURE 13-6
Cindy Rosser competing in a barrel race. (Courtesy Flying U Rodeo Company.)

FIGURE 13-7
Buster Welch cutting cattle on the King Ranch with Mr. Sam Peppy, twice world champion cutting horse. (Courtesy National Cutting Horse Association.)

CUTTING HORSES

A cutting horse separates a given animal from a herd or group of cattle, whether in the contest arena or on the ranch, and prevents its return (Figure 13-7). The cutting horse was born on the open-range ranches of yesteryear, and was a horse born of necessity. Even in this era of modern cattle management with its corrals and systems of pens, chutes, and gates, no superior method has been devised to separate a single animal and prevent its return with a minimum of disturbance to the remaining cattle.

In the early days, ranchers selected a cutting horse only after its ability to handle cattle had been proven. From then on, it was treated with respect and was relieved of other ranch chores. As cowboys from the various ranches got together, they talked of who had the best this or that, and they invariably discussed the merits of the top cutting horse on each ranch. These discussions led to challenges, and the cutting horse contest was on. Probably beginning around 1850, these contests increased in popularity, and on July 22, 1898, the first cutting horse contest with a record of money being posted was held in Haskell, Texas, at the Cowboy Reunion.

On March 14, 1908, the first indoor cutting horse contest was held in the Old North Side Coliseum at Fort Worth, Texas, built to house what is now known as the Southwestern Exposition and Fat Stock Show. Cutting horse contests continued through the years at the Fort Worth show, and when the first indoor rodeo was held there in 1918 a cutting horse contest was held for the first time in connection with a rodeo.

The number of cutting horse contests continued to increase, and in March 1946 the event was placed under a uniform set of rules with the formation of the National Cutting Horse Association (NCHA).

The NCHA has grown from 13 charter members to over 4,000 members. Purses for cutting horses in contests either approved or conducted by NCHA went over the $1 million mark for the first time in 1976. Approved contests occur from Massachusetts in the Northeast to Florida in the South and westward to the Pacific Coast. Contests are also held in six Canadian provinces and in Australia. Cutting horse competitions offer the greatest monetary return available to a horse that does not run on a racetrack. Mr. San Peppy, owned by the King Ranch and ridden by Buster Welch, was the first horse to have lifetime earnings over $100,000 (Figure 13-7), in 1978.

A cutting horse contest is conducted as follows. A herd of usually 15 to 20 cattle is held in one end of an arena by two men called "herd holders." Then, as its name is announced, the cutting horse and rider move into the herd to begin work. One animal is cut out or separated and driven through the herd holders toward the center of the arena. Two more riders, known as "turn back men," turn the separated animal back toward the cutting horse and toward the "safety" of the herd. Now the real action begins.

As the calf or yearling dodges and twists to get by the cutting horse, the horse makes countermoves to hold the animal where it is. Turning on a dime, rolling out swiftly to left and right, and all the while staying head to head with the stock being worked, the cutting horse demonstrates months of training and bred-in cow sense as it does its job with no help from the rider other than the selection of the stock to work in its 2½-minute performance. After getting the desired challenge from one head of stock, the rider signals the cutting horse to stop and to reenter the herd for another demonstration of its ability.

While the cutting horse performs, one or more judges evaluate how hard the horse is challenged by the animal separated from the herd, how the horse handles itself in meeting its challenges, and what mistakes, if any, are made. The judge scores the horse between 60 and 80 points. Where more than one judge is used, the combined scores of the judges determine the contest winner.

FOX HUNTING

Fox hunting has been a recognized sport in Great Britain, Ireland, and America for a number of years. By 1775, it was firmly organized in America, and even before then it was a popular sport. Packs of hounds were imported to America as early as 1650 by Robert Brooke of Maryland. However, they were imported as hunting dogs. The first recorded pack of fox hunting dogs was in 1666 in England. Several packs of hounds were kept in colonial America for fox hunting. Some famous colonists with packs were Thomas Walker, George Calvert, Charles Lee, and George Washington. It was not until 1776 that the first hunting club, Gloucester Foxhunting Club, was organized in America (in Gloucester, Massachusetts). During 1781–1861, the sport was enjoyed by the wealthy landowners of Pennsylvania, Maryland, and Virginia. There were very few organized hunt clubs during this period, but after 1865 many were founded. Since 1907, when the Masters of the Foxhounds Association of America was founded, the sport has been closely regulated and has grown steadily in popularity despite urbanization and modern farming methods. Several hundred packs of hounds are kept by hunt clubs throughout the world so that horsepeople can enjoy a fast cross-country ride behind a pack of hounds when they so desire.

Fox hunting adheres to strict rules of protocol established in the 1900s. Most hunting clubs have a similar type of organization. The master of the foxhounds is in direct command of the field (members of the hunt). He determines if and where a hunt will be held, and makes the necessary arrangements with the farmers. The huntsman maintains the pack and kennels and is responsible for the hounds in the field. He is also responsible for giving various horn signals to the hounds and the field. The huntsman is assisted by whippers-in, who go to the woods where a fox is supposed to be. They watch for the fox to "go away" and signal its escape. The hunt then begins. Once the chase starts, the hounds are in front, followed by the master of the fox hounds and the field. The hunt usually ends when the fox "goes to earth" (disappears in its hole).

Because foxes are getting scarce and because available land is difficult to find, drag hunts are now held. Instead of chasing a live animal, the pursuers and hounds follow a scented trail laid out by touching the ground with a fox's brush or litter from the fox's den. In some areas of America, the hounds chase coyotes instead of foxes.

SOCIAL, ENDURANCE, AND COMPETITIVE TRAIL RIDING

Some horsepeople enjoy the competition of endurance and competitive trail riding, whereas others enjoy social trail riding. Social rides take many forms. Small groups of friends may ride together to discuss common interests and to enjoy scenery. Horseback trips are made into the wilderness areas in and around the national parks. At pack stations, which are either privately owned or a service of the U.S. Forest Service, one may rent guides, horses, and all the necessary gear to travel into the wilderness. Many riding clubs sponsor well-organized rides; one such group is Equestrian Trails, Inc., a California organization made up of many local corrals whose goal is to preserve and develop riding trails. Each corral has many activities, including rides, tack and tog swaps, tours, and educational meetings. Social rides, such as the Desert Caballeros, Old Spanish Trail Riders, Rancheros Visitadores, and Sonoma County

Trail Blazers, are becoming increasingly popular. These rides last for several days and cover 100 to 200 miles (160.9 to 321.8 km).

Competitive trail riding is rapidly becoming one of the nation's most popular horse activities. It is gaining in popularity primarily because of the efforts of the North American Trail Ride Conference (NATRC). This body was organized in 1961 for the purpose of sanctioning competitive trail rides under a uniform judging system. The NATRC publishes an annual rule book, a judges' manual, and a management manual, which are available to interested horsepeople. The objectives of the NATRC are:

1. To stimulate greater interest in the breeding and use of good horses possessed of stamina and hardiness as qualified mounts for trail use

2. To demonstrate the value of type and soundness and the proper selection of horses for a long ride

3. To learn and demonstrate the proper methods of training and conditioning horses for a long ride

4. To encourage horsemanship in competitive trail riding

5. To demonstrate the best methods of caring for horses during and after long rides without the aid of artificial methods or stimulants

The NATRC sponsors both one-day (Class B) and two-day (Class A) rides. Appropriate points are given toward the annual championships, which are awarded to the finishing horses. Daily mileage in the open division is 25 to 40 miles, or 40.2 to 64.4 km (depending on the steepness of the terrain), to be covered in 6½ to 7 hours of riding time. Novice riders cover a shortened course of approximately 20 miles (32.2 km). The main object is to work all of the horses over an identical trail in the same length of time, thereby establishing a basis of fair comparison for determining the horses' soundness, condition, and manners. Although it is not a race, judgment in timing and pacing is important, because the winner usually rode at a consistent pace throughout the ride rather than "hurrying up and waiting."

One attraction of competitive trail riding is that no special type or breed of horse is necessary or favored. The prime requisite is a well-conditioned, calm horse. Riders condition their horses by following a well-planned training schedule that includes careful work over all types of terrain. Straightaway trotting, walking up and down progressively steeper hills, and working in soft sand are a few methods used to maximize development of the muscles, heart, and lungs (see Chapter 6).

The rides are judged by at least two judges, one of whom must be a practicing veterinarian. The open division, for horses over 5 years of age, is divided into junior (riders aged 10 to 17), lightweight (rider and tack weighing less than 190 lb, or 86.2 kg), and heavyweight (rider and tack weighing more than 190 lb) categories. The novice division is primarily for young horses (aged 4 to 5) and for newcomers to the sport.

For purposes of judging, each horse starts the ride with a score of 100, evaluated as follows: soundness, 40 percent; condition, 40 percent; manners, 15 percent; and way of going, 5 percent. Although only the horses are judged, the riders also compete for horsemanship awards and are judged on the care and handling of their mounts during the entire weekend.

One of the older competitive trail rides is sponsored by the Green Mountain Horse Association. It is a 100-mile ride (160.9 km) that is ridden during three days. During the first and second days, 40 miles (64.4 km) are ridden each day between 6½ and 7 hours. During the third day, 20 miles (32.2 km) are ridden between 2¾ and 3 hours. Riders are penalized if they do not finish within the time limits.

Competitive trail riding should not be confused with endurance riding. In contrast to competitive trail riding, where the horse must travel a fixed distance within a fixed period and is judged mainly on its soundness, condition, and manners, the endurance ride is primarily a race. The endurance ride is usually 50 to 100 miles long (80.5 to 160.9 km) and must be completed within a maximum period. The horses are examined to determine that they are sound and are willing to proceed on the trail. There are three 1-hour mandatory rest stops. During the rest stops, each horse is given a thorough veterinary check, which eliminates approximately 40 percent of the entries (Figure 13-8). Awards are made to all horses that complete the ride, to the first horse to finish the ride, and to the horse that finishes in best condition.

FIGURE 13-8
Competitive trail riding horse receiving a veterinary examination at a mandatory rest stop. (Courtesy NATRC.)

FIGURE 13-9

Sharon Saare competing in 102 Bonanza Ox Trail 100-miler on Juaquima Mirage, who won "Best of Condition." (Photo by Barieau, courtesy S. Saare.)

Several 50- to 100-mile endurance rides (80.5 to 160.9 km) are held annually. The "granddaddy" of them all is the Tevis Cup ride, which covers 100 miles in less than 24 hours (Figure 13-9). The Tevis Cup ride runs over the old Pony Express route from Lake Tahoe across the Sierra Nevada to Auburn, California. Horse and rider encounter temperature extremes ranging from freezing in the snow-covered heights to 120°F (48.9°C) in the canyons, as well as altitude extremes. The trail is rough, dusty, and rocky and includes steep inclines.

PACKING AND OUTFITTING

Packing horses and mules to carry cargo have played a role in the history of many countries. The art of packing was developed by Mongolia's Genghis Khan. Pack trains were used in Europe to transport goods in the 1700s. In America, horses were packed with cargo from the time of Cortez' landing in 1519 until the settlement of the West. In the Gold Rush days, ore, supplies, and machinery were packed into the rugged western mountain country.

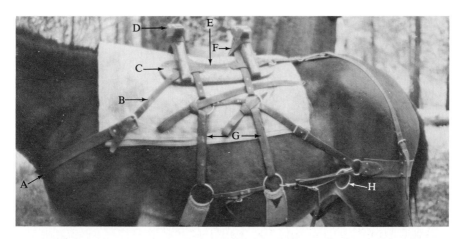

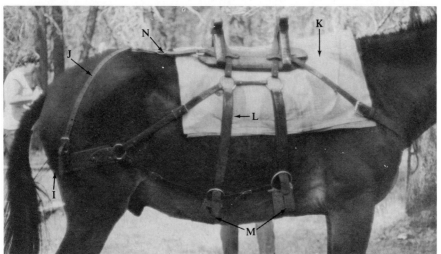

FIGURE 13-10

Sawbuck or humane pack saddle. (A) Breast collar. (B) Connecting strap. (C) Bar of saddle tree. (D) Sawbuck. (E) Humane saddle tree. (F) Rigging. (G) Latigos. (H) Ring to lead mules. (I) Breeching. (J) Hip straps. (K) Pad. (L) Billets. (M) Cinches. (N) Back straps. (Photos by B. Evans.)

Equipment

The basic equipment used for packing consists of the pack saddle, saddle blanket, halter with a 10-foot lead rope (3.05 m), lash rope, lash cinch, pack cover, and panniers.

There are three types of saddles that are used for packing horses, mules, and donkeys (Figures 13-10 and 13-11). The McClellan is seldom used, and the Decker is not as popular as the saw buck. Western stock saddles can be

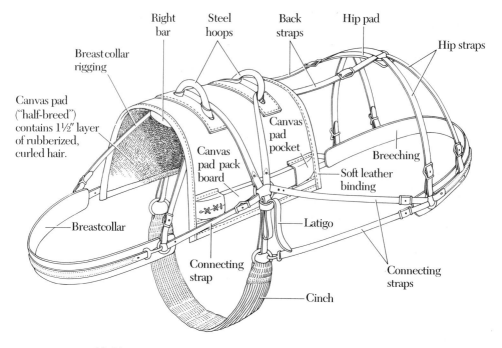

FIGURE **13-11**

Decker pack saddle. (From *Horse Packing in Pictures*, by F. W. Davis. Copyright © 1975 Francis W. Davis. Used by permission of Charles Scribner's Sons.)

used for packing in cases of emergencies or when one is taking a short trip and does not want to buy or rent a pack saddle. Western stock saddles are used by some hunters to pack meat out of the wilderness (Figure 13-12). They ride the horse while hunting and then lead it back to camp carrying the meat. Slings are used to carry dunnage, firewood, or other things, such as bales of hay (Figure 13-13).

The saddle pad used under a pack saddle is larger than a standard western saddle pad. The pack saddle pad should be 30–33 by 36 inches (76.2–83.8 by 91.4 cm). It must be long enough to protect the sides of the pack animal and must extend 4 to 6 inches (10.2 to 15.2 cm) in front of the withers to protect them. If unavailable, pads can be made by placing felt carpet padding between cotton duck material.

Halters should be made of material that will not break easily. Lead ropes are usually made of ¾-inch (19-mm) cotton rope.

Lash ropes are made of ½-inch (12.7-mm) manila rope that has had the splinters singed off and been waxed. Nylon ropes stretch and cannot keep the load tight. Cotton rope becomes very stiff and hard when wet and/or frozen. Manila rope also stiffens when wet but is usually easier to work with. Lash ropes should be 40 to 45 feet long (12.2 to 13.7 m). They are tied to the lash

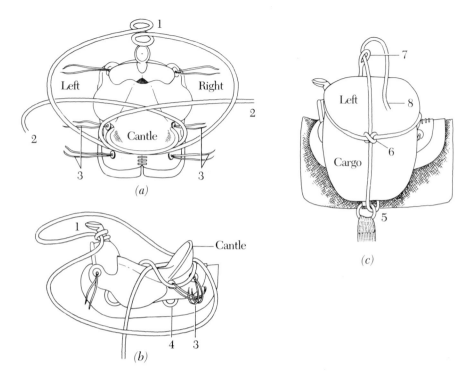

FIGURE **13-12**

Packing on a western stock saddle. *(a)* Top view. *(b)* Side view. *(c)* Side view of pack tied to saddle. Step-by-step instructions: (1) Tie two half hitches on the horn. (2) Run each end of the rope behind the cantle and across the seat. (3) Tie these sets of saddle strings together over the rope to keep the rope from slipping over the cantle. (4) If you are riding a double-rigged saddle, tie the jockey strings over the sling rope and to the back rigging ring. (5) Run the end of the rope through the cinch ring. (6) After placing the pack or cargo on the side of the saddle, pull the rope tight and tie a double half hitch. (7) Tie a hondo in one end of the sling rope and pass the other end through it. (8) Pull the end tight and tie it off. (Adapted from *Horses, Hitches and Rocky Trails.* © 1959, Joe Back. Reprinted by permission of Swallow Press, Chicago.)

FIGURE **13-13**

Set of slings for carrying dunnage, firewood, and bales of hay. (Photo by B. Evans.)

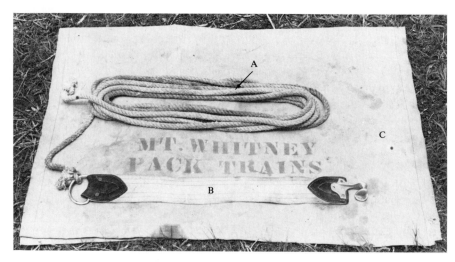

FIGURE **13-14**
(A) Lash rope. (B) Cinch. (C) Tarp. (Photo by B. Evans.)

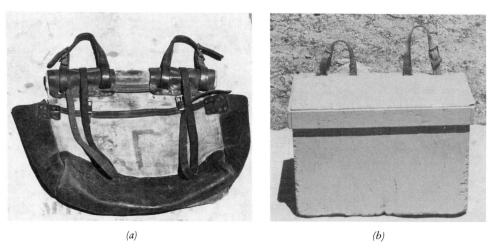

(a)   *(b)*

FIGURE **13-15**
Types of panniers. *(a)* Canvas and leather. *(b)* Box. (Photos by B. Evans.)

cinch. Lash cinches are made of heavy cotton duck webbing with ring and hook ends reinforced with leather (Figure 13-14).

Pack covers (manta covers, manti, or manty) vary in size depending on the type of load being packed. A general-purpose size is 6 by 8 feet (1.8 by 2.4 m), which allows the cover to double as a ground cover for a sleeping bag. They are made of waterproof canvas or rubberized, nylon-reinforced material.

Many types of panniers are used (Figure 13-15). Soft panniers are made of heavy duck webbing with leather-reinforced corners. Leather panniers are difficult to obtain but they wear better and protect the pack animal's sides better than cloth panniers. Rigid panniers are made of plywood and fiberglass, metal, or cowhide stretched over a rigid wood frame. A set of kitchen panniers may have drawers for utensils and staples and a fold-out work area. Canvas panniers are available that will slip over a western stock saddle (Figure 13-16).

In addition to the pack equipment, horse care equipment must be taken on the trip. Grooming equipment, a basic first aid kit, essential shoeing tools and supplies, feed, and salt must be packed. Kitchen and camp supplies may be very elaborate or very simple. Food and personal gear must also be packed. Typical lists of equipment that are packed for trips of a week or longer are given in Table 13-5.

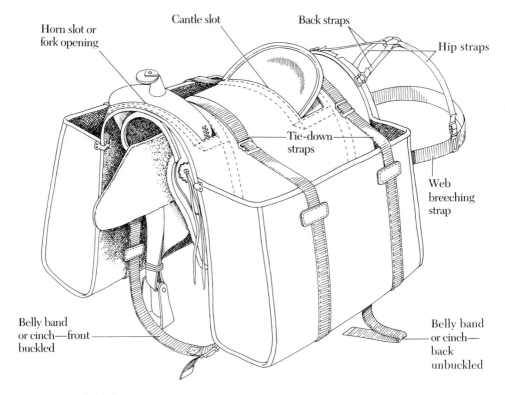

FIGURE 13-16

Canvas saddle panniers. Stirrups can be removed and carried in the panniers so they do not slap the horse.

TABLE 13-5

Supplies and Equipment That May Be Required on a Pack Trip (varies by number of people, length of trip, and number of pack animals)

Personal Checklist	Camp Equipment
To be packed in waterproof bag	Tent and tent stakes
Sleeping bag	Light canvas fly
Moisture-proof ground cover	Coleman lantern (single)
Air mattress or foam hip pad	Extra mantles
	White gas (square-type can)
Pack in duffel bag to put in panniers or on	Axe (cover blade)
slings	Shovel (short handle)
Warm jacket	50 ft ¼-in rope (15.24 m 6.35-mm rope)
Light jacket	Clothespins
Shoes or boots	Saw
Shoes for camp	First aid kit
Shirts or blouses	Toilet tissue
Socks	Grill
Underwear	Coleman stove
Pants	Light collapsible table
Bath towel and washcloth	Folding chairs
Shaving kit or cosmetic kit	Whisk broom
Toothbrush and paste	Wash basin
Flashlight, extra batteries and bulb	Portable camp toilet
Suntan lotion	Pliers
Hair comb	Nails, assortment
Insect bomb for tent	Hammer
Pack in saddle bags	
Camera and film	
Collapsible cup or Sierra-type steel cup	
Sunglasses	
Chapstick	
Insect repellent	
Lunch	
Leather gloves	
Neckerchief	
Pocket knife	
Pack on saddle	
Light rain gear	
Canteen	

Cooking Equipment	
Utensils	*Dishes (Melmac ware):*
Skillets, 10-in and 12-in cast iron (25.4 and 30.5 cm)	Plates (use paper products if you don't like to do dishes)
Pots, combination aluminum set, 3½, 2½, 1½ qt sizes (3.3, 2.4, and 1.4 liters)	Bowls
	Cups
Dishpans	*Other*
Coffee pots	Spatulas, 1 large and 1 small
Biscuit pans	Egg beater (optional)
Washpans, enamel	Can openers
Oven grates, use as grills	Kitchen matches, soaked in wax and stored in waterproof container
Aluminum grill	Plastic quart bottles, with wide mouth
Mixing bowls and baking pan	Pint Mason jars, plastic preferred
Dutch oven, medium size	Plastic bread bags, to pack with and store things, and for fish bags
Silverware	Sandwich and lunch bags
Dipper	Paper towels, napkins
Forks, knives, spoons, in canvas holder	Oilcloth (tablecloth)
Large serving spoons and cooking forks	Dishtowels
Sharp knife in leather case	Dishmop, or dishrags
	Soap and scouring pads
	Potholders

(continued)

TABLE 13-5 *(continued)*

Horse Equipment	
Horseshoeing equipment	Panniers
Bells	Pack covers
Hobbles	Salt
Halters	Feed and grain
Pack saddles	Tack repair items
Saddle pads	

Saddling

To saddle a pack horse, first place the blanket on the horse's back and slide it backward into position where it extends about 4 to 6 inches (10.2 to 15.2 cm) in front of the withers (Figure 13-17). Prepare a saw buck saddle to be placed on the horse's back by folding the breeching and connecting straps on top of the saddle, folding the breastcollar and its connecting straps on top of the breechings, and folding latigos over the breastcollar (Figure 13-18). This prevents the straps and other pieces from falling down and spooking the pack

FIGURE 13-17
Mule with pack saddle and pad in position. (Photo by B. Evans.)

FIGURE 13-18
Pack saddle prepared for storage and saddling. (Photo by B. Evans.)

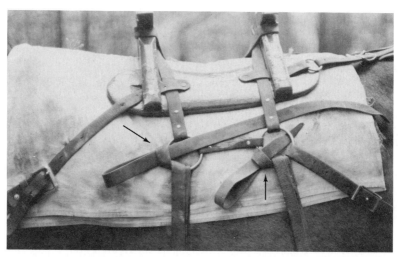

FIGURE 13-19
Quick-release latigo knot (↟). (Photo by B. Evans.)

animal. Place the saddle on the back, and unfold the cinches. Tighten the front cinch with the latigo first and the rear cinch last. Secure latigos with a quick-release knot (Figure 13-19) that does not form a large bump. Tighten them again just before the pack is loaded. Unfold the breastcollar and place it in position. It should be loose enough so that you can run your arm between it and the pack animal. Unfold the breeching and place it about a foot below the pack animal's tail. Pull the tail out and tighten the breeching. After it is tightened, you should be able to slip your hand easily between the breeching and the hindquarters.

Packing

Next, prepare the panniers and slings for loading. A pack animal can carry a load weighing about 12½ percent of its body weight, so a 1,000-pound animal (454 kg) can carry about 125 pounds (56.7 kg) when the animal is used daily. For shorter trips with long layovers, animals may carry a few more pounds. To determine the number of pack animals needed, weigh all the equipment and supplies to be packed. Weigh and balance the load for each pack animal; that is, each pannier in a pair must weigh the same to keep the load balanced. If the load is not balanced, it will shift to one side and make the animal's back sore. A small set of scales is useful. Take special precautions when packing the panniers so that supplies and equipment are not broken or the sides of the pack animal are not injured by protruding objects. Load heavy objects at the bottom of the panniers to maintain a low center of gravity. Place all breakable items in metal containers or wrap them with newspaper. Do not pack lantern fuel and other similar items with food, because a leak will ruin the food. Protect the edges of equipment such as axes, shovels, and saws to prevent cutting the pack animal. Pack personal gear in duffel bags, which are carried in slings. Sleeping bags are carried in waterproof bags and are usually packed between the saw bucks on top of the saddle.

The types of food you pack depend on your appetite and preference. Each meal must be carefully planned ahead of time. If soft-sided panniers are used, place four to five cans of food end to end and rolled in newspaper. Mark their contents in case the paper wrappers come off. Processed and canned meats are used after a few days on the trail. Hams, roast beef, chicken, stews, bacon, Spam, chili, tamales, and spaghetti are often packed. Side bacon keeps for at least 10 days if kept out of the sun. It can be kept cool by hanging it inside a damp burlap bag that is hung in the shade. If you pack frozen meat, wrap it in newspaper for insulation or pack it in a styrofoam cooler inside a pannier. Protect the horse from the cold of the frozen meat. Keep produce cool by packing it on top of the frozen goods. Or pack the bottom of the pannier with solid goods, lay a wet burlap bag on top of them, place the produce on the wet burlap bag, and cover it with another wet burlap bag. All but the most delicate fruits and vegetables keep well if they are kept cool and out of sunlight. Eggs last for 2 weeks if kept cool. They can be packed by leaving them in their cartons and packing them in the top of the pannier. Margarine in squeeze bottles is handy. A variety of liquids, such as instant coffee, tea, and lemonade, offer quick refreshment after arrival at a campsite. Pack powdered or condensed milk instead of fresh milk. Beer, pop, and liquor can also be packed. Use hard-

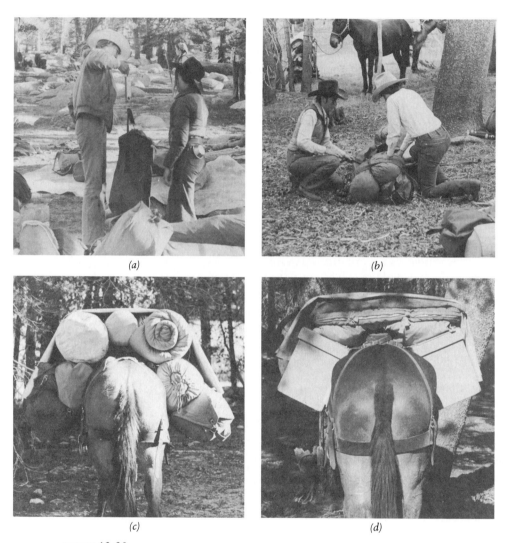

FIGURE 13-20

Packing a load. *(a)* Sorting and weighing loads. *(b)* Making dunnage loads. *(c)* Dunnage load. *(d)* Box load. (Photos by B. Evans.)

stick processed meat, cheese, or canned meat for sandwiches. Cookies, fruit, and candy bars are also useful. Pack loaves of bread close to the top of the panniers and be careful not to mash them. Avoid packing anything in glass containers.

After making up the loads and balancing the panniers, place them on the pack animals. When both panniers are on the saddle, they should be level across the top when viewed from the side. If not, adjust the straps. Place soft dunnage on top of the saddle and panniers (Figure 13-20). Check the balance

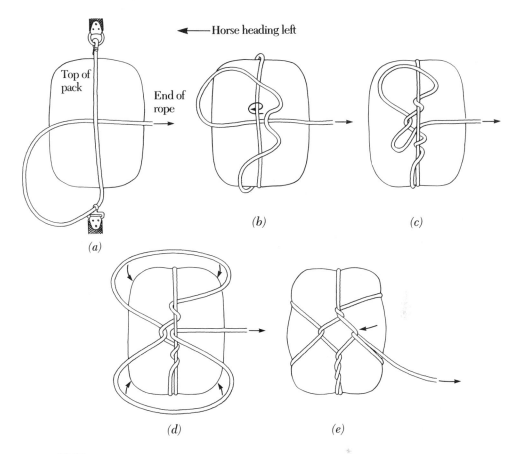

FIGURE 13-21

Tying a diamond hitch. *(a)* Lay most of rope down behind horse and bring up lash rope connected to cinch and then over top of pack and connect it to cinch hook. *(b)* Bring up rope from cinch hook and tuck it under, over, and under the part crossing top of pack. *(c)* Pull part of rope lying down behind horse through loop formed in *(b)*. *(d)* Place rope under corners of pack. *(e)* Tie off rope at point shown by arrow.

of the load by rocking the load. When released, it should stop with the saw bucks to the center of the horse's back. If not, add or remove weight from one side. Cover the load with the pack cover to protect it from rain, tree limbs, and rocks.

A variety of hitches are used to secure a load: the most important are the diamond (Figure 13-21), double diamond (Figure 13-22), basket or box (Figure 13-23), and sling (Figure 13-24) hitches. The diamond (Figure 13-21) and double diamond (Figure 13-22) hitches are used on loads that are packed high and are soft. They draw the load down and close to the saddle. Use a basket hitch to lift a load up and away from a pack animal's sides when the panniers are loaded heavy. Use basket slings to carry items such as bales of hay and

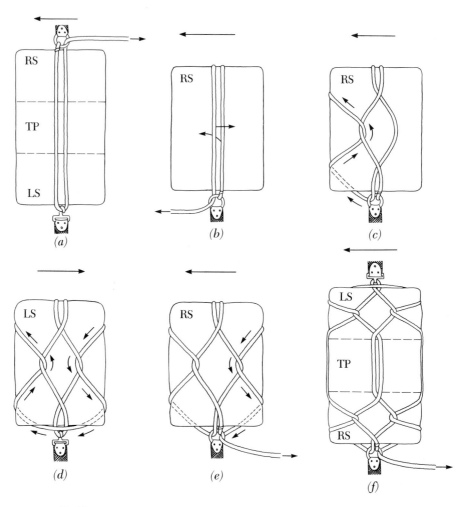

FIGURE 13-22

Tying a double diamond hitch. RS = right side. LS = left side. TP = top of pack. ← = horse heading left. → = horse heading right. *(a)* Pass rope from cinch across top of pack, around hook, back across the top of pack, and through cinch ring. *(b)* On each side of the back, twist the two ropes and then pull them apart to form a small diamond. *(c)* Run rope from cinch ring behind corner of pack and pull it through diamond on side of pack and across front of pack. *(d)* Pass rope from across front of pack through diamond formed on left side of pack, around corner of pack, across hook, around corner, through diamond, and across the back of the pack. *(e)* Pull rope from across back of pack through diamond, around corner of pack, and pass it through cinch ring. *(f)* Pull rope tight around the pack and tie it off.

quarters of a deer or elk carcass (Figure 13-24). After tying the hitch with the lash rope, trace the rope and pull it tight. If the lash rope does not stay tight, the load will shift and/or fall off. Tie off the lash rope with a quick-release knot, and tuck the excess rope securely under the pack cover.

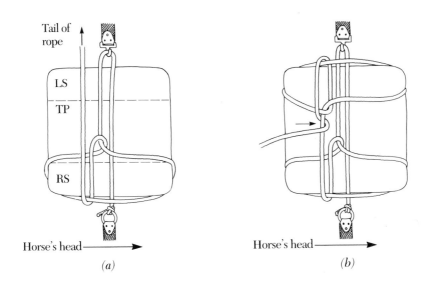

FIGURE 13-23

Tying a basket (box) hitch. RS = right side. LS = left side. TP = top of pack. (a) Pass rope from cinch ring across top of pack, around hook, back across pack, and pull it very tight. Hold the rope with your left hand next to pack. With your right hand, lay tail end of rope across top of your left hand so a loop forms on the left and free end of rope is on the right. With your right hand, take rope extending through left hand and pass it under front corner of pack. Pull rope tight toward free end and pass it across top of pack. (b) Repeat procedure on left side. Pull rope tight and tie it at the arrow.

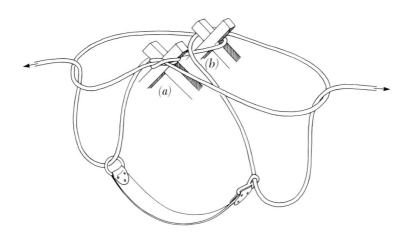

FIGURE 13-24

Tying a sling hitch. Tie a clove hitch on front sawbuck (a) with center of lash rope. Pass each lash rope around back sawbuck (b) and then run through cinch ring (right side) or cinch hook (left side).

FIGURE 13-25

Tying a string of pack animals together. Bowline knot is around lead mule's neck, and rope passes through ring hanging from breechen. (Photo by B. Evans.)

Pack animals are led down the trail by the packer. Three to five pack animals can be tied together and led. The horse that is most difficult to lead follows the packer, and the horse that can be led without any difficulties is the last in the string of horses. The packer should never lead the string by tying the lead rope around the saddle horn but should hold it in his or her hand. Pack animals are tied together by passing the lead rope through a ring (2 inches, or 5.1 cm, in diameter) snapped on the left side of the breeching where the quarter strap connects and then by tying it with a bow-line knot around the neck of the animal in front of it (Figure 13-25).

POLO

Polo developed in Persia, although it was first played in some form in China and Mongolia (Figure 13-26). The earliest historical references to polo were made in connection with Alexander the Great and Darius, King of Persia. The oldest polo club, the Silchar Club, was founded in 1859. The Silchar Club rules of play were the foundation for the modern rules. International matches between Great Britain and the United States began in 1886. The game was played widely throughout the United States until the mid-1930s. Recently, there has been a revival of interest in polo in this country.

FIGURE 13-26
Polo competition between Gulfstream and Delray Beach teams in the T. C. Collin Memorial Match. (Photo by Vogue Photo, courtesy U.S. Polo Association.)

The game requires excellent horsemanship and balance. A successful player must have complete control of the horse, be aware of the position of every other player on the field, relate the ball to the position of teammates, and be able to hit the ball accurately at full speed while competing with a counterpart on the opposite team.

The game is played on a regulation field that is 300 yards long by 160 yards wide (274.3 by 146.3 m). The sidelines of the field have wooden boards to deflect the ball back into play, but these boards are never more than 11 inches high (27.9 cm). If the sides are not boarded, the width of the field is 200 yards (183.9 m). The other essential pieces of equipment include the goal posts, which must be at least 10 feet high (3.05 m) and are located 8 yards apart (7.3 m) at the center of the backlines, a ball made of willow or bamboo root, polo sticks or mallets, the rider's personal equipment, and, of course, horses. Personal equipment includes blunt spurs, protective headgear, brown boots, whip, and polo saddle.

The objective of polo is to get the ball in the opposing team's goal. It is a team effort with established plans and maneuvers. A team consists of four players, of which two are forwards, one is a half-back, and one is a back. The game is played in chukkers of 7 minutes each. There are six chukkers in a match. There is a 4-minute break between chukkers and there is a 10-minute break at half-time. The team scoring the most goals wins the game. Rules governing scoring, penalties, and method of play in America are established by the United States Polo Association.

A variation known as "cowboy polo" is also now enjoying increased popularity. It differs from polo in several respects. First, it is a faster game, is easier to watch, and does not require uniforms or "select" facilities. Cowboy polo can be played both indoors and outdoors. The game is played on a dirt and/or sand field. The field is 260 feet long (79.2 m) and is divided into four zones, each 50 feet long (15.2 m), and a fifth zone 60 feet long (18.3 m). A guard and a forward play in every zone except the 60-foot-long zone, which is patrolled by two centers. Four chukkers of 15 minutes each are played. The ball is 11 inches (27.9 cm) in diameter. Cowboy polo is a more aggressive game and involves considerably more contact than polo.

IV
BREEDING
THE HORSE

14
Breeding
the Stallion

The information presented in this chapter is for those who wish to breed horses and to raise foals. However, an understanding of the reproductive processes in the mare and the stallion is a valuable aid in their care by all horse owners because horse behavior is affected by these processes. This chapter is presented in sufficient detail that the inexperienced horse breeder may conduct a successful breeding program.

REPRODUCTIVE TRACT

A diagram of the reproductive tract of the stallion is shown in Figure 14-1. Specific structures are shown in Figures 14-2, 14-3, and 14-4.

Scrotum

The scrotum contains the two testicles. It is a pouch formed by a diverticulum of the abdomen. The testes lie horizontally in the scrotum so that it is not as pendulous as the ram's or bull's. One testicle is frequently larger than the other,

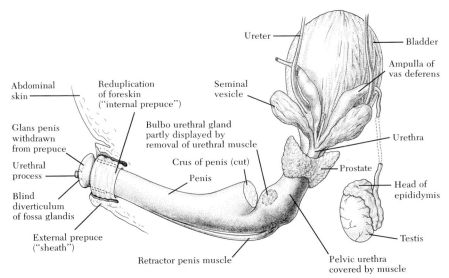

FIGURE 14-1

Reproductive tract of the stallion (lateral view). Bladder and upper urethra are twisted to expose posterior aspects. (From Eckstein and Zucherman, 1956.)

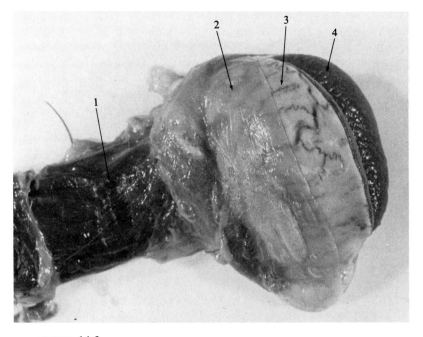

FIGURE 14-2

Testicle of a stallion. (1) Cremaster muscle. (2) Tunica vaginalis propria. (3) Tunica albuginea. (4) Gland substance. (Photo by B. Evans.)

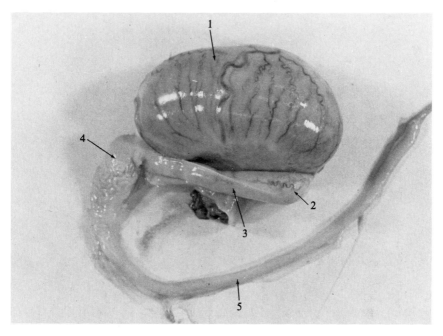

FIGURE 14-3

Testicle, epididymis, and vas deferens of the stallion. (1) Testicle. (2) Head of epididymis. (3) Body of epididymis. (4) Tail of epididymis. (5) Vas deferens. (Photo by B. Evans.)

and one testicle is placed slightly further back than the other, so that the scrotum is asymmetrical. The external cremaster muscle is attached to the testes (Figure 14-2) and serves to draw the testicle close to the abdominal wall for protection or to maintain the proper temperature. When the cremaster muscle is relaxed, the testicles are supported by the scrotum. Their weight tends to stretch the scrotum so that it has a smooth appearance and is constricted above, next to the abdomen. On contraction of the external cremaster and tunica dartos muscles, the testicles are drawn upward and the scrotum becomes wrinkled.

The blood supply to the scrotum derives from branches of the external pudic artery. Blood drains from the scrotum via the external pudic vein. Innervation of the scrotum derives from the ventral branches of the second and third lumbar nerves.

Testes

The testicles are shown in Figures 14-2 and 14-3. They weigh about 225 to 300 gm and frequently the left one is slightly larger. They are about 10 to 12 cm long, 6 to 8 cm high, and 5 cm wide. The outer surface of the testicle is

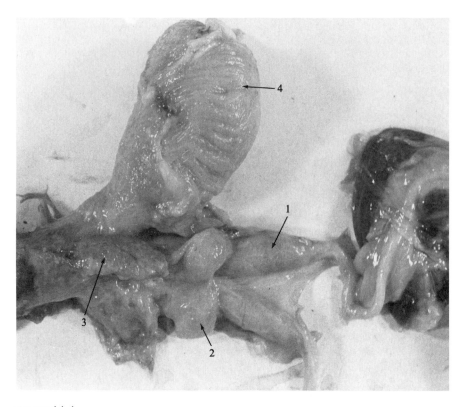

FIGURE 14-4

Accessory sex glands of the stallion. (1) Ampulla. (2) Seminal vesicles. (3) Prostate gland. The bladder is labeled 4. (Photo by B. Evans.)

covered by a serosa, the tunica vaginalis proper, which may or may not be severed during castration. The serosa also envelops the spermatic cord except where the blood vessels and nerves enter. The testicle is divided into lobules, each containing seminiferous tubules, interstitial cells, and loose connective tissue. The interstitial cells produce and secrete the male hormone, testosterone. The seminiferous tubule, the functional unit of the testicle, contains layers of spermatogenic cells in various stages of development, and sustentacular cells extend inward from the basement membrane.

Blood in the testicle circulates via the spermatic artery and spermatic vein.

Epididymis

The epididymis is composed of three areas. The head is the area where the seminiferous tubules from the lobules unite and enter the epididymis. The body is attached to the side of the testis (Figure 14-3). The tail of the epididymis leaves the body of the epididymis and is not attached to the testicle. The epididymis transports spermatozoa and other contributions of the testis to the semen, concentrates the spermatozoa by absorbing water, and provides a place for storage and maturation of the spermatozoa.

Vas Deferens

The vas deferens serves as a transport canal for the spermatozoa between the tail of the epididymis and the urethra. The tail portion of the epididymis merges into the vas deferens, which is characterized by the small diameter of its lumen and by its thick muscular wall. The size of the lumen is constant, but the muscular wall becomes enlarged approximately 15 to 20 cm prior to the entrance into the urethra. The enlarged area is the ampullar gland; many glands are located in this area. The vas deferens decreases in size as it passes under the isthmus of the seminal vesicles to open into the urethra (Figure 14-4). The excretory ducts of the seminal vesicles share the opening into the urethra with the vas deferens.

Penis

The penis of an average-sized stallion is about 50 cm long in the relaxed state. The anterior 15 to 20 cm of it lies in the prepuce. It is divided into three parts: head, body, and glands (Figure 14-1). During erection, it is cylindrical in shape but the lateral sides are compressed slightly. The glans or free end of the penis is bell-shaped. The urethral process extends about 2 to 5 cm from the deep depression or fossa glandis.

Prepuce

The prepuce, commonly called the *sheath*, is a double fold of skin that covers the free portion of the penis, which extends forward from the scrotum area. The external prepuce extends from the scrotum to within 5 to 7 cm of the umbilicus. Before reaching the umbilicus, it is reflected backward to form the preputial orifice. The internal prepuce starts at the preputial orifice and extends backward for 15 to 20 cm. It is reflected forward and extends almost to the preputial orifice before it is reflected backward again to form a secondary tubular invagination that contains the relaxed penis. During an erection, the increased size and length of the penis unfolds the internal prepuce so that it covers the penis. The internal layers of skin contain large sebaceous glands and coil glands or preputial glands whose secretions, together with desquamated epithelial cells form the fatty smegma. The fatty smegma and foreign substances such as dirt or wood shavings accumulate in the fossa glandis to form a round "bean." The presence of a "bean" must be checked when a stallion's or gelding's penis is being washed. The smegma also causes the prepuce and penis to accumulate dirt, so it is necessary to clean them periodically.

Accessory Sex Glands

The accessory sex glands contribute fluids of characteristic composition to stallion semen. There are three accessory sex glands: seminal vesicles, bulbourethral glands, and prostate gland (Figure 14-4). The seminal vesicles are paired glands. Each gland is 15 to 20 cm long and approximately 5 cm in diameter. They are reflected posteriorly so that their long axes parallel the vas deferens. The excretory duct opens into the urethra; often it has passed under the prostate gland. The bulbourethral glands are paired, ovoid-shaped, lobulated glands that lie near the ischial arch. Each gland is approximately 4 × 2.5 cm and has several ducts that open into the urethra behind the prostatic ducts. The prostate gland is composed of two lateral bulbs that are connected by a thin isthmus. Each bulb is approximately 3 × 1.5 × 0.5 cm. The isthmus covers the area where the bladder joins the urethra. Approximately 15 to 20 prostatic ducts perforate the urethra.

SEMEN PRODUCTION AND CHARACTERISTICS

Puberty for stallions varies between 11 and 15 months of age. It is the age when an ejaculate contains 1×10^8 sperm with 10 percent progressive motility. Even though the average stud colt is capable of breeding a mare by the time he is 1 year old, stallions are not usually used at stud until they are 4 to 5 years old. At 2 years, the stallion can be used at stud but it is not recommended. If he is used, the number of services should not exceed two per week, and only 6 or 7 mares should be bred during the breeding season. The 3-year-old stallion can be used four to six times per week but his book should be limited to 10 to 12 mares. The mature stallion can be bred daily but he should get one day of rest every seven to eight days. If necessary, mature stallions can be used twice a day for three or four days each week. The effects of frequency of ejaculation on semen quality are discussed in the section on semen evaluation.

Endocrinology

Hormonal control of the reproductive processes of the stallion is not understood. Currently, it is believed that follicle-stimulating hormone (FSH), secreted by the anterior pituitary gland, is responsible for spermatogenesis in the seminiferous tubule of the testis as well as for maintaining and repairing the tubular epithelium. Luteinizing hormone, known as interstitial cell stimulating hormone (ICSH) in the stallion, stimulates the interstitial cells to produce androgens, one of which is testosterone. Because ICSH is responsible for testosterone production, it is indirectly responsible for stimulating the secondary sex characteristics and accessory sex glands.

No hormonal treatment has been found that stimulates spermatogenesis. Testosterone therapy has been reported to decrease spermatogenesis. However, it is used to increase libido in stallions used for teasing purposes and can be used to elicit stallion behavior in geldings.

Estrogen (female hormone) is produced by the Sertoli cells and interstitial cells in the testes. Stallions secrete large amounts of estrogens in their urine, but their function in the stallion is unknown.

Characteristics

Stallion semen is grayish white in appearance and consists of seminal fluid and spermatozoa. The whole semen is ejaculated in a series of 8 to 10 jets. The first 3 jets are milky in color and consistency. Approximately 80 percent of the total number of spermatozoa is ejaculated in these jets. The fluid in the jets derives mainly from the ampullar glands of the vas deferens. The later jets (4 to 10) contain very few sperm cells. The semen in these jets is mucous in appearance and is mainly derived from the seminal vesicles.

SEMEN COLLECTION, PROCESSING, AND EVALUATION

Evaluating stallion semen helps determine infertility and helps predict fertility. Two methods are used to evaluate a stallion's fertility based on seminal characteristics. The usual procedure is to give the stallion sexual rest for a week. Then his semen is collected two times, an hour apart, with an artificial vagina. Semen is collected twice because there is a 50 percent reduction in total sperm collected between the two collections unless one of the ejaculates was incomplete; high sperm reserves were present, absent, or depleted; or the stallion was immature. If the stallion's semen does not meet minimal standards for quality, he is given at least 60 days' rest, and then the procedure is repeated. The 60-day rest period is essential, because the time required for a spermatogenic cycle is about 60 days.

The second method is used if it is necessary to determine the daily sperm output by a stallion. After sexual rest for a week, the stallion's semen is collected daily for seven or eight days. Extensive investigations in which semen was collected daily for two weeks show that after six to eight days the daily sperm output becomes stable, the extra gonadal reserves have been depleted, and the stallion is ejaculating his daily production of sperm.

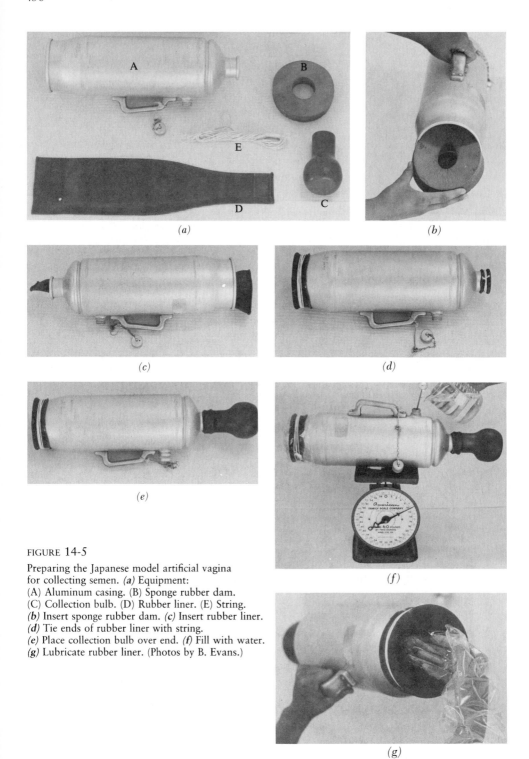

FIGURE 14-5

Preparing the Japanese model artificial vagina
for collecting semen. *(a)* Equipment:
(A) Aluminum casing. (B) Sponge rubber dam.
(C) Collection bulb. (D) Rubber liner. (E) String.
(b) Insert sponge rubber dam. *(c)* Insert rubber liner.
(d) Tie ends of rubber liner with string.
(e) Place collection bulb over end. *(f)* Fill with water.
(g) Lubricate rubber liner. (Photos by B. Evans.)

Collection

The stallion semen is collected with an artificial vagina (AV). Three models can be used: Japanese, Colorado, and Missouri. The Japanese and Colorado models are the most popular. Each model has advantages and disadvantages. However, the preference of the person doing the collecting usually prevails. The Colorado model is heavy and bulky when filled with water. It has the advantage of maintaining the proper temperature for a longer period of time and it is easier to separate out the gel fraction. The Japanese model is much lighter and easier to handle. During cold weather, it loses temperature at a faster rate than does the Colorado model. To collect gel-free semen, the Japanese model must be modified to accommodate a filter.

Preparation of the Colorado model for collection has been described in detail by Pickett and Back (1973). Prepare the Japanese model according to Figure 14-5. First insert the donut-shaped dam made of sponge rubber and position it in the front of the aluminum casing. Then insert the rubber liner, and reflect the small end back over the small end (front end) of the casing. Secure it with cotton string to prevent any water leakage. Keeping the liner straight in the casing, reflect the large end of the liner back over the back end of the casing and secure it with cotton string. The rubber collection bulb is placed over the small (front) end of the AV. Then fill it with water through the valve. Water temperature should be 111 to 118°F (44 to 48°C) at the time of collection. If the water temperature exceeds 120°F, it may cause sperm damage; and if it is below 100°F (37.8°C), it may cause cold shock. If the AV is filled with water that is about 120°F (48.9°C), the AV usually equilibrates at the proper temperature by the time it is used to collect the stallion. The stallion is not very sensitive to temperature but some stallions accept the AV better if it is at the upper or lower end of the temperature range. Rather than guessing at the amount of water in the AV, it is best to weigh the AV. Filled with water, it should weigh 12 to 13 pounds (5.4 to 5.9 kg) for the average stallion. If the penis is smaller than average, more water is required (or vice versa if the stallion has a very large penis). Lubricate the open end of the AV with sterile K-Y lubricant immediately prior to use. Lubricate it about one-half to three-fourths the length of the casing, applying the lubricant with a plastic glove. Use the lubricant sparingly so that it does not contaminate the semen.

Figure 14-6 shows how to collect the stallion's semen. He may be collected off a mare or off a phantom (Figure 14-7). The collector may stand on either side of the mare or phantom, but right-handed people prefer the right side. As the stallion mounts, direct the penis into the AV. Hold the AV steady by pressing it against the mare's hindquarters. Hold the AV at about a 30-degree angle of inclination until ejaculation occurs. As soon as the stallion begins thrusting movements, place your left hand at the base of the penis to feel the ejaculatory pulsations. As the stallion ejaculates, rotate the AV downward so that the semen drains into the collection bulb. After ejaculation is complete, remove the AV from the penis.

FIGURE 14-6
Collecting semen from a stallion with an artificial vagina. (Photo by B. Evans.)

Care of AV Equipment

The AV equipment must be kept clean and sterile to prevent the transmission of infections and diseases. Because rubber components cannot be sterilized with heat without deteroriation, rinse them in warm water and place in a Nolvasan solution (3 oz per gal, or 23.5 ml per liter). Scrub them with a brush in the antiseptic soap solution until they are clean. Next wash them in a mild soap solution (Ivory soap is recommended), rinse them in warm water several times to remove all the soap, rinse them three times in distilled water to remove any residue from the water that would alter the pH of the semen, and finally rinse them in 70 percent ethyl alcohol. Soap is spermicidal, so remove all of it from the rubber components as well as from the glassware that is used. Allow them to air dry in a dust-free cabinet. Rinse glassware in water, wash it in hot, soapy water (Ivory is recommended), rinse it several times to remove all soap residues, rinse it in distilled water two or three times, and dry it in an oven at 167°F (70°C). To prevent dust contamination, seal all glassware with aluminum foil and store it in a dust-free cabinet. If possible, periodically sterilize the glassware with heat. If an infected stallion is used, sterilize all equipment that his semen has touched.

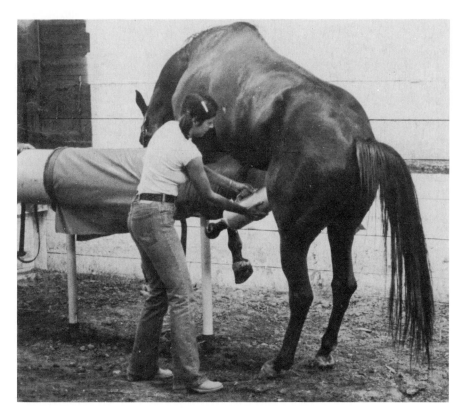

FIGURE 14-7
Collecting semen from a stallion with an artificial vagina and using a phantom.
(Photo by B. Evans.)

Processing

Process the semen by immediately removing the collection bulb and passing the semen through an in-line milk filter (obtainable from Lane Manufacturing Company, Denver, Colorado) to remove the gel (Figure 14-8). Collect the gel-free semen in a prewarmed, sterile plastic bottle and place it immediately in an incubator set at 100°F (37.8°C) (Figure 14-9). Insemination is discussed in Chapter 15.

Evaluation

Evaluate each seminal collection for volume, concentration of spermatozoa per milliliter, motility, morphology, and pH. The evaluation procedure is explained in Appendix 14-1. In a clinical situation, the semen should be cultured for bacteria.

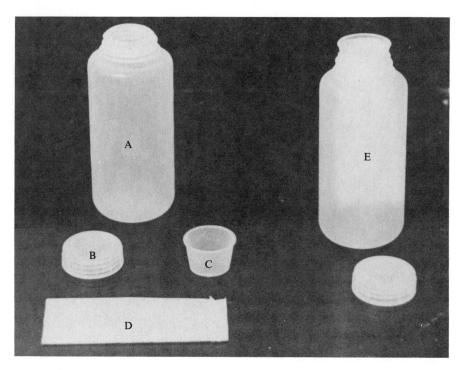

FIGURE 14-8

Filter assembly for obtaining gel-free semen. (A) Sterile plastic bottle. (B) Lid for bottle. (C) Filter ring. (D) In-line milk filter. (E) Assembled components. (Photo by B. Evans.)

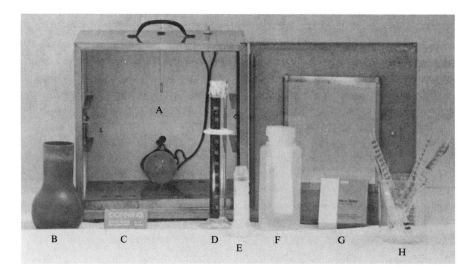

FIGURE 14-9

Incubator for keeping semen and semen evaluation supplies at proper temperature. (A) Incubator. (B) Bulb. (C) Cover slip. (D) Graduated cylinder. (E) Disposable syringe. (F) Filter assembly. (G) Slides. (H) Pipettes. (Photo by B. Evans.)

Volume

Normal values for the ejaculate volume range from 50 to 75 ml but volumes of 300 ml have been observed. Three primary factors influence the volume: stallion, season of year, and frequency of ejaculation. The volume of semen is larger (almost twofold) during the breeding season than during winter. The absence or the amount of gel, which is influenced by stallion variability and season, is an important determinant of volume. Frequency of ejaculation influences the seminal volume. Ejaculate collected 1 hour after the first ejaculate is usually less than the first, largely because of the absence or reduced amount of gel in the second ejaculation. If the ejaculate volumes are less than 50 milliliters, a stallion's fertility may be questioned.

Motility

The percentage of motile spermatozoa in an ejaculate is usually 60 to 100 percent; 70 percent is considered quite good. Motility is not influenced by season nor by collecting two consecutive ejaculates 1 hour apart. When the motility estimate is less than 50 percent, fertility may be questioned.

Concentration

The concentration is the single most important factor affecting fertility. The concentration of sperm per milliliter of gel-free ejaculate ranges between 30 and 800 million. There is a large variation between stallions, between first and second ejaculates, and according to season of the year. One study found that the average of the first ejaculate was 282×10^6 per milliliter (Pickett and Voss, 1973). The average concentration of the second ejaculate was 60 percent of the first ejaculate. During winter, the concentration is about one-half as great as during the breeding season. Stallions ejaculating with a concentration of 8×10^6 spermatozoa per milliliter or less have questionable fertility.

Total Number of Spermatozoa

The total number of spermatozoa collected can be calculated from the volume and concentration determinations. The total is usually about 15 billion per ejaculate. The second ejaculate is about 50 percent less. Season of the year affects sperm production and thus the number collected. During winter, it is about 50 percent of the total collection averaged during the breeding season. Testicle size also influences the total number of spermatozoa collected. Scrotal width is correlated to sperm output. The equation for daily sperm production is

$$-4.88 \times 10^9 + 0.93 \times 10^9 \text{ (scrotal width in millimeters)}$$

Therefore, determine scrotal width with a set of calipers during a breeding soundness evaluation. The average scrotal width is about 90 to 110 mm. If the width is less than 85 mm, the daily sperm output should be determined by collecting the stallion daily for seven to eight days.

Frequency of ejaculation influences the total number of spermatozoa collected. If collected from a stallion one to three times per week, the total number collected per ejaculate is about equal. When collected at more frequent intervals, the total number per ejaculate decreases. To collect the maximum number of spermatozoa each week from a stallion, collect three times per week. If collected daily, the total number collected per ejaculate is about 50 percent of the number collected per ejaculate on an every-other-day basis. Therefore, at the end of a week the number collected for three times a week is the same as for six times per week. This is an important consideration for decreasing labor costs and chances of injury and for preventing overuse of a stallion in an artificial insemination program.

Sexual stimulation of a stallion by teasing does not increase the total number of sperm collected per ejaculate. It does increase the seminal volume. Therefore, it is not necessary to stimulate a stallion beyond the time required for the breeding hygiene procedures.

pH

Determine the pH of the gel-free semen as a precautionary measure if the semen is evaluated. It can be measured routinely if sufficient semen is available after the mares have been inseminated. The pH of first ejaculates should be between 7.4 and 7.6, and the value for second ejaculates should be between 7.5 and 7.7. If an incomplete ejaculation is obtained, the pH will be higher, because the pH of the accessory sex gland fluid without epididymal contributions approaches 8.0. Values for the second ejaculation should be higher than the first ejaculation. Season of the year also affects pH. The lowest pH values are observed during the breeding season, and the highest values are observed during winter.

Morphology

A morphological examination is used to determine the percentage of deformed cells. The percentage of deformed cells usually varies between 20 and 30 percent. The normal spermatozoan is shown in Figure 14-10. It is composed of three principal regions: head, midpiece, and tail. Because the tail is placed abaxially on the head, some normal spermatozoa look as if the tail comes out of the underside of the head. Many types of abnormalities are commonly observed (Figure 14-11). Defective heads may be narrow at the base, pear-shaped, small, abnormally contoured, free heads, double heads, or giant heads. Defective mid-

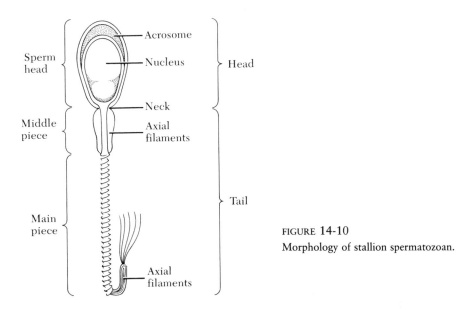

Sperm head

Middle piece

Main piece

Acrosome

Nucleus — Head

Neck

Axial filaments

Tail

Axial filaments

FIGURE 14-10
Morphology of stallion spermatozoan.

pieces may be enlarged and swollen and may have protoplasmic droplets. Tail deformities include free tails and bent, coiled, and double tails. When the percentage of abnormal spermatozoa exceeds 50 percent, fertility is questioned.

BREEDING SOUNDNESS EVALUATION

In addition to a semen evaluation, a breeding soundness evaluation should be completed on stallions before they are bought or before the breeding season starts. If possible, the stallion's past breeding records should be examined to determine his ability to impregnate mares. These records should indicate the number of mares bred, number that conceived, average number of breedings per estrous cycle, average number of estrous cycles mare are bred per conception, and average number of breedings per conception. Fertile stallions breed each mare 1.5 to 2 times per estrous cycle, and they are bred 1 to 2 estrous cycles per conception. The conception percentage for a properly managed and fertile stallion should be greater than 90.

The stallion should be in excellent physical health as shown by a physical examination and by his personal appearance. Injuries to the hindlegs or back may impair or prevent the stallion from mounting mares or phantoms. Overweight stallions tend to lose libido during the latter part of the breeding season, when temperatures are higher. The condition of the respiratory and circulatory systems should be evaluated, because they have a direct bearing on the longevity of the stallion. Problems such as arthritis and melanomas should

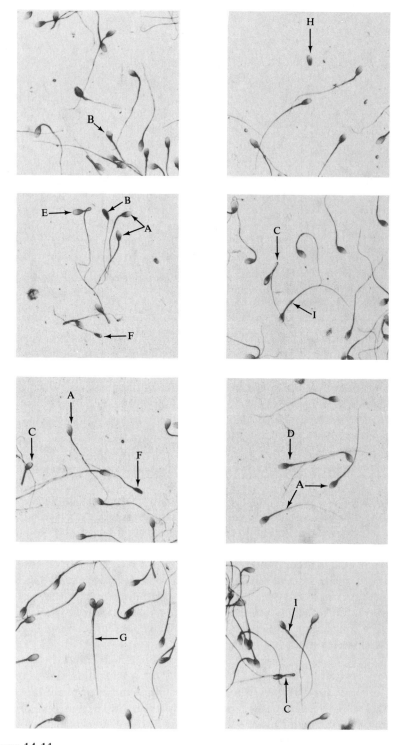

FIGURE 14-11

Normal and abnormal types of spermatozoa. (A) Normal. (B) Abnormal head. (C) Bent tail.
(D) Proximal protoplasmic droplet. (E) Looped tail. (F) Microcephalic head. (G) Twin heads.
(H) Isolated head. (I) Distal protoplasmic droplet. (Photos by B. Evans.)

also be evaluated. Heavy parasite infestations may cause colic and have other long-term effects that reduce the stallion's performance. Check any history of colic. Palpate the reproductive tract for any abnormalities. The scrotum should contain two testicles; cryptorchids should be avoided and castrated. Detect inguinal hernias during a rectal examination. Check for hereditary imperfections such as overshot and undershot jaws.

The temperament of the stallion is important. He should be easy to handle during teasing and breeding procedures. If possible, observe him when he is breeding a mare. He should not have any stable vices. Cribbing, stall kicking, weaving, and other vices detract from the stallion's appearance to mare owners. Masturbation lowers the semen quality. Stallions that severely bite mares during copulation must be muzzled during teasing and breeding. The stallion's libido should be evaluated by determining his reaction time. Normal reaction times (from the time of seeing a mare until an attempted mount) should be about 1 to 5 minutes during breeding season. During winter, it is usually longer and may be as long as 15 minutes. Stallions with depressed libido may take as long as 30 to 60 minutes before they will mount a mare. Breeding these stallions is time consuming. The average number of mounts per ejaculate should be about 1.5. An average of 2 to 3 is also too time consuming. The time of copulation normally lasts from about 30 seconds to 2 or 3 minutes. The stallion usually makes 5 to 10 intravaginal thrusts prior to ejaculation, and each thrust lasts about 11 seconds. The ejaculatory reflex lasts about 15 seconds. About 5 to 12 ejaculation pulses can be felt on the ventral side of the base of the penis. Ejaculation can usually be determined by "tail flagging"—the stallion raises and lowers his tail several times.

Olfaction is one of the fundamental stimuli of the reproductive responses of the stallion. When the stallion smells the external genitalia of the mare or smells voided urine, he displays the olfactory or Flehman reflex, in which he extends his neck upward and curls his lip. During the reflex, he inhales and exhales air in the upper respiratory passages. During courtship or teasing, the stallion also smells the groin of the mare and bites the mare on the croup and neck. At the first approach to a mare or on removal from his stall or paddock to tease mares, the stallion snorts and continues to snort periodically during courtship.

INFERTILITY

There are many possible causes of infertility. A stallion suffering from poor health and/or nutrition, injury, worry, or anxiety sometimes recovers, with proper management. Another common problem is masturbation. It is difficult to observe some stallions masturbating, but the presence of dried semen on the abdomen or on the back of the forelegs and a shrinkage of muscles over the loin are good indicators that the stallion has been masturbating. A couple of

(a)

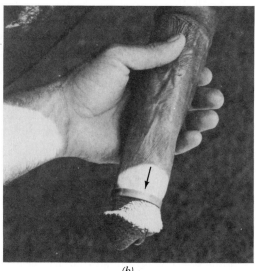

(b)

FIGURE 14-12

(a) Plastic stallion rings.
(b) Stallion ring in position.
(Photos by B. Evans.)

management practices have been tried to eliminate this vice. Turning some stallions outside where they can exercise and see other horses stops masturbating, but others require that a stallion ring be placed behind the glans penis (Figure 14-12). The ring is removed before breeding a mare and is subsequently replaced. Care must be taken that a local inflammation does not develop.

Mismanagement is the most common cause of abnormal sexual behavior leading to infertility. Overuse of a stallion as a 2- or 3-year-old may lead to the development of poor breeding behavior, such as slow reaction time to breed a mare. Mature stallions that are overused during winter and fall may develop a slow reaction time. The overused young stallion often develops a bad habit of savaging (excessively biting) a mare. Disinterest may also result from excessive use of the stallion as a teaser to determine which mares are in estrus.

Infertility may develop in the latter part of the breeding season if a stallion is overused in the early part of the season; that is, before the natural breeding season. During the latter part of winter (early part of breeding season), sperm output and libido decrease. Mares are coming out of anestrus, so their estrous periods are long and irregular. It is difficult to predict ovulation during the first estrous cycle, so these mares require more breedings per conception. Also, barren and problem mares normally are the mares being bred at this time.

Unnecessary roughness while handling a stallion during breeding and teasing may reduce his libido. He may become afraid to mount and copulate. However, stallions should be trained so that they do not endanger personnel or the mare. During teasing, the young stallion should be allowed to be more aggressive than normal. After he learns how to breed mares, any overaggressiveness can be corrected.

TRAINING YOUNG STALLIONS

Training the young, shy stallion takes patience and time. He is usually started by teasing a mare or mares for a few minutes. As soon as he loses interest, or after 5 to 10 minutes, teasing should be stopped. As soon as he teases a mare vigorously, he should be allowed to mount a gentle mare that is in heat. He should be guided into the proper position prior to mounting. A stallion frequently mounts on the mare's side or shoulder until he learns proper position. When he mounts correctly from behind, direct his penis into the vagina. The handler must be careful because the stallion frequently falls backward or off to the side. After bred two to six times, he usually understands the process. At this time, train him to mount the mare by approaching her at the hip.

The stallion must be trained to accept washing procedures before breeding. For the first few times, stand the stallion next to a wall after he drops his penis while being teased. He is washed in this position so that his movements can be controlled.

CASTRATION

A castrated horse is referred to as a *gelding*. Gelding a horse has several advantages. Several geldings may be kept in a paddock, whereas each stallion must be kept by himself. Geldings are easier to care for, less prone to injury, and easier to haul because of their attitude. Many stallions tend to be lazy performers and do not perform consistently.

Age

Most horses chosen to be castrated are usually castrated between birth and 2 years of age. Several factors usually determine the age when they are gelded. Colts that have poor conformation and/or poor pedigrees are gelded as soon as the testicles descend into the scrotum. The testicles are usually in the scrotum at birth or arrive there before the tenth month after birth, but occasionally they are not fully down until the twelfth or fifteenth month after birth.

If a horse has good conformation, a good pedigree, and warrants a performance test, he is kept intact until he fails to meet specific performance criteria. Stallions that are able to perform but do not sire good foals should also be castrated. There are too many good stallions available to keep those that are poorly conformed and lack performance ability or stallions whose progeny are of poor quality.

Procedure for Castration

Because of the possible complications that may result from castration and the need to anesthetize many colts or stallions to prevent injury, it is customary for a veterinarian to perform the castration. At least three methods of castrating a horse have been described.

1. In the *primary closure method,* the veterinarian makes an incision through the skin and vaginal tunic between the scrotum and superficial inguinal ring on each side of the scrotum. He or she ties the vessels and ductus deferens by transfixation ligatures. The testicles are removed, and all dissected planes are closed with chromic gut.

2. In the *closed technique,* the veterinarian frees each testicle and spermatic cord still contained within the parietal layer of the tunica vaginalis by blunt dissection from the surrounding tissue well into the inguinal canal. The cord structures are divided by means of an emasculator, which is left in position for approximately 1 minute.

3. In the *open technique,* the veterinarian frees the tunica vaginalis from the surrounding tissue as in the closed method. Then the tunica vaginalis communis is split with scissors so that all structures to be removed can be identified. The testicles, epididymis, and part of the cord are then removed, as previously indicated.

If the horse has been immunized against tetanus, postoperative care consists of exercise to prevent or control swelling and edema. (Otherwise, immunize immediately after castration.) The stallion libido usually subsides in 4 to 6 months but may last for 1 year.

Failure to remove the epididymis and sufficient vas deferens can cause a gelding to retain stallion behavior. Such geldings are referred to as being "proud cut." Not much can be done to correct their behavior.

Cryptorchids

An animal with one or both testes undescended into the scrotum is a cryptorchid, more commonly referred to as a *ridgeling.* The testes may be descended at birth and are usually descended by the age of 10 months. Some colts may be 12 to 15 months old before the testes descend. After 15 months, colts are considered cryptorchid if the testes are not descended. A positive diagnosis can be made if the vas deferens does not pass through the inguinal canal. These horses are difficult to castrate, and many castrated ridgelings seem to retain their stallion attitude for several months after castration.

15
Breeding
the Mare

Mares have traditionally had a reputation for having low reproductive efficiency compared to other domestic farm animals. On the average, about 50 percent of mares bred have live foals. The low reproductive efficiency has been related to breeding mares with reproductive problems and placing no emphasis on conformation, which is related to reproductive success, or on the reproductive history of the mare' immediate family. Moreover, the mare has a long estrous period, during which it is difficult to determine the appropriate time to breed. Thus it is essential for horse breeders to understand anatomy and the physiology of reproduction in the mare and stallion if they expect high conception rates.

REPRODUCTIVE TRACT

The reproductive tract of the mare consists of two ovaries, two fallopian tubes, uterus, cervix, vagina, clitoris, and vulva. These structures are shown in Figure 15-1. The mammary gland is an accessory sex gland and is shown in Figure 15-2.

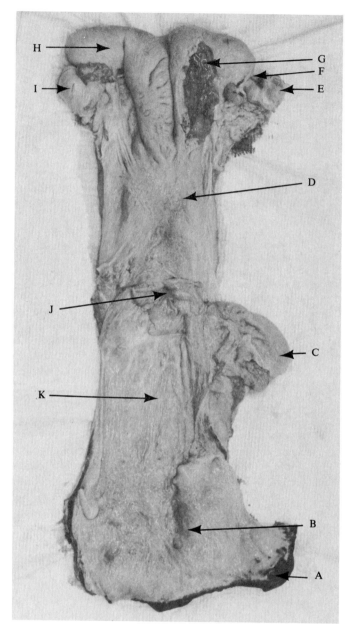

FIGURE 15-1

Reproductive organs of the mare, showing the vagina and a uterine horn opened. (A) Clitoris. (B) Orifice of vestibular glands. (C) Bladder. (D) Body of uterus. (E) Ovary. (F) Fallopian tube. (G) Uterine horn opened to show mucosal folds. (H) Uterine horn. (I) Ovary. (J) External os of cervix. (K) Vagina. (Photo by B. Evans.)

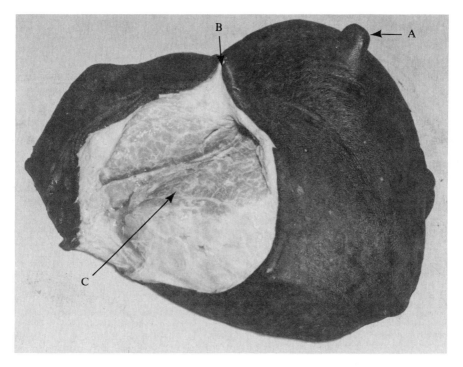

FIGURE 15-2
Mammary gland with one gland complex opened along teat cistern. (A) Teat. (B) Teat opened along teat cistern. (C) Gland complex. (Photo by B. Evans.)

Ovaries

The ovaries of the mare perform two essential functions: the production and release of ova and hormones. In form, the ovary resembles a bean or kidney in that it has an area, the ovulation fossa, that is folded inward (Figure 15-3). This is where ovulation occurs. It is difficult to give average ovary weight or dimensions because they are influenced by many factors, such as age, breed, size, season, and reproductive state of the mare. In mature mares of average size (900 to 1,100 lb), each ovary weighs approximately 50 to 75 gm and is 6 to 7 cm long and 3 to 4 cm wide.

The outer surface of the ovary is covered by a serous coat except at the hilus and ovulation fossa. The hilus is where the blood vessels and nerves enter the ovary. At the ovulation fossa, a layer of primitive germinal epithelium covers the area. The tough, serous coat prevents ovulation from occurring on the surface of the ovary except at the ovulation fossa. The interior of the ovary is composed of connective tissue and follicles or corpora lutea in varying stages of growth and development or degeneration.

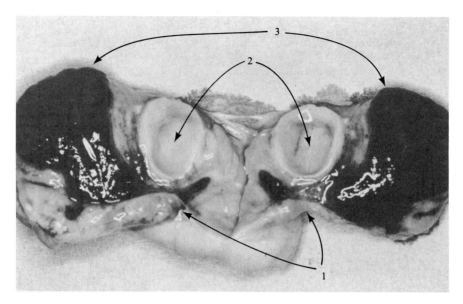

FIGURE 15-3

Cross section of mare's ovary. (1) Ovulation fossa. (2) Graafian follicles. (3) Corpus hemorraghicium. (Courtesy John Hughes.)

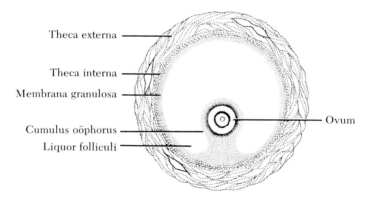

FIGURE 15-4

A Graafian follicle. (Courtesy M. Morris.)

Follicles undergo a series of developmental events leading to the release of the ovum (see Figure 15-4) and formation of the corpus luteum. Immediately after ovulation occurs, the corpus luteum starts to develop in the area from which the follicle was released. By the sixth day after ovulation the corpus luteum is well developed and vascularized. The purpose of the corpus luteum is to produce progestin hormones. It begins secreting progestins in sufficient quantities for their concentration to be measured in the blood about 48 to 72 hours after ovulation. Approximately 14 days after ovulation, the corpus luteum decreases in size, and its progestin-synthesizing capabilities decline quite rapidly if the mare is not pregnant.

In addition to the follicles that are in an active stage of growth in the ovary, there are follicles that do not complete their growth cycle and that start to undergo atresia (a degeneration process). These so-called atretic follicles are found in the ovary during or immediately after the breeding season and are in varying stages of degeneration. Corpora lutea also undergo a degeneration process that may take several months. Therefore, it is common to find degenerating corpora lutea in the ovary, known as *corpus albicans,* resulting from the previous two or three estrous cycles.

The ovarian artery supplies blood to the ovaries (Figure 15-5). Several veins leave the ovary and unite to form the ovarian branch of the ovarian vein. The nerve supply is derived from the renal and aortic plexus of the sympathetic system.

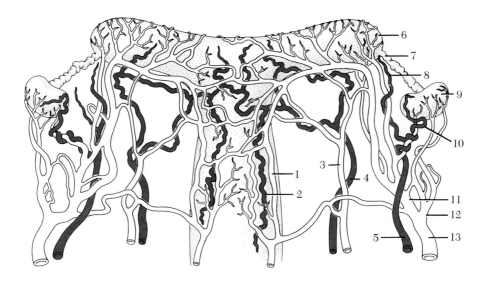

FIGURE 15-5

Composite diagram of dorsal view of arteries and veins (shaded) of uterus and ovaries of the mare. (1) Uterine branch of vaginal vein. (2) Uterine branch of vaginal artery. (3) Uterine vein. (4) Uterine artery. (5) Ovarian artery. (6) Uterine horn. (7) Uterine tube. (8) Uterine branch of ovarian artery. (9) Ovary. (10) Ovarian branch of ovarian artery. (11) Uterine branch of ovarian vein. (12) Ovarian branch of ovarian vein. (13) Ovarian vein. (From Ginther and others, 1972.)

Fallopian Tubes

The fallopian tubes or oviducts are the connecting links between the ovaries and the uterus. They serve two primary functions: they conduct ova from the ovaries to the uterus, and they serve as the usual site of fertilization. Each oviduct is about 30 to 70 mm long and 4 to 8 mm thick and has an inside tubal diameter of 2 to 3 mm. The oviducts are coiled and folded so that the ovary is fairly close to the uterine horn. To prevent the escape of the ovum into the abdominal cavity during ovulation, the end of each oviduct is expanded to form the ampulla and infundibulum. The infundibulum partially surrounds the ovary and has fingerlike projections (fimbrae) extending from its edge. The opening of the oviduct into the peritoneal cavity at the junction of the ampulla is called the *ostium abdominale*. The opening leading into the uterus is the *ostium uterinum* and is located at the uterotubal junction.

Uterus

The uterus of the mare consists of the left uterine horn, right uterine horn, and body. The uterine horns are located in the abdominal cavity. They are approximately 25 cm long (10 inches) and are cylindrical in shape. The horns curl downward and outward from the uterus like rams' horns. The uterine body is approximately 18 to 20 cm long and 10 cm in diameter. It is suspended in position by the broad ligament that attaches to the abdominal and pelvic walls. It is located partially in the abdominal cavity and partially in the pelvic cavity.

The wall of the uterus is composed of three muscle layers. At parturition, these muscles help expel the fetus. An inner layer of loose connective tissue contains the uterine glands.

Blood supply to the uterus is derived from three arteries: the uterine branch of the vaginal artery, the uterine artery, and the uterine branch of the ovarian artery (Figure 15-5). Blood leaves the uterus via three veins whose names correspond to those of the arteries. Nerve supply to the uterus derives from the uterine and pelvic plexus.

Cervix

The cervix is a thick, muscular sphincter that separates the body of the uterus from the vagina. It is about 5 to 7.5 cm long. The muscle layers are well developed and have extensive folds. Because of the extensive folds, there is no well-defined cervical canal. The cervix extends into the vagina, and the angle formed between the cervix and vagina is the formix. The opening into the vagina is the os cervix. Part of the mucus secreted into the vagina is derived from mucosal cells located in the mucosa layer. During estrus, these cells are very active, and a considerable amount of mucus is present in the vagina around the cervix.

Vagina

The vagina is composed of two parts. The vagina proper is located between the hymen and the cervix, whereas the vestibulum vaginae is located between the hymen and the vulva. The hymen is located just forward of where the urethra enters the vestibulum vaginae. The vagina is about 15 to 20 cm long and 10 to 12 cm in diameter. The wall of the vagina is composed of a layer of stratified epithelium, a layer of loose connective tissue, and two layers of muscles.

Blood comes into the vagina through the internal pudic arteries and leaves via the internal pudic veins. The nerve supply is from the pelvic plexus of the sympathetic nervous system.

Vulva

The vulva is the terminal portion of the reproductive tract. It is about 10 to 12 cm in length, and the external orifice is a vertical slit approximately 12 to 15 cm long. The external urethral orifice is located 10 to 12 cm inside the vulva. The glans clitoris, homologue of the male penis, is located about 5 cm within the vulva.

Mammary Gland

The udder of the mare is composed of two gland complexes. Each complex has a broad, flat teat. Each teat has at least two streak canals, and each streak canal is connected to a teat cistern. A canal extends from each teat cistern to a gland cistern, which has a system of ducts leading to it from the secretory gland.

Blood supply to the udder is via the external pudental artery and is drained by the external pudental and subcutaneous abdominal veins. The inguinal nerve innervates the udder.

PHYSIOLOGY OF REPRODUCTION

Estrous Cycle

When the filly attains puberty at the age of 12 to 15 months, she starts to come into estrus (heat) periodically. The estrous cycle is the length of time from the first day of behavioral estrus until the first day of behavioral estrus of the next estrous cycle. For practical purposes, the estrous cycle is divided into two phases, estrus and diestrus (Figure 15-6). The average length of the estrous cycle is 21 to 23 days. Estrus lasts for about 4 to 6 days, and diestrus lasts for

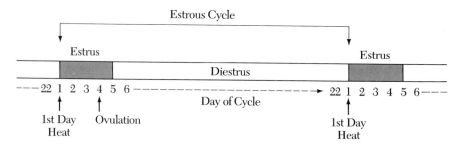

FIGURE 15-6
Estrous cycle components and time relationship.

about 17 to 19 days. Ovulation usually occurs during the third to fifth days of the cycle. During the two phases, definite changes occur in the mare's behavior, in the hormone concentrations in her blood, and in her reproductive tract.

Behavioral Changes

During diestrus, the mare shows complete resentment of the stallion. She will kick, bite, and chase him. One or two days before estrus, she becomes somewhat more passive to the stallion. She may or may not object to his presence. During estrus, she is receptive to the stallion. She flexes her pelvis, raises her tail, frequently urinates, spreads her hindlegs apart, contracts and relaxes the lips of the vulva with eversion of the clitoris, and allows the stallion to bite and chew on her neck and flanks (Figures 15-7 and 15-8). The behavioral signs of the jinny (donkey) are a little different: she holds her head low, and her jaws make a chewing motion.

The end of the estrous period is closely correlated to ovulation. About 50 percent of all mares are out of estrus 24 hours after ovulation; 80 percent are out after 48 hours.

To detect when mares are in estrus, they must be teased with a stallion. The mares should be teased daily or at least every other day if they are to be bred. Several methods are used to tease mares, depending on the number of mares to be bred, physical facilities, and available labor. Each mare can be brought to the teaser, or vice versa. The teasing-rail method (Figure 15-9) and teasing-wall method (Figure 15-10) require a stallion handler and a mare handler. The rail or wall should be about 4 feet high, and the wall should be of solid construction. Because both the mare and stallion kick and strike, the wall should be padded. The personnel are protected by standing behind the two short padded walls. The stallion can be kept in a stall (Figure 15-11) or an outside paddock (Figure 15-12) where the mares can be brought to him. This eliminates the need for a stallion handler. The chute keeps the mare close to

FIGURE 15-7

Mare in estrus, showing her acceptance of the stallion. (Photo by B. Evans.)

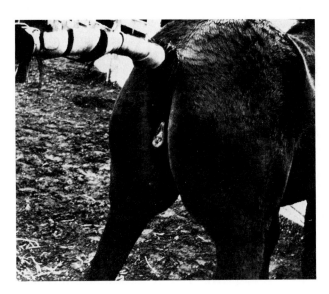

FIGURE 15-8

Estrous mare with hindlegs spread apart, pelvis flexed, raised tail, and eversion of clitoris. (Photo by B. Evans.)

FIGURE 15-9

Teasing rail separates mare and stallion.

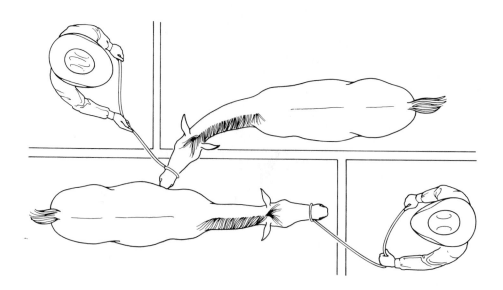

FIGURE 15-10

Teasing a mare with a stallion across a teasing wall. Both handlers are protected from the mare and stallion kicking and striking. (Illustration by M. Morris.)

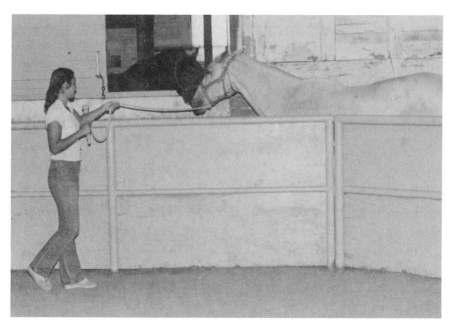

FIGURE 15-11
Stallion teasing a mare through a window in the stall wall that is adjacent to a teasing chute. (Photo by B. Evans.)

FIGURE 15-12
Paddock designed so that mares can be efficiently teased individually. (Photo by B. Evans, courtesy Taylor Ranch, Davis, California.)

the stallion. By using a long lead shank, the handler can get in a position to observe the mare's perineal area. For teasing large numbers of mares, a teasing mill can be used (Figure 15-13). By the time the last mare is in her small pen, the first mare can be returned to her paddock. To tease 50 to 75 mares in a short period of time, two stallions can be placed in teasing boxes between pens

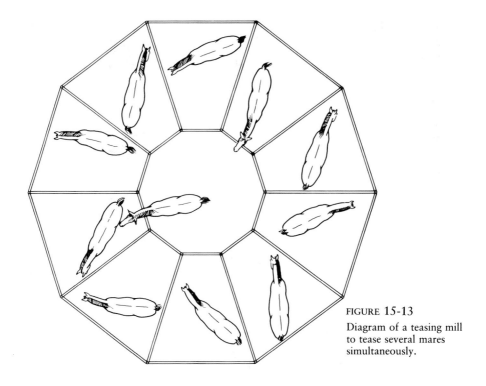

FIGURE 15-13

Diagram of a teasing mill to tease several mares simultaneously.

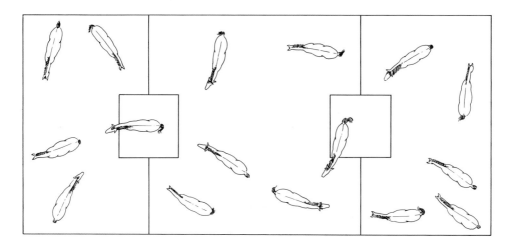

FIGURE 15-14

Large-scale teasing operation in which 50 to 75 mares can be teased simultaneously. A teaser stallion is placed in each of the two small pens.

(Figure 15-14). Keep exact records on each mare because some mares are timid and will not approach the teaser when they are in estrus. Any mare that does not show estrus within three weeks of teasing should be teased individually. The stallion can be taken to each mare in her paddock, pens of mares, or each pasture of mares. Teasing across a fence is dangerous for the horses. They can easily get a foot caught in the fence unless a teasing wall is built in the fence. A teasing wall is 4 feet high, about 10 to 12 feet long, and of solid construction. Mares will approach the stallion when he is brought to the teasing wall. A pony stallion can be turned into the pasture or large paddock, because he is not tall enough to breed the mares if they are in estrus.

The teaser stallion is an important part of the breeding program because he gets the mare to show behavioral signs of estrus. Most farms use a low-quality stallion for a teaser. Cryptorchid stallions whose descended testicle has been removed and vasectomized stallions make excellent teasers. A vasectomized stallion cannot impregnate a mare if he gets loose and mounts a mare. Stallions with a genetic deformity in which the penis is reflected back between the legs are aggressive teasers. Some geldings will tease a mare and if they are given testosterone they may be suitable to use during a breeding season.

Hormonal Changes

The endocrine changes that occur during the two phases of the estrous cycle are diagrammed in Figure 15-15. Although the changes in concentration of the hormones in the blood have been determined, the interrelationships between

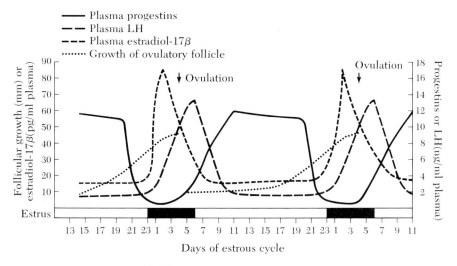

FIGURE 15-15
Endocrine patterns during the estrous cycle of a mare.

the hormones are not fully understood. The hormones controlling the estrous cycle originate in the brain, pituitary, and ovary.

The pituitary hormones are follicle-stimulating hormone (FSH) and luteinizing hormone (LH). FSH is responsible for the growth of follicles. Generally, FSH in the blood increases during the mid-diestrous phase and again during estrus. It is believed that the mid-diestrous surge initiates the growth of follicles, and the estrous surge aids in their final growth.

Luteinizing hormone causes ovulation and is responsible for the formation of the corpus luteum. LH concentration starts to increase 2 to 3 days prior to ovulation and reaches a maximum 1 to 2 days after ovulation. Several days after ovulation, it reaches a minimal concentration during diestrus.

FSH and LH work in a synergistic manner in the processes of follicular growth, maturation, and ovulation. Their secretion by the pituitary is probably controlled by other hormones known as "releasing factors" that are secreted by the hypothalamus in the brain.

Two ovarian hormones—estrogens and progestins—participate in the control of the estrous cycle (Figure 15-15). Estradiol-17β is produced by the granulosa cells and by the theca interna of the Graafian follicles. As the follicle enlarges, the amount of estradiol-17β in the blood starts to increase. The increased concentration reaches a peak just prior to ovulation and then declines to a minimal concentration at 5 to 8 days after ovulation. Estrogen may facilitate the surge in LH, because it does reach a peak prior to the LH.

Within 24 hours after ovulation, the corpus luteum begins to secrete progestins, primarily progesterone. By the sixth day after ovulation, the corpus luteum is producing maximal amounts of progestins. This continues until about 12 to 14 days after ovulation, when the blood concentration of progestins rapidly declines. The cessation of corpus luteum function is caused by the release of a luteolysin, prostaglandin, by the lining of the uterus. The interval between cessation of corpus luteum function and the beginning of estrus is approximately 3 days. The blood concentration of progesterone must decline below 1 ng/ml plasma before estrus is manifested.

Reproductive Tract Changes

Characteristic changes occur in the reproductive tract of the mare during the different phases of the estrous cycle. These changes can help determine the best time to breed a mare. For mares that have silent estrous periods, these changes are the only indicators used to determine when to breed the mare. The changes in weight and size of the different parts are listed in Table 15-1.

The most important changes occur on the ovary. They are determined by palpating the ovaries via the rectum (Figure 15-16). During the winter anestrous period, the ovaries are usually small and inactive. However, the ovaries of

TABLE 15-1
Changes in Mare's Reproductive Tract During the Estrous Cycle

	Day of Estrous Cycle				
Structure	2	4	7	11	17
Anterior portion of vagina (gm)	188	216	173	194	230
Cervix (gm)	141	168	153	133	164
Uterus (gm)	696	674	558	669	849
Uterine horn (left)					
Diameter (mm)	68	55	61	58	62
Length (mm)	151	146	152	154	137
Uterine horn (right)					
Diameter (mm)	58	60	44	61	64
Length (mm)	148	155	144	160	152
Oviduct (left)					
Weight (gm)	5.57	5.72	4.85	6.17	5.20
Length (mm)	147	140	149	162	152
Oviduct (right)					
Weight (gm)	7.12	6.83	6.02	6.61	6.61
Length (mm)	154	155	147	166	143
Ovaries (gm)	136.7	128.4	122.0	111.7	116.4
Extraluteal tissue (gm)	71.9	69.6	70.5	69.0	81.8
Extraluteal fluid (gm)	62.7	54.4	42.9	25.6	22.4
Follicles 10–30 mm (no.)	6.3	6.1	7.6	4.6	3.8
Follicles 20–30 mm (no.)	3.4	2.0	1.4	1.7	1.3
Follicles > 30 mm (no.)	1.3	0.9	0.5	0.2	0.1

SOURCE: Adapted from Warszawsky and others (1972).

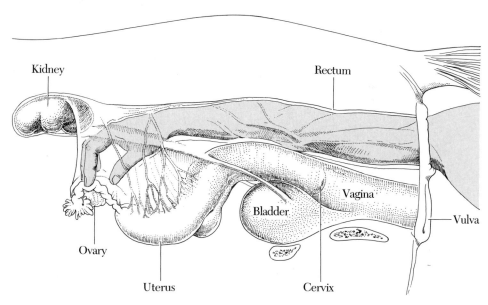

FIGURE 15-16
Palpating (feeling) the ovaries of a mare.

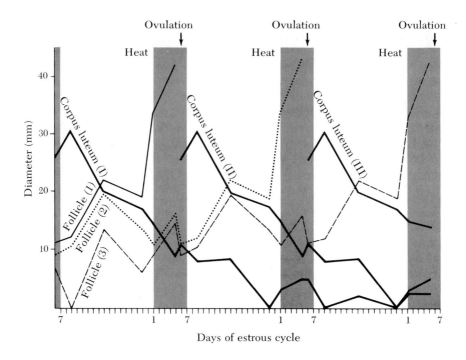

FIGURE 15-17

Curves of growth of follicle and corpus luteum during the estrous cycle. (From Hammond and Wodzicki, 1941.)

some anestrous mares continually develop follicles that may or may not ovulate. As the breeding season approaches, the ovaries begin to develop. Although several follicles start to develop each estrous cycle, only one or two reach ovulatory size (Figure 15-17). The follicle that will ovulate during estrus usually becomes prominent just before estrus. It starts a rapid growth phase about 5 days prior to ovulation. Maximum follicular size attained before ovulation is about 40 to 50 mm, and the maximum size is attained the day before ovulation. Immediately before ovulation, some follicles get softer, while others do not. Therefore, a change in softness cannot be used to predict ovulation. Follicles present at the first ovulation may continue to develop, and about 25 percent of the estrous cycles have a second ovulation. Some follicles become atretic while others may remain for a couple of estrous cycles before they become the ovulatory follicles (Figure 15-17).

Definite changes that occur in the cervix during the estrous cycle are also quite useful in determining estrus. During diestrus, the cervix is tightly constricted, pale in color, and covered with a dry, gummy mucus. As the mare approaches estrus, the cervix begins to relax, the blood supply increases, and

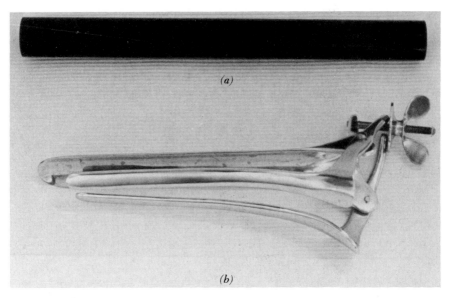

FIGURE 15-18

Vaginal speculums to visually examine the vagina and cervix. *(a)* A plastic speculum. *(b)* A "Caslick" speculum. (Photo by B. Evans.)

a clear mucus is secreted. During estrus, vascularity has increased so that the cervix is a reddish color. The cervix has relaxed so that it may lie on the floor of the vagina, and two or three fingers can be easily passed through it. A considerable amount of clean, clear, slimy mucus is secreted, covering the cervix and vagina. These changes are determined visually through a speculum (Figure 15-18).

The changes in the vagina are definite but not as dramatic as the cervical changes. During diestrus, the walls are dry and sticky and are pale in color. As the mare comes into estrus, the walls of the vagina become rosy red in color and covered with the slimy cervical mucus.

Changes that occur in the uterus are more difficult to determine and to predict. Therefore, from a practical viewpoint, they are not very useful in determining when to breed a mare. During estrus, the uterus has some muscle tone but not as much as during diestrus. During estrus, it has slight tone; during diestrus, it has good tone. Changes in vascularity also change the feel of the uterus. During estrus, it is turgid but flaccid. During diestrus, tissues are not edemic, so that the uterus starts to resemble a rigid tube.

During winter anestrus, the cervix and uterus lose muscle tone and usually become completely flaccid in most mares that have no follicular activity. In mares with considerable ovarian activity, the uterus may have some muscle tone and the cervix may be firm and tightly constricted.

TABLE 15-2
Length of Estrous Cycle and Estrous Period of Mares with One or Two Ovulations per Cycle

No. of Ovulations per Estrous Cycle	Length of Estrous Cycle (days)	Length of Estrus (days)	Length of Diestrus (days)	Corpus Luteum Life Span (days)
One	20.7 ± 0.2 (range 13–34)	5.4 ± 3.0 (range 1–24)	15.4 ± 3.1 (range 6–25)	12.6 ± 2.9 (range 8–18)
Two	20.2 ± 4.5 (range 13–27)	6.1 ± 2.7 (range 1–21)	14.6 ± 3.5 (range 8–25)	12.2 ± 3.3 (range 5–17)

Note: All deviations are standard deviations. Same 11 mares were studied to obtain data for Tables 15-2 through 15-5.

SOURCE: Figures in last column from Stabenfeldt, Hughes, and Evans (1970). All other figures from Hughes, Stabenfeldt, and Evans, 1972b.

Variability in the Estrous Cycle

One of the most important aspects of managing reproduction in the mare is the extreme variability in all aspects of the estrous cycle. There is considerable variation from cycle to cycle in the same mare, and there is considerable variation from mare to mare. An understanding of these variations, of management techniques to detect variations, and of how to breed mares during the variations is necessary to prevent stallion overuse and to achieve a high conception rate.

The average length of the estrous cycle is 21 to 23 days (Table 15-2). Only about 50 percent of all mares have a normal estrous cycle length. A few are shorter but most are longer.

Winter Anestrus

A primary reason for the variability in estrous cycle is season of the year, and estrous cycle patterns are classified according to seasonal variation. Many years ago, the wild mare probably had only two or three estrous cycles per year. This restricted estrous season enabled foals to be foaled during favorable environmental conditions. The second general pattern is seasonal polyestrus, characterized by a period of time when the mare does not cycle (Figure 15-19). The anestrous period is usually during winter and is called *winter anestrus*. The period of winter anestrus will vary depending on the climate. Some mares exhibit winter anestrus one year and not the following year. Normally, the more definite the seasons of the year, the higher the percentage of mares displaying winter anestrus. The third pattern is polyestrus—the mare cycles throughout the year (Figure 15-20).

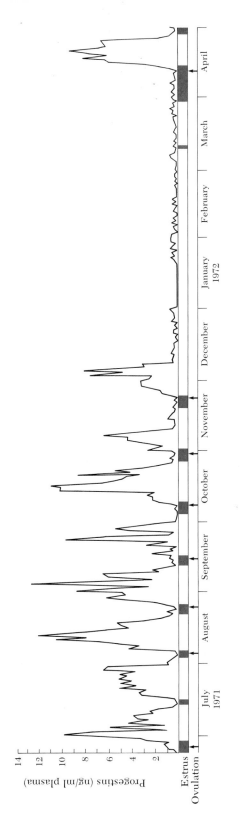

FIGURE 15-19

Reproductive history of Hi Aggie. (From Hughes, Stabenfeldt, and Evans, 1972b.)

488

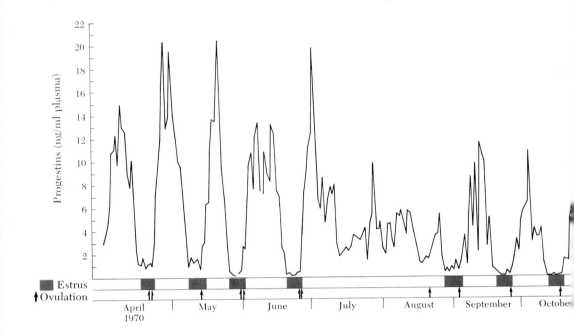

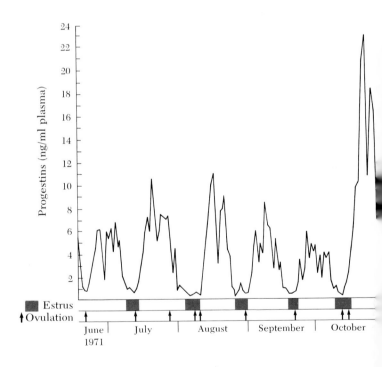

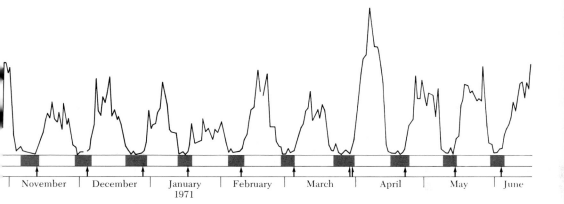

November | December | January 1971 | February | March | April | May | June

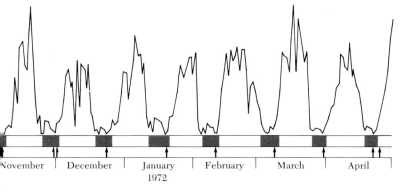

November | December | January 1972 | February | March | April

FIGURE 15-20

Reproductive history of Paisley Star. (From Hughes, Stabenfeldt, and Evans, 1972b.)

Prolonged Corpus Luteum

Another type of anestrus—or, really, a period of irregularity in the rhythmical pattern of the estrous cycles—is the result of a spontaneously prolonged corpus luteum life span (Figure 15-20). A spontaneously prolonged corpus luteum continues to secrete enough progestins to maintain plasma concentrations above 1 ng/ml. When the progestins are maintained above this level, signs of estrus are suppressed. The exact cause is unknown but is believed to be a failure of the uterus to synthesize or release sufficient prostaglandin to lyse the corpus luteum (make it stop functioning) at the proper stage of the cycle or a failure of the corpus luteum to respond to prostaglandin. There are two methods of treatment for the anestrous condition. A warm, 0.9 percent saline solution (500 ml) can be infused into the uterus. The mare will come into estrus in about 5 to 6 days and will ovulate approximately 9 to 10 days after the infusion. Prostaglandin or an analogue of prostaglandin is the usual treatment. The mare comes into heat approximately 3 to 4 days after the injection and will ovulate during the 6- to 12-day postinjection period (Figure 15-21).

Prolonged corpus luteum function can occur at any time and lasts for 35 to as long as 100 days or more. Therefore, the infertile period may last beyond the length of the normal breeding season. A maiden, open, or barren mare that is not pregnant and does not come into heat at the expected time could possibly have a spontaneously prolonged corpus luteum. Actual diagnosis can be made by determining the concentration of progesterone in the blood. Concentrations over 1 ng/ml confirm the condition. It is possible for a mare to develop the condition following the estrous period in which she was bred (Figure 15-22).

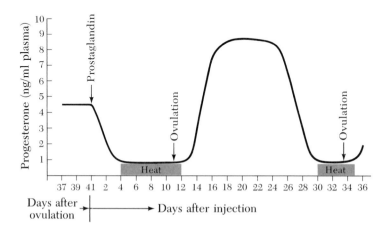

FIGURE 15-21

Effect of administering prostaglandin to a mare with a prolonged corpus luteum.

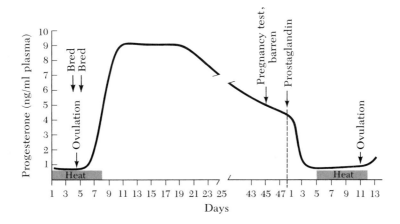

<small>FIGURE 15-22</small>

Effect of administering prostaglandin to a pseudopregnant mare that did not conceive or resorbed the fetus prior to 36 days of gestation.

Because she does not come back into heat, she is assumed to be pregnant until diagnosed nonpregnant by a manual pregnancy check. The mare will respond to prostaglandin treatment or intrauterine saline infusion. However, be careful to determine that the mare did not conceive and abort (see section on pregnancy problems).

Diestrus Length

Aside from winter anestrus, season of the year influences the length of the estrous cycles in other ways (Table 15-3). The average length of the diestrous period (in cycling mares) tends to be shorter in the early part of the natural breeding season and longer during winter.

Estrus Length

The estrous period is normally 5 to 6 days long (Table 15-2), but only about 50 percent of the cycles have a normal length. Some mares tend to have uniform estrous periods, while others have quite variable estrous periods. Seasonal variability of the estrous period, excluding anestrus, is responsible for part of the variation. During later winter and early spring, the average estrous period is 7.6 days rather than the normal 5 to 6 days. The longest estrous periods occur when mares are going into or coming out of winter anestrus (Figure 15-19).

TABLE 15-3
Effects of Seasonal Changes on Reproduction in the Mare

	Jan.	Feb.	Mar.	Apr.	May	Jun.	Jul.	Aug.	Sep.	Oct.	Nov.	Dec.
Average estrous cycle length (days)	32.3	34.4	26.1	21.9	19.5	21.1	24.6	20.4	20.6	20.5	24.8	30.4
Number of cycles observed	22	20	25	28	39	31	28	27	32	35	25	21
Number of periods of estrus	18	13	26	29	28	28	20	24	27	31	27	22
Average length of estrus (days)	5.06	6.69	8.38	7.76	5.72	4.47	4.70	4.75	4.44	4.53	5.20	7.23
Total ovulations for each month for all 11 mares	20	23	46	45	44	54	48	41	42	44	43	27
Average number of ovulations per month	0.9	1.1	2.1	2.1	2.0	2.5	2.2	1.9	1.9	2.0	2.0	1.2
Total number of diestrous periods without anestrus	16	13	21	27	24	24	21	21	26	27	22	20
Average length of diestrus without anestrus	16.3	16.3	14.2	11.8	14.0	15.8	16.2	15.4	16.1	15.3	16.0	14.3

SOURCE: Adapted from Hughes, Stabenfeldt, and Evans (1972b).

Split Estrus

Another variation in the estrous period is the split estrus: The mare is in estrus for a few days, goes out of estrus for one or more days, and returns to estrus for a few days. The total time period may last 12 to 14 days or even longer or may be only 4 to 6 days long. There is considerable variation in the split estrus condition. Fortunately, they only occur in about 5 percent of estrous periods. They are frequently associated with estrous cycles just before or after winter anestrus.

Silent Heat

"Silent heats" are another variation of the estrous period that causes management problems in getting the mare in foal. Some mares never show estrus when teased with a stallion (Figure 15-23) or may show estrus for a few cycles and then have a few silent heat periods (Figure 15-20). Internally, the mare is having normal estrous cycles: She ovulates and has normal plasma progesterone profiles. To breed such mares, palpate them two to three times per week to determine follicular growth. At the appropriate time before ovulation, they can be artificially inseminated.

Mares are sensitive to strange places, people, and horses and alter their behavior accordingly. A mare may be teased and found to be in estrus, but when she is transported to a stallion to be bred she may not show the behavioral signs of estrus. In this situation, the mare must be palpated through the rectum and artificially inseminated just before ovulation. Few stud managers handle this common problem properly.

A mare that is cycling and normally shows estrus can fool a stud manager just long enough to miss breeding her because she may not show estrus during one cycle (Figure 15-24). A stud manager must keep good records to detect silent heats in time for the mare to be bred. As soon as the records indicate she should be in estrus but is not, she should be palpated. If necessary, she can be artificially inseminated.

Ovulation

During estrus, the mare should ovulate one follicle on the third or fourth day of estrus. Ovulation is related to the end of estrus, not the onset (see the earlier section on "Behavioral Changes"). Approximately 75 percent of ovulations occur when expected. About 12 percent occur 3 days before the end of estrus. About 10 percent occur after the mare is out of estrus (Figure 15-25). Occasionally, however, a mare will ovulate before she shows signs of estrus or on the first day of estrus (Figure 15-24).

494

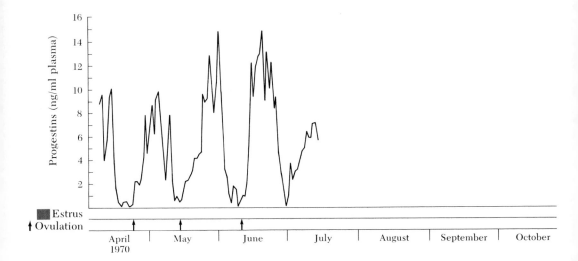

495

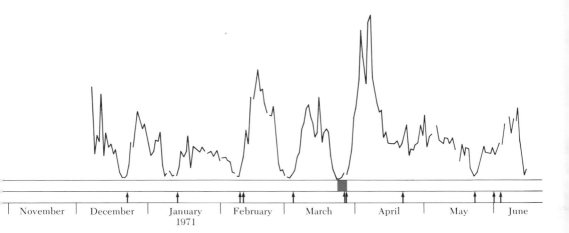

FIGURE 15-23

Reproductive history of Bashful Boots. Notice the "silent heat" periods and double ovulations.
(Courtesy G. H. Stabenfeldt, J. P. Hughes, and J. W. Evans.)

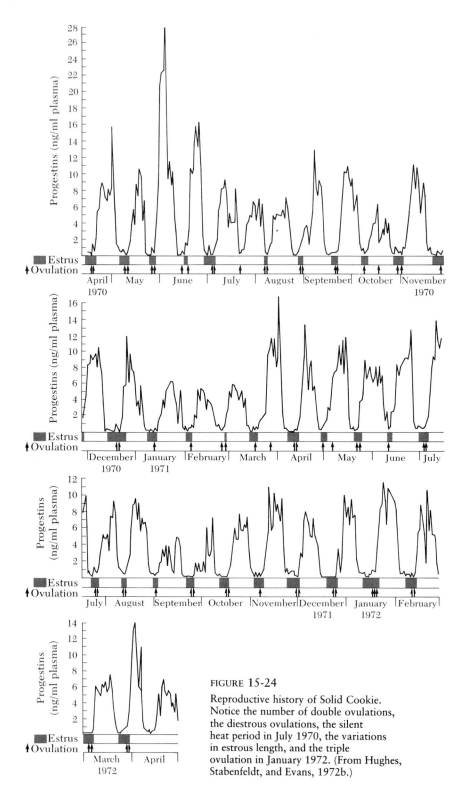

FIGURE 15-24

Reproductive history of Solid Cookie.
Notice the number of double ovulations,
the diestrous ovulations, the silent
heat period in July 1970, the variations
in estrous length, and the triple
ovulation in January 1972. (From Hughes,
Stabenfeldt, and Evans, 1972b.)

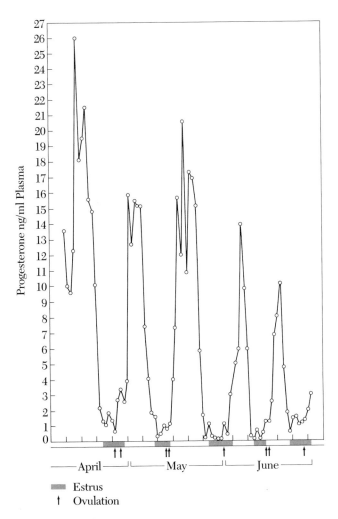

FIGURE 15-25

Reproductive history of Tule Goose. Notice ovulation in June
after mare went out of estrus. (Courtesy J. W. Evans, J. P. Hughes,
and G. H. Stabenfeldt.)

Multiple ovulations occur in 20 percent of the estrous cycles. The percentage of twin conceptions is much lower (1 to 5 percent). Usually multiple
ovulations occur in all mares if they are open for a period of 2 to 3 years and
if they are being closely observed. Some mares tend to have multiple ovulations.
The mare in Figure 15-24 is an example of one that has frequent multiple
ovulations. Most multiple ovulations are twin ovulations, but triple ovulations
during a heat period have been observed (Figure 15-24). Fortunately, multiple

TABLE 15-4
Incidence of Multiple Ovulation by Mare and by Ovaries, During Two-Year Period

Mare	No. of Triple Ovulations	No. of Double Ovulations	Ovulations				Average Interval Between Ovulations (days)
			No. LL	No. RR	No. LR[a]	No. RL[a]	
Aggie Tess	0	7	1	1	2	3	0.7
Bashful Boots	0	9	2	2	1	3	1.5
Double Dark	0	5	2	2	1	0	1.6
Hi Aggie	0	1	0	0	1	0	0.0
Mr. Bruce's Miss	0	2	1	1	0	0	1.0
Paisley Star	0	6	2	1	3	0	1.2
Psychedelic Miss	0	4	0	2	0	2	1.2
Solid Charm	0	4	2	0	1	1	1.0
Solid Cookie	1	25	3	9	10	3	0.7
Solimistic	0	10	2	2	3	3	1.5
Tule Goose	1	11	1	3	6	1	0.5
Total	2	83	16	23	28	16	1.0

[a]First letter indicates ovary in which first ovulation occurred.

SOURCE: Hughes, Stabenfeldt, and Evans (1972b).

TABLE 15-5
Multiple Ovulations, by Month

Month	No. per Month	Percentage per Month
January	3	23.5
February	4	22.2
March	10	40.6
April	9	25.7
May	10	32.3
June	10	25.5
July	9	25.0
August	7	21.2
September	8	24.2
October	7	20.0
November	6	17.1
December	4	18.2

SOURCE: Hughes, Stabenfeldt, and Evans (1972b).

ovulations do not upset the rhythmicity of the estrous cycle. The total length of the cycle as well as the length of estrus and diestrus are not affected. Double ovulations are rare in pony mares.

The average interval between the multiple ovulations is about 1 day. However, they may occur on the same day or as much as 5 days apart. The ovulations may occur on one ovary or on both ovaries. The distribution between the two ovaries is about equal (Table 15-4). Multiple ovulations occur more frequently during late spring and early summer (Table 15-5).

From a management viewpoint, a mare having two ovulations is not bred before the first ovulation. If at least 48 hours elapse between ovulations, the mare can be bred immediately before the second ovulation. If necessary, she can be bred shortly after the second ovulation. This procedure is based on the length of time the ovum is fertile. The normal period is 8 to 24 hours after ovulation. To prevent twins, one must wait until the first ovum is no longer fertile before the mare is bred.

At the opposite extreme, sometimes a mare goes through a heat period and fails to ovulate (Figure 15-19). About 25 percent of all open mares do not ovulate during at least one heat period in a year's time. Almost all the estrous periods without ovulation occur during the winter months—the mare is usually going into or coming out of the anestrous period. Others occasionally show heat at any time during the anestrous period but the heat period is not accompanied by ovulation. Even though they show heat, very few of the mares' ovaries have any significant follicular activity during winter anestrus. Fortunately, the incidence of estrus without ovulation is not common. It has been observed in about 3 percent of the estrous cycles, almost always in the winter anestrous period.

Mares normally ovulate only during the estrous period, but occasionally they ovulate during the diestrous period (Figure 15-24). They do not show any behavioral signs of estrus. It is doubtful that most such mares conceive if bred because the reproductive tract is probably not prepared to receive the fertilized egg. However, one mare has been reported to have conceived during a diestrous ovulation.

Foal Heat

A mare usually has her first estrous period about 7 to 9 days after foaling. However, there is considerable variation; she may show signs of estrus 1 or 2 days after foaling or as long as 50 days. Very seldom is the period longer than 12 days. She is in estrus normally for about 3 days but often for as long as 12 days. Ovulation is expected to occur on the eighth to tenth day after foaling.

Mares can be successfully bred during the foal heat but the breeder must be very careful. Only those mares that had a normal delivery should be bred. Seven days after foaling, the mare should be examined by a veterinarian. The mare should not be bred if she has a bruised cervix, lacerations or tears in cervix or vagina, vaginal discharge, placenta retained more than 3 hours, lack of tone in the uterus or vagina, or presence of urine in the tract. If there are any other questions about breeding her, it is best to wait. In many instances, the mare can be bred if she ovulated 2 or 3 days later than the normal, expected time. The reproductive tract needs that extra time to return to a breedable condition. If time of conception is extremely important, one may use prostaglandins on the sixth day after the foal heat ovulation (see section on estrous cycle manipulations).

During lactation, some mares do not show estrus to a stallion but do ovulate at the expected times. They may or may not exhibit foal heat and then may not exhibit estrus while suckling their foals. Several things may cause this lactation anestrus. Some mares are very possessive or protective of their foals so that they take the foal away from the teaser and will not approach the teaser. Such mares are usually ovulating on a regular schedule. A spontaneously prolonged corpus luteum may have developed after the foal heat ovulation. If so, the mares respond to prostaglandin treatment.

Estrous Cycle Manipulation

In several circumstances, the estrous cycle or some phase of it must be manipulated. Mares in anestrus must be stimulated to start cycling. Winter anestrus has been previously discussed, as has the spontaneously prolonged corpus luteum. The use of prostaglandins and saline infusions was discussed for the spontaneously prolonged corpus luteum condition.

Human Chorionic Gonadotropin

Human chorionic gonadotropin (HCG) is a hormone used to help control ovulation. HCG mimics LH activity and is commonly used to get a follicle to ovulate after it has reached ovulatory size but has failed to ovulate as expected. Following administration of HCG, most follicles ovulate within 24 to 48 hours. Some farms use HCG on an extensive basis. Each mare receives HCG on the second day of estrus and is bred the third day, and ovulation occurs 24 to 48 hours after the injection. This procedure can reduce the number of times that a mare needs to be bred. If used in conjunction with palpation, most mares are bred once each estrous cycle. Extensive use of HCG may cause antibodies to develop against it. The antibodies are specific for HCG and do not cross react with endogenous LH. Therefore, one may not get the full benefit of an HCG injection but it will not interfere with the activity of the endogenous LH. In one study of 12 mares receiving HCG on the day when a 35-mm follicle was present for six consecutive estrous cycles, 5 mares developed antibodies against HCG but all 12 mares ovulated each time within 48 hours after the injection (Figure 15-26). The effects of the antibodies on pregnancy are still being studied.

Prostaglandins

Prostaglandins can be used to control the time of ovulation by controlling the life span of the corpus luteum. Within 48 hours after the injection, the progesterone concentration in the blood is minimal. The mare comes into estrus in about 3 to 4 days and ovulates 6 to 12 days after treatment (Figure 15-27). Prostaglandins can be used to breed a mare earlier than the next estrous period following the foal heat period (Figure 15-28). If treated on the sixth day after foal heat ovulation, she comes into estrus in 3 to 4 days and ovulates within 6 to 12 days after treatment. This method is useful for mares that were not quite ready to be bred during their foal heat. The time of treatment after ovulation is important, because, from the sixth day after ovulation to the endogenous release of prostaglandin, the corpus luteum is sensitive to prostaglandin treatment. Less success is achieved if prostaglandins are used on the fourth day or fifth day after ovulation. After the fifth day, the success rate is approximately 65 to 80 percent. After the prostaglandin treatment, six types of responses arise of which one must be aware. Complete luteolysis (cessation of corpus luteum function) occurs for about 65 percent of the treated cycles; incomplete luteolysis occurs for 26 percent. Incomplete luteolysis, with subsequent recovery by the corpus luteum, occurs in about 6 percent of treated cycles. Very few (about 3 percent) corpora lutea fail to show any type of response. Prostaglandin treatment of a prolonged corpus luteum life span is about 90 percent effective. Therefore, a mare may fail to respond to a treatment

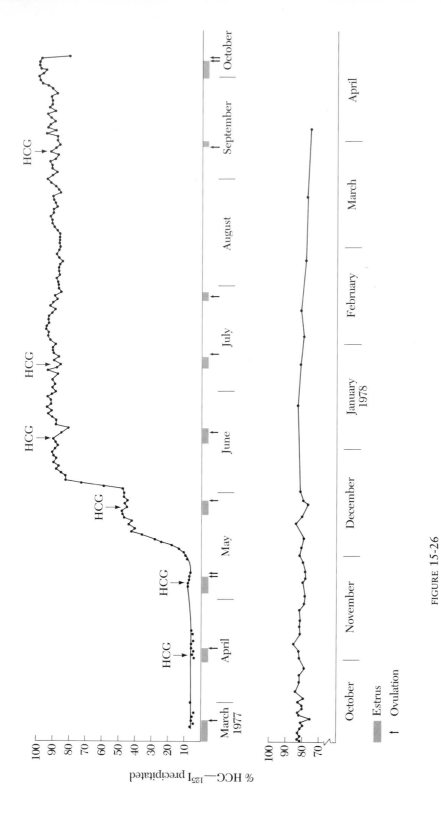

FIGURE 15-26

Time course of anti-HCG antibody development following treatments with HCG. (Courtesy J. Roser and J. W. Evans.)

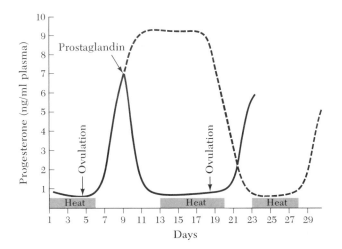

FIGURE 15-27

Effect of administering prostaglandin to shorten the length of the estrous cycle.

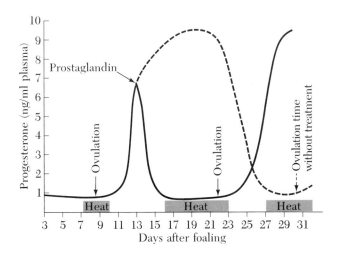

FIGURE 15-28

Effect of administering prostaglandin to breed the foaling mare before her normal second estrous period.

but usually responds to a second treatment. Prostaglandins provoke ovulation within 4 days after treatment in about 32 percent of treated cycles (Figure 15-29), and about 25 percent of ovulations occur without behavioral signs of estrus (Figure 15-29). These two responses occur primarily when prostaglandin is given in the presence of a large follicle on the ovary. Therefore, it is very important to know the condition of the ovary before prostaglandin treatment and to follow very closely ovarian changes by palpation.

For an artificial insemination program, prostaglandins can be used to synchronize the estrous cycle of mares so that they can all be bred in a short

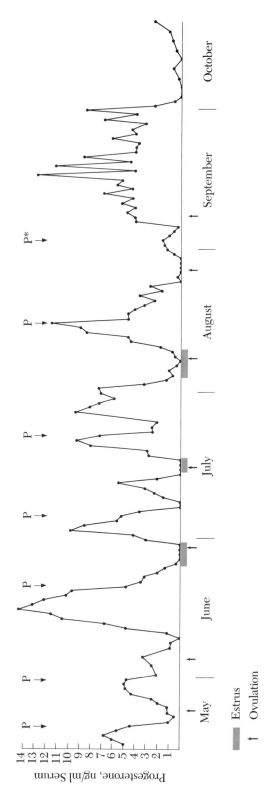

FIGURE 15-29

Luteolytic responses of a mare to sequential administration of a prostaglandin analogue (prostalene). Incomplete luteolytic responses followed prostalene injections on May 31 and August 15. Incomplete luteolysis with subsequent luteal recoveries occurred after prostalene injections on July 5 and July 22. Failure to manifest estrus was observed after prostalene treatments on May 21, May 31, and August 15. Ovulation within 4 days after treatment followed the prostalene injection on May 21. The posttreatment control cycle contained a prolonged luteal phase. P* is an injection of prostalene (4 mg). (From Kiefer and others, 1979.)

period of time. The treatment program is to give all the mares a prostaglandin treatment on Day 0, HCG on Day 6, prostaglandin on Day 14, HCG on Day 20, and breed on Days 19 and 21. About 75 percent of the mares ovulate during the 3- to 4-day period.

Prostaglandins may not prompt the pseudopregnant mare to cycle again if the pseudopregnancy is caused by an abortion. If the mare aborts soon after the endometrial cups are stimulated to form and secrete pregnant mare serum gonadotropin (PMSG), the PMSG suppresses ovarian activity in this type of pseudopregnancy and the corpus luteum of pregnancy may regress. The mare will not come into estrus until after the PMSG disappears from the blood—about 120 days. Saline infusions or prostaglandin treatments are not successful because no corpus luteum is present. If the corpus luteum is still present, prostaglandin or saline treatment will cause it to lyse but the mare may remain anestrous.

Light and Hormone Treatments for Winter Anestrus

The winter anestrous mare can be managed to have rhythmic estrous cycles before the normal breeding season by using an artificial light regime. The artificial light regime is started approximately 90 days before the start of the breeding season. The mares are exposed to 16 hours of light and 8 hours of darkness each day. Turn lights on at 6 A.M. and off at 10 P.M. Use an automatic timer to prevent control failure. Lights can be fluorescent or incandescent. A 200-watt incandescent bulb provides about the right amount of light in a 12-foot × 12-foot stall. It is not necessary to control the temperature so the mare does not need to be kept in a box stall. For outside paddocks, lights can be mounted on poles or a lighted shed can be used. About 50 percent of the mares that go into winter anestrus thus become ready to breed at the start of the breeding season. Because a few mares are polyestrous, a large percentage of the open, barren, and maiden mares are thus ready to breed at the start of the breeding season.

Another method has been developed to induce one estrous cycle in the anestrous mare, but further work is necessary before it has widespread practical use. Mares are treated with gonadotropin-releasing hormone and progesterone to mimic the changes in FSH, LH, and progesterone during a normal cycle. Following treatment, the mare will have one fertile estrous cycle.

Records

One of the most important aspects of a stud manager's duties is to keep an accurate set of records for the breeding farm. It is critical for the stud manager to know the reproductive status of each mare each day. Each breed and registry

Breeding Farm Name
Address
Phone Number

ARRIVAL

Time _____   Date _____

HORSE _____   Color _____ Sex _____ Age _____

OWNER _____   Address _____

_____

Method of Arrival _____   From _____

_____

Appearance of Horse _____   Tatoo Number _____

Equipment _____

Instructions _____

_____

Remarks _____

_____

Signed _____   Signed _____
          (Owner or Agent)   (Stud Manager)

Copy 1–Owner   Copy 2–Horse Barn   Copy 3–Accounting   Copy 4–Farm Manager

FIGURE 15-30

"Arrival" form, to be filled out on arrival of mare to the breeding farm.

association has established a specific set of records that the breeding farm must keep. The information required to complete the forms can be collected from a basic set of farm records.

When a mare is delivered to the breeding farm, fill out an arrival form and ask the person delivering the mare to sign it (Figure 15-30). The form shows information concerning the mare's name, description, age, owner's name and address, how she arrived, appearance, tattoo number, equipment with her, and any other pertinent remarks. Place a numbered neck band around the mare's neck, and enter the ID number on the arrival form. Some farms take a Polaroid snapshot of the mare for identification purposes and to record the mare's physical condition on arrival. Send a notice of the arrival to the mare's owner (Figure 15-31), and use the same form to obtain any further needed information about the mare. When the mare leaves after the breeding season, fill out a departure form (Figure 15-32), to be signed by the person picking

Breeding Farm Name
Address
Phone Number

NEW ARRIVAL REPORT

TO: _____   Date: _____

We have received _____ , a _____ , _____ .
                            (Name)   (Color)   (Sex)

To facilitate our records, please fill out applicable information below and return promptly.

Age _____ Breeding: Sire/Dam _____

Sire of Foal _____ Foaling Date _____

Date of Last Service _____ Booked to _____

Shots	Yes	No	Date
Encephalomyelitis	☐	☐	Date Last Wormed _____
Virus Abortion	☐	☐	
Tetanus Toxoid	☐	☐	Date Last Cultured _____
Flu Vaccine	☐	☐	

Insurance _____ (Yes/No) Who to contact? _____

Bills and reports should be mailed to: _____

_____ Phone No. _____

Manager's or trainer's instructions _____

_____

All due precautions will be taken, but under no circumstances will stable accept any responsibility for accident or disease.

FIGURE 15-31

"New Arrival Report," to notify owner of receipt of mare and to obtain additional information about her.

Breeding Farm Name
Address
Phone Number

DEPARTURE

Time _____

Date _____

HORSE _____ Color _____ Sex _____ Age _____

OWNER _____ Address _____

_____

Shots	Yes	No	Date
Encephalomyelitis	☐	☐	_____
Virus Abortion	☐	☐	_____
Tetanus Toxoid	☐	☐	_____
Flu Vaccine	☐	☐	_____
VEE	☐	☐	_____

Date Last Wormed _____

Oiled for Trip _____ (Yes/No)

Tranquilized for Trip _____ (Yes/No)

Veterinarian _____

Method of Departure _____ To _____

Appearance of Horse _____ Tatoo Number _____

Remarks _____

Signed _____ Signed _____
(Owner or Agent) _(Stud Manager)_

Copy 1–Owner   Copy 2–Horse Barn   Copy 3–Accounting   Copy 4–Farm Manager

FIGURE 15-32
"Departure" form, to be filled out when mare departs from breeding farm.

Name _____ # _____   Sire _____   Dam _____
Reg. No. _____   Reg. No. _____   Reg. No. _____
Breed _____   Year Foaled 19 _____   Markings _____
Characteristics _____

Date Foal	Sex	Name	No.	Sire	No.

ESTROUS CYCLE RECORD

Year 19   Bred to:

Mo. Day	J	F	M	A	M	J	J	A	S	O	N	D
1												
2			O									
3			O/H									
4			H/O									
5			H									
6			H–B									
7			H–X									
8		O	H/O									
9		O/H	O/H									
10		O/H	O	pg								
11		H										
12		H										
13		H										
14		H										
15		H										
16												
17		H–X										
18		H										
19		O										
20												
21												
22												
23												
24			O									
25			O/H									
26			H									
27			H–B									
28			H–X									
29			O									
30												
31												

Year 19   Bred to:

Mo. Day	J	F	M	A	M	J	J	A	S	O	N	D
1												
2												
3												
4												
5												
6												
7												
8												
9												
10												
11												
12												
13												
14												
15												
16												
17												
18												
19												
20												
21												
22												
23												
24												
25												
26												
27												
28												
29												
30												
31												

O—out of heat.   H—in heat.   pg—positive pregnancy test.   B—bred to stallion.   O/H—coming into heat.   H/O—going out of heat.   X—ovulated.

FIGURE 15-33

"Estrous Cycle Record," to keep a daily record of mare's reproductive status.

509

Date _____ 3/5 _____ Date _____

Cervix ____ slightly relaxed _____ Cervix _____

Uterus _____ Uterus _____

Left ovary ___ active _____ Left ovary _____

Right ovary __ 25mm anterior pole ____ Right ovary _____

Spec. _____ Spec. _____

Treatment _____ Treatment _____

Date _____ 3/6 _____ Date _____

Cervix _____ relaxed _____ Cervix _____

Uterus _____ slight tone _____ Uterus _____

Left ovary ___ 25mm follicle _____ Left ovary _____

Right ovary __ 45mm follicle _____ Right ovary _____

Spec. _____ Spec. _____

Treatment _____ Treatment _____

Date _____ 3/7 _____ Date _____

Cervix _____ relaxed _____ Cervix _____

Uterus _____ slight tone _____ Uterus _____

Left ovary ___ 25mm _____ Left ovary _____

Right ovary __ corpus luteum–ovulated Right ovary _____

Spec. _____ Spec. _____

Treatment _____ Treatment _____

FIGURE 15-34

Form to record changes in reproductive tract observed during female genital organ examinations.

up the mare. It also shows information regarding her health care and treatment. Maintain a daily record of the mare's reproductive status (Figure 15-33). Record each day the mare is in estrus. When the mare is coming into estrus, enter the symbol *O/H* (out of heat/heat) and when the mare is in estrus enter *H*. As the mare goes out of heat enter *H/O* (heat/out of heat). Each day the mare is bred, enter *B*. Mark the ovulation date with *X*. Enter the results of all palpations on another form (Figure 15-34). Keep a similar set of records for the ranch-owned mares. When the mare is bred, make an entry on the stallion record form (Figure 15-35). Keep one of these forms for each stallion. To follow the health status and to maintain an accurate record of preventative medical procedures, keep a clinical record and a log of procedural activities for each mare (Figure 15-36). For ranch-owned mares, use a different form

STALLION RECORD   Page _____

Name _____   Breed _____   Markings _____

No. _____   Date Foaled _____

DATE	MARE	REG. NO. of MARE	BREED	COMMENTS

FIGURE 15-35

"Stallion Record," to record mares bred to a stallion.

for each mare (Figure 15-37) as well as a general record (Figure 15-38). These records are useful when veterinary services are required and increase breeding efficiency. After the mare is bred and had a positive pregnancy examination, send a postcard to notify the mare owner (Figure 15-39). If a mare is foaled out on the farm for a client, use a foal report form to notify the owner (Figure 15-40).

CLINICAL RECORD

Name _____ Date Foaled _____ Markings _____

No. _____ Sex _____ _____

DATE	EXAMINATION	TREATMENT AND COMMENTS

FIGURE 15-36

"Clinical Record," to record veterinary medical services required by the horse.

Breeding Farm Name
Address
Phone Number

FOALING RECORD for

_____

(Mare's Name)

Year _____ : Sex _____ Color _____ Sire _____

Comments:

Year _____ : Sex _____ Color _____ Sire _____

Comments:

Year _____ : Sex _____ Color _____ Sire _____

Comments:

Year _____ : Sex _____ Color _____ Sire _____

Comments:

Year _____ : Sex _____ Color _____ Sire _____

Comments:

FIGURE 15-37
"Foaling Record" for an individual brood mare.

19____ FOALING RECORD

MARE	DUE	FOALED	GESTATION LENGTH	SEX	WEANED	MARKINGS	STALLION

FIGURE 15-38

"Foaling Record" for all foals foaled during the year.

Breeding Farm Name
Address
Phone Number

Date _____ , 19 _____

HORSE OWNER: _____

The mare _____ bred to _____

last service date _____ was examined for

pregnancy on _____ and has been pronounced:

☐ IN FOAL   ☐ BARREN

Signed: _____

FIGURE 15-39

Notice informing owner that mare is pregnant or barren.

Breeding Farm Name
Address
Phone Number

FOAL REPORT

TIME FOALED _____   **DATE** _____

☐ Bay   ☐ Black   ☐ Chestnut   ☐ Dr. Bay or Br.   ☐ Colt   ☐ Filly

Sire/Dam _____

Owner _____

Address _____

Condition After Foaling _____

Approximate Markings _____

_____

_____

Medical Treatment _____

_____

Equipment _____

Signed _____

FIGURE 15-40

"Foal Report," informing owner that mare has foaled.

BREEDING

The procedures for breeding a mare are basically the same for natural breeding as for artificial insemination except for the actual act of breeding. Tease the mare to determine her receptivity and then clean her external genitalia. Clean the stallion's penis. If the mare is to be bred by artificial insemination, collect the stallion's semen with an artificial vagina.

Mare Preparation

After determining that the mare is ready to be bred, the first consideration is hygiene. Immediately before she is washed, tease her so that she will urinate and empty her bladder. Restrain the mare in a palpation or breeding chute to

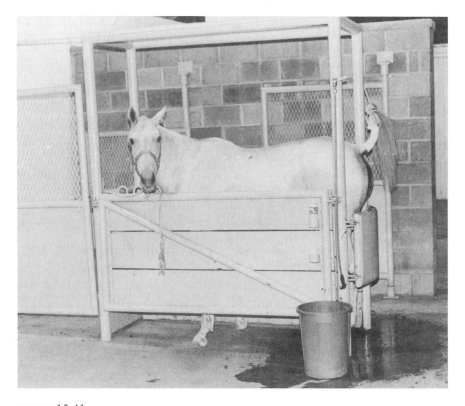

FIGURE 15-41

Restraint chute used during reproduction examinations, breeding hygiene procedures, and medical treatments. (Photo by B. Evans.)

be washed (Figure 15-41). Wrap the mare's tail with a roll of 3-inch gauze or some other disposable wrapping such as a disposable, long-sleeved plastic glove, to prevent hair from being taken into the vagina by the penis during breeding. The hair may be contaminated with infectious contaminants that may cause uterine infections or abortions. Hair also causes hair burns on the penis, making it sore enough to stop the stallion from breeding. Wash the mare's external genitalia and the surrounding area with warm water and a mild soap (Figure 15-42). Use a disposable plastic bag to line the inside of the bucket, to prevent the spread of infectious agents to other mares. Use Ivory soap or other mild antiseptic soap solutions to wash the mare. After the area is clean, rinse off all the soap. Soap is spermicidal, and it causes irritation if it is not removed. Dry off the washed area with a disposable paper towel.

To protect the stallion during natural breeding, the mare is usually restrained. A lip twitch may be used if the mare will not stand still when the stallion is approaching. If the mare is prone to kick, she can be restrained with a set of breeding hobbles (Figure 15-43). The hobbles should have a quick-release snap so that if the stallion steps over the rope it can be released without injuring the horse. Another method of restraint is to tie up a front leg and release it as the stallion is mounting. A mare's hindshoes should be removed if she is shod, or padded boots can be placed over her hindfeet.

Some stallions bite the neck of mares when they are breeding. A horse blanket and hood will protect the mare from being scratched and scraped. This is important if the mare is to be shown later in horse shows.

If a small maiden mare is being bred, the stallion may injure the cervix or vagina. To help prevent tears, lacerations, and bruises, a breeding roll may be used to restrict entry by the stallion (Figure 15-44). A breeding roll is a padded roll about 4 to 6 inches in diameter (10.2 to 15.2 cm) and about 18 inches long (457.2 mm). A handle on one end allows easy insertion of the roll between the mare's rump and the stallion just above the penis as the stallion mounts. To prevent the spread of infection, the roll should be covered with a disposable wrapper such as a plastic examination sleeve.

As a safety precaution when there is a marked difference in the physical size of the mare and stallion, the mare may need to stand with her hindlegs on a small mound or in a slight depression. If a stallion is too small, he may repeatedly mount and dismount the mare, which increases the chance of injury.

Stallion Preparation

Allow the stallion to tease a mare until he drops his penis. Then take him to an area a short distance away to be cleaned. Hold his penis with one hand and wash it with a mild soap and warm water (Figure 15-45). A terry washcloth or strip of sheet cotton can be used to lightly scrub the penis. The prepuce must

FIGURE 15-42

Breeding hygiene equipment and procedures. *(a)* Equipment: (A) A bucket lined with a disposable plastic bag and containing warm water. (B) Septi-Soft® soap. *(C)* Squeeze bottle with Septi-Soft® soap solution. (D) Roll of cotton. (E) Plastic cup to rinse off area with water. (F) Plastic liners for bucket. (G) Tail wrap. (H) Paper towels. *(b)* Wrap tail with 3-inch gauze (0.52 cm). *(c)* Tie up tail out of the way. *(d)* Rinse buttocks with water from hose or with a cup of water from a bucket. *(e)* Wash buttocks with soap several times depending on how dirty the mare is. *(f)* Rinse off soap. *(g)* Remove debris from lips of the vulva. *(h)* Dry buttocks with a paper towel. *(i)* Mare is clean. (Photos by B. Evans.)

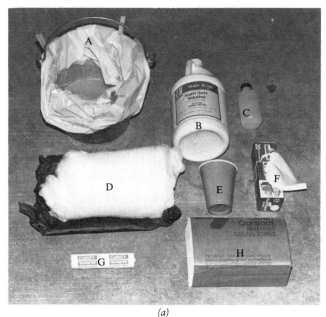

(a)

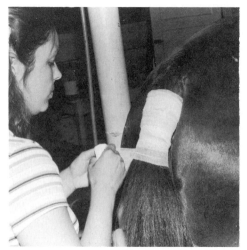

(b)

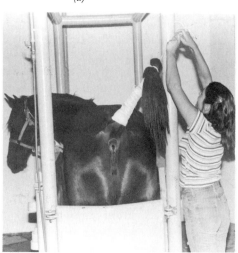

(c)

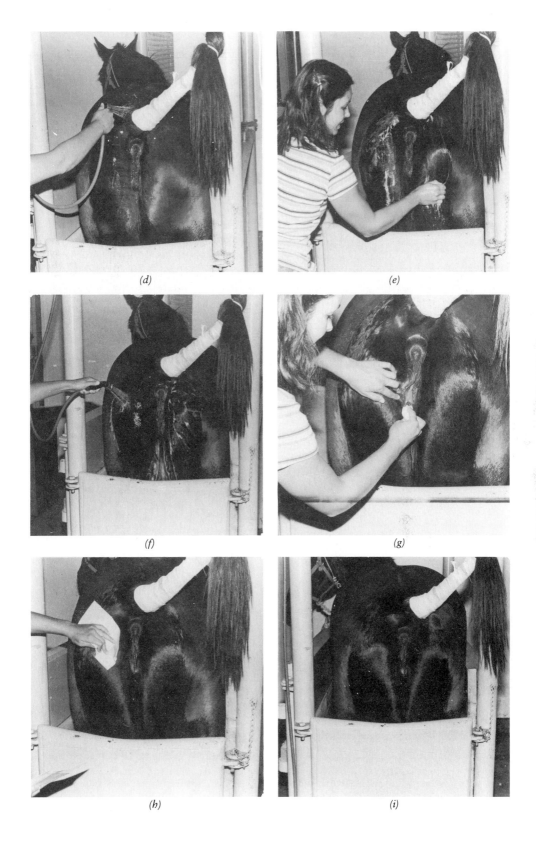

(d)

(e)

(f)

(g)

(h)

(i)

FIGURE 15-43

Preparing the mare for breeding. Hobble the mare with a set of breeding hobbles that prevent her from kicking the stallion. Use a lip twitch to prevent the mare from trying to kick or move about. Wash the area around her genital organs and wrap her tail with sterile gauze to prevent transmission of venereal diseases or infections. (Photo by B. Evans.)

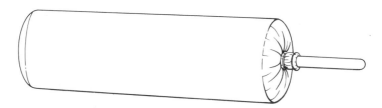

FIGURE 15-44

Breeding roll used to protect maiden mares from excessive penetration by the stallion's penis. (Illustration by M. Morris.)

FIGURE 15-45 *(facing page)*

Hygiene procedures for stallion. *(a)* Equipment: (A) Paper towels. (B) Squeeze bottle with mild soap solution (Septi-Soft®). (C) Plastic water cup. (D) Plastic liners for water bucket. (E) Roll of cotton. (F) Bucket lined with disposable plastic bag and filled with warm water. *(b)* Apply soap to penis with squeeze bottle. *(c)* Wash penis with wet cotton. *(d)* Wash end of penis carefully. *(e)* Rinse off soap. *(f)* Blot penis dry with paper towels. (Photos by B. Evans.)

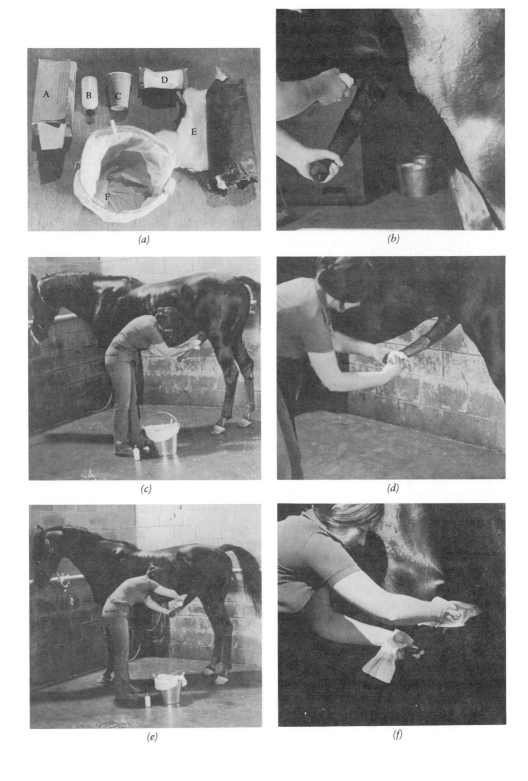

(a)

(b)

(c)

(d)

(e)

(f)

be thoroughly cleaned. Be careful to remove all dirt, debris, and smegma. The fatty secretions from the glands in the prepuce, along with epithelial cells, form the smegma. The urethral diverticulum in the glans penis must be thoroughly cleaned, because it is common to find a small "bean" of smegma in this area. After the penis is thoroughly clean, remove all soap by rinsing with clean, warm water. Then dry the penis with disposable paper towels or a Turkish towel. After the stallion has bred a mare, rinse his penis with a mild antiseptic solution.

Before the breeding season, clean the stallion several times. If this is not done, he will usually be very dirty, and the penis will become irritated if he is thoroughly cleaned the first time. Because of the soreness, he may refuse to breed mares. The three or four cleaning sessions before breeding season also train the stallion to behave himself while he is being cleaned.

Breeding Area

The breeding area should be free from dust and have a surface that allows good footing. It can be outside, on a grassy area, or enclosed. It should be at least 40 × 40 feet in size (12.2 × 12.2 m) so that if problems develop the mare and stallion can be safely separated. It is advisable to have the palpation or examination chute and the wash area located in another adjacent area. This prevents the breeding area from becoming too wet from the water used to wash the mare and stallion.

Natural Breeding

After preparing the stallion, mare, and breeding area, position the mare in the middle of the area. The mare's handler usually stands on the mare's left side opposite the head and facing her hindquarters (Figure 15-46). The stallion is led so that he approaches the mare at a slight angle to her hindquarters on the left side. If he is not ready to mount, he can tease the mare in the flank for a moment. This way, the mare can easily see his approach and is not startled by his presence. As the stallion mounts, the stallion handler can help guide the penis into the mare's vagina. The stallion handler should observe the stallion's tail for the characteristic "flagging movements" during ejaculation or hold the base of the penis to feel the ejaculatory pulses. After ejaculation, the stallion will dismount, and he should be immediately separated from the mare.

If a good cover was obtained, the semen was deposited in the uterus of the mare. It is not necessary to walk the mare to prevent the semen from draining out. If a considerable volume of semen runs out of the vagina after the stallion dismounts, a good cover was not obtained and the mare should be bred again in about 1 hour.

FIGURE 15-46

Natural breeding of mare by a well-mannered stallion. Note the position of the mare and stallion handlers. (Courtesy William E. Jones, D.V.M.)

Artificial Insemination

Artificial insemination has some advantages in a breeding program. Venereal disease control is accomplished by diluting the semen with antibiotics and by decreasing the number of disease-producing organisms placed in the mare. The chance of injury to horses and personnel is also reduced, particularly if the stallion's semen is collected with the aid of a phantom. And overuse of the stallion can be prevented, because each collection is usually enough to breed several mares. Artificial insemination also permits evaluation of the semen for quality at periodic intervals.

When the mare is to be bred by artificial insemination, collect the semen as previously discussed. It can be processed two ways for insemination after it is collected with an artificial vagina. First determine the volume of semen to inseminate that contains sufficient spermatozoa for conception. Each mare should be inseminated with at least 100×10^6 and preferably with 500×10^6 actively motile spermatozoa. To determine the volume containing 500×10^6 actively motile spermatozoa, use the following equation:

$$\text{Milliliters} = \frac{500 \times 10^6}{\text{Concentration per ml} \times \frac{\% \text{ motility}}{100}}$$

A typical example for an ejaculate containing 250×10^6 spermatozoa per milliliter that is 80 percent motile is as follows:

$$\text{Milliliters} = \frac{500 \times 10^6}{250 \times 10^6 \times \frac{80}{100}} = 1.5$$

Fresh semen without any diluters gives excellent results even when only 1.5 to 2.0 ml is used, providing there are 500×10^6 actively motile spermatozoa.

Diluters or extenders are added to the semen to permit antibiotic and antibacterial treatment and to enhance sperm survival. One of the most popular diluters or extenders is called cream-gel (see Appendix 15-1 for preparation instructions).

Prepare the mare by wrapping her tail and washing the perineal area. The fresh extended semen is contained in a disposable syringe maintained at 100°F (38°C). The syringe is connected to a sterile insemination catheter (Figure 15-47). The inseminator places his or her hand and arm in a long-sleeved plastic glove. Over the plastic glove, a sterile surgeon's glove is placed over the hand. A small amount of sterile lubricant (K-Y jelly) is placed on the gloves. The hand and arm is inserted in the vagina, and one finger is passed through the cervix. The hand is withdrawn slightly and the insemination catheter is inserted into the vagina under the palm of the hand. If necessary, a speculum can be used to find the cervix. The tip of the catheter is covered by the tip of a finger and passed through the cervix. The semen is deposited into the uterus and the catheter withdrawn.

Insemination or breeding should be done just before ovulation. Spermatozoa retain their capacity for fertilization for at least 72 hours after deposition in the mare's reproductive tract. There are documented cases in which breeding has preceded ovulation by 5 days and conception still occurred. However, after

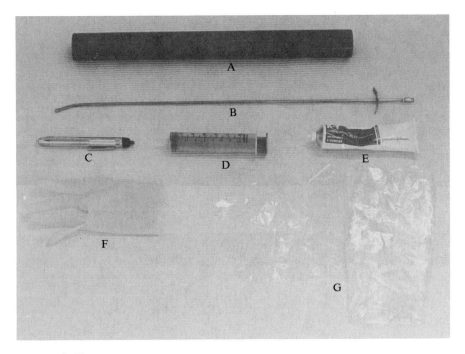

FIGURE 15-47

Artificial insemination equipment. (A) Vaginal speculum. (B) Chamber's catheter. (C) Pen light. (D) Disposable syringe. (E) Sterile K-Y lubricant. (F) Disposable sterile surgeon's glove. (G) Disposable plastic sleeve. (Photo by B. Evans.)

72 hours the conception rate decreases rapidly. The ovum remains viable for conception for a period of 8 to 24 hours after ovulation. Therefore, from a practical standpoint the mare should be bred every other day during estrus if she is not palpated. Breeding should begin on the second day of estrus. If the mare is bred according to palpation and follicle growth, she usually needs to be bred one time just before ovulation. If the mare ovulates just prior to breeding, her chances of conception are reduced but she still may conceive.

Infertility Problems

The pneumovagina or "windsucker" condition is one of the most frequently observed causes of infertility. In this conformation of the perineal area, the anus is recessed or sunken. The lips of the vulva are not almost vertical but are more horizontal (Figure 15-48). In extreme cases, the lips of the vulva are almost horizontal. This type of conformation facilitates contamination of the vagina with fecal material. During estrus when the vulva is relaxed, air is aspirated into the vagina and uterus. These conditions are conducive for bac-

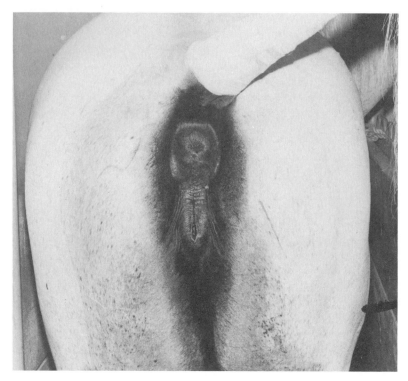

FIGURE 15-48
Perineal conformation that is prone to cause "wind sucking." (Photo by B. Evans.)

terial invasion of the vagina and uterus, and genital infections are common. Approximately 90 percent of mares that have a poor prognosis for a successful pregnancy have the pneumovagina condition. Therefore, evaluate perineal conformation before purchasing brood mares. To help prevent infections, the lips of the vulva can be sutured part of the way down to prevent fecal contamination and the aspiration of air. If the mare is artificially inseminated, the vulva does not have to be opened up for breeding. Care must be taken, however, to be sure the vulva is opened up before foaling.

Other infertility problems can be determined during a female genital organ examination to evaluate breeding soundness. Inside the vagina, one may observe scars, abrasions, ulcers, and other defects of the mucosa that may cause infertility. Infections, vaginitis, and cervicitis are frequently the result of urine pooling in the anterior vagina. When urine is being pooled, there is a marked downward and forward slope of the vaginal floor, which prevents the urine from flowing out. In older mares, this occurs as the broad ligament loses elasticity. Surgical correction to allow the urine to drain out is successful in a few cases.

Damage to the cervix area can also cause infertility. If the cervical muscles are torn during foaling and if the tear does not heal properly, the cervix does not seal during pregnancy to prevent bacterial invasions, which can cause abortion. Adhesions may develop that permanently close the vagina and prevent spermatozoa from entering the uterus.

Endometritis (infection of the uterus) is frequently found in mares that aspirate air. The air contains infectious contaminants that eventually overcome tissue resistance. The condition may become chronic, and in severe cases pyometria (pus in the uterus) may result. The uterus may contain as much as a gallon or more of purulent discharge. In these cases, the lining of the uterus is destroyed, and no prostaglandin is released to cause the corpus luteum to stop secreting progesterone. Therefore, such mares tend to have long periods of anestrus due to prolonged corpus luteal function. The most prevalent organisms causing uterine infections are Beta-hemolytic streptococcus, *E. coli*, and *K. pneumoniae*. They are detected by swabbing the cervix or uterus and culturing the specimen. The streptococci are sensitive to penicillin treatment, and the *E. coli* and *K. pneumoniae* are sensitive to chloraphenicol, gentamicin, and the polymyxins. A veterinarian should select and use the appropriate pharmacological agent.

The ovaries of mares may be underdeveloped or have become atrophied (small in size). The problem can be congenital. Ovarian cysts are numerous follicles on an enlarged ovary. The follicles fail to ovulate, so the mare is in constant estrus. Their presence requires diagnosis and treatment by a veterinarian. Tumors may be present in the uterus and block the pathway for spermatozoa. Granulosa cell tumors in the ovary cause different types of estrous behavior: anestrus, continuous sexual receptivity, or male behavior. Male behavior results from increased testosterone secretion by the tumors. Treatment is usually surgical removal of the affected ovary. The time between surgical removal of the ovary with the tumor and normal ovulation by the other ovary varies from 90 days to 2 years.

Embryo Transfer

Surgical and nonsurgical methods for recovering and transferring fertilized eggs are well established by several laboratories and allow good pregnancy rates. The nonsurgical method is easy to perform. The recovery of fertilized eggs is approximately 75 percent, and an overall pregnancy rate is about 45 percent. The initial step is to synchronize the donor and recipient mares with prostaglandins and HCG as previously described. On Days 6 to 8, the embryo is collected by flushing the uterus with physiological saline (TCM-199, available from BDH Chemicals, Ltd.). The embryo is then transferred to the uterus of the recipient mare.

Embryos can be collected from mares several times during the early part of the breeding season, and the mare can be allowed to carry her own fetus later in the season. The technique offers several advantages. First, a large number of foals can be obtained from genetically superior mares. Second, the embryos from mares that chronically abort can be transferred to recipient mares. Third, mares with injuries that would endanger their lives if they remained pregnant can be used as donors. The development of methods for inducing superovulation in the mare would enhance the value of the technique.

GESTATION

Length of Gestation Period

The average length of gestation usually ranges between 335 to 350 days for all breeds. The length varies considerably because it is influenced by sex of the foal, month of conception, and individual traits of the mare. A range of 300 to 385 days has been observed for Morgan horses. The average was 340 days, and the median was 342 days. Colts are usually carried 2 to 7 days longer than fillies, indicating that the sire influences gestation length through the genotype. The gestation period of mares carrying a mule fetus (jack and mare) is approximately 10 days longer than the gestation period of the same mare carrying a horse fetus. Mares foaling during the spring usually have a 4- to 5-day shorter gestation. This is probably related to nutrition, because mares on a high plane of nutrition foal approximately 4 days before mares on a low plane of nutrition. Therefore, the rate of fetal development could be affected by the nutritional status of the mare.

A foal born at 300 to 320 days is classified as premature. Those born after 360 days are classified as postmature.

Hormonal Changes

After the mare conceives, a sequence of changes occurs in the blood hormone concentrations (Figure 15-49). The corpus luteum of pregnancy continues to secrete progesterone until it regresses at 150 to 210 days. Because the accessory corpora lutea developed during the second and third month and the placenta produce progesterone, the concentration is high until 120 to 150 days. Then it declines to rather low concentrations until just before parturition. About 30 days before parturition, the blood concentration increases. After the placenta is passed, it declines very rapidly. The ovaries are not essential for progesterone production after 100 days, because they can be removed without subsequent abortion.

One of the most significant hormone changes is the secretion of large quantities of PMSG between 40 and 130 days of gestation. The PMSG is

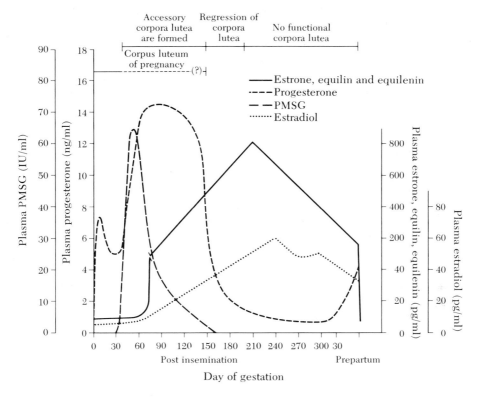

FIGURE 15-49

Endocrine pattern during gestation of a mare.

detectable as early as 35 days after conception. Maximum concentrations are attained at 55 to 60 days. After 130 to 170 days, it is undetectable in the blood. PMSG is secreted by the endometrial cups, but its function in the mare is not known. It predominantly stimulates follicles in effect, but it also possesses a luteinizing fraction. Therefore, it is also used to stimulate follicular growth and ovulation in other species.

Estrogen starts to increase in the blood at approximately 45 days of gestation. At 80 days of gestation, it starts to increase rapidly, and maximum concentrations are attained at 200 to 210 days. A gradual decline starts after 200 to 210 days until parturition, at which time a rapid decline occurs.

Because of these hormonal changes, management problems can arise. Pregnant mares may come into estrus and ovulate. This normally occurs between 30 and 50 days after conception. Before breeding the mare, it is advisable to give her a manual pregnancy examination if she skipped the expected estrus at 18 to 23 days after she was bred.

Placentation

Shortly after fertilization (4 to 6 days), the ovum migrates to the uterus. However, implantation does not occur for approximately 8 weeks. During the first 8 weeks, the chorion (Figure 15–50) is held next to the uterine mucosa by fetal fluid pressure. By 10 weeks, the uterine mucosa is penetrated by the chorionic villi, and by 14 weeks the attachment is complete. The type of attachment is epitheliochorial; that is, the chorion of the fetus is in contact with the epithelium of the mare's uterus.

The fetal placenta is formed by three membranes (Figure 15-50): chorion, allantois, and amnion. Villi, or fingerlike projections, are scattered over the surface of the chorion (outermost membrane), and for this reason the attachment is said to be "diffuse." The second membrane, allantois, lines the inside of the chorion and is fused with the amnion, thus forming the first waterbag, or *allantoic cavity*. The allantoic cavity is connected to the bladder by the urachus, which passes through the umbilical cord. If the urachus fails to close after the umbilical cord is broken after parturition, bacteria may invade the foal. The amnion is the innermost membrane and forms the second water bag, or *amnionic cavity*.

The umbilical arteries and veins form the umbilical cord. When they reach

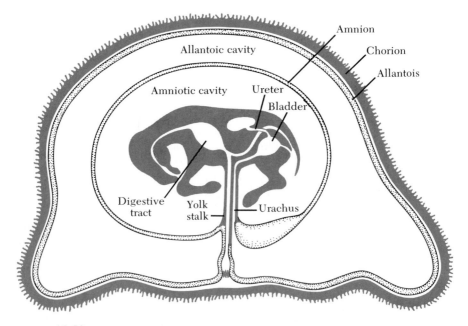

FIGURE 15-50

Fetus of horse within the placenta. The chorion and allantois make up the chorioallantois, also often called the allantochorion. (From Frandson, 1974; after Witchi, 1956.)

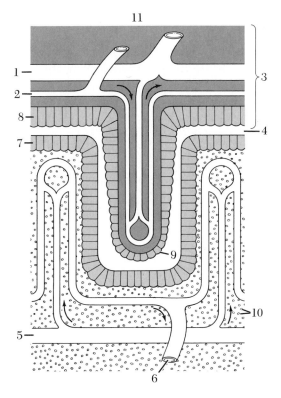

FIGURE 15-51

Diagrammatic representation of relationship between maternal
and fetal tissues in the epitheliochorial type of placenta (for horse).
(1) Fetal artery. (2) Fetal vein. (3) Allantochorion. (4) Uterine lumen.
(5) Maternal artery. (6) Maternal vein. (7) Uterine epithelium.
(8) Cytotrophoblast. (9) Villous. (10) Stroma. (11) Allantoic vesicle.
(From Harvey, 1959.)

the chorion, they spread out in the connective tissue between the allantois and chorion. The umbilical arteries and their tributaries carry unoxygenated blood and waste products from the fetus to the mare. The umbilical veins carry oxygenated blood and nutrients from the mare to the fetus. In general, blood from the two circulations do not mix, because at least six layers of tissue separate them (Figure 15-51).

The mare has a peculiar specialization: Endometrial cups begin to form on the third to sixth day of pregnancy. They form from fetal cells, and they start development prior to implantation. They form cuplike depressions in the chorion, which becomes filled with coagulum. Coagulum is an accumulation of degenerate epithelial cells, erythrocytes, leucocytes, and PMSG. The endometrial cups secrete PMSG.

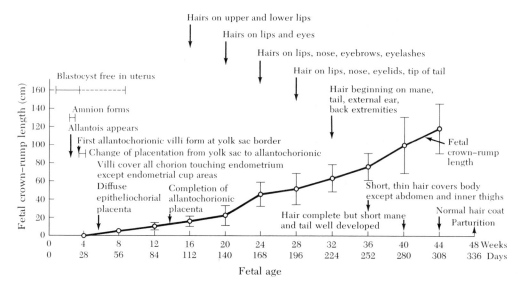

FIGURE 15-52

Fetal growth. (From Bitteridge and Laing, 1970.)

Fetal Growth

The growth and development of the fetus is illustrated in Figure 15-52. During the first two trimesters of pregnancy, fetal growth is slow. About 50 percent of the fetal growth (weight) occurs during the last 110 days of pregnancy.

Pregnancy Examination

Pregnancy in the mare can be determined by palpating the ovaries and uterus through the rectum and by determining the presence of PMSG or high concentrations of estrogen in the blood.

The quickest and most efficient method is by palpation. This is usually done by a veterinarian, but stud farm managers can learn the technique. Position of the ovaries and presence of the fetus help determine pregnancy by palpation. During the early stages of pregnancy and during a nonpregnant state, the ovaries are high in the sublumbar area. As pregnancy advances, the ovaries are displaced downward, forward, and medially. The most important indicators are the form, size, consistency, and position of the pregnant uterus and the presence of the fetus. The earliest stage that pregnancy can be positively determined is 28 to 30 days (Figure 15-53). At this stage, a bulge about 2 to 3 cm in diameter is present in the uterus, usually located at the base of one

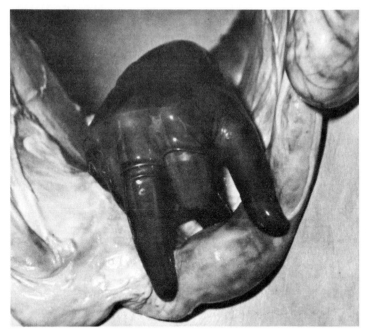

FIGURE 15-53

Left horn pregnancy, 28 to 30 days. Note the slight enlargement between the thumb and the middle fingers. (From R. Zemjanis, *Diagnostic and Therapeutic Techniques in Animal Reproduction*, Second Edition. The Williams & Wilkins Co., Baltimore. Copyright © 1970.)

horn. Transmigration of the fetus is common in the mare, so the fetus may be in the opposite horn of the uterus than the ovary in which ovulation occurred. The uterus has good muscle tone, and the cervix is tightly closed. At 5 weeks, the bulge is about 3 to 4 cm (size of a golf ball) (Figure 15-54). By 6 weeks, the bulge is 5 to 7 cm long and about 5 cm in diameter (Figure 15-55). The enlargement begins to involve the body of the uterus during the seventh week. It is 7 to 8 cm long and 6 to 7 cm in diameter. At this time, fluctuation of the fetal fluids is clearly palpable. The uterus is still in the pelvic area. At 2 months, the bulge is in the form of a football, 12 to 15 cm long, and 8 to 10 cm in diameter. About one-half of the enlargement is in the body of the uterus (Figure 15-56). The fetus continues to enlarge so that the entire uterus is involved by 3 months. The uterus starts to descend over the pelvic brim. The bulge is 20 to 25 cm long and 12 to 16 cm in diameter (Figure 15-57). To detect the fetus, use your hand with a blotting-type motion. During the 3- to 5-month period, the uterus is still partially located in the pelvic area, so the fetus can be felt without difficulty. Toward the end of the 5- to 7-month period, the uterus has

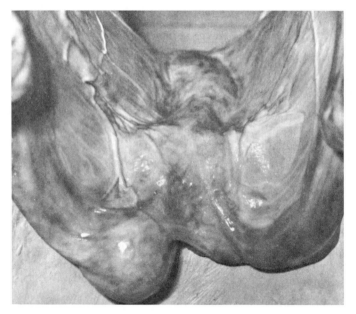

FIGURE 15-54

Right horn pregnancy, 35 days. (From R. Zemjanis, *Diagnostic and Therapeutic Techniques in Animal Reproduction,* Second Edition. The Williams & Wilkins Co., Baltimore. Copyright © 1970.)

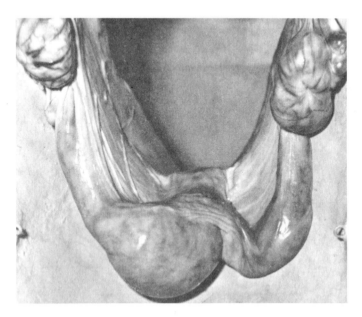

FIGURE 15-55

Right horn pregnancy, 42 to 45 days. (From R. Zemjanis, *Diagnostic and Therapeutic Techniques in Animal Reproduction,* Second Edition. The Williams & Wilkins Co., Baltimore. Copyright © 1970.)

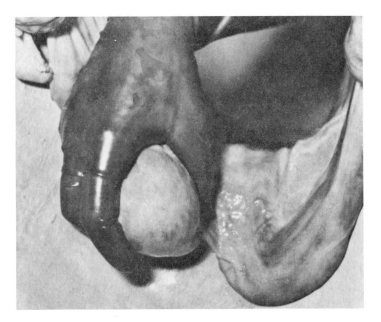

FIGURE 15-56

Right horn pregnancy, 48 to 50 days. (From R. Zemjanis, *Diagnostic and Therapeutic Techniques in Animal Reproduction*, Second Edition. The Williams & Wilkins Co., Baltimore. Copyright © 1970.)

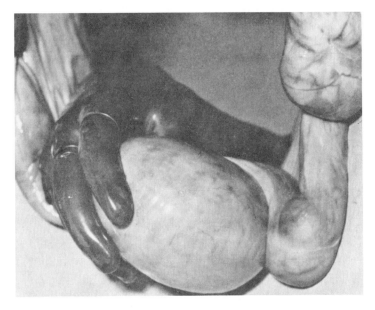

FIGURE 15-57

Right horn pregnancy, 60 to 65 days. (From R. Zemjanis, *Diagnostic and Therapeutic Techniques in Animal Reproduction*, Second Edition. The Williams & Wilkins Co., Baltimore. Copyright © 1970.)

completely descended. After 7 months, the pregnant uterus starts to ascend, and the fetus can be easily palpated.

The secretion of PMSG during the 35- to 40-day period to about 100 to 150 days can be used for a positive pregnancy test. The MIP-Test, manufactured for Diamond Laboratories, is a 2-hour immunological test for PMSG and is easy to perform. The test procedure is given in Appendix 15-2.

Estrogen concentration in the blood is determined by a chemical assay. However, it is very seldom used.

Abortions

Abortion is expulsion of the fetus before the time of normal foaling. There are many causes of abortions, and in some instances the cause cannot be determined. The common causes are bacteria, viruses, fungi, genetic abnormalities, and injuries.

Viral abortions are caused by viruses such as herpes, which can cross the placenta and infect the fetus. The fetus dies because of extensive damage to the liver and other organs. Herpes virus is discussed as rhinopneumonitis in Chapter 10.

Bacteria and fungi normally do not cross the placenta but cause abortions by disrupting the function of the placenta.

Bacterial abortions resulting from salmonella and streptococci occur during the latter half or during early pregnancy, respectively. Streptococci are usually introduced during breeding, so hygienic procedures are important for prevention. These bacteria account for about 25 percent of all abortions.

Genetic abortions occur when the fetus is malformed. For reasons unknown, the mare's body usually rejects the malformed fetus, and abortion occurs.

Injury-related abortions are rare, because the fetus is well cushioned by the structure of the fetal membranes. However, it is advisable to prevent injuries caused by kicking, bumping, or squeezing.

Signs

Abortions may be sudden in onset and may occur before any premonitory signs are manifest, or they may occur over a prolonged period of time. Prolonged abortions are usually accompanied by readily recognized symptoms. There is usually mammary development, and milk may stream from the udder for several days. Some mares may go through the normal symptoms exhibited prior to a normal foaling, including Stage 1 labor. The symptoms disappear for 1 to 2 days and then reappear. After two to three cycles, the fetus is expelled. Most fetuses are expelled without any difficulties but the placenta may be retained.

If so, the mare requires veterinary assistance for its removal. All abortions should be followed by a female genital organ examination by a veterinarian.

Precautions

After an abortion has occurred, the fetus should be examined by a veterinarian to determine the possible cause of abortion. During the period of time required for the analysis, the mare should be isolated to prevent the spread of a virus that may have caused the abortion. The mare's stall should be disinfected with steam or a tamed iodine solution, all contaminated bedding removed, and the afterbirth removed and disposed.

16
Care During
and After
Pregnancy

Proper care of the mare during and after pregnancy is critical if she is to produce healthy and well-developed foals. Improper care can result in abortion or the presentation of a dead or deformed foal. Poor care can also result in infertility or sterility caused by damage to or infection of the reproductive tract.

FOALING

Pregnant Mare Care Immediately Before Foaling

During the last 90 to 120 days prior to foaling, the increased nutrient requirements of the pregnant mare must be met. Failure to do so will result in excessive loss of condition, and the foal's birth weight will be reduced. Her feet and teeth also need proper care, and she should be treated for external parasites.

At about 30 days before foaling, the mare should be given a booster shot for tetanus (tetanus toxoid). This results in high antibody titers, so the foal receives adequate antibodies in the colostrum for protection against tetanus. Check the mare to determine if her vulva has been sutured. If it has, the vulva should be opened. Early removal of the sutures (opening the vulva) prevents an unexpected parturition in which the vulva is torn. Exercise should be continued if she is used to being exercised. However, exercise should not be strenuous during this period. Exercise maintains muscle tone.

When the mare approaches her time to foal, remove her shoes and bring her into a foaling stall or into a large, clean, grassy paddock. Mares foaled on clean grassy areas seldom have complications. They are brought into foaling stalls to receive close attention during foaling. The stall should be at least 12 × 14 feet (3.66 × 4.27 m). It should be bedded with straw or have a floor covered with synthetic material so that it can be kept clean. Straw that is clean, free of dust, and cut in long lengths is preferable. Wood shavings and other similar materials may stick to the wet nostrils of a newborn foal and suffocate it. Before the mare is placed in the stall, disinfect it with a tamed iodine disinfectant solution. The stall should be equipped with good lights and be free of any projections such as nails or broken boards.

Immediately before the mare foals, reduce her ration. This decreases the amount of material in the digestive tract and may prevent a foaling problem.

It is advisable to keep a foaling kit assembled and ready for use (Figure 16-1). It should contain two buckets, to be used for warm, clean water if

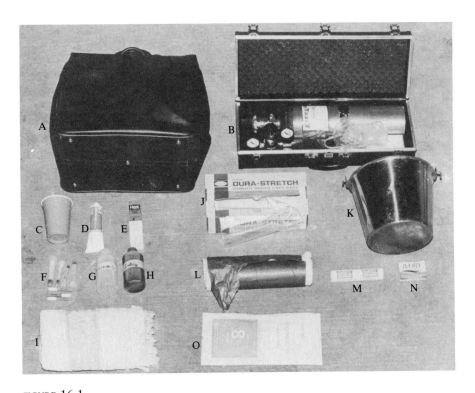

FIGURE 16-1
Foaling kit. (A) Storage bag. (B) Portable oxygen tank. (C) Water cup. (D) Sterile K-Y lubricant. (E) Fleet enema. (F) Needles and syringes. (G) Soap solution in squeeze bottle. (H) Iodine solution. (I) Turkish towel. (J) Plastic sleeves. (K) Bucket. (L) Roll of cotton. (M) Gauze roll (tail wrap for mare). (N) Ivory soap. (O) Sterile gloves. (Photo by B. Evans.)

trouble arises. A mild soap such as Ivory for cleaning the mare and for cleaning the attendant's hands if the mare requires help should be available. A mild disinfectant solution such as Nolvasan is desirable to keep instruments clean. Tail bandages such as 4-inch (10.2-cm) gauze bandages or leg wraps are used to wrap the tail prior to foaling during Stage 1 labor. A cup and a solution of 10 percent iodine tincture are used to treat the navel cord. Supplies (Fleet enema) for administering an enema are essential. Towels for drying the foal are handy. Obstetrical sleeves and a lubricant are used when it is necessary to reposition or examine the foal. Syringes and needles are used to administer antibiotics and/or tetanus antitoxin if necessary. A small tank of oxygen that fits in the palm of the hand, obtained from an emergency medical supply store, is valuable if artificial respiration is necessary. The flow should be 2 liters per minute for 5 minutes. One must be careful to avoid lung damage. Keep the supplies in a box that is easy to carry to the foaling stall.

Signs of Approaching Parturition

The signs of approaching parturition are very evident in some mares, but others show few, if any, signs. Some mares merely lie down and have their foals. The general indicators are as follows. Distension of the udder is usually the first indicator, and it occurs about 2 to 6 weeks before foaling. Approximately 7 to 10 days before foaling, the muscles in the croup area shrink markedly because the muscles and ligaments in the pelvic area generally relax. The teats fill out their nipples 4 to 6 days before foaling. A waxy secretion oozes out of the end of the nipples, building up in the nipples 2 to 4 days before foaling. Within 24 hours of foaling, the wax may drop off, and milk may start to drip. Finally, the mare becomes quite nervous, frequently raises her tail and passes a small amount of urine, appears to have cramps, breaks out in a sweat, and may walk anxiously around the stall. At this time, wash the perineal area with warm water and a mild soap. Wrap the tail to help maintain cleanliness and prevent the tail from becoming entangled with the foal.

Foaling

Foaling is divided into three stages (Figure 16-2). During Stage 1, the muscles in the uterine wall contract rhythmically. The contraction waves start at the anterior end of the uterus and move toward the cervix. As the contraction wave moves toward the cervix, it increases in strength. These contractions are responsible for the mare's anxious behavior. As the contraction forces get stronger, they force the allantois into the cervical canal and the vagina. Then it may appear through the vulva. Increasing pressure causes it to rupture, and 2 to 5 gallons of water (7.6 to 18.9 liters) are released. When it ruptures,

Stage 1 is completed. The time period for Stage 1 varies from a few minutes to several hours.

Stage 2 is the actual expulsion of the fetus from the uterus. After the allantochorial membrane ruptures, the contractions continue. At this time the

FIGURE 16-2

Delivery of a normal foal. Stage 1: *(a)* Mare rolling. *(b)* Mare looking at her side. *(c)* Mare sweating and kicking at her belly. *(d)* First appearance of the placenta. Stage 2 (delivery of foal): *(e)* Appearance of forelegs. *(f)* Body of foal passing through the vulva. Stage 3: *(g)* Forelegs tearing placenta. *(h)* Placenta torn. *(i)* Mare getting up. *(j)* Mare up and navel cord broken. (Photos by B. Evans.)

(a)

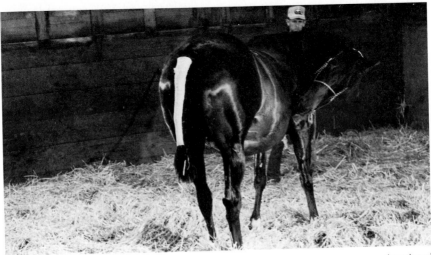

(b)

(continued)

FIGURE **16-2** *(continued)*

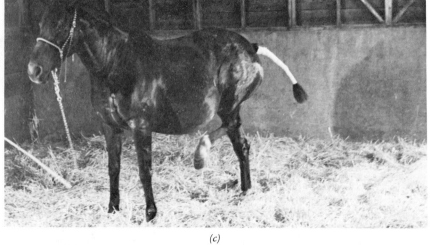

(c)

(d)

(e)

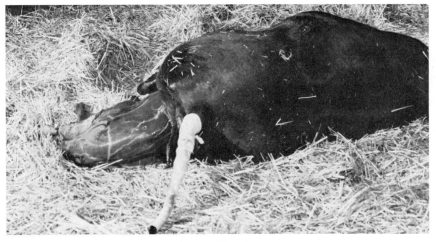

(f)

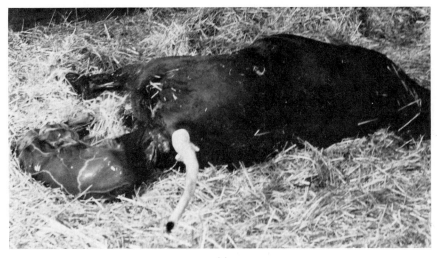

(g)

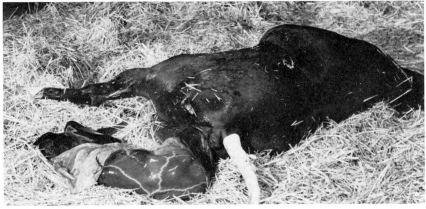

(h)   *(continued)*

FIGURE 16-2 *(continued)*

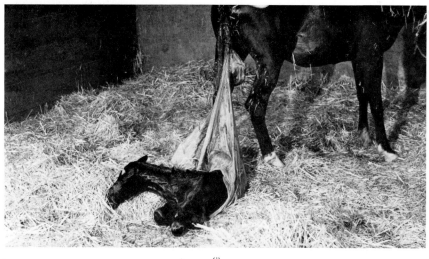

(i)

(j)

mare may lie down, roll, and get up several times. This is important to get the foal in the proper position for delivery. The two front feet appear, and the nose is resting just behind them (Figure 16-3). The bottoms of the feet should point downward. The forequarters pass rather easily. As the hips start to pass, the mare may have some difficulty. After a few more strong contractions, the foal is completely delivered. Stage 2 takes 10 to 20 minutes if no difficulties are encountered.

After the foal has been delivered, make sure that the membranes do not

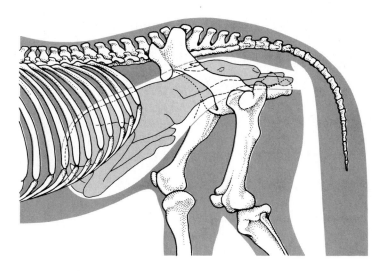

FIGURE 16-3

Normal position of a foal during foaling. The forelegs are extended,
and the head and neck rest on the forelegs. The hindlegs become extended
backward as the body passes out of the mare.

cover the foal's nostrils. If they do, the foal may suffocate. Do not disturb the
mare and foal when reflecting the membranes back over the foal's body.

Stage 3 begins after the foal is completely delivered and ends with the
expulsion of the afterbirth. After the foal is delivered, the mare should lie
undisturbed for as long as she wants—the longer, the better. During this time,
blood from the placenta returns to the foal. If the umbilical cord breaks too
soon, 1 or 2 pints (0.47 or 0.94 liter) of the foal's blood may be lost, resulting
in an anemic condition. The foal will struggle to rise and will break the cord.
Soon after the cord breaks, treat it with a solution of iodine (see the section
on postnatal care).

The afterbirth can be tied up in a ball with a piece of string after the mare
stands. This prevents her from stepping on it and tearing it off so that part of
it remains in the uterus. A young mare may try to kick it, because it will slap
against her legs. The weight of the placenta provides a steady pull that is
usually sufficient to aid in its expulsion. It should be expelled within 30 minutes
to 3 hours after foaling. If retained over 4 hours, an abnormal situation exists,
requiring treatment by a veterinarian. A retained placenta can cause complications such as laminitis, septicemia, and/or colic. Spread the placenta out and
examine it to be sure it is intact (Figure 16-4). If a piece of it is torn off and
retained in the uterus, complications arise. Also, the placenta should be
weighed. A normal placenta weighs approximately 13 to 14 pounds (5.9 to
6.4 kg). If it weighs about 18 to 20 pounds (8.2 to 9.1 kg) and is very bloody,
the uterus may be infected, requiring veterinary treatment.

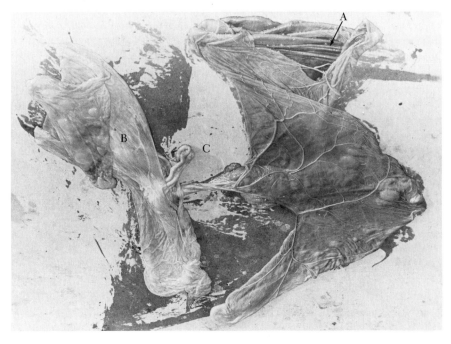

FIGURE 16-4

Placenta that has been spread out on the ground to be sure it is complete and that part of it has not been left in the mare. (A) Point of rupture of the allantochorion. (B) Allantois and amnion, which form the second water bag that immediately surrounds the foal. (C) Navel cord. (Photo by H. Stokes.)

Dystocia

Dystocia is defined as any difficulty during parturition. Fortunately, mares seldom experience dystocia. If they do, they require immediate veterinary assistance. Otherwise, the mare and/or foal will usually be critically injured. The mare should be kept up and walking around until help arrives.

When a mare starts labor, do not interfere unless she starts to have trouble. After the first water bag ruptures, people tend to try to help the mare. But if one or both feet are visible, she requires no help at this stage. In a few minutes and if both feet and the muzzle are visible, all will be well. Do not rush the mare, but let her continue to foal naturally. When the hips pass through the pelvis, she may encounter a slight difficulty. Let her finish and do not disturb her. If she must be helped to deliver a foal that is in the proper position, pull the foal's feet outward and downward when her abdominal muscles contract. When the mare relaxes, the attendant should also relax.

The following situations may be corrected if veterinary help is unavailable. Determine them by passing an arm into the vagina and feeling the fetus' forelegs and head. Most problems result from improper positioning of the limbs, body, or head: The forelegs may be bent at the knee, one foreleg bent backward, head bent backward toward the chest, head bent to the side, head and neck bent backward, buttocks presented first, hocks flexed with hindlegs under body, back against cervix, or hindleg present with forelegs. To correct these situations, push the fetus back in and reposition it with the forelegs and muzzle in the birth canal. The foal can be delivered hindlegs first without too much difficulty. However, if you are not skilled in delivering foals and do not understand fetal positions, you can easily make the conditions worse. Veterinary help is preferred.

Inducing Parturition

Mares can be induced to foal but the procedure is generally not recommended. It may be indicated for mares that have sustained an injury, preparturient colic, prolonged gestation, or colostrum loss. Induction can alleviate the mare's condition and can allow delivery to occur when veterinary assistance is available. It may be useful on a farm in which personnel are not skilled at foaling mares or when veterinary assistance may not be easily available for urgent assistance.

Before a mare is induced, she must meet three criteria: (1) the gestational period must be greater than 320 days; (2) colostrum must be present in the udder, and the cervix should be relaxed; and (3) the fetus should be in the proper position. If the mare fails to meet these criteria, serious consequences usually develop.

To induce the mare, give her oxytocin or prostaglandin. Following treatment, signs of Stage 1 labor are observed within 30 minutes. The foaling process is completed by the end of 40 to 200 minutes.

Postnatal Foal and Mare Care
Foal

As soon as the foal is delivered, examine it to see if it is breathing. If no movement can be observed from a distance, inspect it closely. Artificial respiration can be given by blowing into the foal's nostrils, by pressing and releasing the rib cage, by pulling the forelegs forward and then backward, or by the administration of oxygen. Vigorously rubbing the foal with a towel may stimulate it to breathe. If all else fails, picking up the foal and dropping it a couple of feet may start respiration.

The navel cord should be dipped in a tincture of 10 percent iodine to prevent bacterial invasions. Failure to treat the navel cord frequently results in navel ill. Examine the stump of the cord for a few days to ensure that no urine is present. If the urachus does not close off, urine will pass from the bladder to the stump. The condition is referred to as "pervious or persistent urachus." If a silver nitrate application does not cause it to close, it should be sutured, because it can easily become a route for bacterial invasions.

The foal must be protected against tetanus. This can be provided by two means. As mentioned earlier, the mare can be given a tetanus toxoid booster shot about 30 days before foaling. The foal thus receives high antibody titers against tetanus in the colostrum. These antibodies or "passive immunity" protect the foal until it is 3 to 5 months old. The second method, which is less effective, is to give the foal a tetanus antitoxin injection. Tetanus antitoxin protects the foal for 2 to 3 weeks. The immune system does not function very well until the foal is 3 to 5 months old. One must wait to give the tetanus toxoid sequence to the foal until it is 3 to 5 months old. Therefore, it is not protected for a period of about 2½ to 3 months.

A veterinarian should decide whether or not to administer a prophylactic dose of antibiotics. The practice is controversial. It is usually not required on farms that practice good hygiene procedures.

Normal foals usually nurse within 30 minutes to 2 hours after foaling. Before nursing, it is helpful to clean the udder with warm water and a mild soap. All soap must be washed off. If necessary, help the foal up and guide it to the mare's udder. Foals that are too weak to nurse should be fed by stomach tube. It is important for the foal to receive colostrum because it contains antibodies for disease protection, contains high levels of vitamin A, and has a laxative effect. Mares produce colostrum for about 48 hours. Therefore, if a mare streams milk from the udder for a couple of days before foaling, the foal may not get any colostrum. A colostrum bank should be kept for such emergencies. Colostrum can be milked from mares that produce a large quantity of milk. It can be stored frozen in pint containers for up to 5 years. When necessary, a pint can be warmed to body temperature and given to a foal. Place it in a soft-drink bottle fitted with a lamb's rubber nipple. The foal must receive the colostrum within 36 hours of birth and should receive it as soon as possible. The large protein molecules (antibodies) are absorbed intact from the digestive tract for up to 36 hours. Thereafter, the protein is digested and provides no benefit.

Meconium, the dark-colored fetal excrement, should be passed within 12 hours of birth. If it is not passed, the foal is constipated and should be given an enema. Give 1 or 2 pints (0.47 or 0.94 liter) of warm, soapy water. Pass the end of the hose 3 to 4 inches (7.6 to 10.2 cm) into the rectum. A 4-oz prepackaged Fleet enema is easy to obtain and use, and it gives excellent results.

Foals that are foaled in a pasture without close supervision should be watched to see if they are constipated. If they are, they frequently stop moving, squat, raise their tails, and try to defecate. These same symptoms are observed in foals that are impacted high in the colon. A veterinarian should treat them with a lubricant. Once yellow feces are passed, the digestive track is no longer blocked.

Some foals' eyes start to water within 1 to 2 days after birth. The eyelids and lashes may be turned in (a condition called *entropion*) and should be rolled out. In such cases, rub an eye ointment in the eye.

Mare

Very few mares have complications after they foal. However, foal colic is common. This usually occurs within 10 to 15 minutes after foaling but may not occur for several hours. Foal colic is caused by continuing uterine contractions as the uterus decreases in size. The usual signs of mild colic are observed: The mare looks back at her abdomen, gets up and down, rolls, kicks at her abdomen, and bites her flanks and abdomen. Treatment is usually not necessary, but a mild sedative is helpful.

If part of the placenta is retained, the next day the mare has colic symptoms, increased temperature, increased heart and respiration rates, and possible lameness. Veterinary assistance is necessary.

Internal hemorrhaging usually results in death. It is caused by damage to the uterus or by rupture of the uterine artery, and the mare bleeds internally. At first the condition resembles mild colic but gets worse until death occurs.

Some mares are hesitant to defecate after foaling because of the pain involved. If you observe this, give the mare a wet bran mash containing mineral oil or give her a gallon of mineral oil by stomach tube.

FOAL DISEASES

It is crucial to recognize sick foals early enough so that treatment will be effective. During the first few days, observe the foal frequently. It should move about normally and nurse every few minutes. Most foals nurse two to three times each hour. Failure to nurse causes the mare's udder to become swollen and sore. The presence of diarrhea or constipation, nasal discharge, difficult breathing, swelling or fluid discharges at the navel, swelling of the joints, or increased temperature (normal is 100 to 101°F, or 37.8 to 38.3°C) usually indicate a disease problem.

Diarrhea or scours may appear at any age and may or may not be serious. It is a sign of disease, not a disease itself. Diarrhea may be caused by improper management. Some common causes are foal heat in the mare, dietary upsets, coprophagy, parasites, bacterial infections, and stress (too much exercise, excitement, or exposure). The foal can become dehydrated very quickly and die, so serious cases require immediate attention. Avoid dietary upsets such as separation of mare and foal for a few hours. When diarrhea results from coprophagy (eating feces), the foal may have to be muzzled except when it is allowed to nurse. This usually occurs when the foal's digestive tract is not ready for solid materials. Digestive upsets occur when the foal is forced to exercise too long or if it runs and plays too long after it is turned out to pasture. Foals are very susceptible to ascarid and intestinal threadworm infestations, which cause diarrhea. Foals should be routinely wormed every 2 months for the first year.

Navel ill or joint ill is a serious bacterial disease that affects foals. The organism gains entrance through the unhealed navel stump or through the placenta before birth. Symptoms of the disease include listlessness, swollen joints, reluctance to nurse, temperature, and tucked-up appearance in flanks.

About one-half of all foal deaths can be attributed to septicemias. The bacteria *Actinobacillus equuli* is usually responsible if the foal dies within a few days of birth. Death is the main symptom, but the foal may be drowsy and weak and may sleep a great deal before dying.

Convulsive syndrome is another disease of newborn foals. It is also called "barkers," "dummies," or "wanderers." The condition is caused by extreme lack of oxygen just prior to, during, or immediately after birth. Premature separation of the placenta from the uterus may be the cause in some cases. Usually, it results from unwarranted assistance at parturition. The overanxious assistant causes premature rupture of the umbilical cord, thus losing blood that is vital to the foal. The placental blood makes up about 30 percent of the foal's potential blood volume. This deficit reduces blood pressure in the lung capillaries and causes an improper inflation of the lungs, so that insufficient oxygen is transported to the brain.

The terms "barker," "dummy," and "wanderer" are descriptive of the signs of this disease. The foal may appear normal for a time after birth. Signs are often detected within minutes after birth, but they may be delayed as long as 2 hours. The barker syndrome usually is preceded by a rapid onset of signs. The signs consist of jerking the head up and down in a strange fashion, aimless leg movements, colic, and muscle spasms of the limbs, body, neck, and head. These initial signs are followed by marked convulsions during which the foal may make a peculiar noise resembling the bark of a small dog.

Dummies and wanderers have less severe signs and are so-called because of apparent blindness and failure to respond to external stimuli. The dummy tends to stand or lie quietly in one place with a drowsy attitude. The wanderer moves about aimlessly, unconscious of its surroundings. If the barker survives the convulsions, it passes into the dummy or wanderer stage.

ORPHAN FOALS

Orphan foals are not difficult to raise if one has the time and patience and does not mind work and loss of sleep. Within 2 to 3 hours of birth, the foal must be fed. It is better to feed it as soon as possible. It should receive about 1 pint

(0.47 liter) of colostrum, which can be freshly milked from another mare or which has been stored frozen. Follow the other standard procedures for foal care. Make arrangements for a steady supply of milk to the foal. Very seldom is a nurse mare available, but if so the foal should be grafted to her. If no mare is available, the foal can be taught to suckle a goat. It is easy to train the goat to stand on a small stand while the foal suckles. The stand prevents the foal from standing on its knees while suckling. Goat milk is very similar to mare's milk and can be fed unaltered. Most orphan foals must be hand fed. A variety of formulas can be used. Milk substitutes such as Foal-Lac are excellent and easy to use. Cow's milk can be used but must be altered, because it is higher in fat and protein and lower in water and sugar. Therefore, dilute it with water (½ pint, or 0.24 liter, water to 2 pints, or 0.94 liter, whole milk), and add 1 ounce (28 g) of sugar and a tablespoon, or 14.3 ml, of saturated lime water. The following feeding schedule has been used successfully. During the first 2 days, the foal should be fed ½ pint of formula (warmed to 100°F, or 37.7°C) at hourly intervals around the clock. For the next 5 days, it can be fed ½ to 1 pint (0.24 to 0.47 liter) at 2-hour intervals. Therefore, the foal should be receiving about 8 to 9 pints (3.8 to 4.27 liters) each 24-hour period toward the end of the first week. Some foals may be able to take 10 to 12 pints (4.73 to 5.7 liters). Feed the formula in a bottle fitted with a lamb's nipple. At the end of the week or sooner, it can be trained to drink out of a bucket. Get the foal to lick the milk in the bucket by placing some milk on its lips. As soon as it starts to lick, it will start to drink. Introduce the foal to a creep formula containing milk protein and a legume hay at the end of the first week. Foals will start eating it because they like to nibble on things. During the second week, feed the milk formula at 4-hour intervals. It should receive 1½ to 2 pints at each feeding (0.71 to 0.94 liter). Maintain the schedule during the third week. Intake may increase to 3 pints (1.42 liters) each feeding during the 3 weeks.

It is better to underfeed the foal than to overfeed. Overfeeding causes diarrhea. At the end of the first month, the foal can be fed twice daily and will consume about 4 pints (1.9 liters) each feeding. At this time, the foal should be receiving (daily) about 1 pint per 10 pounds of body weight (0.47 liter per 4.5 kg). It should be eating about 3 to 4 pounds of hay (1.4 to 1.8 kg) and about 2 pounds (0.9 kg) of creep ration. When the foal is 6 weeks old, the amount of formula can be decreased to 2 pints per feeding (0.94 liter). At 7 weeks, the formula intake can be limited to 1½ pints per feeding (0.71 liter), and the foal can be weaned when it is 8 weeks old. By this time, the foal will be eating 6 pounds of hay (2.7 kg) and an equal amount of creep ration.

WEANING

The stress of weaning can be reduced significantly if the foal is used to eating hay and grain. By the time it is 4 months old, it is receiving an insignificant proportion of its nutrient intake from the mare and can be weaned without a nutritional problem. If possible, wean two or more foals at the same time. A foal frets less if kept with a "buddy." Keep them in a box stall or paddock in which there is little chance of injury. Separate the mares and foals so that they are unable to see or hear each other. By the end of a week, the process should be completed.

Separate colts and fillies soon after they are 6 to 8 months of age. Puberty occurs as early as 10 months in both colts and fillies. Failure to separate them may result in an unwanted pregnancy.

The nutritional program at weaning is important (see Chapter 5).

TRAINING THE FOAL

Young foals should become accustomed to being handled if circumstances permit. They should be taught how to be led, to allow their feet to be handled, and to be groomed. This makes the horse manageable when serious training starts later in life.

If the mare is gentle, some foals let people scratch them on the back and hindquarters while nursing. While handling the mare, take every opportunity to scratch and rub the foal. Foals that do not allow this must be caught and held. To catch the foal, crowd it against the mare or a wall. Hold it with one arm under its neck and grip the tail with the opposite hand (Figure 16-5). In this position, you can hold the foal safely or move it about. Holding the foal adjacent to the mare for a while usually calms it down. Then, after someone scratches and brushes it a few times while it is being held, it will not be afraid. It is important not to frighten the foal. As soon as it stands relaxed, place a halter on it so one person can groom it. As soon as it accepts its legs being groomed, pick them up and handle them. By gradually increasing the time the feet are held up, the feet can be cleaned with a hoof pick in a few days.

The foal should be taught to be led at an early age. There are several ways of teaching the foal to be led before it is weaned. One method is to encourage the foal (while being led) to follow its mare while she is being slowly led away. If the foal fights being led, have an assistant walk behind it to make it move forward. A butt rope can be used to encourage it to move forward (Figure 16-6). As soon as it will follow the mare, lead the foal in front of the mare, with her following behind. A few lessons and tugs on the butt rope are usually sufficient to train a foal to be led.

After a foal is weaned, it is more difficult to train to lead, because it is larger and stronger. A gentle foal can be taught by the use of a butt rope. If it is not very gentle, the foal can be taught by tying a large cotton rope around the throatlatch with a bow-line knot, passing it through the halter ring, and tying the rope off about 6 feet (1.8 m) above the head. The rope should hang down about 6 to 8 inches (15.2 to 20.3 cm) below the level of normal head

FIGURE 16-5
Restraint of a foal. (Photo by B. Evans)

carriage. A foal soon learns that it must stand tied without pulling back. As soon as it stands quietly, it can be taught to lead very quickly. A rubber tire tube tied into the cotton rope teaches a foal not to pull back because it learns not to resist the tension. If the foal is tied to a stout post, tie the rope at the level of the withers and tie a piece of tire tube in the rope to prevent the neck from dislocating. The foal will vigorously fight being restrained, but it must remain tied until it no longer fights. At this time, it is advisable to "sack out" the weanling by rubbing a gunny sack or raincoat over it. As soon as it accepts the rubbing, slap the gunny sack or raincoat lightly all over its body until the weanling stands still and does not resist being tied. This type of training teaches the horse to accept being handled, not to be a halter puller, and not to "spook."

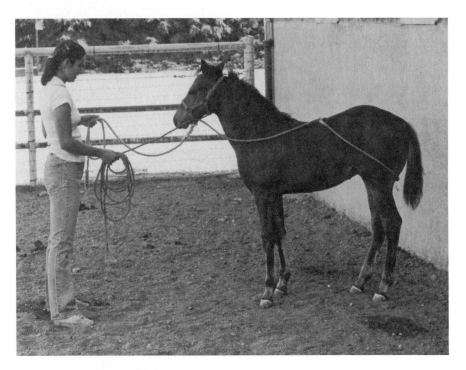

FIGURE 16-6
Butt rope used to teach a foal to lead. (Photo by B. Evans.)

CONGENITAL DEFECTS IN FOALS

Congenital defects are abnormalities of structure or function present at birth. The defects are the result of disruptive events in fetal development that are genetic or environmental in origin. Many defects have no clearly established cause, while the causes of some are understood. At least 87 congenital defects have been reported for foals. Some common defects are as follows.

Malformations

Cleft palate and parrot mouth are defects of the mouth. Cleft palate is a fissure in the soft or soft and hard palates. It leads to secondary pneumonia due to inhalation of fluids into the lungs. Some can be surgically repaired. Parrot mouth is a malformation or underdevelopment of the lower jaw that hinders eating. It is believed to be caused by a dominant gene.

Atresia ani is an anus without an opening. It can be surgically corrected. Atresia coli exists when the colon ends in two blind ends. The foal develops colic shortly after birth and dies in a few days.

Interventricular septal defect is a hole or passageway between the ventricles (lower chamber) of the heart. It is the most common heart defect seen in young horses, and they usually die.

There are several eye defects. Complete, mature cataracts are the most common cause of blindness in young horses (the lens of the eye is opaque). It can be surgically corrected. Iridal heterochromia is an abnormal coloration of the iris. Glass eye is a nearly all-white iris (or combinations of white, brown, and blue). It causes no problems for the horse.

Wobbles or equine incoordination may be inherited by a recessive process that is not understood. It causes ataxia of the rear legs. Wobblers are discussed in Chapter 2.

Umbilical hernia is an opening in the abdominal wall at the umbilicus. The hernia is present at birth or occurs within a few days. It usually corrects itself by 1 year of age. If it does not, it can be repaired surgically.

Scrotal hernia is similar to umbilical hernia except the gut protrudes through the inguinal canal into the scrotum. It also usually corrects itself.

Cryptorchidism is retention of the testicles in the inguinal canal or abdomen. It may be a unilateral or bilateral retention. The mode of inheritance is not understood but is probably a dominant trait.

Incompatible Blood Groups

Neonatal isoerythrolypsis (jaundice foal, isohemolytic icterus, or NI) is a condition caused by incompatibility of blood groups. The incompatibility causes the destruction of the foal's red blood cells by serum antibodies absorbed from the colostrum. Horse blood contains factors that are inherited. If a horse fails to inherit one of the factors and receives the blood of another horse containing the factor, antibodies are formed against the factor. These antibodies destroy the red blood cells that have the factor. When a mare that lacks one of the blood factors (usually A or O) is bred to a stallion that has the blood factor, the foal's blood type will be positive for the factor, because it inherits the factor from the stallion. Some of the foal's blood containing the factor passes to the mare through a faulty placenta. The mare has an antigenic response to the foal's blood and forms antibodies, which build up in the colostrum. The foal is born normal and becomes affected after it suckles and receives the colostrum. Antibodies in the colostrum are absorbed and destroy the foal's red blood cells. The foal becomes progressively anemic and dies within a few days unless treated with blood transfusions. During the first one or two pregnancies, the foal's blood sensitizes the mare's system, and subsequent pregnancies cause antibody formation. The absence of one or more of the blood factors does not occur very frequently, so the chances of the condition are not very great. During late pregnancy, the mare's blood can be tested for the presence of the antibodies.

After a mare produces one NI foal, she and the foal require special management during foaling to prevent the foal from receiving its dam's colostrum. Milk the mare by hand several times daily for 3 days. Feed the foal a milk formula or place it on a nurse mare until the fourth day. The foal must be muzzled and kept with the mare during this period to prevent being rejected by her after her milk becomes safe. After 3 days, the foal can be allowed to nurse the mare.

Combined Immunodeficiency Disease

Combined immunodeficiency disease (CID) is a deficiency of the B- and T-lymphocytes found in Arabian foals. The T-lymphocytes are responsible for all cell-mediated immunity, and B-lymphocytes secrete antibodies. Because of the deficiency in the immune system, the foal dies by the time it is 5 months old, when the passive immunity is lost. CID is transmitted as an autosomal recessive genetic defect (Figure 16-7). Therefore, both the sire and dam of a CID foal carry the recessive gene. The condition has been found in Arabian and part-Arabian horses. It occurs in about 2 to 3 percent of Arabian foals. It has been estimated that about 25 percent of Arabian horses are carriers. Death prior to

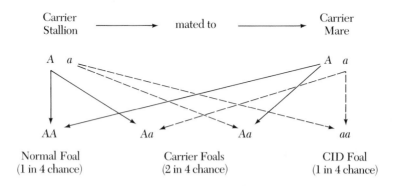

FIGURE 16-7

Inheritance of combined immunodeficiency disease (CID).

5 months is usually caused by an adenoviral infection in conjunction with secondary bacterial infection. Any young Arabian foal that dies should be tested for CID.

Other Immunity Failures

CID should not be confused with three other types of failures of the immune system. Agammaglobulinemia is a deficiency of the B-lymphocytes, but the T-lymphocytes are functional. Only one case involving a Thoroughbred foal has been described. Selective deficiency of IgM (specific immunoglobin) can occur and has been observed in Arabians, Thoroughbreds, and Quarter Horses. These foals may survive but they are plagued with respiratory infections. A foal may fail to receive enough antibodies in the mare's colostrum to provide adequate immunity from diseases. It may have disease problems until its immune system becomes functional at 3 to 5 months of age. If given colostrum at birth or protected with antibodies, it usually suffers few, if any, consequences.

17
Understanding
Basic Genetics

Genetic principles are important to horse breeders because they are concerned about inheritance of lethal traits and coat color and about how to improve their horses for conformation and performance ability. Also, occasionally there are parentage disputes. Although our knowledge of the genetics of horses is incomplete, the basic principles used for genetic improvement of horses were developed by Mendel in the middle 1800s.

BASIC GENETICS

Genes and Chromosomes

Genetic material is passed from generation to generation in chromosomes, which are threadlike structures in the cell nucleus. Each chromosome contains a number of genes, and the location of a gene on a chromosome is a locus. Genes are responsible for controlling body functions because they control the synthesis of all body proteins. Each member of a gene series is an allele; there is one allele at a locus on the chromosome. There may be several alleles for

TABLE 17-1
Numbers of Chromosomes for Some Equine Species

Scientific Name	Common Name or Names	Diploid No. of Chromosomes
Equus caballus, przewalskii	Przewalski's horse	66
Equus caballus	Horse	64
Equus asinus	Donkey, ass	62
Equus hemionus	Onager, kiang, Asiatic wild ass	56
Equus grevyi	Grevy's zebra	46
Equus burchelli	Burchell's zebra, Damara zebra, Chapman's zebra, Grant's zebra, Boehm's zebra	44
Equus zebra	Mountain zebra, true zebra, Hartmann's zebra	32

each gene, as is the case for some genes for coat color inheritance. Each equine species has a different number of chromosomes (Table 17-1). The horse has 64 chromosomes, in pairs. One chromosome from each pair is inherited from each parent. The chromosome number is reduced from 64 to 32 during the formation of ovum and spermatozoa. This reduction in chromosome number is called *meiosis*. During meiosis, a chromosome pair becomes entwined and exchanges chromosome material. The alleles are mixed randomly in the pair of chromosomes, and then the chromosome pair divides. Each germ cell receives one member of the pair of chromosomes. This prevents all the genes in a chromosome inherited from a parent from being passed intact to future generations. Therefore, each germ cell has a mixture of genes from the previous parents. Because of the mixing, the foal receives genes from all its grandparents. With the union of the ovum and spermatozoa (germ cells) at fertilization, the embryo receives 32 chromosomes from each germ cell so that it has 64 chromosomes in all its body cells.

Sex Determination

One pair of chromosomes, the sex chromosomes, determines the sex of the foal. In the female, the chromosomes are matched (same size) and called XX. Therefore, during meiosis the mare's germ cells contain only the X chromosome. In males, the sex chromosomes are unmatched (unequal size) and are termed X and Y, the Y chromosome being smaller than the X. During meiosis, half the sperm cells receive an X chromosome, and the other half receive a Y chromosome. Therefore, the male sperm cell determines the sex of the foal. Because of this situation and the unequal size of the X and Y chromosomes, a filly receives more genetic material from her sire than does a colt, and colts receive more genetic material from their dams than do fillies.

Hybrid Crosses

Hybrid crosses have been obtained among most equine species—horses, asses, and zebras. Although many crosses have been made, the most popular have been between the ass and the horse. The jackass mated to a horse mare produces a mule, and the stallion mated to the jinny produces a hinny. These two crosses produce superior, rugged work animals. The mule and the hinny are infertile. They have 63 chromosomes, because they received 32 from the horse and 31 from the donkey. During meiosis, the chromosomes are not in pairs, and as their number is halved the sex cells usually end up without a complete set. This unbalanced condition does not allow the sex cells to function, resulting in sterility.

Genetic Infertility

Chromosomal abnormalities occur in the horse and can cause infertility. These abnormalities are verified by karyotyping, a process in which the chromosomes are spread out, viewed with a microscope, and counted. The abnormalities usually cause abortions or the development of an abnormal foal. The foal may be phenotypically normal but may fail to reproduce after reaching maturity. Some of these abnormalities involve the sex chromosomes. The most common cases involve mares that failed to receive one X chromosome (XO genotype). Their ovaries are small and firm and do not have germ cells. The result is infertility. Other causes of infertility include the failure to receive any sex chromosomes. This condition results from a failure of the X and Y chromosomes to separate out during meiosis, so one germ cell does not receive sex chromosomes. The same condition can produce a horse with an extra sex chromosome (XXY genotype). The embryo that does not receive an X chromosome (YO genotype) fails to develop.

Parentage Disputes

Some of the major breed registries require that stallions standing at stud or all breeding animals be blood typed. Other registries require that stallions used in an AI program be blood typed.

At least eight blood antigen systems (blood factors) characterize the red blood cell membrane of horses. Each system has alleles that are codominant (each allele causes production of a specific antigen), except for one recessive gene (the dominant allele does not allow production of the antigen specific for recessive alleles). (See Table 17-2.) In addition to the antigen systems, 13 loci control the formation of blood proteins. At each of these loci, there are co-

TABLE 17-2
Genetic Loci That Control Blood Antigen Systems

Blood Antigen Locus Symbol	Known Alleles	No. of Phenotypes	Comments
A	a^{A1}, $a^{A'}$, a^{H}, $a^{A'H}$, a	8	All are dominant to a; remainder are codominant to each other. $a^{A'H}$ will give the same phenotype as $a^{A'}a^{H}$.
C	c^{C}, c	2	Dominant.
D	$d^{Sw10,E}$, d^{Sw10}, $d^{Sw14,D}$ $d^{Sw10,J}$, $d^{Sw14,J}$, d^{Sw14}	21	The J, E, and D types are codominant but dominant to their absence. The Sw10 and Sw14 characters are codominant.
K	k^{K}, k	2	Dominant.
P	p^{P1}, $p^{P'}$, p	4	All are dominant to p; remainder are codominant.
Q	q^{Q}, q^{R}, q^{S}, q^{QR}, q^{RS}, q	8	All are dominant to q; others are codominant but different combinations can give same phenotype—not very useful.
T	t^{T}, t	2	Dominant.
U	u^{U}, u	2	Dominant.

dominant alleles, or each allele causes production of a specific protein. Both the blood antigens and the proteins can be used to settle parentage disputes The tests are based on the exclusion principle. The blood antigens and proteins in the foal must come from the stallion and mare. If the foal has one or more alleles in the antigen system or blood proteins that are not present in the mare, they must be inherited from the stallion. All stallions that do not have the antigen(s) or protein(s) that are present in the foal and absent in the mare can be excluded as the sire. A stallion cannot be proven to be the sire because there are other stallions with the same blood type that could possibly be the sire. There are no tests to prove parentage. A typical example of a questionable parentage and the results of the blood-type tests are presented in Table 17-3. Parentage questions also occur when two mares foal in the same paddock or pasture and there is a possibility that they switched foals. Blood typing may resolve the parentage question before the owner applies for a foal's registration certificate.

Coat Color Inheritance

The interest in coat color inheritance is shared by most horsepeople, and most have a favorite color pattern. In fact, several registries based on particular coat colors have been formed over the years. In most instances, the knowledge of

TABLE 17-3
Typical Example of Questionable Parentage, with Results of Blood-Type Tests

Lab. Code No.	Name and Registration Number of Animal	Sex	Blood Group Systems												Unclassified
			A	C	D	K	P	Q	T	U	Tf	Alb	Es	Pa	CA
A	Copper Coed	F	A₁/A'	C	—	—	P₁	RS	T	U₁	DD	B	I	F	M
B	Foal of Copper Coed		A₁/A'	C	—	—	P₁	Q	T	—	DF	AB	I	?	?
C	Bullfighter	M	A₁	C	—	—	P₁	Q	T	—	FF	AB	I	F	M
D	The Searcher	M	A₁	C	—	—	P'	Q	T	—	FR	B	I	F	M
E	Solidarity	M	A₁	C	J	—	—	Q	T	—	DF	B	FI	F	M

The foal of Copper Coed possesses Q, Tf-F, and Alb-A, which must have been inherited from the sire. Stallions The Searcher and Solidarity lack Alb-A and are excluded as possible sires. Bullfighter possesses all the critical factors and qualifies as a sire.

coat color inheritance is sufficient to formulate certain guidelines for breeding horses with specific color patterns.

The inheritance of coat color in horses is very complex and is not completely understood. It is difficult to obtain the same color description for the same coat color pattern from all investigators, genetic trials are expensive, and the number of genes involved is large. Therefore there are several theories on coat color inheritance. One of the more popular theories is discussed here.

The basic color pattern of horses is believed to be controlled by five loci. Several other loci may modify the basic color pattern.

The first step in the development of color is controlled by the alleles of the color gene (C). There are three alleles: C, c, c^{cr}. If the dominant gene (C) or the dilution gene (c^{cr}) is present, the hair coat, eyes, and skin will be pigmented. Any horse that is homozygous recessive (cc) would be a true albino. It would have no pigment in the hair, eyes, or skin regardless of the other genes present for coat color. There are no true albino horses. Apparently, the homozygous recessive genotype (cc) is lethal in utero. Several types of horses are classified as albinos by horsepeople, but they are not true albinos. They are white horses resulting from the action of the white gene (W) or from the c^{cr} allele. The recessive allele (c) has undergone a mutation (c^{cr}) to cause the dilution of the color pattern. The c^{cr} is codominant with the dominant allele (C), so both genes exert their effect. The effects of the c^{cr} gene will be discussed in the section on dilution.

There are two basic colors of horses, black and brown, controlled by two alleles at one locus on the chromosome. The allele for black (B) is dominant (masks the effect of the other allele) to the allele for brown (b). Therefore, black horses are homozygous dominant (BB) or heterozygous (Bb). The black pigment is also present in the eyes and skin. The homozygous recessive (bb) genotype causes the coat color, eyes, and skin to have brown pigment.

The genes at the other three loci regulate the distribution of the pigments to specific areas of the coat and body.

The agouti or wild pattern gene is named A. Four alleles of the A gene are present in horses, and their order of dominance is A, A^+, a^t, and a. The allele A is the bay pattern gene. It affects the distribution of black pigment by allowing normal production of black pigment in the mane and tail and on the legs, nose, and ears. These peripheral parts of the body are referred to as "the points." The hair on the main body parts is lighter in color; normal pigment production is restricted in these areas and becomes reddish in color. The overall effect is the bay color pattern. In brown horses, its effect is not understood.

A mutation (change in the effect of the allele) of the A allele has been postulated and is termed A^+. Its existence is controversial, but the color pattern of Prezwalski's horse is believed to be due to A^+. The A^+ allele causes a bay color pattern to have a dark spinal stripe, a dark vertical bar on each shoulder, and horizontal bars on the front legs. A is dominant over A^+. The A and A^+

alleles probably have no effect in the brown pigmented horse, or at least there is no evidence for any effect.

Allele *a* in the homozygous condition (both alleles are the same, *aa*) is responsible for the uniform color pattern. The horse is a uniform black or brown color. The color of the uniform black horse fades when the horse is exposed to sun.

Allele *aᵗ* is a mutation of the *a* gene and is responsible for the seal-brown color in the black-pigmented horse. It causes production of brown pigment on the muzzle, under the eyes, on tips of the ears, behind the elbows, and in the flanks. Frequently, it is referred to as "the gene for tan points." The action of the *aᵗ* allele in brown horses is not understood. However, it may be required for production of the so-called claybank dun pattern by one of the dilution genes.

Alleles of the *E* gene act in conjunction with the *B* and *C* genes to govern the intensity of pigment production. There are three alleles: E^D, *E*, and *e*, the order of dominance being E^D, *E*, and *e*. Allele *E* permits the normal display or intensity of dark or light pigment production throughout the body; that is, a uniform coat color. Allele *e* is the recessive of *E* (*E* masks the effects of *e*). When present in the homozygous condition *(ee)*, the intensity of dark pigments (black and brown) is suppressed, and light (yellow-red) pigment intensity is enhanced. The E^D allele increases the intensity of the dark pigment. When homozygous $(E^D E^D)$, it will hide the presence of the *A* gene and when heterozygous, $(E^D E$ or $E^D e)$, the bay pattern will be slightly visible.

The effects of the three alleles that are believed to occur in black- and brown-pigmented horses can be summarized as follows:

1. Black horses with the *A* or A^+ allele and the E^D allele are jet black at birth and do not fade in the sun. The *A* or A^+ allele with an *E* allele produces a normal bay pattern, whereas the *A* or A^+ allele with an *e* allele produces a light (reddish) bay pattern.

2. Black horses with the *aᵗaᵗ* or *aᵗa* genotype in conjunction with E^D__ (__ means any other allele), *EE,* or *Ee* and *ee* genotypes produce jet black (nonfading), normal seal brown, and seal brown (light areas difficult to see) color patterns, respectively.

3. Brown horses with the *A* or A^+ allele are chestnuts regardless of the *E* allele. The E^D may cause a dark liver chestnut, and the *ee* genotype produces a lighter chestnut.

4. The effects in combination with *aᵗaᵗ* or *aᵗa* are not known.

5. Brown horses that are homozygous *aa* and have the E^D allele are darker chestnuts compared with those having the *E* allele. Those with the homozygous *ee* genotype are light chestnuts and are called *sorrel* horses.

The various combinations of these genes are shown in Table 17-4.

"Dilution Genes"

The effects of dilution genes on the basic coat color patterns are very specu-
lative, and there is little definite evidence to support the current theories. One
theory is that two loci have alleles that dilute the basic patterns. At one locus,
the dominant dilution allele is D and the recessive is d. The homozygous recessive, dd, allows a normal color pattern. The D allele in the heterozygous
(Dd) or homozygous (DD) condition causes the dilution of all colors: black
color patterns to produce the grulla, mouse, or smoky black color; brown color
patterns to redden; chestnuts and sorrels to produce yellow dun with dun mane
and tail; and bay color patterns to produce yellow dun with dark mane and
tail. The homozygous condition does not cause further dilution of the color
patterns. The A allele is not necessary for the D allele to exert its effect.

The c^{cr} allele at the C locus also causes the dilution of color patterns and
is partially dominant (one allele has one effect, whereas the homozygous condition has a different effect). It is referred to as the palomino gene. When
heterozygous (cc^{cr}), the bay pattern $(A \_ B \_)$ becomes characterized by a
yellow body and black mane and tail ("buckskin" pattern). The seal-brown
pattern $(a^{t}a^{t}B\_$ or $a^{t}aB\_)$ would not be changed very much and would be
hard to recognize, but may be changed to bay or dark bay. Chestnut and sorrel
horses would become palominos. There is no effect on black horses. When
homozygous $(c^{cr}c^{cr})$, the brown-pigmented horses are diluted to cremellos, and
the black-pigmented horses are diluted to perlinos. Cremellos and perlinos are
white but may be distinguished because the mane and tail of the perlino is a
light rust color.

Before the coat color patterns presented in Table 17-4 are known to be
correct, more evidence must be accumulated. Currently, no hypothesis and
observation agree completely.

Modifying Genes

Several genes modify the basic coat color pattern. These genes are usually
codominant to the basic coat color genes just described. Some of the modifying
genes control the presence or absence of white on the face and legs.

The graying gene (G) is responsible for a progressive silvering of the coat
(see Figure 3-4). The foal is born with its basic coat color pattern. After the
first hair coat is shed, white hairs start to appear. As the horse gets older, the
number of white hairs increases. Some horses become gray faster and are white
by the time they are old. Some never get white but remain gray because they
are turning gray slowly. The homozygous condition (GG) may cause a faster
rate of graying. The mane and tail may or may not become gray. Black horses
with the gray gene are born black. While there is still a larger proportion of
black hairs, the color pattern is called an "iron" or "steel" gray. After there
is a larger proportion of white hairs, the color pattern is simply referred to as

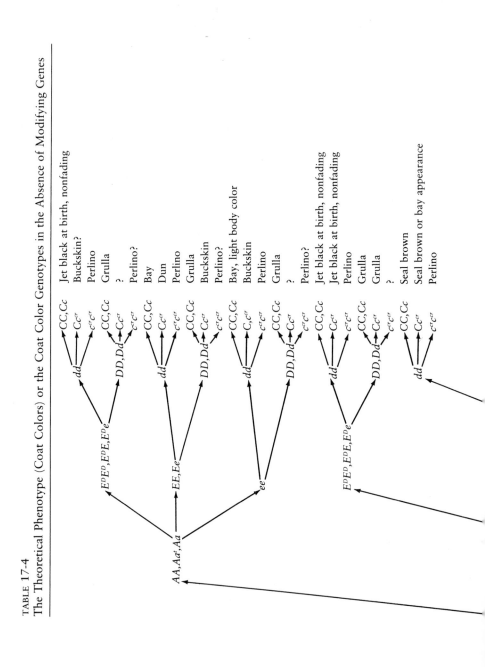

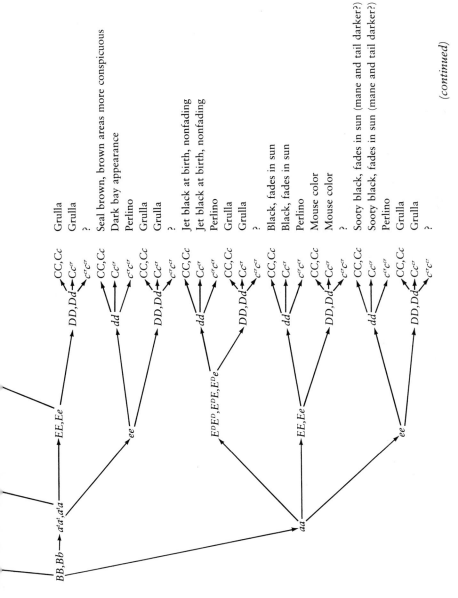

BB,Bb → a^t a^t; a^t a

EE,Ee → DD,Dd → CC,Cc — Grulla
                Cc^cr — Grulla
        → Cc^cr — ?

ee → dd → CC,Cc — Seal brown, brown areas more conspicuous
              Cc^cr — Dark bay appearance
          Cc^cr — Perlino
      DD,Dd → CC,Cc — Grulla
              Cc^cr — Grulla
          c^cr c^cr — ?

E^D E^D, E^D E, E^D e → dd → CC,Cc — Jet black at birth, nonfading
                               Cc^cr — Jet black at birth, nonfading
                           Cc^cr c^cr — Perlino
                       DD,Dd → CC,Cc — Grulla
                               Cc^cr — Grulla
                           c^cr c^cr — ?

aa → EE,Ee → dd → CC,Cc — Black, fades in sun
                    Cc^cr — Black, fades in sun
                Cc^cr c^cr — Perlino
              DD,Dd → CC,Cc — Mouse color
                      Cc^cr — Mouse color
                  c^cr c^cr — ?

ee → dd → CC,Cc — Sooty black, fades in sun (mane and tail darker?)
              Cc^cr — Sooty black, fades in sun (mane and tail darker?)
          Cc^cr c^cr — Perlino
      DD,Dd → CC,Cc — Grulla
              Cc^cr — Grulla
          c^cr c^cr — ?

(continued)

TABLE 17-4 (continued)

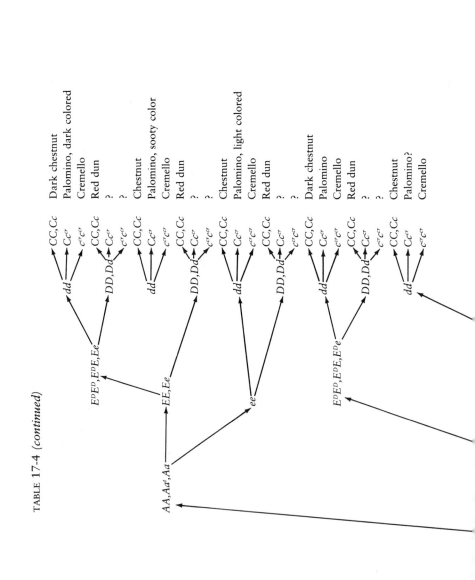

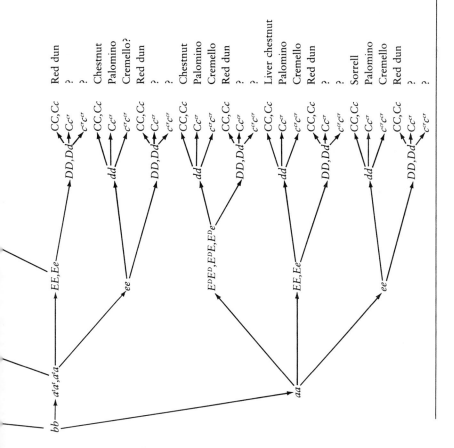

gray. Bay horses with the gray pattern are called "red grays," and chestnuts with gray pattern are called "chestnut grays."

Gray horses can always be distinguished from white horses because the skin of gray horses contains brown or black pigment. However, one must be careful not to confuse the roan gene *(Rn)* with the gray gene *(G)*. The *Rn* gene is a simple dominant gene that causes a mixture of white and colored hairs. The white hairs are present and scattered throughout the hair coat at birth. The pattern does not change. If the white hairs are mixed with black hairs, the coat color is called "blue roan." The strawberry roan is a sorrel or chestnut color with the white hairs. A red roan is a bay with white hairs.

The homozygous dominant *(RnRn)* is believed to be lethal while the fetus is in utero. If so, when roans are mated to roans, the fertility rate should not exceed 75 percent, and 66.6 percent of the foals born will be roan. Other genes may control the amount of white in the coat, because it varies from horse to horse.

Breeders that are selecting for coat color patterns such as Appaloosa, paint, pinto, and palomino should avoid breeding animals that have the roan and/or gray genes. Their presence detracts from the desired color pattern. The gray gene causes the pattern to fade, and both genes cause the pattern to be without sharp definition.

The white gene *(W)* is epistatic (masks the effect of other genes) to all other coat color genes (Figure 3-17). Therefore, the presence of the dominant allele causes the horse to be white. The homozygote *(WW)* is lethal at an early stage of pregnancy, because no true-breeding white horses have been reported. Mating white horses results in two *Ww* white foals and one *ww* nonwhite foal. White horses are not albinos, because their eyes are pigmented. These white horses are not to be confused with the perlino and cremello horses resulting from the $c^{cr}c^{cr}$ genotype. The true white horse *(Ww)* has a much whiter coat than the off-white color of perlinos and cremellos.

A progressive silvering gene *(Sl)* has been postulated to cause a silvering effect ranging from a few white hairs to a completely white horse.

Shetland ponies have a dominant silvering gene *(S)* that causes a silver dapple color and a white mane and tail. The homozygote *(SS)* may have more white hairs than the heterozygote *(Ss)*. If the pony has both the *S* and *G* genes, it will turn almost white within 1 to 2 years.

One of two possible genotypes are responsible for paint and pinto horses. The tobiano pattern is white spotting that crosses over the top of the horse and extends downward (see Figure 3-20). It is caused by the dominant allele at the *T* locus. The homozygote *(TT)* breeds true for the tobiano pattern. When mating a heterozygote *(Tt)* to a solid-color horse, one-half of the foals should have the tobiano pattern. When one solid-colored foal is sired by a stallion or produced by a mare, the parents are proven to be heterozygotes. The overo pattern (see Figure 3-19) was once thought to be caused by the homozygous recessive genotype *(oo)*. Recent evidence indicates it is more complicated. However, the overo pattern occasionally crops up when horses are bred that are registered with a breed association that does not register paint or pinto horses. The pattern is characterized by white that crosses under the belly and extends upward. Frequently, the horse has excessive white on its face. When some overo horses are mated and a foal with excessive white is produced, it may have atresia coli (a colon that ends in two blind ends). The amount of white for both patterns is probably controlled by several genes. When the tobiano and overo color patterns are on a black horse, it is called a "piebald," and when on a brown horse it is called a "skewbald."

The inheritance of the Appaloosa color patterns is not very well understood. One hypothesis is that the dominant gene *(Ap)* must be present for the appaloosa pattern to be expressed. The pattern may be a white blanket extending over the hindquarters, or a combination of a blanket with dark spots (see Figure 3-21). The gene *wb* is necessary for the blanket, and the gene *sp* is necessary for the dark spots in the blanket. Both genes are influenced, because males may be homozygous or heterozygous for each gene. Mares must be homozygous for the pattern to be expressed. The amount of white and size of the blanket as well as the characteristics of the spots are determined by the modifier gene.

Based on the preceding hypothesis, the following genotypes may result in the corresponding phenotypes:

1. *Ap__ wbwb spsp* results in blanket with spots in male and female.

2. *Ap__ wbwb Spsp* results in blanket with no spots in female and blanket with spots in male.

3. *Ap__ WbWb spsp* results in no blanket with spots in female and blanket with spots in male.

4. *Ap__ WbWb Spsp* results in no blanket with spots in female and blanket with spots in male.

The leopard pattern, white with dark spots over entire or most of the body, may be controlled by a dominant gene (Figure 17-1).

Appaloosa horses also have particolored skin around the nostrils, nose, and anus (Figures 15-1 and 3-21). This is apparently caused by a dominant gene *(M)*. The sclera around the eye is white and more exposed than on nonappaloosa horses, giving the impression of a white ring around the eye. The hooves have vertical light and dark stripes. A sparse mane and tail are sometimes present (Figure 3-21). An Appaloosa horse always has one or more of these characteristics. If the color pattern is absent, the horse may still show these minor characteristics.

The basic coat color pattern is also modified by genes that control the expression of white on the face and legs. Very little is known about the genes controlling the expression of white on the face and legs. A gene *(F)* has been proposed, and the dominant *F* is for solid color of the face and legs. The homozygous *ff* allows varying degrees of white on the face and legs. The

FIGURE 17-1

Tot's Hama's Moolah, California State Champion Appaloosa Stallion, has the leopard color pattern. (Photo by Tallaw, courtesy Sandy Leoni, owner.)

amount of white on the face may be controlled by at least four modifying genes. A white spot on the chin is the recessive genotype, *chch*. The genes for star *(st)*, stripe *(sr)*, and snip *(sn)* may be dominant or recessive. Genes controlling the white stocking effect are not understood. One theory is that a gene *(Pl)* is for white stockings. It is known that the spots located in the white above the coronary band are dominant.

The flaxen manes and tails of chestnuts may be due to a recessive genotype. The gene probably has no effect in horses carrying the black *(B)* gene.

Curly Hair

The hair on the body and tail can be curly. A recessive genotype *(ff)* is said to result in the kink found in the tail of many Appaloosas. The curly coat of horses in the Buckskin registry is caused by a homozygous recessive *(ctct)*. The hair is in small curls over the entire body of the horse.

BREEDING SUPERIOR HORSES

There are two general types of inheritance, qualitative and quantitative. In qualitative inheritance, the phenotype (physical appearance and attributes) is controlled by one or a few pairs of genes. Phenotypic differences are easy to distinguish. In quantitative inheritance, the phenotype is controlled by many pairs of genes. One phenotype tends to blend into the next one so that there is no easily recognizable, clear-cut distinction among phenotypes.

Qualitative Inheritance

In any breeding program, horsepeople try to improve their horses by eliminating easily recognizable undesirable traits (phenotypes). The dominant and recessive genotype causing the phenotype may be selected to improve horses, or the breeder may select to eliminate them. Selection for or against a dominant gene is easy because the phenotype allows recognition of the genotype. The phenotype is recognized, and the individual is kept or sold. The major difficulty in a breeding program is eliminating unwanted recessive genes. Animals with the recessive allele in combination with a dominant allele are termed "carrier animals," and the phenotype does not express the genotype (that is, the breeder does not know that the recessive allele is present). The procedures used to detect carriers require time, effort, and money. The simplest method is to recognize and eliminate all animals that have both recessive genes (homozygous recessive). Progress by this method is slow and never completely eliminates all carriers. For example, if the frequency of the undesirable recessive gene in the population is 0.25 (one of every four is the recessive allele), the frequency will be reduced to 0.07 after 10 generations. The generation interval in horses is 4 to 5 years; therefore, elimination would take 40 to 50 years.

The rate of progress by other methods is faster but requires that the stallion or mare be tested. If a stallion is bred to a group of mares and one foal is produced that is homozygous recessive for the unwanted gene, the stallion is proven to carry the gene. A quick way to detect a carrier stallion is to breed him to 4 mares that are homozygous for the unwanted gene. If the foals are normal, the probability that the stallion is not a carrier is 94 percent. If 8 mares are bred without siring a homozygous recessive foal, the probability is about 100 percent that he is not a carrier. If the mares used in the test program are known carriers, more matings are needed. When 8 mares do not produce a homozygous recessive foal, the probability that the sire is not a carrier is 90 percent. When bred to 20 carrier mares without producing an undesirable foal, the probability of him not being a carrier is essentially 100 percent. However, there is always a slim chance that the stallion could be a carrier.

To determine if a mare is a carrier, she must be bred to known carriers or homozygous recessive stallions. Because the number of foals she must produce is the same as the number a stallion must sire, it is not feasible to test mares unless the embryo transfer technique is used.

To detect all the unwanted recessive genes that a stallion carries, he is bred to his own daughters. It requires 20 daughters to reach a 93 percent probability for each recessive gene, and he would have to be bred to 50 daughters to obtain almost a 100 percent probability. This would be a time-consuming and costly effort requiring at least 4 years if the mares are bred in one year and foal the next and if the fillies were bred when they were 2 years old and foaled the next year. To obtain 50 fillies to breed, the stallion would have to initially breed about 125 mares, assuming an 80 percent foal crop of which 50 percent were fillies.

Quantitative Inheritance

Quantitative traits are difficult to recognize, so a deliberate effort must be made to measure quantitative traits. Body weights are measured in pounds, pulling power is measured by the load the horse can pull, and speed is measured by time to race a given distance. After the measurements are made, superior animals may be identified. To identify a superior animal for a quantitative trait, one must know the average value and the variation of the horse population for that trait. The standard deviation describes the variation of a trait about the average for that trait. A standard deviation for a trait of a population is calculated by finding the deviation (difference) between the average for the population and each individual value, squaring each deviation, adding all the squared deviations, dividing the sum of the squared deviations by 1 less than the number of individuals, and taking the square root. In a normal population, 68 percent of all the values should fall in the range of the average ±1 standard deviation. About 95 percent of all individuals are included in the range of the average ±2 standard deviations. A horse would be superior to 95 percent of the general population if its value for the trait were greater than the value of the average +2 standard deviations. If a smaller value for the trait is desired, the horse's value would be less than the average −2 standard deviations.

Each individual has a genetic limit for each trait, and this limit is set at fertilization. Rarely does an individual reach its genetic potential, because environmental factors influence performance. To eliminate environmental influence from performance ability, heritability estimates have been made for some traits. Heritability estimates are expressed in percentages. Some heritability estimates are given in Table 17-5. The heritability estimates for speed range from 25 to 50 percent. For an actual value of 47 percent, that percent of the variability in speed is probably caused by heredity and 53 percent of

TABLE 17-5
Heritability Estimates for Some Traits

Trait	Heritability
Speed	0.25–0.50
Wither height	0.25–0.60
Body weight	0.25–0.30
Body length	0.25
Breast circumference	0.34
Cannon bone circumference	0.19
Pulling power	0.25
Walking speed	0.40
Trotting speed	0.40
Points for movement	0.40
Points for temperament	0.25
Reproductive traits	Low?
Cow sense	Medium to high?
Conformation characteristics	Medium to high?

the variability is caused by environmental influences. Low heritability estimates indicate that most of the variability is caused by the environment; that is, by training, feeding, care, and so on. Therefore, a high heritability estimate indicates to the breeder that considerable genetic progress can be made in improving a trait by mating the best to the best. A low heritability estimate means that attention should be paid to environmental influences to obtain good performances and that not much progress can be made in improving the trait by a carefully planned breeding program.

Locating Animals with Superior Genotypes

The location of horses with superior genotypes depends on one's ability to eliminate or minimize environmental influences on performance. The environmental influence can never be eliminated, but it can be minimized by the examination of accurate performance records. Statistical services for some breed associations provide extensive performance records. Some records attempt to eliminate environmental influences such as racetrack and condition of track for the race. Show ring performance records for horses shown under several different judges eliminate some environmental variation. These records may be most important for the horse being considered; with older animals, the progeny record is most important. If the horse is very young, the pedigree and the appearance of the individual may be the most important. Performance records of closely related relatives are important, because superior breeding animals usually come from superior families. Selection is to increase the frequency of desirable genes and to increase the homozygosity of desirable gene pairs.

Linebreeding and Inbreeding Relationships

Many successful long-term horse breeding programs have used extensive cull-ing and the breeding of closely related individuals (inbreeding) or of individuals with a common ancestor (linebreeding). Inbreeding or linebreeding raises the percentage of genes inherited from the common ancestor. For successful line-breeding or inbreeding, the common ancestor must be genetically superior.

Inbreeding increases the homozygosity of genes. There is an increased number of dominant and recessive pairs of genes. Inbreeding is accomplished by the mating of closely related individuals. The percentage of inbreeding is an estimate of the increased homozygosity of genes in the offspring as compared to the noninbred population (Table 17-6). There are advantages and disad-vantages of the increased homozygosity. When inbreeding is accompanied by selection, individuals that are homozygous for desired traits are selected and used as breeding animals. They are prepotent because they pass desirable genes to the next generation. Their offspring usually resemble the prepotent parent more than the other parent. The disadvantage is that inbreeding uncovers undesirable recessive genes if they are present in the population. Homozygous recessive genes are expressed phenotypically.

There are phenotypic effects of inbreeding. Genetic defects are seen so one must constantly be aware of the possibility of defects. The thriftiness and vigor of the animals are usually reduced. The less vigor the animal has, the more undesirable homozygous recessive genes it probably possesses. Horses that maintain vigor probably have an increased number of heterozygous or homo-zygous dominant genes that cover up or mask the detrimental recessive allele. Because of the reduced vigor, they may not look as good as the noninbred population, but one must not forget that the genetically superior ones produce better offspring. Therefore, the inbreeding program must have enough animals

TABLE 17-6

Percent of Inbreeding in the Offspring When Various Kinds of Relatives Are Mates (assuming that the parents themselves are not inbred)

Kinds of Relatives Mated	Percent of Inbreeding in Offspring
Parent and progeny	25.0
Full brother and full sister	25.0
Half brother and half sister	12.5
First cousins	6.3
Double first cousins	12.5
Uncle and niece or aunt and nephew	12.5
Grandparent and grandchild	12.5

Note: The percent of inbreeding is an estimate of the increased homozygosity of genes (both dominant and recessive) in the offspring as compared to their noninbred parents.

SOURCE: Adapted from Lasley (1970).

involved to be able to select a sufficient number of desirable animals and be able to cull the undesirable animals.

Linebreeding is a special form of inbreeding. In a linebreeding program, an attempt is made to keep a high relationship to a common ancestor in the pedigree. The common ancestor usually has been proven to be a genetically superior animal by its progeny. The common ancestor is usually at least three to four or more generations removed from the individual. However, some linebreeding programs use intensive inbreeding of closely related individuals. The purpose of linebreeding is the same as inbreeding except that it attempts to fix in the population the desirable characteristics of one or two superior ancestors. The inbreeding percentage is not as high, and the vigor is not reduced as much.

The purpose of linebreeding and inbreeding is to develop a genetically superior population of animals that can execute a particular type of performance. In many instances, the inbred or linebred families may be crossed with other similar families. If the appropriate niche is found, very superior offspring are produced. This is called *outcrossing,* and its genetic effect is to increase the number of heterozygous genes. In fact, outcrossing or outbreeding is a tool that many breeders use to improve their animals. Because of the increased number of heterozygous genes, many undesirable recessive alleles are masked. This system is used by most breeders because they cannot afford to develop linebred or inbred families.

Stallion Selection to Complement the Mare

There is no perfect horse, so horse owners are constantly trying to improve the quality of the horses they raise. In most instances, the mare owner tries to select a stallion to breed to a mare with the hope that the stallion will sire a foal with better conformation or performance potential than the mare has. Stallion owners try to select mares that will complement their stallion so that he will sire well-conformed foals.

The first step in the selection process is to find well-conformed individuals. The owner may quickly eliminate mares or stallions that have serious conformation faults, such as calf knees and offset knees, or that have too many faults. The next step is to decide which of the well-conformed stallions or mares are particularly strong where the horse to be bred is weak. Horses with poor hindleg conformation should be bred to those with very correct hindlegs and have a reputation for siring or producing foals with correct hindlegs. Most breeds have sires or families that are known for certain conformation characteristics. These facts should be taken into account.

Selection is less risky if one avoids breeding to stallions during their first year as studs. To prove a stallion, do not select maiden mares but use older, proven mares.

In addition to trying to eliminate conformation faults, one must breed horses of similar size. If large stallions are bred to small mares, the foal will inherit characteristics that do not match, and the foal will be out balance. The front end must match the hind end. The slope of the hip must complement the slope of the shoulder. Tall horses have longer necks to maintain balance. As a result of improper crosses, foals may inherit long legs from a tall parent and a short neck from a short, stocky parent.

Appendixes

ADDRESSES OF BREED ASSOCIATIONS AND REGISTRIES, BUSINESS ASSOCIATIONS, AND MAGAZINES

Breed Associations and Registries

American Association of Owners and Breeders of Peruvian Paso Horses. P.O. Box 371L, Calabasas, CA 91302.

American Bashkir Curly Registry. Sunny Martin, secretary. Box 453, Ely, NE 89301.

American Bay Horse Association. D. M. Cordes, vice-president. P.O. Box 884F, Wheeling, IL 60090.

American Buckskin Registry Association. Randi Spears, executive secretary. P.O. Box 1125, Anderson, CA 96007.

American Connemara Pony Society. Mrs. John E. O'Brien, secretary. Hoshiekon Farm, R.D. #1, Goshen, CT 06756.

American Fox Trotting Horse Breed Association. Box 666, Marshfield, MO 65706.

American Gotland Horse Association. Bob Lee, president. R.R. #2, Box 181, Elkland, MO 65644.

American Hackney Horse Society. Mrs. L. M. Hanks, executive secretary. P.O. Box 174, Pittsville, IL 62363.

American Hunter and Jumper Association. Herb Glass, executive secretary. P.O. Box 1174, Fort Wayne, IN 46801.

American Indian Horse Registry. Nancy Falley, president. Route 1, Box 64, Lockhart, TX 78644.

American Miniature Horse Registry. P.O. Box 468, Fowler, IN 47944.

American Morgan Horse Association. Box 1, Oneida County Airport, Westmoreland, NY 14390.

American Mustang Association. P.O. Box 338, Yucaipa, CA 92399.

American Paint Horse Association. Ed Roberts, executive secretary. P.O. Box 18519, Fort Worth, TX 76116.

American Part-Blooded Horse Registry. Mrs. Barbara J. Bell, secretary. 4120 South East River Drive, Portland, OR 97222.

American Paso Fino Pleasure Horse Association. Mellon Bank Building, Room 3018, 525 William Penn Place, Pittsburgh, PA 15219.

American Quarter Horse Association. 2736 West Tenth, P.O. Box 200, Amarillo, TX 79168.

American Quarter Pony Association. Edward Ufferman, secretary. Route 1, Box 585, Marengo, OH 43334.

American Saddlebred Pony Association. Irene Zane, president. 801 South Court Street, Scott City, KS 67871.

American Saddle Horse Breeders' Association. Charles J. Cronan, Jr., secretary. 929 South Fourth Street, Louisville, KY 40203.

American Shetland Pony Club. P.O. Box 468, Fowler, IN 47944.

American Shire Horse Association. Heather D. Erskine, secretary. 14410 High Bridge Road, Monroe, WA 98272.

American Suffolk Horse Association. Mary Read, secretary. 15B Roden, Wichita Falls, TX 76311.

American Trakehner Association. David Gribbons, secretary. P.O. Box 268, Norman, OK 73070.

American Walking Pony Association. Mrs. Joan H. Brown, executive secretary. Route 5, Box 88, Upper River Road, Macon, GA 31201.

Americana Pony. Vern Brewer. 926 Summit Avenue, Gainesville, TX 76240.

Andalusian Horse Registry of the Americas. Glenn O. Smith, registrar. P.O. Box 1290, Silver City, NM 88061.

Appaloosa Horse Club. Charles Nuber, executive secretary. P.O. Box 8403, Moscow, ID 83843.

Arabian Horse Registry of America. 3435 South Yosemite Street, Denver, CO 80231.

Belgian Draft Horse Corporation of America. Rollin Christner, secretary-treasurer. P.O. Box 335, Wabash, IN 46992.

Canadian Arabian Horse Registry. R.R. 1, Bowden, Alberta, Canada T0M 0K0.

Capitol Quarter Horse Association. 25302 Edison Road, Box 594, South Bend, IN 46624.

Chickasaw Horse Association. Mrs. J. A. Barker, Jr., secretary. Box 607, Love Valley, NC 28677.

Cleveland Bay Horse Society of America. Mrs. W. R. L. Dorman, secretary. P.O. Box 182, Hopewell, NJ 08525.

Clydesdale Breeders Association of the United States. Betty Groves, secretary. Route 1, Pecatonica, IL 61063.

Colorado Ranger Horse Association. John E. Morris, president. 7023 Eden Mill Road, Woodbine, MD 21797.

Cream Horse Registry. Ruth Thompson, secretary. P.O. Box 79, Crabtree, OR 97335.

Cross-Bred Pony Registry. Eric Caleca, registrar. Box M, Andover, NJ 07821.

Galiceño Horse Breeders Association. Mary E. Stubblefield, secretary. 111 East Elm Street, Tyler, TX 75701.

Hungarian Horse Association. Bitterroot Stock Farm, Hamilton, MT 59840.

Hunter Club of America. Box 274, Washington, MI 48094.

Icelandic Pony Club and Registry. Mrs. Judith Hassed, secretary. 56 Alles Acres, Greeley, CO 80631.

International Arabian Horse Association. 224 East Olive Avenue, Burbank, CA 91503. (Half-Arab and Anglo-Arab Registries.)

International Buckskin Horse Association. Richard E. Kurzeja, executive secretary. P.O. Box 357, St. John, IN 64373.

International Miniature Horse Registry. 3 Malaga Cove Plaza, Palos Verdes Peninsula, CA 90274.

Jockey Club. 380 Madison Avenue, New York, NY 10017. (Thoroughbred horses.)

Missouri Fox Trotting Horse Breed Association. L. Barnes, secretary. P.O. Box 637, Ava, MO 65608.

Model Quarter Horse Association. Mrs. Lavonne Foster. P.O. Box 127, Monument, OR 97864.

Morab Horse Registry of America. P.O. Box 143, Clovis, CA 93612.

Morgan Horse Club. Seth P. Holcombe, secretary. P.O. Box 2957, Bishop's Corner Branch, West Hartford, CT 06117.

Morocco Spotted Horse Cooperative Association of America. Lowell H. Rott, secretary-treasurer. Route 1, Ridott, IL 61067.

National Appaloosa Pony. Eugene Hayden, executive secretary. P.O. Box 206, Gaston, IN 47342.

National Association of Paso Fino Horses of Puerto Rico. Aptdo. 253, Guaynabo, PR 00657.

National Chickasaw Horse Association. Mrs. Duane Sunderman, secretary. R.R. #2, Clarinda, IA 51632.

National Palomino Breeders Association. Mrs. Nora Lee Howard, secretary. P.O. Box 146, London, KY 40741.

National Quarter Horse Registry. Cecilia Connell, secretary. Box 235, Raywood, TX 77582.

Original Half Quarter Horse Registry. I. M. Hunt, secretary. Hubbard, OR 97032.

Palomino Horse Association. P.O. Box 324, Jefferson City, MO 65101.

Palomino Horse Breeders of America. Mrs. Melba Lee Spivey, secretary. P.O. Box 249, Mineral Wells, TX 76067.

Paso Fino Owners and Breeders Association. P.O. Box 1579, Tryon, NC 28782.

Percheron Horse Association of America. Lucille Gossett, secretary. Rural Route 1, Belmont, OH 43718.

Peruvian Paso Horse Registry of North America. Janetta Michael, president. P.O. Box 816, Guerneville, CA 95776.

Pinto Horse Association of America. 7525 Mission George Road, Suite C, San Diego, CA 92103.

Pony of the Americas Club. George A. Lalonde, executive secretary. 1452 North Federal Street, P.O. Box 1447, Mason City, IA 50401.

Racking Horse Breeders Association of America. Helena, AL 35080.

Shetland Pony Identification Bureau. 1108 Jackson Street, Omaha, NB 68102.

Spanish-Barb Breeders Association. Peggie Cash, secretary. Box 7479, Colorado Springs, CO 80907.

Spanish Mustang Registry. Mrs. Leana Rideout, secretary-treasurer. Route 2, Box 74, Marshall, TX 75670.

Standard Quarter Horse Association. 4390 Fenton Street, Denver, CO 80212.

Tennessee Walking Horse Breeders' Association of America. John Price, executive secretary. P.O. Box 286, Lewisburg, TN 37091.

United States Trotting Association. William R. Hilliard, executive vice-president. 750 Michigan Avenue, Columbus, OH 43215. (Standardbred horses.)

United States Trotting Pony Association. P.O. Box 468, Fowler, IN 47944.

Welsh Pony Society of America. Gail Hamilton, secretary. P.O. Box 2977, Winchester, VA 22601.

White Horse Registry. Ruth Thompson, secretary. P.O. Box 79, Crabtree, OR 97335.

Ysabella Saddle Horse Association. L. D. McKenzie, secretary. McKenzie Rancho, R.R. #2, Williamsport, IN 47993.

Business Associations

American Horse Council. 1776 K Street Northwest, Washington, DC 20006.

American Horse Shows Association. 527 Madison Avenue, New York, NY 10022.

American Thoroughbred Breeders and Owners Association. 36 Ozone Park, Jamaica, NY 11417.

Arabian Horse Racing Association of America. 66 South Riversdale Drive, Batavia, OH 45103.

International Society for the Protection of Mustangs and Burros. Helen A. Reilly. Badger, CA 93603.

National Cutting Horse Association. P.O. Box 12155, Fort Worth, TX 76116.

National Reining Horse Association. William E. Garvey. R.R. #2, Greenville, OH 45331.

National Trotting and Pacing Association. Ronald R. Moul, executive secretary. 575 Broadway, Hanover, PA 17331.

National Trotting Pony Association. Ronald R. Moul, executive secretary. 575 Broadway, Hanover, PA 17331.

Thoroughbred Racing Associations of the United States. Five Dakota Drive, Lake Success, Hyde Park, NY 12538.

United States Pony Clubs. Mrs. John Reidy, secretary. Pleasant Street, Dover, MA 02030.

Magazines

A.B.R.A. Newsletter. American Buckskin Registry Association. P.O. Box 1125, Anderson, CA 96007.

American Connemara Pony Society News. American Connemara Pony Society. Hoshiekon Farm, R.D. #1, Goshen, CT 06756.

American Fox Trotting Horse. American Fox Trotting Horse Breed Association. Route 2, Box 200, Marshfield, MO 65706.

American Horseman. Countrywide Communications. 257 Park Avenue South, New York, NY 10010.

American Shetland Pony Journal. American Shetland Pony Club. 218 East Fifth Street, P.O. Box 468, Fowler, IN 47944.

Appaloosa News. Appaloosa Horse Club. P.O. Box 8403, Moscow, ID 83843.

Arabian Horse. 1777 Wynkoop Street, Suite 1, Denver, CO 80202.

Arabian Horse News. P.O. Box 2264, Fort Collins, CO 80522.

Arabian Horse World. 2650 East Bayshore, Palo Alto, CA 94303.

Arizona Horseman. 2517 North Central Avenue, Phoenix, AZ 85004.

Arizona Thoroughbred. 3223 Pueblo Way, Scottsdale, AZ 85251.

Backstretch. 19363 James Couzens Highway, Detroit, MI 48235.

Belgian Review. Belgian Draft Horse Corporation of America. P.O. Box 335, Wabash, IN 46992.

Blood-Horse. P.O. Box 4038, Lexington, KY 40504.

British Columbia Thoroughbred. 4023 East Hastings Street, North Burnaby, British Columbia, Canada.

California Horse Review. P.O. Box D, Acampo, CA 95220.

Chronicle of the Horse. Middleburg, VA 22117.

Equestrian Trails. 10723 Riverside Drive, North Hollywood, CA 91602.

Equus. 665 Quince Orchard Road, Gaithersburg, MD 20760.

Florida Horse. P.O. Box 699, Ocala, FL 32670.

Harness Horse. Telegraph Press Building, P.O. Box 1831, Harrisburg, PA 17105.

Hoofbeats. United States Trotting Association. 750 Michigan Avenue, Columbus, OH 43215.

Hoof and Horn. 3900 South Wadsworth Boulevard, Denver, CO 80235.

Horse and Horseman. P.O. Box HH, Capistrano Beach, CA 92624.

Horse and Rider. 41919 Moreno Road, Temecula, CA 92390.

Horse and Show. Box 386, Northfield, OH 44067.

Horse Show. American Horse Shows Association. 527 Madison Avenue, New York, NY 10022.

Horse Show World. P.O. Box 39848, Los Angeles, CA 90039.

Horse World. P.O. Box 588, Lexington, KY 40501.

Horseman. 5314 Bingle Road, Houston, TX 77018.

Horseman & Fair World. 904 North Broadway, P.O. Box 11688, Lexington, KY 40511.

Horseman's Gazette. Box 202, Badger, MN 56714.

Horseman's Journal. Suite 317, 6000 Executive Boulevard, Rockville, MD 20852.

Horseman's Review. P.O. Box 116, Roscoe, IL 61073.

Horsemen's Yankee Pedlar. Wilbraham, MA 01095.

I.B.H.A. News Report. International Buckskin Horse Association. P.O. Box 357, St. John, IN 46373.

Lariat. 12675 Southwest First Street, Beaverton, OR 97005.

Maryland Horse. P.O. Box 4, Timonium, MD 21093.

Morgan Horse. American Morgan Horse Association. Box 1, Westmoreland, NY 13490.

Mr. Longears. 100 Church Street, Amsterdam, NY 12010.

National Horseman. 933 Baxter Avenue, Louisville, KY 40204.

Northeast Horseman. P.O. Box 131, Hampden Highlands, ME 04444.

Oregon Thoroughbred Review. 1001 North Schmeer Road, Portland, OR 97217.

Owners and Breeders Registry. Drawer XX, Livingston, AL 35470.

Paint Horse Journal. American Paint Horse Association. P.O. Box 18519, Fort Worth, TX 76118.

Palomino Horses. Palomino Horse Breeders Association. P.O. Box 249, Mineral Wells, TX 76067.

Peruvian Horse Review. P.O. Box 816. Guerneville, CA 95446.

Peruvian Horse World. Box 2035, California City, CA 93505.

Pinto Horse. 7525 Mission George Road, Suite C, San Diego, CA 92103.

Pony of the Americas. Pony of the Americas Club. 1452 North Federal Street, P.O. Box 1447, Mason City, IA 50401.

Practical Horseman. 225 South Church, West Chester, PA 19380.

Quarter Horse Digest. Route 2, Box 14, Gann Valley, SD 57341.

Quarter Horse Journal. P.O. Box 9105, Amarillo, TX 79105.

Quarter Horse Racing World. P.O. Box 1597, Roswell, NM 88201.

Que Paso? American Association of Owners and Breeders of Peruvian Paso Horses. P.O. Box 371, Calabasas, CA 92302.

Rangerbred News. Colorado Ranger Horse Association. 7023 Eden Mill Road, Woodbine, MD 50677.

Rodeo News. 703 North Cedar, P.O. Box 587, Pauls Valley, OK 73075.

Saddle and Bridle. 2333 Bretwood Boulevard, St. Louis, MO 63144.

Side-Saddle News. R.D. #2, Box 2096, Mt. Holly, NJ 08060.

Southern Horseman. P.O. Box 5735, Meridian, MS 39301.

Spanish-Barb Quarterly. Box 7479, Colorado Springs, CO 80907.

Spanish Mustang News. 2005 Ridgeway, Colorado Springs, CO 80906.

Tack 'n Togs. Miller Publishing Company. P.O. Box 67, Minneapolis, MN 55440.

Thoroughbred of California. 201 Colorado Place, Arcadia, CA 91006.

Thoroughbred Record. P.O. Box 11788, Lexington, KY 40578.

Trail Rider. Trail Rider Publications. Route 1, Box 387, Chatsworth, GA 30705.

Trottingbred. 575 Broadway, Hanover, PA 17331.

Turf and Sport Digest. 511-513 Oakland Avenue, Baltimore, MD 21212.

Voice of the Tennessee Walking Horse. Voice Publishing Company. 3710 Calhoun Avenue, P.O. Box 6009, Chattanooga, TN 37401.

Washington Horse. P.O. Box 88258, Seattle, WA 98188.

Welsh News. Welsh Pony Society of America. P.O. Drawer A, White Post, VA 22663.

Western Horseman. P.O. Box 7980, Colorado Springs, CO 80933.

Whip. American Driving Society. Kings Hill Road, Sharon, CT 06069.

Your Pony. Box 125, Barboo, WI 53913.

RATION FORMULATION

Rations will be used to demonstrate the method, using rules of thumb refined by the National Research Council recommendations.

Example 1

The first sample ration formulated will be for a 500-kg (1,100-lb) gelding that is ridden occasionally for ½ hour at a time; that is, a gelding fed a maintenance ration. Calculations are given in both the metric and avoirdupois weight systems.

Step 1. Determine the daily nutrient requirements. These requirements are given in Appendix Table 5-3.

Digestible energy	16.39 Mcal
Total protein	630 gm, or 1.39 lb
Digestible protein	290 gm, or 0.64 lb
Calcium	23 gm, or 0.05 lb
Phosphorus	14 gm, or 0.031 lb
Vitamin A	12,500 IU

Step 2. From Appendix Table 5-4 on rules of thumb, we can make a tentative decision to feed the horse a roughage ration. The limiting factor for its ration should

be the energy content. Therefore, the amount of a particular roughage that will provide the necessary energy is determined by the equation:

$$\frac{\text{Mcal digestible energy required by horse}}{\text{Mcal digestible energy per kg (or lb) of dry matter in roughage}} = \begin{array}{l}\text{weight of roughage}\\\text{on a dry matter}\\\text{basis that is}\\\text{necessary}\end{array}$$

If we decide to use alfalfa hay (hay, s-c, early-bloom), its nutrient content is given in Appendix Table 5-6, Line 3. This alfalfa hay contains 90 percent dry matter. The dry matter contains 2.42 Mcal/kg, or 1.1 Mcal/lb; 17.2 percent crude protein; 13.4 percent digestible protein; 1.75 percent Ca; and 0.26 percent P. This hay contains about 26.1 mg carotene/kg, or 11.86 mg carotene/lb (Appendix Table 5-7). Therefore, from Appendix Table 5-7 we can calculate that the vitamin A content is 10,440 IU per kg, or 4,744 IU per pound.

The amount of energy required by the horse is contained in 6.77 kg (14.9 lb) dry matter:

$$\frac{16.39 \text{ Mcal}}{2.42 \text{ Mcal/kg}} = 6.77 \text{ kg of dry matter}$$

or

$$\frac{16.39 \text{ Mcal}}{1.1 \text{ Mcal/lb}} = 14.9 \text{ lb of dry matter}$$

Step 3. The ration must be checked to see if the remainder of the nutrient requirements is met by feeding 6.77 kg (14.9 lb) of dry matter (alfalfa hay). The nutrient content for alfalfa hay is given in Appendix Table 5-6.

Total protein = weight × CP content:
$$6.77 \text{ kg} \times .172 = 1.164 \text{ kg}$$
or
$$14.9 \text{ lb} \times .172 = 2.56 \text{ lb}$$

Digestible protein = weight × DP content:
$$6.77 \text{ kg} \times .134 = 0.907 \text{ kg}$$
or
$$14.9 \text{ lb} \times .134 = 2.0 \text{ lb}$$

Calcium = weight × calcium content:
$$6.77 \text{ kg} \times .0175 = 0.118 \text{ kg}$$
or
$$14.9 \text{ lb} \times .0175 = 0.261 \text{ lb}$$

Phosphorus = weight × phosphorus content:
$$6.77 \text{ kg} \times .0026 = 0.018 \text{ kg}$$
or
$$14.9 \text{ lb} \times .0026 = 0.039 \text{ lb}$$

Vitamin A = weight × Vitamin A content:
$$6.77 \text{ kg} \times 10,440 = 70,679 \text{ IU}$$
or
$$14.9 \text{ lb} \times 4,744 = 70,686 \text{ IU}$$

Then compare the daily requirements with the ration contents:

	Requirements	Ration Contents	Comments
Digestible energy	16.39 Mcal	16.39 Mcal	Ration meets requirement.
Digestible protein	290 gm *or 0.64 lb*	1,164 gm *or 2.56 lb*	Ration has fourfold requirement.
Total protein	630 gm *or 1.39 lb*	907 gm *or 2.0 lb*	Ration has excessive protein, and it may be possible to reduce amount of ration since some protein will be used for energy.
Calcium	23 gm *or 0.05 lb*	118 gm *or 0.263 lb*	Ration has fivefold requirement.
Phosphorus	14 gm *or 0.031 lb*	18 gm *or 0.039 lb*	Ration meets requirement.
Vitamin A	12,500 IU	70,679 IU	Ration has fivefold requirement.

Therefore the ration meets the daily nutrient requirements. The other necessary questions are (1) Will the horse consume the ration? and (2) Is the Ca:P ratio acceptable? The horse will consume up to 1.5 percent of its body weight, which is 7.5 kg (16.5 lb) and is greater than the weight of the ration. The Ca:P ratio is 6.6 (118 gm Ca/18 gm P, or 0.263 lb Ca/0.039 lb P), which is more than 6:1. Therefore, it would be advisable to mix a phosphorus supplement with the trace-mineralized salt.

Step 4. The final step is to convert the ration to an "as fed" basis, because the calculations were based on 100 percent dry matter. Alfalfa is 90 percent dry matter; therefore, the equation for conversion is

$$\frac{\text{Weight in dry matter}}{\text{\% dry matter}} = \text{weight "as fed"}$$

$$\frac{6.77 \text{ kg}}{0.9} = 7.52 \text{ kg} \quad or \quad \frac{14.9 \text{ lb}}{0.9} = 16.55 \text{ lb}$$

Example 2

The calculations for a more difficult ration will be used to demonstrate how to use a combination of roughage and grain. The calculations are taken from the National Academy of Sciences, National Research Council (1978) bulletin, *Nutrient Requirements for Horses.*

The basic idea is to compare the animal's needs to the amount supplied and, if the nutrients are not adequate, to determine what feedstuffs should be added and how much is needed to correct the deficiency. All of the calculations in the example are on a dry matter basis.

Calculate ration for yearling, 12 months old, weighing 325 kg (715 lb), with expected mature weight of 500 kg (1,100 lb). The hay available is timothy, sc prehead (Appendix Table 5-6, line 64).

Step 1. Compare the nutrient content of the forage with the requirements of the horse. Use actual analysis if available; if not, use Appendix Tables 5-6 and 5-7 for the forage content. The approximate horse requirements for the ration are from Table 5-4 in the text.

	Digestible energy		Crude Protein (%)	Calcium (%)	Phosphorus (%)	Vitamin A	
	(Mcal/kg)	(Mcal/lb)				(IU/kg)	(IU/lb)
Forage supplies	2.2	1.0	11.5	0.5	0.25	3,600	1,636
Horse needs	2.8	1.3	13.5	0.55	0.40	2,000	900

Timothy hay alone lacks energy, protein, and phosphorus concentration. Therefore, a grain supplement rich in these nutrients must be added. According to Table 5-3 in the text, the yearling will eat 6.0 kg (13.2 lb) of a ration containing 2.8 Mcal of DE per kg (1.3 Mcal/lb). A mixture of equal parts of hay containing 2.2 Mcal/kg (1.0 Mcal/lb) and grain with 3.4 Mcal/kg (1.54 Mcal/lb) would provide 2.8 Mcal/kg (1.27 Mcal/lb); that is,

$$\frac{2.2 + 3.4}{2} = 2.8$$

Corn and oats in equal parts would provide 3.6 Mcal per kilogram (1.64 Mcal/lb); that is,

$$\frac{3.87 + 3.34}{2} = 3.6$$

(See Appendix Table 5-6, lines 36 and 47.)

Step 2. Calculate the crude protein. The yearling required 13.5 percent crude protein in the total ration. One part of hay with 11.5 percent crude protein plus 1 part of grain (corn and oats, equal parts) with 12.25 percent crude protein will contain 11.87 percent protein.

$$11.87\% = \frac{11.5 + \left(\dfrac{13.6 + 10.9}{2}\right)}{2}$$

in the total ration. The "square" method can be used to calculate what level of protein is required in the grain ration in order to correct the protein deficiency of the hay and corn–oats ration:

Grain = ?   2.0 = 1 part

                    13.5% desired in
                    total ration

Hay = 11.5%   2.0 = 1 part
                                            Total   2 parts

The difference between the percent crude protein in the hay and the desired level (13.5 percent) is 2.0 percent, which is equivalent to 1 part of the ration. Therefore, 2.0

percent (1 part) added to 13.5 percent gives 15.5 percent, the required percentage of crude protein in the grain ration. That is, 15.5 percent plus 11.5 percent divided by 2 = 13.5 percent in the total ration.

The same "square" method can be used to determine how much protein supplement should be added to the grain in order to attain a mixture with 15.5 percent protein. Soybean meal contains 50.9 percent protein.

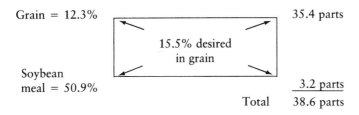

Grain = 12.3%   35.4 parts

15.5% desired
in grain

Soybean
meal = 50.9%   3.2 parts
                                                    Total   38.6 parts

Subtracting diagonally, we determine that a mixture of 35.4 parts of grain and 3.2 parts of soybean meal would result in a concentrate containing 15.5 percent protein or, in other words, 91.7 percent (35.4 parts grain/38.6 parts total) corn–oats and 8.3 percent (3.2 parts soybean meal/38.6 parts total) soybean meal is needed. Check to be certain that the protein level is adequate. According to Appendix Table 5-3, the yearling needs 0.76 kg (1.67 lb) of crude protein daily. Multiply the amount fed by the concentration:

$$3.0 \text{ kg (6.6 lb) of hay} \times 11.5\% = 0.345 \text{ kg (0.76 lb) of protein}$$
$$3.0 \text{ kg (6.6 lb) of grain} \times 15.5\% = 0.465 \text{ kg (1.02 lb) of protein}$$

The total of 0.81 kg (1.78 lb) is a little more than the requirement.

Step 3. Now determine the calcium and phosphorus status. Again multiply amount fed by concentration. The grain concentration of 0.09 percent Ca was determined by corn containing 0.05 percent Ca; oats, 0.07 percent Ca; and soybean meal, 0.31 percent Ca. Therefore, $(0.46 \times 0.05) + (0.46 \times 0.07) + (0.08 \times 0.31) = 0.083$ percent Ca in the mixture.

For total calcium,

$$3.0 \text{ kg of hay} \times 0.5\% \text{ Ca} = 0.015 \text{ kg} \times 1{,}000 = 15 \text{ gm}$$

or

$$6.6 \text{ lb of hay} \times 0.5\% \text{ Ca} = 0.033 \text{ lb}$$

$$3.0 \text{ kg of grain} \times 0.083\% \text{ Ca} = 0.0025 \times 1{,}000 = 2.5 \text{ gm}$$

or

$$6.6 \text{ lb of grain} \times 0.083\% \text{ Ca} = 0.005 \text{ lb}$$

The total is 17.5 g (or 0.038 lb).

The grain concentration of 0.37 percent P was determined by corn containing 0.60 percent P; oats, 0.37 percent P; and soybean meal, 0.70 percent P. Therefore, $(0.46 \times 0.31) + (0.46 \times 0.37) + (0.08 \times 0.71) = 0.37$ percent P in the mixture.

For total phosphorus,

$$3.0 \text{ kg of hay} \times 0.25\% \text{ P} = 0.0075 \text{ kg} \times 1{,}000 = 7.5 \text{ gm}$$

or

$$6.6 \text{ lb of hay} \times 0.25\% \text{ P} = 0.016 \text{ lb}$$

$$3.0 \text{ kg of grain} \times 0.37\% \text{ P} = 0.011 \text{ kg} \times 1{,}000 = 11.0 \text{ gm}$$

or

$$6.6 \text{ lb of hay} \times 0.37\% \text{ P} = 0.024 \text{ lb}$$

The total is 18.5 g (0.04 lb).

Subtract the amount of calcium and phosphorus supplied from the amount needed:

Yearling requires	31 gm (0.068 lb) of Ca	22 gm (0.048 lb) of P
Hay and grain supply	− 17.5 gm (0.038 lb)	− 18.5 gm (0.04 lb)
Additional needed	13.5 gm (0.030 lb)	3.5 gm (0.008 lb)

When phosphorus is supplied by dicalcium phosphate containing 24 percent calcium and 19 percent phosphorus (Table 5-2 in the text), divide amount needed by concentration: 3.5 gm P needed/0.19 = 18.4 gm, *or* 0.008 lb/0.19 = 0.042 lb, of dicalcium phosphate needed. Therefore, there is 0.6 percent dicalcium phosphate in the grain mixture (0.0184 kg/3.0 kg grain, *or* 0.042 lb/6.6). The 18.4 gm (0.042 lb) of dicalcium phosphate would also supply 4.4 gm (18.4 gm × 0.24, *or* 0.042 lb × 0.24 = 0.010 lb) of calcium. Therefore, 14.4 − 3.5 = 0.9 gm (*or* 0.042 − 0.030 = 0.12 lb) of calcium more than needed.

APPENDIX TABLE 5-1
Nutrient Requirements of Horses (Daily Nutrients per Horse) and Ponies, 200 kg Mature Weight

	Weight		Daily Gain		Digestible Energy (Mcal)	TDN		Crude Protein		Digestible Protein		Calcium (gm)	Phosphorus (gm)	Vitamin A Activity (1,000 IU)	Daily Feed[a]	
	kg	lb	kg	lb		kg	lb	kg	lb	kg	lb				kg	lb
Mature ponies, maintenance	200	440	0.0	—	8.24	1.87	4.12	0.32	0.70	0.14	0.31	9	6	5.0	3.75	8.2
Mares, last 90 days gestation	—	—	0.27	0.594	9.23	2.10	4.62	0.39	0.86	0.20	0.44	14	9	10.0	3.70	8.1
Lactating mare, first 3 months (8 kg milk per day)	—	—	0.0	—	14.58	3.31	7.29	0.71	1.56	0.54	1.19	24	16	13.0	5.20	11.5
Lactating mare, 3 months to weaning (6 kg milk per day)	—	—	0.0	—	12.99	2.95	6.50	0.60	1.32	0.34	0.75	20	13	11.0	5.00	11.0
Nursing foal (3 months of age)	60	132	0.70	1.54	7.35	1.67	3.68	0.41	0.90	0.38	0.84	18	11	2.4	2.25	5.0
Requirements above milk	—	—	—	—	3.74	0.85	1.87	0.17	0.37	0.20	0.44	10	7	0.0	1.20	2.7
Weanling (6 months of age)	95	209	0.50	1.10	8.80	2.0	4.40	0.47	1.03	0.31	0.68	19	14	3.8	2.85	6.3
Yearling (12 months of age)	140	308	0.20	0.44	8.15	1.85	4.07	0.35	0.77	0.20	0.44	12	9	5.5	2.90	6.4
Long yearling (18 months of age)	170	374	0.10	0.22	8.10	1.84	4.05	0.32	0.70	0.17	0.37	11	7	6.0	3.10	6.8
2-year-old (24 months of age)	185	407	0.05	0.11	8.10	1.84	4.05	0.30	0.66	0.15	0.33	10	7	5.5	3.10	6.8

[a]Dry matter basis.

SOURCE: National Research Council (1978).

593

APPENDIX TABLE 5-2
Nutrient Requirements of Horses (Daily Nutrients per Horse), 400 kg Mature Weight

	Weight		Daily Gain		Digestible Energy (Mcal)	TDN		Crude Protein		Digestible Protein		Calcium (gm)	Phosphorus (gm)	Vitamin A Activity (1,000 IU)	Daily Feed[a]	
	kg	lb	kg	lb		kg	lb	kg	lb	kg	lb				kg	lb
Mature horses, maintenance	400	880	0.0	—	13.86	3.15	6.93	0.54	1.19	0.24	0.53	18	11	10.0	6.30	13.9
Mares, last 90 days gestation	—	—	0.53	1.17	15.52	3.53	7.76	0.64	1.41	0.34	0.75	27	19	20.0	6.20	13.7
Lactating mare, first 3 months (12 kg milk per day)	—	—	0.0	—	23.36	5.31	11.68	1.12	2.46	0.68	1.50	40	27	22.0	8.35	18.4
Lactating mare, 3 months to weanling (8 kg milk per day)	—	—	0.0	—	20.20	4.59	10.10	0.91	2.00	0.51	1.12	33	22	18.0	7.75	17.1
Nursing foal (3 months of age)	125	275	1.00	2.2	11.51	2.62	5.76	0.65	1.43	0.50	1.10	27	17	5.0	3.55	7.8
Requirements above milk	—	—	—	—	6.10	1.39	3.05	0.40	0.88	0.30	0.66	15	12	0.0	1.95	4.3
Weanling (6 months of age)	185	407	0.65	1.43	13.03	2.96	6.51	0.66	1.45	0.43	0.95	27	20	7.4	4.20	9.2
Yearling (12 months of age)	265	583	0.40	0.88	13.80	3.14	6.91	0.60	1.32	0.35	0.77	24	17	10.0	4.95	10.9
Long yearling (18 months of age)	330	726	0.25	0.55	14.36	3.26	7.17	0.59	1.30	0.32	0.70	22	15	11.5	5.50	12.2
2-year-old (24 months of age)	365	803	0.10	0.22	13.89	3.16	6.95	0.52	1.14	0.27	0.59	20	13	11.0	5.35	11.8

[a]Dry matter basis.

SOURCE: National Research Council (1978).

APPENDIX TABLE 5-3
Nutrient Requirements of Horses (Daily Nutrients per Horse), 500 kg Mature Weight

	Weight		Daily Gain		Digestible Energy (Mcal)	TDN		Crude Protein		Digestible Protein		Calcium (gm)	Phos-phorus (gm)	Vitamin A Activity (1,000 IU)	Daily Feed[a]	
	kg	lb	kg	lb		kg	lb	kg	lb	kg	lb				kg	lb
Mature horses, maintenance	500	1,100	0.0	—	16.39	3.73	8.20	0.63	1.39	0.29	0.64	23	14	12.5	7.45	16.4
Mares, last 90 days gestation	—	—	0.55	1.21	18.36	4.17	9.18	0.75	1.65	0.39	0.86	34	23	25.0	7.35	16.2
Lactating mare, first 3 months (15 kg milk per day)	—	—	0.0	—	28.27	6.43	14.14	1.36	2.99	0.84	1.85	50	34	27.5	10.10	22.2
Lactating mare, 3 months to weaning (10 kg milk per day)	—	—	0.0	—	24.31	5.53	12.16	1.10	2.42	0.62	1.36	41	27	22.5	9.35	20.6
Nursing foal (3 months of age)	155	341	1.20	2.64	13.66	3.10	6.83	0.75	1.65	0.54	1.19	33	20	6.2	4.20	9.2
Requirements above milk	—	—	—	—	6.89	1.57	3.45	0.41	0.90	0.31	0.68	18	13	0.0	2.25	4.9
Weanling (6 months of age)	230	506	0.80	1.76	15.60	3.55	7.80	0.79	1.74	0.52	1.14	34	25	9.2	5.00	11.0
Yearling (12 months of age)	325	715	0.55	1.21	16.81	3.82	8.41	0.76	1.67	0.45	0.99	31	22	12.0	6.00	13.2
Long yearling (18 months of age)	400	880	0.35	0.77	17.00	3.90	8.58	0.71	1.56	0.39	0.86	28	19	14.0	6.50	14.3
2-year-old (24 months of age)	450	990	0.15	0.33	16.45	3.74	8.23	0.63	1.39	0.33	0.72	25	17	13.0	6.60	14.5

[a]Dry matter basis.
SOURCE: National Research Council (1978).

APPENDIX TABLE 5-4
Nutrient Requirements of Horses (Daily Nutrients per Horse), 600 kg Mature Weight

	Weight		Daily Gain		Digestible Energy (Mcal)	TDN		Crude Protein		Digestible Protein		Calcium (gm)	Phos-phorus (gm)	Vitamin A Activity (1,000 IU)	Daily Feed[a]	
	kg	lb	kg	lb		kg	lb	kg	lb	kg	lb				kg	lb
Mature horses, maintenance	600	1,320	0.0	—	18.79	4.27	9.40	0.73	1.61	0.33	0.73	27	17	15.0	8.50	18.8
Mares, last 90 days gestation	—	—	0.67	1.47	21.04	4.78	10.52	0.87	1.91	0.46	1.01	40	27	30.0	8.40	18.5
Lactating mare, first 3 months (18 kg milk per day)	—	—	0.0	—	33.05	7.51	16.53	1.60	3.52	0.99	2.18	60	40	33.0	11.80	26.0
Lactating mare, 3 months to weanling (12 kg milk per day)	—	—	0.0	—	28.29	6.43	14.15	1.29	2.84	0.73	1.61	49	30	27.0	10.90	23.9
Nursing foal (3 months of age)	170	374	1.40	3.08	15.05	3.42	7.53	0.84	1.85	0.78	1.72	36	23	6.8	4.65	10.2
Requirements above milk	—	—	—	—	6.93	1.58	3.47	0.51	1.12	0.38	0.84	18	15	0.0	2.25	4.9
Weanling (6 months of age)	265	583	0.85	1.87	16.92	3.85	8.47	0.86	1.89	0.57	1.25	37	27	10.6	5.45	12.0
Yearling (12 months of age)	385	847	0.60	1.32	18.85	4.28	9.42	0.90	1.98	0.50	1.10	35	25	14.0	6.75	14.8
Long yearling (18 months of age)	475	1,045	0.35	0.77	19.06	4.33	9.53	0.75	1.65	0.43	0.95	32	22	13.5	7.35	16.2
2-year-old (24 months of age)	540	1,188	0.20	0.44	19.26	4.38	9.64	0.74	1.63	0.39	0.86	31	20	13.0	7.40	16.3

[a]Dry matter basis.

SOURCE: National Research Council (1978).

APPENDIX TABLE 5-5
Dietary Minerals and Vitamins for Horses

	Adequate Levels		
	Maintenance of Mature Horses	Growth	Toxic Levels[a]
Calcium	—[b]	—[b]	
Phosphorus	—[b]	—[b]	
Sodium, %	0.35	0.35	
Potassium, %	0.4	0.5	
Magnesium, %	0.09	0.1	
Sulfur, %	0.15	0.15	
Iron, mg/kg	40	50	
Zinc, mg/kg	40	40	9,000
Manganese, mg/kg	40	40	*
Copper, mg/kg	9	9	*
Iodine, mg/kg	0.1	0.1	4.8
Cobalt, mg/kg	0.1	0.1	
Selenium, mg/kg	0.1	0.1	5.0
Fluorine, mg/kg	—	—	50+
Lead, mg/kg	—	—	80
Vitamin A	—[b]	—[b]	*
Vitamin D, IU/kg	275	275	150,000
Vitamin E, mg/kg	15	15	
Thiamin, mg/kg	3	3	
Riboflavin, mg/kg	2.2	2.2	
Pantothenic acid, mg/kg	15	15	

[a]Nutrients known to be toxic to other species but without adequate information on the horse are indicated by *.
[b]See Appendix Tables 5-1 to 5-4.

SOURCE: National Research Council (1978).

APPENDIX TABLE 5-6

Composition of Feeds Commonly Used in Horse Diets—Dry Basis (moisture free)

Line No.	Short Feed Name Scientific Name	International Feed Number[a]	Dry Matter (%)	DE (Mcal/ kg)	DE (Mcal/ lb)	TDN (%)	Crude Protein (%)	Digestible Protein (%)	Lysine (%)	Crude Fiber (%)	Cell Walls (%)	ADF (%)
	Alfalfa											
	Medicago sativa											
1	grazed, prebloom	2-00-181	21	2.51	1.14	57	21.2	15.6	1.06	22	—	—
2	grazed, full bloom	2-00-188	25	2.29	1.04	52	16.3	11.4	0.65	33	—	—
3	hay, s-c, early bloom	1-00-059	90	2.42	1.10	55	17.2	13.4	0.94	31	48	38
4	hay, s-c, midbloom	1-00-063	89	2.29	1.04	52	16.0	11.6	0.90	32	50	40
5	hay, s-c, full bloom	1-00-068	89	2.16	0.98	49	15.0	10.1	0.64	34	52	42
6	meal, dehy, 15% protein	1-00-022	91	2.42	1.10	55	16.3	11.8	0.66	33	51	41
7	meal, dehy, 17% protein	1-00-023	92	2.46	1.12	56	19.7	13.9	0.96	27	45	35
	Bahiagrass											
	Paspalum notatum											
8	grazed	2-00-464	30	2.11	0.96	48	7.9	4.2	—	32	—	—
9	hay, s-c	1-00-462	91	1.89	0.86	43	5.8	2.5	—	30	—	—
	Barley											
	Hardeum vulgare											
10	grain	4-00-549	89	3.61	1.64	82	13.9	11.4	0.48	6	19	7
11	grain, Pacific Coast	4-00-939	90	3.48	1.58	79	10.7	7.0	0.35	7	21	9
12	hay, s-c	1-00-495	89	1.89	0.86	44	8.5	4.7	—	27	—	—
13	straw	1-00-498	90	1.63	0.74	37	4.0	0.9	—	42	80	59
	Beet, Sugar											
	Beta vulgaris,											
	B. saccharifera											
14	pulp, dehy	4-00-669	91	2.86	1.30	65	8.0	5.0	0.66	22	59	34
	Bermudagrass											
	Cynodon dactylon											
15	grazed	2-00-712	39	2.20	1.00	50	9.1	5.2	—	28	—	—
16	hay, s-c	1-00-716	91	1.98	0.90	45	7.0	4.2	—	34	80	35
	Bluegrass, Kentucky											
	Pao pratensis											
17	grazed, early	2-00-777	31	2.46	1.12	56	17.0	12.4	—	26	—	—
18	grazed, posthead	2-00-782	35	2.20	1.00	50	11.6	7.4	—	27	—	—
19	hay, s-c	1-00-776	90	2.20	1.00	50	11.0	5.1	—	30	—	—
	Brewer's											
20	grains, dehy	5-02-141	92	2.99	1.36	68	27.0	20.9	0.95	16	42	23
	Brome											
	Bromus spp											
21	grazed, vegetative	2-00-892	32	3.00	1.36	68	18.3	12.6	—	24	60	31
22	hay, s-c, late bloom	1-00-888	90	2.38	1.08	54	7.4	5.0	—	40	72	44
	Canarygrass, Reed											
	Phalaris arundinacea											
23	grazed	2-01-113	27	2.38	1.08	54	12.0	7.5	—	29	—	—
24	hay	1-01-104	91	2.16	0.98	49	12.3	7.6	—	33	—	—
	Citrus											
25	pulp wo fines, dehy	4-01-237	90	2.99	1.36	68	6.9	3.6	—	14	23	23
	Clover, Alsike											
	Trifolium hybridum											
26	hay, s-c	1-01-313	89	2.11	0.96	48	14.8	10.1	—	29	—	—
	Clover, Crimson											
	Trifolium incarnatam											
27	grazed	2-01-336	17	2.42	1.10	55	17.2	12.1	—	27	—	—
28	hay, s-c	1-01-328	89	2.16	0.98	49	18.0	13.1	—	32	—	—

[a]First digit is class of feed: 1, dry forages and roughages; 2, pasture, range plants, and forages fed green; 3, silages; 4, energy feeds; 5, protein supplements; 6, minerals; 7, vitamins; 8, additives.

Cellulose (%)	Lignin (%)	Calcium (%)	Copper		Iron		Magnesium (%)	Manganese		Phosphorus (%)	Potassium (%)	Sodium (%)	Sulfur (%)	Zinc	
			(mg/kg)	(mg/lb)	(mg/kg)	(mg/lb)		(mg/kg)	(mg/lb)					(mg/kg)	(mg/lb)
—	—	2.26	10	4.54	200	90.9	0.25	28	12.7	0.35	2.35	0.20	0.50	18	8.2
—	—	1.53	9	4.09	330	150.0	0.27	25	11.4	0.27	2.15	0.15	0.31	15	6.8
28	10	1.75	15	6.82	200	90.9	0.30	32	14.5	0.26	2.55	0.15	0.29	17	7.7
29	11	1.50	13	5.91	180	81.8	0.29	29	13.2	0.25	1.90	0.14	0.28	17	7.7
30	12	1.29	12	5.45	170	77.3	0.31	27	12.3	0.24	1.80	0.14	0.26	17	7.7
29	12	1.40	11	5.00	330	150.0	0.30	31	14.1	0.24	2.50	0.10	0.20	22	10.0
24	11	1.50	10	4.54	400	181.8	0.39	31	14.1	0.26	2.70	0.10	0.26	22	10.0
—	—	0.45	—	—	60	27.3	0.25	—	—	0.19	1.45	—	—	—	—
—	—	0.45	—	—	60	27.3	0.19	—	—	0.22	1.45	—	—	—	—
—	—	0.05	9	4.09	90	40.9	0.15	19	8.6	0.37	0.45	0.03	0.18	17	7.7
—	—	0.05	9	4.09	80	36.4	0.13	18	8.2	0.37	0.58	0.02	0.17	17	7.7
—	—	0.21	4	1.82	300	136.4	0.19	39	17.7	0.31	1.49	0.14	0.17	—	—
37	12	0.24	10	4.54	300	136.4	0.15	17	7.7	0.05	2.01	0.14	0.17	—	—
—	—	0.75	14	6.36	330	150.0	0.30	38	17.3	0.10	0.20	0.23	0.22	10	4.5
—	—	0.49	—	—	—	—	0.19	—	—	0.27	—	—	—	—	—
23	12	0.40	—	—	—	—	0.17	—	—	0.19	1.57	0.44	—	20	9.1
—	—	0.56	10	4.54	—	—	0.20	79	35.9	0.40	2.20	—	—	—	—
—	—	0.46	9	4.09	—	—	0.18	68	30.9	0.39	2.01	—	—	—	—
—	—	0.30	9	4.09	260	118.2	0.16	93	42.3	0.29	1.70	0.14	0.13	—	—
18	5	0.30	24	10.90	270	122.7	0.17	42	19.1	0.58	0.09	0.28	0.34	30	13.6
27	4	0.55	5	2.27	100	45.4	0.18	—	—	0.35	2.32	0.02	0.20	—	—
36	8	0.32	7	3.18	100	45.4	0.13	106	48.2	0.22	2.00	0.02	0.20	—	—
—	—	0.42	9	4.09	150	68.2	—	—	—	0.35	3.64	—	—	—	—
—	—	0.37	9	4.09	150	68.2	0.31	106	48.2	0.25	1.86	0.39	0.41	—	—
—	—	2.07	6	2.73	170	77.3	0.16	7	3.18	0.13	0.77	0.10	0.07	16	7.3
—	—	1.32	6	2.73	260	118.2	0.41	69	31.4	0.29	2.46	0.46	0.17	—	—
—	—	1.33	—	—	250	113.6	0.29	317	144.1	0.32	2.51	0.40	0.28	—	—
—	—	1.39	—	—	300	136.4	0.29	200	90.9	0.20	2.00	0.39	0.28	—	—

(continued)

APPENDIX TABLE 5-6 *(continued)*

Line No.	Short Feed Name Scientific Name	International Feed Number[a]	Dry Matter (%)	DE (Mcal/ kg)	DE (Mcal/ lb)	TDN (%)	Crude Protein (%)	Digestible Protein (%)	Lysine (%)	Crude Fiber (%)	Cell Walls (%)	ADF (%)	Cellulose (%)
	Clover, Ladino *Trifolium repens*												
29	hay, s-c	1-01-378	90	2.24	1.02	51	21.0	15.6	—	20	36	32	25
	Clover, Red *Trifolium pratense*												
30	grazed, early bloom	2-01-428	20	2.51	1.14	57	21.1	12.0	—	19	—	—	—
31	grazed, late bloom	2-01-429	26	2.42	1.10	55	14.5	9.8	—	30	—	—	—
32	hay, s-c	1-01-415	89	2.16	0.98	49	14.9	10.0	—	30	56	41	30
	Corn *Zea mays*												
33	cobs, ground	1-02-782	90	1.36	0.62	31	2.8	0.5	—	36	89	35	28
34	distillers grains, dehy	5-02-842	92	3.08	1.40	70	29.8	21.0	0.87	12	43	—	—
35	ears, grnd	4-02-849	87	3.26	1.48	74	9.1	5.6	0.20	10	—	—	—
36	grain	4-02-985	88	3.87	1.76	88	10.9	8.5	0.30	2	—	—	—
	Cotton *Gossypium* spp												
37	hulls	1-01-599	91	1.45	0.66	33	4.2	1.1	—	50	90	71	48
	Fescue, Meadow *Festuca elatior*												
38	grazed	2-01-920	27	2.29	1.04	52	11.5	7.3	—	29	—	—	—
39	hay, s-c	1-01-912	88	2.02	0.92	46	10.5	5.8	—	33	65	43	37
	Flax *Linum usitatissimum*												
40	seeds, meal, solv extd (Linseed meal)	5-02-048	91	3.04	1.38	69	38.9	27.6	1.34	10	—	—	—
	Lespedeza *L. striata, L. stipulacea*												
41	grazed	2-02-568	31	2.20	1.00	50	14.9	10.2	—	38	—	—	—
42	hay, s-c	1-08-591	91	2.07	0.94	47	13.9	9.3	—	32	—	—	—
	Linseed—see Flax Milk *Bos taurus*												
43	skimmed, dehy	5-01-175	94	4.05	1.84	92	36.0	30.3	2.69	0.3	—	—	—
	Molasses												
44	beet, sugar, mn 48% invert	4-00-668	78	3.17	1.44	72	8.7	5.3	—	—	—	—	—
45	sugarcane, molasses, dehy	4-04-695	94	3.17	1.44	72	9.3	5.8	—	5.0	—	—	—
46	sugarcane, molasses, mn 48% invert	4-04-696	75	3.26	1.48	74	4.3	2.0	—	—	—	—	—
	Oats *Avena sativa*												
47	grain	4-03-309	89	3.34	1.52	76	13.6	10.5	—	12	31	17	14
48	grain, Pacific Coast	4-07-999	91	3.34	1.52	77	10.1	6.5	—	12	—	—	—
49	hay, s-c	1-03-280	90	2.07	0.94	47	8.9	5.1	—	32	—	36	30
50	straw	1-03-283	92	2.11	0.96	40	4.3	2.5	—	40	70	47	34
	Orchardgrass *Dactylis glomerata*												
51	grazed	2-03-439	19	2.42	1.10	55	18.4	13.2	—	27	55	31	28
52	hay, s-c	1-03-438	89	2.07	0.95	47	10.1	6.1	—	36	—	—	—
	Pangolagrass *Digitara decumbens*												
53	grazed	2-03-493	19	2.24	1.02	51	12.5	8.1	—	29	—	—	—
54	hay, s-c	1-09-459	88	1.98	0.90	45	9.6	5.7	—	27	—	—	—
	Prairie												
55	midwest, hay, s-c	1-03-191	90	2.02	0.92	46	6.7	3.2	—	33	—	—	—

Lignin (%)	Calcium (%)	Copper (mg/kg)	Copper (mg/lb)	Iron (mg/kg)	Iron (mg/lb)	Magnesium (%)	Manganese (mg/kg)	Manganese (mg/lb)	Phosphorus (%)	Potassium (%)	Sodium (%)	Sulfur (%)	Zinc (mg/kg)	Zinc (mg/lb)
7	1.32	9	4.09	600	272.7	0.29	200	90.90	0.24	2.80	0.39	0.18	17	7.7
—	2.26	—	—	300	136.4	0.51	—	—	0.38	2.49	0.22	0.17	—	—
—	1.01	—	—	306	139.1	0.43	—	—	0.27	1.96	0.20	0.17	—	—
10	1.49	11	5.00	310	140.9	0.45	73	33.20	0.25	1.66	0.18	0.17	17	7.7
7	0.12	7	3.18	230	104.5	0.07	6	2.73	0.04	0.91	—	0.47	—	—
—	0.11	48	21.80	200	90.9	0.08	20	9.09	0.44	0.20	0.10	0.46	35	15.9
—	0.05	8	3.64	80	36.4	0.16	6	2.73	0.26	0.56	0.05	0.22	18	8.2
—	0.05	4	1.82	30	13.6	0.03	6	2.73	0.60	0.35	0.01	0.14	21	9.5
23	0.15	13	5.91	150	68.2	0.14	10	4.54	0.08	0.87	0.02	—	16	7.3
—	0.60	4	1.82	—	—	0.37	27	12.30	0.43	2.34	—	—	—	—
6	0.57	4	1.82	—	—	0.59	24	10.90	0.37	1.74	—	—	—	—
—	0.43	28	12.70	360	163.6	0.67	42	19.10	0.90	1.53	0.15	0.44	—	—
—	1.10	—	—	310	140.9	0.29	154	70.70	0.28	1.26	0.31	—	—	—
—	1.15	—	—	330	150.0	0.25	184	83.60	0.25	1.03	0.30	—	—	—
—	1.30	1	0.45	10	4.54	0.13	2	0.90	1.09	1.66	0.50	0.34	68	30.9
—	0.21	22	10.00	100	45.4	0.30	6	2.70	0.03	6.20	1.52	0.61	18	8.2
—	0.87	73	33.20	240	109.1	0.43	52	23.60	0.20	3.68	0.19	0.46	33	15.0
—	1.05	80	36.40	250	113.6	0.47	57	25.90	0.15	3.80	0.22	0.46	30	13.6
3	0.07	7	3.18	80	36.4	0.19	43	19.50	0.37	0.44	0.18	0.38	33	15.0
—	0.11	6	2.73	90	40.9	0.19	42	19.10	0.34	0.44	0.16	0.23	—	—
6	0.30	4	1.82	400	181.8	0.75	120	54.50	0.26	1.23	0.17	0.30	—	—
13	0.25	10	4.54	200	90.9	0.19	37	16.80	0.07	2.37	0.40	0.23	—	—
3	0.57	7	3.18	170	77.3	0.19	40	18.20	0.54	3.27	0.04	0.21	17	7.7
—	0.35	14	6.36	110	50.0	0.20	40	18.20	0.31	3.01	—	0.26	18	8.2
—	0.45	—	—	—	—	0.14	—	—	0.35	—	—	—	—	—
—	0.37	—	—	—	—	0.13	—	—	0.23	—	—	—	—	—
—	0.41	23	10.50	100	45.4	0.28	48	21.80	0.15	1.01	0.04	—	—	—

(continued)

APPENDIX TABLE 5-6 *(continued)*

Line No.	Short Feed Name Scientific Name	International Feed Number[a]	Dry Matter (%)	DE (Mcal/ kg)	DE (Mcal/ lb)	TDN (%)	Crude Protein (%)	Digest-ible Protein (%)	Lysine (%)	Crude Fiber (%)	Cell Walls (%)	ADF (%)	Cellu-lose (%)
	Rye *Secale cereale*												
56	grain	4-04-047	88	3.52	1.60	80	13.8	9.9	0.48	3	—	—	—
	Sorghum *Sorghum vulgare*												
57	grain	4-04-383	90	3.52	1.60	80	12.6	8.8	0.28	3	—	—	—
	Soybean *Glycine max*												
58	hay, s-c	1-04-558	89	2.11	0.96	48	15.9	11.0	—	34	—	—	—
59	hulls	1-04-560	92	2.64	1.20	60	12.0	7.7	1.61	40	67	46	44
60	seeds	5-04-610	91	4.05	1.84	92	43.2	31.7	2.93	6	—	—	—
61	seeds, meal, solv extd	5-04-604	90	3.60	1.64	82	50.9	35.7	3.28	7	14	10	8
	Sunflower *Helianthus* spp												
62	seeds wo hulls, meal, solv extd	5-04-739	92	3.12	1.42	71	50.3	—	1.85	12	—	—	—
	Timothy *Phleum pratense*												
63	grazed, midbloom	2-04-905	30	2.15	0.98	49	9.6	5.2	—	31	—	—	—
64	hay, s-c, pre-head	1-04-881	89	2.20	1.00	50	11.5	7.2	—	31	64	37	33
65	hay, s-c, head	1-04-883	88	1.98	0.90	45	9.0	4.8	—	32	70	45	34
	Trefoil, Birdsfoot *Lotus corniculatus*												
66	hay, s-c	1-05-044	91	2.20	1.00	50	16.0	12.5	—	30	44	34	25
	Wheat *Triticum* spp												
67	bran	4-05-190	89	2.94	1.34	67	17.0	14.4	0.68	11	45	12	8
68	grain, hard red winter	4-05-268	89	3.83	1.74	87	14.4	10.7	0.42	3	40	—	—
69	grain, soft red winter	4-05-294	89	3.83	1.74	87	13.0	9.2	0.57	3	30	—	—
70	grain, soft white winter	4-05-337	89	3.83	1.74	87	11.5	7.5	0.35	3	14	4	—
71	hay, s-c	1-05-172	89	1.89	0.86	43	8.7	4.9	—	29	68	41	- -
72	straw	1-05-175	89	1.50	0.68	34	4.2	1.0	—	41	85	54	39
	Yeast *Saccharomyces cerevisiae*												
73	Brewer's, dehy	7-05-527	93	3.30	1.50	75	48.3	32.8	3.33	3	—	—	—

Lignin (%)	Calcium (%)	Copper		Iron		Magnesium (%)	Manganese		Phosphorus (%)	Potassium (%)	Sodium (%)	Sulfur (%)	Zinc	
		(mg/kg)	(mg/lb)	(mg/kg)	(mg/lb)		(mg/kg)	(mg/lb)					(mg/kg)	(mg/lb)
—	0.07	8	3.64	70	31.8	0.14	62	28.2	0.36	0.52	0.03	0.17	36	16.4
—	0.03	11	5.00	50	22.7	0.20	17	7.7	0.33	0.39	0.03	0.16	16	7.3
—	1.22	9	4.09	290	131.8	0.79	101	45.9	0.28	1.02	0.09	0.24	24	10.9
2	0.45	18	8.18	320	145.4	—	14	6.4	0.15	1.03	0.05	—	24	10.9
—	0.28	17	7.73	90	40.9	0.31	32	14.5	0.66	1.77	0.13	0.24	18	8.2
2	0.31	30	13.60	130	59.1	0.30	32	14.5	0.70	2.19	0.31	0.48	48	21.8
—	0.41	4	1.82	40	18.2	0.81	25	11.4	1.10	1.10	0.44	—	—	—
—	0.28	11	5.00	200	90.9	0.15	190	86.4	0.25	2.40	0.19	0.13	—	—
4	0.50	6	2.73	200	90.9	0.15	—	—	0.25	1.92	0.18	0.13	—	—
11	0.41	5	2.27	140	63.6	0.16	46	20.9	0.19	1.60	0.18	0.13	—	—
9	1.75	9	4.09	230	104.5	0.51	15	6.8	0.22	1.80	0.18	—	77	35.0
4	0.12	14	6.37	190	86.4	0.59	130	59.1	1.43	1.60	0.04	0.25	120	54.5
—	0.05	5	2.27	40	18.2	0.17	44	20.0	0.48	0.45	0.03	0.18	43	19.5
—	0.05	7	3.18	30	13.6	0.11	36	16.4	0.46	0.46	0.02	0.12	48	21.8
—	0.05	8	3.64	40	18.2	0.11	40	18.2	0.45	0.41	0.02	0.13	30	13.6
—	0.15	—	—	200	90.9	0.12	40	18.2	0.19	1.00	0.28	0.24	—	—
15	0.21	3	1.36	200	90.9	0.12	40	18.2	0.08	1.10	0.14	0.19	—	—
—	0.50	36	16.40	100	45.4	0.25	6		1.52	1.86	0.08	0.41	42	19.1

Abbreviations Used			
dehy	dehydrated	mech extd	mechanically extracted, expeller extracted, hydraulic extracted, or old process
extd	extracted		
extn	extraction		
extn unspecified	extraction unspecified	μg	microgram
gm	gram	mg	milligram
gr	grade	mt	more than
grnd	ground	s-c	sun-cured
IU	International Units	solv extd	solvent extracted
kcal	kilocalories	spp	species
kg	kilogram(s)	w	with
lt	less than	wo	without

Stage of Maturity Terms Used in Appendix Table 5-6

Preferred Term	Definition	Comparable Terms
For Plants That Bloom		
Germinated	Stage in which the embryo in a seed resumes growth after a dormant period	Sprouted
Early vegetative	Stage at which the plant is vegetative and before the stems elongate	Fresh new growth, before heading out, before inflorescence emergence, immature, prebud stage, very immature, young
Prebloom	Stage at which stems are beginning to elongate to just before blooming; first bud to first flowers	Before bloom, bud stage, budding plants, heading to in bloom, heads just showing, jointing and boot (grasses), prebloom, preflowering, stems elongated
Early bloom	Stage between initiation of bloom and stage in which $1/10$ of the plants are in bloom; some grass heads are in anthesis	Early anthesis, first flower, headed out, in head, up to $1/10$ bloom
Midbloom	Stage in which $1/10$ to $2/3$ of the plants are in bloom; most grass heads are in anthesis	Bloom, flowering, flowering plants, half-bloom, in bloom, mid-anthesis
Full bloom	Stage in which $2/3$ or more of the plants are in bloom	Three-fourths to full bloom, late anthesis
Late bloom	Stage in which blossoms begin to dry and fall and seeds begin to form	Fifteen days after silking, before milk, in bloom to early pod, late to past anthesis
Milk stage	Stage in which seeds are well formed but soft and immature	After anthesis, early seed, fruiting, in tassel, late bloom to early seed, past bloom, pod stage, post anthesis, post bloom, seed developing, seed forming, soft, soft immature
Dough stage	Stage in which the seeds are of doughlike consistency	Dough stage, nearly mature, seeds, dough, seeds well developed, soft dent
Mature	Stage in which plants are normally harvested for seed	Dent, dough to glazing, fruiting, fruiting plants, in seed, kernels ripe, ripe seed
Post ripe	Stage that follows maturity; seeds are ripe and plants have been cast and weathering has taken place (applies mostly to range plants)	Late seed, over ripe, very mature
Stem-cured	Stage in which plants are cured on the stem; seeds have been cast and weathering has taken place (applies mostly to range plants)	Dormant, mature and weathered, seeds cast
Regrowth early vegetative	Stage in which regrowth occurs without flowering activity; vegetative crop aftermath; regrowth in stubble (applies primarily to fall regrowth in temperate climates); early dry season regrowth	Vegetative recovery growth

APPENDIX TABLE 5-7
Vitamin Content of Common Feeds

	International Feed Number	Carotene[a]		Riboflavin		Thiamin		Vitamin E	
		(mg/kg)	(mg/lb)	(mg/kg)	(mg/lb)	(mg/kg)	(mg/lb)	(mg/kg)	(mg/lb)
Alfalfa hay, midbloom	1-00-063	26.1	11.90	10.6	4.8	3.0	1.36	90.0	40.9
Alfalfa meal, 17% protein	1-00-023	131.1	59.60	14.4	6.5	3.7	1.68	134.8	61.3
Barley, grain	4-00-549	1.0	0.45	1.8	0.8	5.0	2.27	18.0	8.2
Brewer's grains	5-02-141	—	—	1.4	0.6	0.6	0.27	28.1	12.8
Clover hay, ladino	1-01-378	20.0	9.10	15.0	6.8	4.2	1.91	70.0	31.8
Clover hay, red, sun-cured	1-01-115	19.8	9.00	17.8	8.1	2.2	1.00	60.0	27.3
Corn, grain	4-02-935	2.5	1.10	1.5	0.7	2.3	1.04	25.7	11.7
Linseed meal	5-02-048	—	—	3.1	1.4	8.6	3.91	18.0	8.2
Oats, grain	4-03-309	0.1	0.04	1.7	0.8	7.2	3.27	18.5	8.4
Oat hay, sun-cured	1-03-280	15.0	6.80	5.3	2.4	3.3	1.50	12.5	5.7
Rye, grain	4-04-383	1.0	0.45	1.8	0.8	3.4	1.55	17.2	7.8
Skim milk, dehydrated	5-01-175	—	—	20.4	9.3	4.0	1.82	10.0	4.5
Soybean meal	5-04-604	0.2	0.09	3.3	1.5	6.2	2.82	2.3	1.0
Timothy hay, sun-cured	1-04-893	9.0	4.10	12.4	5.6	1.7	0.77	63.1	28.7
Wheat bran	4-05-190	2.9	1.30	5.4	2.5	7.3	3.32	13.4	6.1
Brewer's yeast	7-05-527	—	—	39.2	17.8	100.0	45.45	2.4	1.1

[a]1 mg = 400 IU; 1 IU = 0.3 µg crystalline vitamin A alcohol; 0.344 µg vitamin A acetate, or 0.55 µg vitamin A palmitate.

SOURCE: National Research Council (1978).

APPENDIX TABLE 5-8
Weight-Unit Conversion Factors

Units Given	Units Wanted	For Conversion, Multiply by	Units Given	Units Wanted	For Conversion, Multiply by
lb	gm	453.6	μg/kg	μg/lb	0.4536
lb	kg	0.4536	Mcal	kcal	1,000.
oz	gm	28.35	kcal/kg	kcal/lb	0.4536
kg	lb	2.2046	kcal/lb	kcal/kg	2.2046
kg	mg	1,000,000.	ppm	μg/gm	1.
kg	gm	1,000.	ppm	mg/kg	1.
gm	mg	1,000.	ppm	mg/lb	0.4536
gm	μg	1,000,000.	mg/kg	%	0.0001
mg	μg	1,000.	ppm	%	0.0001
mg/g	mg/lb	453.6	mg/gm	%	0.1
mg/kg	mg/lb	0.4536	gm/kg	%	0.1

SOURCE: National Research Council (1978).

APPENDIX TABLE 5-9
Weight Equivalents

1 lb = 453.6 gm = 0.4536 kg = 16 oz
1 oz = 28.35 gm
1 kg = 1,000 gm = 2.2046 lb
1 gm = 1,000 mg
1 mg = 1,000 μg = 0.001 gm
1 μg = 0.001 mg = 0.000001 gm
1 μg per gm or 1 mg per kg is the same as ppm

SOURCE: National Research Council (1978).

RACING TABLES

Past performance information for racehorses gives a detailed analysis of performance in previous races. The information is used by bettors to handicap the horses and determine which one should win a race. It is used by trainers, owners, and breeders to compare horses' racing abilities.

APPENDIX TABLE 12-1
How to Read Past Performances of Racehorses

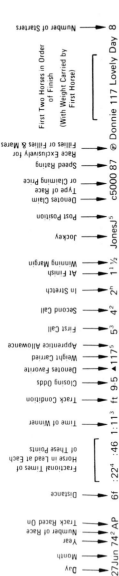

Day	Month	Year	Number of Race	Track Raced On	Distance	Fractional Times of Horse in Lead at Each of These Points	Time of Winner	Track Condition	Closing Odds	Denotes Favorite	Weight Carried	Apprentice Allowance	First Call	Second Call	In Stretch	At Finish	Winning Margin	Jockey	Post Position	Denotes Claim	Type of Race or Claiming Price	Speed Rating	Race Exclusively for Fillies or Fillies & Mares	First Two Horses in Order of Finish (With Weight Carried by First Horse)	Number of Starters
27Jun	74²	AP		6f	:22⁴	:46	1:11³	ft	9-5		▲117⁵		5³	4²	2ʰ		1¹½	JonesJ⁵			c5000	87		ⓔ Donnie 117 Lovely Day	8

EVERYTHING YOU WANT TO KNOW IS IN THE P.P.'s!

27Jun 74² AP   6f   :22⁴   :46   1:11³   ft   9-5   ▲117⁵   5³   4²   2ʰ   1¹½   JonesJ⁵   c5000 87   ⓔ Donnie 117 Lovely Day   8
DATE RACE WAS RUN—The day, month and year. This race was run June 27, 1974.

27Jun 74² AP   6f   :22⁴   :46   1:11³   ft   9-5   ▲117⁵   5³   4²   2ʰ   1¹½   JonesJ⁵   c5000 87   ⓔ Donnie 117 Lovely Day   8
NUMBER OF RACE AND TRACK RACED ON—This was the second race at Arlington Park (AP). See the past performances section of Daily Racing Form for a complete list of track abbreviations.

27Jun 74² AP   6f   :22⁴   :46   1:11³   ft   9-5   ▲117⁵   5³   4²   2ʰ   1¹½   JonesJ⁵   c5000 87   ⓔ Donnie 117 Lovely Day   8
DISTANCE OF RACE—The race was at six furlongs or ¾ of a mile (there are 8 furlongs to a mile). An "a" before the distance denotes an "about" or inexact distance ("about six furlongs"). A circled ⓣ before the distance denotes the race was run on the main turf course; a squared ⊤ a race run on track's inner turf course.

27Jun 74² AP   6f   :22⁴   :46   1:11³   ft   9-5   ▲117⁵   5³   4²   2ʰ   1¹½   JonesJ⁵   c5000 87   ⓔ Donnie 117 Lovely Day   8
FRACTIONAL TIMES—The first fraction (.22 4/5) is the time of the horse in front at the ¼; the second fraction (.46) is the time of the horse in front at the ½. For a listing of the fractions at other distances, see Fractional Times explanation below.

27Jun 74² AP   6f   :22⁴   :46   1:11³   ft   9-5   ▲117⁵   5³   4²   2ʰ   1¹½   JonesJ⁵   c5000 87   ⓔ Donnie 117 Lovely Day   8
FINAL TIME OF FIRST HORSE TO FINISH—This is the winner's final time (the six furlongs were run in 1:11 3/5). In all cases, this is the time of the first horse to finish (when the winner is disqualified, this is HIS time, not the time of the horse awarded first money).

27Jun 74² AP   6f   :22⁴   :46   1:11³   ft   9-5   ▲117⁵   5³   4²   2ʰ   1¹½   JonesJ⁵   c5000 87   ⓔ Donnie 117 Lovely Day   8
TRACK CONDITION—The track was fast (ft). See reverse side for all other track condition abbreviations.

27Jun 74² AP   6f   :22⁴   :46   1:11³   ft   9-5   ▲117⁵   5³   4²   2ʰ   1¹½   JonesJ⁵   c5000 87   ⓔ Donnie 117 Lovely Day   8
APPROXIMATE CLOSING ODDS—The horse was approximately 9-5 in the wagering. A triangle ▲ following the odds indicates the horse was the favorite; an "e" that he was part of an entry (two or more horses coupled in the wagering); an "f" that he was in the mutuel field.

27Jun 74² AP   6f   :22⁴   :46   1:11³   ft   9-5   ▲117⁵   5³   4²   2ʰ   1¹½   JonesJ⁵   c5000 87   ⓔ Donnie 117 Lovely Day   8
WEIGHT CARRIED IN THIS RACE—The horse carried 117 pounds. The small figure following the weight indicates the amount of the apprentice allowance claimed. When an apprentice allowance is claimed the exact amount of the claim is listed.

27Jun 74² AP   6f   :22⁴   :46   1:11³   ft   9-5   ▲117⁵   5³   4²   2ʰ   1¹½   JonesJ⁵   c5000 87   ⓔ Donnie 117 Lovely Day   8
FIRST CALL—The horse was running fifth, three lengths behind the leader, at this stage of the race (at the ¼ mile in this instance). The larger figure indicates the horse's running position; the small one his total margin behind the first horse. If he was in front at this point (12) the superior figure would indicate the margin by which he was leading the second horse.

27 Jun 74² AP 6f :22⁴ :46 1:11³ ft 9-5 ▲117⁵ 5³ 4² 2ʰ 1¹½ JonesJ⁵ c5000 87 ℗ Donnie 117 Lovely Day 8
SECOND CALL—The horse was fourth at this stage of the race (at the ½-mile in this case) a total of two lengths behind the leader.

27 Jun 74² AP 6f :22⁴ :46 1:11³ ft 9-5 ▲117⁵ 5³ 4² 2ʰ 1¹½ JonesJ⁵ c5000 87 ℗ Donnie 117 Lovely Day 8
STRETCH CALL—The horse was second at this stage of the race, a head behind the leader.

27 Jun 74² AP 6f :22⁴ :46 1:11³ ft 9-5 ▲117⁵ 5³ 4² 2ʰ 1¹½ JonesJ⁵ c5000 87 ℗ Donnie 117 Lovely Day 8
FINISH—The horse finished first, 1½ lengths in front of the second horse. If second, third or unplaced, the superior figure indicates his margin behind the winner.

27 Jun 74² AP 6f :22⁴ :46 1:11³ ft 9-5 ▲117⁵ 5³ 4² 2ʰ 1¹½ JonesJ⁵ c5000 87 ℗ Donnie 117 Lovely Day 8
JOCKEY AND POST POSITION—J. Jones rode the horse, who started from post position number 5.

27 Jun 74² AP 6f :22⁴ :46 1:11³ ft 9-5 ▲117⁵ 5³ 4² 2ʰ 1¹½ JonesJ⁵ c5000 87 ℗ Donnie 117 Lovely Day 8
CLAIMING PRICE OR TYPE OF RACE—The horse was entered to be claimed for $5,000 and the "c" indicates he was claimed. See reverse side for a complete list of the race classification abbreviations used in this portion of the past performance line.

27 Jun 74² AP 6f :22⁴ :46 1:11³ ft 9-5 ▲117⁵ 5³ 4² 2ʰ 1¹½ JonesJ⁵ c5000 87 ℗ Donnie 117 Lovely Day 8
SPEED RATING—The horse's speed rating was 87. See reverse side for an explanation of speed ratings.

27 Jun 74² AP 6f :22⁴ :46 1:11³ ft 9-5 ▲117⁵ 5³ 4² 2ʰ 1¹½ JonesJ⁵ c5000 87 ℗ Donnie 117 Lovely Day 8
FIRST TWO FINISHERS—These are the first two finishers in the race, with the weight carried by the first horse. The ℗ preceding the first horse's name indicates the race was exclusively for fillies or mares.

27 Jun 74² AP 6f :22⁴ :46 1:11³ ft 9-5 ▲117⁵ 5³ 4² 2ʰ 1¹½ JonesJ⁵ c5000 87 ℗ Donnie 117 Lovely Day 8
NUMBER OF STARTERS—Eight horses started in this race.

POINTS OF CALL IN PAST PERFORMANCES

The points of call in the past performances vary according to the distance of the race. The points of call of the running positions for the most frequently raced distances are:

Distance	1st Call	2nd Call	3rd Call	4th Call
2 Furlongs	Start	—	Stretch	Finish
5/16 Mile	Start	—	Stretch	Finish
3 Furlongs	Start	—	Stretch	Finish
3½ Furlongs	Start	¼ mile	Stretch	Finish
4 Furlongs	Start	¼ mile	Stretch	Finish
4½ Furlongs	Start	¼ mile	Stretch	Finish
5 Furlongs	⅛ mile	⅜ mile	Stretch	Finish
5½ Furlongs	¼ mile	⅜ mile	Stretch	Finish
6 Furlongs	¼ mile	½ mile	Stretch	Finish
6½ Furlongs	¼ mile	½ mile	Stretch	Finish
7 Furlongs	¼ mile	½ mile	Stretch	Finish

Distance	1st Call	2nd Call	3rd Call	4th Call
1 Mile	½ mile	¾ mile	Stretch	Finish
1 Mile, 70 Yds	½ mile	¾ mile	Stretch	Finish
1 1/16 Miles	½ mile	¾ mile	Stretch	Finish
1⅛ Miles	½ mile	¾ mile	Stretch	Finish
1¼ Miles	½ mile	1 mile	Stretch	Finish
1 3/16 Miles	½ mile	1 mile	Stretch	Finish
1⅜ Miles	½ mile	1¼ miles	Stretch	Finish
1½ Miles	½ mile	1¼ miles	Stretch	Finish
1⅝ Miles	½ mile	1⅜ miles	Stretch	Finish
1¾ Miles	½ mile	1½ miles	Stretch	Finish

NOTE: The second call in most races is made ¼ mile from the finish; the stretch call ⅛-mile from the finish.

FRACTIONAL TIMES

In races at less than one mile, fractional times are given for the horse in the lead at the quarter-mile and half-mile. In races at one mile or more, fractional times are given for the half-mile and three-quarter mile.

NOTE: In races at the shorter distances, only the fraction for the quarter-mile is used, except for races at three and one-half furlongs, in which the fractions are given for the quarter-mile and three-eighths mile.

KEY TO SYMBOLS, ABBREVIATIONS IN PAST PERFORMANCES

FOREIGN-BRED HORSES

An asterisk(*) preceding the name of the horse indicates foreign-bred. (No notation is made for horses bred in Canada and Cuba).

MUD MARKS

✱ – Fair mud runner X–Good mud runner
⊗–Superior mud runner

COLOR

B.–Bay	Blk.–Black	Br.–Brown	Ch.–Chestnut

Gr.–Gray Ro.–Roan Wh.–White
Dk. b. or br.–Dark bay or brown
(Lt. before color denotes Light)

SEX

c–colt g–gelding rig–ridgling
f–filly m–mare

PEDIGREE

Each horse's pedigree lists, in the order named, color, sex, year foaled, sire, dam and grandsire (sire of dam).

BREEDER

Abbreviation following breeder's name indicates the state, place of origin or foreign country in which the horse was foaled.

TODAY'S CLAIMING PRICE

Claiming prices for horses entered today appear in bold face type to right of trainer's name.

RECORD OF STARTS AND EARNINGS

The horse's racing record for his most recent two years of competition appears to the extreme right of the name of the breeder. This lists the year, number of starts, wins, seconds, thirds and earnings. The letter "M" in the win column of the upper line indicates the horse is a maiden. If the letter "M" is in the lower line only, it indicates the horse was a maiden at the end of that year.

RACE CLASSIFICATIONS

10000–Claiming race (eligible to be claimed for $10,000). Note: The letter c preceding claiming price (c10000) indicates horse was claimed.

M10000–Maiden claiming race (non-winners–eligible to be claimed).

10000H–Claiming handicap (eligible to be claimed).

o10000–Optional claiming race (entered **NOT** to be claimed).

10000o–Optional claiming race (eligible to be claimed).

Mdn–Maiden race (non-winners).

AlwM–Maiden allowance race (for non-winners with special weight allowance for those having started in a claiming race).

Alw–Allowance race.

HcpO–Overnight handicap race.

SplW–Special weight race.

Wfa–Weight-for-age race.

Mtch–Match race.

A10000–Starter allowance race (horses who have started for claiming price shown, or less, as stipulated in the conditions).

H10000–Starter handicap (same restriction as above).

S10000–Starter special weight (restricted as above). Note: Where no amount is specified in the conditions of the "starters" race dashes are substituted, as shown below:

A—— H—— S——

STEEPLECHASE AND HURDLE RACES

[S–Steeplechase   [H–Hurdle race

STAKES RACES

AlwS–Allowance stakes.

HcpS–Handicap stakes.

ScwS–Scale weight stakes.

SpwS–Special weight stakes.

WfaS–Weight-for-age stakes.

50000S–Claiming stakes (eligible to be claimed).

DISTANCE

a – preceding distance (a6f) denotes "about" distance (about 6 furlongs in this instance.)

TURF COURSES

Ⓣ – before distance indicates turf (grass) course race.

Ⓣ – before distance indicates inner turf course.

TRACK CONDITIONS

ft–fast	fr–frozen	gd–good	sl–slow
sy–sloppy	m–muddy	hy–heavy	
hd–hard	fm–firm	gd–good	yl–yielding	sf–soft

Turf course races, including steeplechase and hurdles:

Note: SC in place of track condition indicates Synthetic Course in use at various tracks. The condition of this synthetic course is fast at all times.

CLOSING ODDS

2▲–favorite   2e–entry   15f–mutuel field

APPRENTICE OR RIDER WEIGHT ALLOWANCES

Allowance (in pounds) indicated by superior figure following weight–117⁵.

ABBREVIATIONS USED IN POINTS OF CALL

no–nose   h–head   nk–neck

POST POSITION

Horse's post position appears after jockey's name–Wilson R⁴

EACH HORSE'S MOST RECENT WORKOUTS APPEAR DIRECTLY UNDER THE PAST PERFORMANCES

For example, July 30 Bel 3f ft :38b indicates the horse worked on July 30 at Belmont Park. The distance of the trial was 3 furlongs over a fast track and the horse was timed in 38 seconds, breezing. The following abbreviations are used to describe how each horse worked:

b–breezing	e–easily	bo–bore out
d–driving	g–worked from stall gate	
h–handily	tc–turf course	

trt following track abbreviation indicates horse worked on training track.   TR-Trial race

HC–Hillside Course

INVITATIONAL RACES

InvH–Invitational handicap.
InvSc–Invitational scale weight
InvSp--Invitational special weights.
InvW–Invitational weight-for-age.

RACES EXCLUSIVELY FOR FILLIES, MARES

Ⓕ – Immediately following the speed rating indicates races exclusively for fillies or fillies and mares.

SPEED RATINGS

This is a comparison of the horse's time with the track record established prior to the opening of the meeting. The track record is given a rating of 100 and one point is deducted for each one-fifth second slower than the record. When a horse breaks the track record one point is added to the par 100 for each one-fifth second faster than the record. One-fifth of a second is considered the equivalent of one length.

No ratings are given for hurdle or steeplechase events, for races of less than three furlongs, for races where the horse's speed rating is less than 25.

NOTE: Speed ratings for new distances are computed and assigned when adequate time standards are established. At Caliente, Mexico, which operates throughout the year, rating changes are made whenever a track record is broken.

Source: Copyright © 1974 Triangle Publications

APPENDIX TABLE 12-2

Leading Money-Earning Quarter Horses, 1949–1979 (capital letters denote Register of Merit Qualifiers)

	Earnings
MOON LARK, s.s. 76 by TOP MOON out of PAN O LAN	$859,356.25
EASY DATE, b. m. 72 by EASY JET out of Spot Cash (TB)	849,709.45
TOWN POLICY, b. g. by Reb's Policy (TB) out of CAMPTOWN GIRL	629,743.62
TIMETO THINKRICH, br. s. 71 by Aforethought (TB) out of CHRONOMETER	612,858.12
BUGS ALIVE IN 75, s.s. 73 by TOP MOON out of RALPH'S LADY BUG	550,969.51
REAL WIND, br. m. 74 by GO WITH THE WIND out of REAL NEW	528,161.37
PIE IN THE SKY, b. s. 77 by EASY JET out of Miss Jelly Roll	521,418.00
PASS OVER, s.m. 71 by Pass 'Em Up (TB) out of Revision	521,172.70
DASH FOR CASH, s.s. 73 by ROCKET WRANGLER out of Find A Buyer (TB)	507,687.22
MY EASY CREDIT, s.s. 74 by EASY JET out of KITS CHARGE ACCT	502,504.08

Note: Money earned at official distances only. m. = mare, s. = stallion, g. = gelding, br. = brown, s. = sorrel, b. = bay, TB = Thoroughbred, 76 (etc.) = birthdate.

SOURCE: American Quarter Horse Association (1980).

APPENDIX TABLE 12-3

Leading Quarter Horse Sires of Money Earners, 1949–1979 (capital letters denote Register of Merit Qualifiers)

	Sire	Earnings
EASY JET, s.s. 67 by JET DECK	557	$9,330,234
GO MAN GO, ro.s. 53 by Top Deck (TB)	774	7,468,123
JET DECK, b. s. 60 by MOON DECK	486	6,779,535
TOP MOON, bk. s. 60 by MOON DECK	915	5,501,669
Azure Te (TB), b. s. 62 by Nashville (TB)	510	4,527,029
THREE OH'S, br. s. 66 by THREE CHICKS	584	3,992,787
ROCKET BAR (TB), ch.s. 51 by Three Bars (TB)	360	3,815,328
ALAMITOS BAR, b. s. 59 by Three Bars (TB)	425	3,687,575
LADY BUG'S MOON, s.s. 66 by TOP MOON	613	3,218,799
Three Bars (TB), ch.s. 40 by Percentage (TB)	424	3,207,856

Note: m. = mare, s. = stallion, br. = brown, b. = bay, s. = sorrel, ro. = roan, bk. = black, ch. = chestnut, TB = Thoroughbred, 76 (etc.) = birthdate.
*a*Includes every horse that has been indexed since 1945.
*b*From January 1, 1949.

SOURCE: American Quarter Horse Association (1980).

APPENDIX TABLE 12-4

Leading Quarter Horse Maternal Grandsires of Register of Merit Qualifiers
(capital letters denote Register of Merit Qualifiers)

Broodmare Sire	Mares	Performers	Starts	Qualifiers	Earnings
LEO	269	1,269	17,437	686	$5,746,660
Three Bars (TB)	237	1,033	15,227	625	5,469,602
GO MAN GO	278	784	10,405	503	6,213,319
SUGAR BARS	257	703	8,283	359	1,692,654
Top Deck (TB)	162	585	7,855	320	2,277,799
JET DECK	219	445	5,513	310	3,375,545
VANDY	141	560	8,306	301	2,387,861
ROCKET BAR (TB)	146	478	6,502	299	3,753,695
TONTO BARS GILL	207	659	8,707	292	1,666,392
MR BAR NONE	126	442	5,996	250	2,138,811

SOURCE: American Quarter Horse Association (1980).

APPENDIX TABLE 12-5

Leading Quarter Horse Dams of Register of Merit
Qualifiers, 1945–1979 (capital letters denote Register
of Merit Qualifiers)

Dam	Total Qualifiers
MISS NIGHT BAR by BARRED	13
ROSA LEO by LEO	12
Chicaro Sue by Chicaro Bill	12
Star Brandy by FIREBRAND REED	12
Super's Baby by SUPER CHARGE	12
Ginger Baker by Harmon Baker B	12
LITTLE CAPRI by Little Request (TB)	11
Little Peach by Beggar Boy (TB)	11
89'ER by King	11
MISS OVERTIME by OVERTIME LEO	11
ALFARETTA by ROBIN REED	11
Barbara Lam (TB) by Little Request (TB)	11
BARBARA L by Patriotic (TB)	11
HI B HIND by Bob Kerr	11
BARBLO TWIST by Sabre Twist	10
F1 Lady Bug by Sergeant	10
SUGAR TIME GAL by OVERTIME LEO	10
MISS OLENE by LEO	10
Lady Me by JOE REED II	10
Freda Ruby by TONY RED	10
Vera Vay by Tommy	10
RANALDA REED by RUSTY WENTZ	10

SOURCE: American Quarter Horse Association (1980).

Champion Quarter Running Horses, 1940–1979

Year	World Champion	2-Year-Old Champion	3-Year-Old Champion	Aged Champion
1979	Moon Lark	Easy Angel	Moon Lark	Mr. Doty Bars
1978	Miss Thermolark	Moon Lark	Miss Thermolark	Azure Three
1977	Dash for Cash	Town Policy	My Easy Credit	Dash for Cash
1976	Dash for Cash	Real Wind	Dash for Cash	She's Precious

Year	World Champion	Champion Stallion	Champion Mare	Champion Gelding
1975	Easy Date	Bugs Alive in 75	Easy Date	Wanta Go
1974	Tiny's Gay	Tiny's Gay	Charger Bar	Flight 109
1973	Truckle Feature	Truckle Feature	Charger Bar	Come Six
1972	Mr Jet Moore	Mr Jet Moore	Charger Bar	Kaweah Bar
1971	Charger Bar	Bunny Bid	Charger Bar	Kaweah Bar
1970	Kaweah Bar	Easy Jet	Go Together	Kaweah Bar
1969	Easy Jet	Easy Jet	Go Derussa Go	Kaweah Bar
1968	Kaweah Bar	Mighty Deck	Top Rockette	Kaweah Bar
1967	Laico Bird	Duplicate Copy	Laico Bird	Brad Len
1966	(No Award)	(Duplicate Copy) (Tiny Watch)	Go Josie Go	Cee Bar Deck
1965	Go Josie Go	Tiny Watch	Go Josie Go	Joe Sherry
1964	Goetta	Pasamone Paul	Goetta	Dari Star
1963	Jet Deck	Jet Deck	Anna Dial	Dariman
1962	No Butt	Jet Deck	No Butt	Caprideck
1961	Pap	Breeze Bar	First Call	Pap
1960	Vandy's Flash	Tonto Bars Hawk	Triple Lady	Vandy's Flash
1959	(No Award)	Double Bid	(Miss Louton) (Triple Lady)	Pap
1958	Mr Bar None	Mr Bar None	Vanetta Dee	Vandy's Flash
1957	Go Man Go	Go Man Go	Vanetta Dee	Vannevar
1956	Go Man Go	Go Man Go	Vanetta Dee	Vannevar
1955	Go Man Go	Go Man Go	Posey Vandy	Arizonan
1954	Josie's Bar	Palleo Pete	Josie's Bar	Brigand (TB)
1953	Miss Meyers	Rukin String	Miss Meyers	Brigand (TB)
1952	Johnny Dial	Johnny Dial	Stealla Moore	Brigand (TB)
1951	(Monita) (Maddon's Bright Eyes)	(Bart B S) (Hard Twist) (Clabbertown G)	(Monita) (Maddon's Bright Eyes)	
1950	Blob Jr	Blob Jr	Maddon's Bright Eyes	
1949	Maddon's Bright Eyes	Diamond Bob	Maddon's Bright Eyes	
1948	Woven Web (TB)	Scotter W	Woven Web (TB)	
1947	Woven Web (TB)	Pelican	Woven Web (TB)	
1946–1947	Woven Web (TB)	Hard Twist	Woven Web (TB)	
1945–1946	Queenie	Piggin String (TB)	Queenie	
1944–1945	(No Award)	See Dee	Queenie	
1943–1944	Shue Fly	Piggin String (TB) (Texas Lad)	Shue Fly	
1942–1943	Shue Fly	Joe Reed II	Shue Fly	
1941–1942	Shue Fly	Nobodies Friend	Shue Fly	
1940–1941	Clabber	Clabber		

SOURCE: *Quarter Horse Racing Chart Book* (1977) and American Quarter Horse Association (1980).

APPENDIX TABLE 12-7

Breakdown of the Total Gross Distribution in Stakes and Purses in 1979 for
Thoroughbreds by States, including Canada and Mexico (Caliente and Juarez)

Locality	Racing Days	Races Run	Gross Distribution
Arizona	200	1,822	$ 4,571,295
Arkansas	50	477	6,625,650
California	512	4,583	55,069,374
Colorado	119	1,075	2,753,352
Delaware	73	672	4,962,990
Florida	361	3,694	25,320,818
Georgia	1	6	34,000
Idaho	81	625	550,092
Illinois	482	4,493	29,301,277
Kentucky	276	2,440	15,762,091
Louisiana	534	4,734	25,421,413
Maryland	304	2,715	23,000,739
Massachusetts	236	2,280	11,593,175
Michigan	235	2,161	12,125,315
Montana	39	406	421,954
Nebraska	180	1,604	8,251,020
New Hampshire	64	612	3,594,200
New Jersey	264	2,415	25,900,010
New Mexico	290	2,275	6,406,105
New York	469	4,233	57,497,174
North Carolina	3	18	71,800
Ohio	380	3,601	15,478,694
Oregon	128	1,085	2,166,810
Pennsylvania	543	5,130	25,945,290
South Carolina	3	15	117,600
South Dakota	36	336	337,121
Tennessee	1	5	27,000
Virginia	8	45	134,275
Washington	255	2,250	7,854,974
West Virginia	534	5,403	11,880,545
Canada	759	6,865	29,729,900
Mexico (Caliente and Juarez)	125	1,322	3,482,416
Total	7,515	69,407	$415,414,224

SOURCE: *American Racing Manual* (1980).

APPENDIX TABLE 12-8
Annual Statistics for Thoroughbred Racing, 1907–1979

Year	Racing Days	No. Races	Starters	Gross Distribution
1907	1,004	6,252	5,662	$ 5,375,554
1908	921	5,699	5,405	4,351,691
1909	724	4,510	4,890	3,146,695
1910	1,063	6,501	4,180	2,942,333
1911	1,037	6,289	4,038	2,337,957
1912	926	5,806	3,553	2,391,625
1913	969	6,136	3,541	2,920,963
1914	906	5,849	3,632	2,994,525
1915	839	5,454	3,700	2,853,037
1916	1,035	6,098	3,754	3,842,471
1917	902	5,899	4,200	4,066,253
1918	610	3,968	3,575	3,425,347
1919	686	4,408	3,531	4,642,865
1920	1,022	6,897	4,032	7,773,407
1921	1,074	7,250	4,623	8,435,083
1922	1,182	8,045	5,049	9,096,215
1923	1,319	8,991	5,437	9,675,811
1924	1,456	10,007	5,906	10,825,446
1925	1,656	11,579	6,438	12,577,270
1926	1,713	12,065	7,218	13,884,820
1927	1,680	11,832	7,794	13,935,610
1928	1,613	11,465	8,171	13,332,361
1929	1,599	11,133	8,332	13,417,817
1930	1,653	11,477	8,791	13,674,160
1931	1,660	11,690	9,128	13,084,154
1932	1,518	10,835	9,017	10,082,757
1933	1,746	12,680	9,176	8,516,326
1934	1,959	14,261	9,470	10,443,495
1935	2,133	15,830	10,544	12,794,418
1936	2,033	15,344	10,757	12,994,605
1937	2,140	16,250	11,515	14,363,562
1938	2,140	16,243	12,185	14,946,609
1939	2,199	16,967	12,804	15,312,839
1940	2,096	16,401	13,257	15,911,167

Note: The following number of horses started in 1902, 5,271; 1903, 5,525; 1904, 5,962; 1905, 6,232; and 1906, 5,962. All statistics include United States, Canadian, Mexican, Cuban, and Puerto Rican racing with the following exceptions:

1. Cuba: not included after 1957.

2. Mexico: Caliente, only, through 1943; 1958 through 1960. Including Mexico City, 1944 through 1957; 1961 through 1970.

3. Puerto Rico: 1963 and 1964 only.

SOURCE: *American Racing Manual* (1980).

Year	Racing Days	No. Races	Starters	Gross Distribution
1941	2,162	16,912	13,683	17,987,225
1942	2,228	17,593	12,614	18,136,118
1943	2,052	16,094	11,258	18,555,680
1944	2,396	19,228	12,959	29,159,099
1945	2,480	19,587	14,307	32,300,060
1946	3,020	23,940	17,601	49,291,024
1947	3,134	24,884	19,063	53,932,141
1948	3,183	25,388	20,254	54,436,063
1949	3,309	26,832	21,616	52,317,078
1950	3,290	26,932	22,554	50,102,098
1951	3,394	27,856	22,819	55,551,124
1952	3,515	29,051	23,813	63,950,236
1953	3,635	30,069	24,417	72,870,819
1954	3,685	30,467	25,294	74,255,611
1955	3,827	31,757	26,056	76,643,696
1956	3,979	33,445	26,507	81,311,581
1957	4,120	34,982	27,355	85,300,966
1958	3,910	33,325	28,099	85,467,082
1959	4,218	36,579	28,623	92,848,541
1960	4,304	37,661	29,773	93,741,552
1961	4,641	40,744	30,381	98,846,843
1962	4,772	41,766	33,579	103,525,712
1963	5,203	45,449	35,828	113,122,209
1964	5,326	46,922	37,812	121,777,847
1965	5,284	47,335	38,502	126,463,984
1966	5,254	46,814	39,604	130,653,813
1967	5,344	47,811	41,853	139,170,738
1968	5,553	49,777	43,710	150,644,478
1969	5,825	52,315	45,808	168,713,911
1970	6,242	56,676	49,087	185,625,110
1971	6,394	57,467	50,470	201,435,894
1972	6,624	59,410	52,488	210,607,082
1973	6,888	62,272	54,813	233,936,153
1974	7,201	65,312	56,520	263,951,376
1975	7,488	68,210	58,816	292,583,798
1976	7,593	69,480	61,084	318,963,255
1977	7,525	68,822	61,970	336,022,495
1978	7,579	69,499	62,950	367,637,193
1979	7,515	69,407	70,914	415,414,224

APPENDIX TABLE 12-9
Thoroughbred Horses That Were Annual Money-Winning Leaders Since 1902

Year	Horse	Age	Starts	1st Place	2nd Place	3rd Place	Amount Won
1902	Major Daingerfield	3	7	4	2	1	$ 57,685
1903	Africander	3	15	8	3	1	70,810
1904	Delhi	3	10	6	2	0	75,225
1905	Sysony	3	9	9	0	0	144,380
1906	Accountant	3	13	9	1	1	83,570
1907	Colin	2	12	12	0	0	131,705
1908	Sir Martin	2	13	8	4	0	78,590
1909	Joe Madden	3	15	5	9	1	44,905
1910	Novelty	2	16	11	2	2	72,630
1911	Worth	2	13	10	1	0	16,645
1912	Star Charter	4	17	6	2	5	14,655
1913	Old Rosebud	2	14	12	2	0	19,057
1914	Roamer	3	16	12	1	2	29,105
1915	Borrow	7	9	4	1	1	20,195
1916	Campfire	2	9	6	2	0	49,735
1917	Sun Briar	2	9	5	1	2	59,506
1918	Eternal	2	8	6	1	0	56,173
1919	Sir Barton	3	13	8	3	2	88,250
1920	Man O' War	3	11	11	0	0	166,140
1921	Morvich	2	11	11	0	0	115,234
1922	Pillory	3	7	4	1	1	95,654
1923	Zev	3	14	12	1	0	272,008
1924	Sarzen	3	12	8	1	1	95,640
1925	Pompey	2	10	7	2	0	121,630
1926	Crusader	2	15	9	4	0	166,033
1927	Anita Peabody	2	7	6	0	1	111,905
1928	High Strung	2	6	5	0	0	153,590
1929	Blue Larkspur	3	6	4	1	0	153,450
1930	Gallant Fox	3	10	9	1	0	308,275
1931	Gallant Flight	2	7	7	0	0	219,000
1932	Gusto	3	16	4	3	2	145,940
1933	Singing Wood	2	9	3	2	2	88,050
1934	Calvacade	3	7	6	1	0	111,235
1935	Omaha	3	9	6	1	2	142,255
1936	Granville	3	11	7	3	0	110,295
1937	Seabiscuit	4	15	11	2	2	168,580
1938	Stagehand	3	15	8	2	3	189,710
1939	Challedon	3	15	9	2	3	184,535

SOURCE: *American Racing Manual* (1980).

Year	Horse	Age	Starts	1st Place	2nd Place	3rd Place	Amount Won
1940	Bimelech	3	7	4	2	1	110,005
1941	Whirlaway	3	20	13	5	2	272,386
1942	Shutout	3	12	8	2	0	238,872
1943	Count Fleet	3	6	6	0	0	174,055
1944	Pavot	2	8	8	0	0	179,040
1945	Busher	3	13	10	2	1	273,735
1946	Assault	3	15	8	2	3	424,195
1947	Armed	6	17	11	4	1	376,325
1948	Citation	3	20	19	1	0	709,470
1949	Ponder	3	21	9	5	2	321,825
1950	Noor	5	12	7	4	1	346,940
1951	Counterpoint	3	15	7	2	1	250,525
1952	Crafty Admiral	4	16	9	4	1	277,225
1953	Native Dancer	3	10	9	1	0	513,425
1954	Determine	3	15	10	2	3	328,700
1955	Nashua	3	12	10	1	1	752,550
1956	Needles	3	8	4	2	0	440,850
1957	Round Table	3	22	15	1	3	600,383
1958	Round Table	4	20	14	4	0	662,780
1959	Sword Dancer	3	13	8	4	0	537,004
1960	Bally Ache	3	15	10	3	1	455,045
1961	Carry Back	3	16	9	1	3	565,349
1962	Never Bend	2	10	7	1	2	402,969
1963	Candy Spots	3	12	7	2	1	604,481
1964	Gun Bow	4	16	8	4	2	580,100
1965	Buckpasser	2	11	9	1	0	568,096
1966	Buckpasser	3	14	13	1	0	669,078
1967	Damascus	3	16	12	3	1	817,941
1968	Forward Pass	3	13	7	2	0	546,674
1969	Arts and Letters	3	14	8	5	1	555,604
1970	Personality	3	18	8	2	1	444,049
1971	Riva Ridge	2	9	7	0	0	503,263
1972	Droll Role	4	19	7	3	4	471,633
1973	Secretariat	3	12	9	2	1	860,404
1974	Chris Evert	3	8	5	1	2	551,063
1975	Foolish Pleasure	3	11	5	4	1	716,278
1976	Forego	6	8	6	1	1	491,701
1977	Seattle Slew	3	7	6	0	1	641,370
1978	Affirmed	3	11	8	2	0	901,541
1979	Spectacular Bid	3	12	10	1	1	1,279,334

APPENDIX TABLE 12-10

Leading Money-Winning Thoroughbred Horses of North America with Earnings of at Least $1 Million, Through 1979

Horses	Years Raced	Starts	1st Place	2nd Place	3rd Place	Earnings
Affirmed (c., 1975; Exclusive Native—Won't Tell You, by Crafty Admiral)	2-5	29	22	5	1	$2,393,818
Kelso (g., 1957; Your Host—Maid of Flight, by Count Fleet)	2-8	63	39	12	2	1,977,896
Forego (g., 1970; *Forli—Lady Golconda, by Hasty Road)	3-8	57	34	9	7	1,938,957
Round Table (c., 1954; *Princequillo—*Knights Daughter, by Sir Cosmo)	2-5	66	43	8	5	1,749,869
Spectacular Bid (c., 1976; Bold Bidder—Spectacular, by Promised Land)	2-4	21	7	2	1	1,663,817
Exceller (c., 1973; *Vaguely Noble—Too Bald, by Bald Eagle)	2-6	33	15	5	5	1,654,002
Dahlia (f., 1970; *Vaguely Noble—Charming Alibi, by Honeys Alibi)	2-6	48	15	3	7	1,544,139
Buckpasser (c., 1963; Tom Fool—Busanda, by War Admiral)	2-4	31	25	4	1	1,462,014
Allez France (f., 1970; *Sea-Bird-Priceless Gem, by Hail to Reason)	2-5	21	13	3	1	1,386,146
Secretariat (c., 1970; Bold Ruler—Somethingroyal, by *Princequillo)	2-3	21	16	3	1	1,316,808
*Nashua (c., 1952; *Nasrullah—Segula, by Johnstown)	2-4	30	22	4	1	1,288,565

Note: Horses still in training listed in boldface type. c. = colt, g. = gelding, f. = filly, * = imported.

SOURCE: *American Racing Manual (1980).*

Horses	Years Raced	Starts	1st Place	2nd Place	3rd Place	Earnings
Ancient Title (g., 1970; Gummo—Hi Little Gal, by Bar Le Duc)	2-8	57	24	11	9	1,252,791
Susan's Girl (f., 1969; Quadrangle—Quaze, by *Quibu)	2-6	63	29	14	11	1,251,668
Carry Back (c., 1958; Saggy—Joppy, by Star Blen)	2-5	62	21	11	11	1,241,165
Foolish Pleasure (c., 1972; What a Pleasure—Fool-Me-Not, by Tom Fool)	2-4	26	16	4	3	1,216,705
Seattle Slew (c., 1974; Bold Reasoning—My Charmer, by Poker)	2-4	17	14	2	0	1,208,726
Damascus (c., 1964; Sword Dancer—Kerala, by *My Babu)	2-5	32	21	7	3	1,176,781
*Cougar II (c., 1966; Tale of Two Cities—*Cindy Lou II, by Madara)	2-7	50	20	7	17	1,162,725
Riva Ridge (c., 1969; First Landing—Iberia, by *Heliopolis)	2-4	30	17	3	1	1,111,497
Fort Marcy (g., 1964; *Amerigo—Key Bridge, by *Princequillo)	2-7	75	21	18	14	1,109,791
Citation (c.,1945; Bull Lea—*Hydroplane II, by Hyperion)	2-3, 5-6	45	32	10	2	1,085,760
Native Diver (g., 1959; Imbros—Fleet Diver, by Devil Diver)	2-8	81	37	7	12	1,026,500
Royal Glint (g., 1970; Round Table—Regal Gleam, by Hail to Reason)	3-6	52	21	9	4	1,004,815
Dr. Fager (c., 1964; Rough'n Tumble—Aspidistra, by Better Self)	2-4	22	18	2	1	1,002,642

APPENDIX TABLE 12-11
Annual Thoroughbred Champions

Year	2-Year-Old Colt or Gelding	2-Year-Old Filly	Best 2-Year-Old	3-Year-Old Colt or Gelding	3-Year-Old Filly	Best 3-Year-Old
1957	Nadir	Idun	Idun	Bold Ruler	Bayou	Bold Ruler
1958	First Landing	Quill	First Landing	Tim Tam	Idun	Tim Tam
1959	Warfare	My Dear Girl	Warfare	Sword Dancer	Royal Native	Sword Dancer
1960	Hail to Reason	Bowl of Flowers	Hail to Reason	Kelso	Berlo	Kelso
1961	Crimson Satan	Cicada	Crimson Satan	Carry Back	Bowl of Flowers	Carry Back
1962	Never Bend	Smart Deb	Never Bend	Jaipur	Cicada	Jaipur
1963	Hurry to Market	Tosmah	Hurry to Market	Chateaugay	Lamb Chop	Chateaugay
1964	Bold Lad	Queen Empress	Bold Lad	Northern Dancer	Tosmah	Northern Dancer
1965	Buckpasser	Mocassin	Buckpasser	Tom Rolfe	What A Treat	Tom Rolfe
1966	Successor	Regal Gleam	Successor	Buckpasser	Lady Pitt	Buckpasser
1967	Vitriolic	Queen of the Stage	Vitriolic	Damascus	Furl Sail	Damascus
1968	Top Knight	Gallant Bloom	Top Knight	Stage Door Johnny	Dark Mirage	Dark Mirage
1969	Silent Screen	Fast Attack	Silent Screen	Arts and Letters	Gallant Bloom	Arts and Letters
1970	Hoist the Flag	Forward Gal	Hoist the Flag	Personality	Office Queen	Personality

Year	2-Year-Old Colt or Gelding	2-Year-Old Filly	3-Year-Old Colt or Gelding	3-Year-Old Filly	Older Colt, Horse, or Gelding
1971	Riva Ridge	Numbered Account	Canonero II	Turkish Trouser	Ack Ack
1972	Secretariat	La Prevoyante	Key to the Mint	Susan's Girl	Autobiography
1973	Protagonist	Talking Picture	Secretariat	Desert Vixen	Riva Ridge
1974	Foolish Pleasure	Ruffian	Little Current	Chris Evert	Forego
1975	Honest Pleasure	Dearly Precious	Wajima	Ruffian	Forego
1976	Seattle Slew	Sensational	Bold Forbes	Revidere	Forego
1977	Affirmed	Lakeville Miss	Seattle Slew	Our Mims	Forego
1978	Spectacular Bid	Candy Eclair and It's In the Air	Affirmed	Tempest Queen	Seattle Slew
1979	Rockhill Native	Smart Angle	Spectacular Bid	Davona Dale	Affirmed

Note: Beginning with 1971, the annual champions were determined in a combined poll of the staff members of the *Daily Racing Form*, the Board of Selection of the Thoroughbred Racing Associations, and members of the National Turf Writers Association.
SOURCE: *American Racing Manual* (1980).

Year	Handicap Horse	Handicap Mare	Grass Horse	Sprinters	Steeplechaser or Hurdler	Horse of Year
1957	Dedicate	Pucker Up	Round Table	Decathalon	Neji	Bold Ruler
1958	Round Table	Barnastar	Round Table	Bold Ruler	Neji	Round Table
1959	Sword Dancer	Tempted	Round Table	Intentionally	Ancestor	Sword Dancer
1960	Bold Eagle	Royal Native			Benguala	Kelso
1961	Kelso	Airmans Guide	TV Lark		Peal	Kelso
1962	Kelso	Primonetta			Barnabys Bluff	Kelso
1963	Kelso	Cicada	Mongo		Amber Driver	Kelso
1964	Kelso	Tosmah			Bon Nouvel	Kelso
1965	Roman Brother	Old Hat	Parka	Affectionately	Bon Nouvel	Roman Brother
1966	Buckpasser	Open Fire	Assagai	Impressive	Mako	Buckpasser
1967	Damascus	Straight Deal	Fort March	Dr. Fager	Quick Pitch	Damascus
1968	Dr. Fager	Gamely	Dr. Fager	Dr. Fager	Bon Nouvel	Dr. Fager
1969	Arts and Letters	Gallant Bloom	Hawaii	Ta Wee	L'Escargot	Arts and Letters
1970	Fort Marcy	Shuvee	Fort Marcy	Ta Wee	Top Bid	Fort Marcy

Year	Older Filly or Mare	Grass Horse	Sprinter	Steeplechaser or Hurdler	Horse of Year
1971	Shuvee	Run the Gauntlet	Ack Ack	Shadow Brook	Ack Ack
1972	Typecast	Cougar II	Chou Croute	Soothsayer	Secretariat
1973	Susan's Girl	Secretariat	Shecky Greene	Athenian Idol	Secretariat
1974	Desert Vixen	Dahlia	Forego	Gran Kan	Forego
1975	Susan's Girl	Snow Knight	Gallant Bob	Life's Illusion	Forego
1976	Proud Delta	Youth	My Juliet	Straight and True	Forego
1977	Cascapedia	Johnny D.	What A Summer	Cafe Prince	Seattle Slew
1978	Late Bloomer	Mac Diarmida	Dr. Patches and J. O. Tobin	Cafe Prince	Affirmed
1979	Waya	Bowl Game (M) and Trillion (F)	Star De Naskra	Martie's Anger	Affirmed

APPENDIX TABLE 12-12
Annual Leading Thoroughbred Sire, Money Won, 1955–1979

Year	Horse	Performers Sired	Races Won	Amount
1955	Nasrullah	40	69	$1,433,660
1956	Nasrullah	50	106	1,462,413
1957	Princequillo	75	147	1,698,427
1958	Princequillo	65	110	1,394,540
1959	Nasrullah	69	141	1,434,543
1960	Nasrullah	64	122	1,419,683
1961	Ambiorix	73	148	936,976
1962	Nasrullah	62	107	1,474,831
1963	Bold Ruler	26	56	917,531
1964	Bold Ruler	44	88	1,457,156
1965	Bold Ruler	51	90	1,091,924
1966	Bold Ruler	51	107	2,306,523
1967	Bold Ruler	63	135	2,249,272
1968	Bold Ruler	57	99	1,988,427
1969	Bold Ruler	59	90	1,357,144
1970	Hail to Reason	53	82	1,400,839
1971	Northern Dancer	44	93	1,288,580
1972	Round Table	65	98	1,199,933
1973	Bold Ruler	41	74	1,488,622
1974	TV Lark	98	121	1,242,000
1975	What a Pleasure	90	101	2,011,878
1976	What a Pleasure	85	108	1,622,159
1977	Dr. Fager	79	124	1,593,079
1978	Exclusive Native	63	106	1,969,867
1979	Exclusive Native	71	104	2,872,605

SOURCE: *American Racing Manual* (1980).

APPENDIX TABLE 12-13

Annual Stakes, Winning Jockey Riding Thoroughbreds, Money Won (total winners' value)

Year	Jockey	Stakes Won	Total Win Value
1935	Wright, W. D.	18	$ 122,150
1936	Richards, H.	14	173,470
1937	Kurtsinger, C.	16	259,820
1938	Wall, N.	19	286,170
1939	Stout, J.	15	231,985
1940	Arcaro, E.	16	170,165
1941	Arcaro, E.	15	229,975
1942	Woolf, G.	23	341,680
1943	Longden, J.	20	290,222
1944	Woolf, G.	14	338,135
1945	Longden, J.	24	528,220
1946	Arcaro, E.	21	404,380
1947	Dodson, D.	35	899,915
1948	Arcaro, E.	35	1,082,585
1949	Brooks, S.	26	728,335
1950	Arcaro, E.	30	689,035
1951	Arcaro, E.	18	531,250
1952	Arcaro, E.	40	1,172,404
1953	Arcaro, E.	20	839,734
1954	Arcaro, E.	29	895,690
1955	Arcaro, E.	34	1,226,657
1956	Hartack, W.	26	1,037,077
1957	Hartack, W.	43	1,718,231
1958	Shoemaker, W.	36	1,600,503
1959	Shoemaker, W.	29	1,034,422
1960	Ussery, R.	20	931,441
1961	Shoemaker, W.	25	1,156,470
1962	Shoemaker, W.	38	1,474,516
1963	Shoemaker, W.	32	996,785
1964	Shoemaker, W.	34	1,243,827
1965	Baeza, B.	24	1,085,661
1966	Shoemaker, W.	37	1,303,787
1967	Shoemaker, W.	38	1,629,874
1968	Baeza, B.	25	1,181,745
1969	Rotz, J. L.	20	931,565
1970	Rotz, J. L.	24	1,031,548
1971	Shoemaker, W.	46	1,567,295
1972	Turcotte, R.	21	1,226,282
1973	Turcotte, R.	16	1,087,397
1974	Pincay, L., Jr.	38	1,427,407
1975	Shoemaker, W.	29	1,353,655
1976	Shoemaker, W.	30	1,569,960
1977	Shoemaker, W.	30	1,590,920
1978	McHargue, D. G.	37	1,662,595
1979	Pincay, L., Jr.	44	2,720,006

SOURCE: *American Racing Manual* (1980).

APPENDIX TABLE 12-14
Annual Leading Trainer of Thoroughbreds, Money Won

Year	Trainer	Starts	Number of Wins	Amount Won
1908	Rowe, J.		50	$ 284,335
1909	Hildreth, S. C.		73	123,942
1910	Hildreth, S. C.		84	148,010
1911	Hildreth, S. C.		67	49,418
1912	Schorr, J. F.		63	58,110
1913	Rowe, J.		18	45,936
1914	Benson, R. C.		45	59,315
1915	Rowe, J.		19	75,596
1916	Hildreth, S. C.		39	70,750
1917	Hildreth, S. C.		23	61,698
1918	Bedwell, H. G.		53	80,296
1919	Bedwell, H. G.		63	208,728
1920	Feustal, L.		22	186,087
1921	Hildreth, S. C.		85	262,768
1922	Hildreth, S. C.		74	247,014
1923	Hildreth, S. C.		75	392,124
1924	Hildreth, S. C.		77	255,608
1925	Tompkins, G. R.		30	199,245
1926	Harlan, S. P.		21	205,681
1927	Bringloe, W. H.		63	216,563
1928	Schorr, J. F.		65	258,425
1929	Rowe, J., Jr.		25	314,881
1930	Fitzsimmons, J.		47	397,335
1931	Healy, J. W.		33	297,300
1932	Fitzsimmons, J.		68	266,650
1933	Smith, R. A.		53	135,720
1934	Smith, R. A.		43	249,938
1935	Stotler, J. H.		87	303,005
1936	Fitzsimmons, J.		42	193,415
1937	McGarvey, R.		46	209,925
1938	Sande, E. H.		15	226,495
1939	Fitzsimmons, J.		45	266,205
1940	Smith, T.		14	269,200
1941	Jones, B. A.		70	475,318
1942	Gaver, J. M.		48	406,547

Note: No "starts" records exist prior to 1963.
SOURCE: *American Racing Manual* (1980).

Year	Trainer	Starts	Number of Wins	Amount Won
1943	Jones, B. A.		73	267,915
1944	Jones, B. A.		60	601,660
1945	Smith, T.		52	510,655
1946	Jacobs, H.		99	560,077
1947	Jones, H. A.		85	1,334,805
1948	Jones, H. A.		81	1,118,670
1949	Jones, H. A.		76	978,587
1950	Burch, P. M.		96	637,754
1951	Gaver, J. M.		42	616,392
1952	Jones, B. A.		29	662,137
1953	Trotsek, H.		54	1,028,873
1954	Molter, W.		136	1,107,860
1955	Fitzsimmons, J.		66	1,270,055
1956	Molter, W.		142	1,227,402
1957	Jones, H. A.		70	1,150,910
1958	Molter, W.		69	1,116,544
1959	Molter, W.		71	847,290
1960	Jacobs, H.		97	748,349
1961	Jones, H. A.		62	759,856
1962	Tenney, M. A.		58	1,099,474
1963	Tenney, M. A.	192	40	860,703
1964	Winfrey, W. C.	287	61	1,350,534
1965	Jacobs, H.	610	91	1,331,528
1966	Neloy, E. A.	282	93	2,456,250
1967	Neloy, E. A.	262	72	1,776,089
1968	Neloy, E. A.	212	52	1,233,101
1969	Burch, E.	156	26	1,067,936
1970	Whittingham, C.	551	82	1,302,354
1971	Whittingham, C.	393	77	1,737,115
1972	Whittingham, C.	429	79	1,734,028
1973	Whittingham, C.	423	85	1,865,385
1974	Martin, F.	846	166	2,408,411
1975	Whittingham, C.	487	93	2,437,244
1976	Van Berg, J.	2,362	496	2,976,196
1977	Barrera, L. S.	781	127	2,715,848
1978	Barrera, L. S.	592	100	3,314,564
1979	Barrera, L. S.	492	98	3,608,517

GLOSSARY OF RACING TERMS

Acey-Deucey   Jockeys with their stirrups at uneven levels, the inside or left iron lower than the right, ride *acey-deucey*.

Age   The age of a racehorse is recorded as beginning on January 1 of the year in which the horse is foaled (born). Even if a horse is foaled December 31, it is considered 1 year old on January 1 (breeding is planned to avoid this).

Backside   The stable and training area of a racetrack.

Bleeder   A horse that bleeds after or during a workout or race; the result of a nasal hemorrhage caused from a ruptured throat vein.

Blowout   A brief (usually 3 to 4 furlongs, or 0.6 to 0.8 km) final drill two days or the day before a race designed to further sharpen or maintain an edge of conditioning for the impending contest.

Bolt   When a horse swerves sharply from the regular course, it is said to have *bolted*.

Brace Bandages   Resilient bandages on the legs of horses worn in some cases in an effort to support lame legs, worn in other cases to protect a horse from cutting and skinning its legs while racing.

Breaking   When a horse leaves its gait and breaks into a gallop, it is *breaking*. A trotter or pacer must remain on that gait in a race. If it makes a *break*, the driver must immediately pull it back to its gait.

Breeder   The breeder of a horse is considered to be the owner of its dam, at the time of service.

Breeze, Breezing   An easy workout under stout restraint by the exercise rider to stabilize an already sharp horse's condition between engagements.

Bug Boy   An apprentice jockey, so-called because of the "bug" or asterisk denoting the 5-pound weight allowance in the official program.

Chalk   The favorite or most heavily played horse in a race. The term originated in the days of bookmakers, when the odds were written on slates with chalk.

Check Rein   A line running from the bit to the top of the horse's head, then to the saddle hook to keep a horse's head up. Trotters and pacers commonly race with heads high to maintain a balanced, reaching stride.

Chute   The straightaway entering onto the main oval track for races at 6 furlongs, (1.2 km) (starting on backstretch) and 10 furlongs (2 km) (beginning at left of grandstand).

Climbing   A Thoroughbred is said to be *climbing* when striding in an unnatural, upward fashion, not reaching out forward as in a coordinated gait.

Clocker   The clocker times a horse's workouts, and these times are published for the benefit of the public. Almost all workouts are taken early in the morning during training hours.

Colors   The jockey's silk or nylon jacket and cap provided by the owner.

Colt (c)   A male horse between 2 and 3 years of age.

Coupled Entry   Two or more horses belonging to the same owner or trained by the same person are said to be *coupled*, and they run as an *entry* comprising a single betting unit. Their program number, regardless of post position, would be "1" and "1A." A second entry in the race would be listed in the program as "2" and "2B." A bet on one horse of a coupled entry is a bet on both.

Crab Bit   A bit with prongs extending at the horse's nose. Its purpose is to tip the horse's head up and help prevent it from ducking its head, bowing its neck, and pulling hard on the rein.

Cross   Where a jockey joins the reins is his or her "cross." American riders generally ride with a shorter "cross" than elsewhere in the world.

Cushion   The loose top surface of the racetrack.

Dash   A race decided in a single trial.

Dead Heat   When the photofinish camera shows two horses inseparable at the finish, the race is declared a *dead heat* or tie.

Declaration   Withdrawing an entered horse from a race before the closing of overnight entries. A horse that has been withdrawn is said to have been *declared*.

Derby   A stakes race exclusively for 3-year-olds.

Driving   An all-out effort by horse and/or jockey.

Elbow Boots   Sheepskin-lined pads worn high on front legs to protect elbows (points at rear and bottom of shoulders) from the front feet as they are folded back in top stride. Needed on high-gaited trotters.

Field   The entire group of starters in a race is known collectively as the *field*.

Field Horses   Two or more horses coupled as one betting interest. A *field* appears when there are more than ten entries in a race. Horse number 10 on the program and over are the *field*. A bet on one field horse is bet on all field horses.

Filly (f)   A female horse between 2 and 3 years of age.

Foul   Any action by any jockey that tends to hinder another jockey or any horse in the proper running of the race.

Free-for-All   Horses (or races for such horses) that have won considerable money and must race in fast classes. Means "free for all to enter and open to all horses, regardless of earnings."

Free-Legged Pacer   A pacer that races without hobbles.

Futurity   A stakes race exclusively for 2-year-olds.

Gaiting Strap   Strap strung inside shafts of sulky to keep the horse from swinging the rear end to the right and left and traveling sideways on its gait.

Green Horse   A horse that has never trotted or paced in a public race or against time.

Handle   The aggregate amount of money passing in and out of the parimutuel machines for a given period—a race, a day, a meeting, or a season.

Head Pole   A cue (usually a billiard cue with a hole bored in the handle to accommodate a leather thong) fastened alongside horse's head and neck to keep its head straight.

Heat   One trip in a race that will be decided by winning two or more trials.

Hind Shin Boots   Leather protecting guards on the lower hindlegs prevent cuts and bruises from the front shoes, which graze the hindlegs of some trotters.

Hobbles   Leather straps encircling the front and hind legs of a pacer on the same side to keep those legs moving in unison and to help the horse maintain gait. The straps running over the horse's back and to the haunches are known as *hobble hangers*.

Horse (h)   A male horse 4 years of age or older.

Impost   Weight assessment.

In the Money   A horse finishing first, second, or third is *in the money*. To horsepeople, fourth and fifth are also included because the fourth- and fifth-place finishers also receive purse money.

Inquiry   In some races, even though the numbers of the first four horses to cross the finish line are posted on the tote (totalisator) board, the *official sign* (located on the tote board) is delayed due to the posting of the *inquiry sign*. The inquiry can originate from the stewards or an objection can be lodged by a jockey. Until the objection is resolved by the stewards, the result of the race is not official. The order of finish may stay the same or be revised to penalize a finisher that interfered with another horse during the running of a race. Once a decision is reached by the stewards, the *official sign* will replace the *inquiry sign*.

Irons   Stirrups.

Jog Cart   A cart longer and heavier than a racing sulky, used in warmup miles because it is more comfortable for the rider than a sulky.

Jogging   A slow warmup or exercise for several miles, with the horse going the wrong way on the track.

Knee Boots   Boots fitted around the knees and held up by suspenders (usually white) over the horse's withers. They protect the knees from blows from the opposite foot, especially on turns where a horse is more likely to break stride if it "brushes its knees."

Lead Pad   A piece of equipment under the saddle in which thin slabs of lead may be inserted to bring the rider's weight up to that assigned a horse, in a specific race.

Lugging and Pulling   Some horses pull on the reins, "lug" on one rein, or bear out or in with the driver or jockey, making it hard to drive or ride them and rate the race.

Lugging In or Out   A horse is said to be "lugging in" or "lugging out" if it is pulling toward the inside or outside. A jockey or driver must make every effort to keep his mount on a straight course to prevent interference with other entrants.

Maiden   Horse, mare, or gelding that has never won a heat or race.

Mare (m)   A female horse 4 years of age or older.

Martingale   A strap running from the girth between the horse's front legs to the reins, which are threaded through the rings at the end of the martingale. Helps prevent the horse from tossing and raising its head.

Minus Pool   When an outstanding horse is so heavily played that, after the deduction of the state tax and commission, not enough money remains in the pool to pay off the legally prescribed minimum, it is called a *minus pool,* and the racetrack or racing association makes up the difference.

Morning Glory   A horse that performs well in morning workouts but fails to race to its potential in the afternoon.

Morning Line   The approximate odds usually printed in the program and posted on the totalisator board prior to any betting. The morning line is a forecast of how it is believed the betting will go on a particular race.

Mutuel Pool   The total amount wagered on a race in each ticket category (win-place-show). The total number of winning tickets in the *win* category share the entire pool equally. The same is true of the daily double and quinella pool. The *place* pool is divided into two parts, and the *show* pool into three parts, and divided among the holders with winning tickets on the horse involved.

Official   The designation given to the result of a race by the stewards when no occurrence has happened during the running that in their judgment would revise the actual order of finish, in terms of purse distribution to owners, and parimutuel payoffs paid to winning bettors.

Open Bridle   A bridle without blinds or blinkers covering the eyes. Some bridles are rigged with blinds that shut off vision to the rear and side, and a few horses are raced with goggles or "peekaboo" blinds.

Overlay or Underlay   A horse is called an *overlay* or an *underlay* in wagering parlance when its odds are greater or less than those estimated by the track's official morning line maker.

Overnight   A race for which entries close 72 hours or less before the post time for the race on the day the race is to be run. Also, the mimeographed sheet available to horsepeople at the racing secretary's office showing the entries for the following day.

Overweight   Depending on the conditions of a race, each horse carries an assigned weight. When the jockey cannot make the weight, overweight is allowed, but not more than 5 pounds.

Owner   This includes sole owner, part owner, or lessee of a horse. An interest in the winnings of a horse does not constitute part ownership. The owner makes one of the greatest contributions to the sport of horse racing—he or she pays the bills! In all states, the owner must be licensed, of good reputation, and financially responsible.

Paddock   The area where the horses are saddled and viewed prior to the race. The paddock is always adjacent to the jockeys' quarters.

Parked Out   Lapped on horses at the pole or rail so that there is no chance to get in. A horse parked out, or "on the limb," has farther to go and usually tires and falls back, unless it is far superior to the other horses in the race. The parked-out symbol is used in many past performance charts.

Post   The starting point for the race.

Post Parade   The time before the race when the horses leave the paddock, come on the racetrack, and walk in front of the stands in order for everyone to have a look at them. The duration of the post parade is usually 10 minutes.

Post Position   A horse's position in the starting gate from the inside rail outward is decided by a drawing at the close of entries, before the race. In Quarter Horse racing, every horse must maintain its position as nearly as possible in the lane in which it starts, from the gate down the straightaway to the finish.

Post Time   The official time set by the stewards at which a race will start and at which horses are required to be at the post and ready to start.

Quarter Boots   Close-fitting boots on the front heel to protect the tender quarter (heel of the foot) from being cut by the hind shoes.

Racing Secretary   The official who writes the conditions for the races and assigns the weights for handicap races.

Racing Strip   The track surface itself, which is made up of two parts. The bottom layer, called the *base,* is a mixture of sand and (mostly) clay. It supports the *cushion,* which is generally a 3-inch (76-mm) layer of topsoil mixed with sand. The cushion serves as a shock absorber to ease the intense striking force to which a horse's forelegs are subject when in full stride.

Router   A horse at its best over a "distance of ground" or more than a mile.

Scoring   Preliminary warming up of horses before the start. The horses are turned near the starting point and hustled away as they will be in the race.

Scratch   Racing's idiom for indicating that a horse is declared out of a race in which it was entered, after the entries were drawn.

Shadow Roll   A large sheepskin-type roll worn just above a horse's nose and just below its eyes. It cuts off the horse's view of the track so that it will not shy at shadows, pieces of paper, or other objects.

Short   A Thoroughbred not advanced enough in its conditioning to run the complete distance of a race at its fastest often is termed a "short" horse.

Silks   Jockey's or driver's silks. *See* Colors.

Standardbred   Pure-bred trotting or pacing horses. A nonstandard horse is a crossbred horse or one that cannot be traced in breeding far enough to qualify for standard registration.

Starter   The starter is one of the track officials. He or she has complete jurisdiction over the starting of the horses and has the authority to give orders necessary to ensure a fair start. He or she directs the assistant starters, who lead each horse into the stall gate according to the post position corresponding to the draw made when the horses were entered. These people must know the horses and do any special handling required to get each horse safely into the gate.

Starting Gate   Generally, an electromechanical device in which Quarter Horses and Thoroughbred horses are loaded. When they are all in and standing in an alert manner with their heads up, the starter presses a button that throws open all of the stall doors simultaneously, a bell clangs, and the field is dispatched. Opening the gate activates the electric timer, which records the time of the race in 1/100ths of a second.

Stewards   Racetrack officials. A minimum of three stewards preside over the racing at all tracks, one of whom is a representative of the respective state racing commission. The word of the stewards in all matters is law. It is their duty to see that the race meeting is run according to the rules of racing. They rule on claims of foul or any protests and impose fines and suspensions. All fines and suspensions are reported by them to the state racing commission and in turn to the National Association of State Racing Commissioners so that offenders may be barred from taking part in other race meetings in that state or any other during the period of suspension. The stewards have complete jurisdiction over a race meeting.

Stick   The jockey's whip (sometimes called a *bat.*)

Sulky or Bike   A light racing rig with bicycle-type wheels used in harness races. The sulkies weigh from 29 to 37 pounds (13.2 to 16.8 kg) and usually have hardwood shafts. Aluminum and steel sulkies have been introduced recently.

Tack   The saddle and other equipment worn by a horse during a race or exercise.

Toe Weights   Brass or lead weights weighing from 2 to 4 ounces (56.7 to 113.4 gm). They are clipped to the edge of the front hoofs to extend a horse's stride.

Totalisator   An intricate piece of electronic equipment that records each wager and total wagering in the win, place, show pools and other pools as the parimutuel tickets are sold by a manually operated vending machine. This equipment also calculates the odds on each horse at given intervals, according to the amount wagered.

Totalisator or "Tote" Board   A display board in the infield, on which is posted, electronically, data essential to the racegoer such as the approximate odds, total amount bet in each pool, track condition, post time, time of day, result of race, official sign or inquiry sign (if a foul is claimed), running time of each race, and the mutuel payoff prices, after each race is declared official.

Track Conditions

Fast   A track that is thoroughly dry and at its best. Footing is even.

Sloppy   During or immediately after a heavy rain; water has saturated the cushion and there may be puddles but the base is still firm. Footing is splashy but even, and the running remains fast.

Muddy   Water has soaked into the base and it is soft and wet. Footing is deep and slow.

Heavy   A drying track that is muddy and drying out. Footing is heavy and sticky.

Slow   Still wet, between heavy and good. Footing is heavy.

Good   Rated between slow and fast. Moisture remains in the strip but footing is firm.

Off   An off track is anything other than fast. The usual progression of track conditions before and after a heavy rain is fast—sloppy—muddy—heavy—slow—good—fast.

Trainer   The person who conditions and prepares horses for racing; the coach.

Valet   An employee who takes care of a jockey's equipment, sees to it that the correct silks are at his or her locker, that the rider has the proper weight in his or her lead pad, carries the saddle and equipment to the paddock, helps the trainer saddle the horse, meets the rider after the race, and carries the saddle and equipment back to the jockeys' room.

Walkover   A race from which all entrants but one have been withdrawn. Thus, the remaining horse need but "walk over" the prescribed course to win.

Weigh Out   The procedure in which the racing official known as the clerk of the scales, before the race, checks the weights of jockeys and their riding equipment (saddle, pommel pad, girth) against the officially assigned weights to be carried by each horse in the race. These weights must agree. After the race, the jockeys and their riding equipment are weighed in. This procedure assures that the assigned weights are carried to the post and throughout the race.

Wheel   A horse, turning around suddenly, without guidance by its rider, is said to *wheel*.

GLOSSARY OF RODEO TERMS

Added Money The purse, or the part of the total money put up by the committee.

All-Around World Champion To be eligible for the title All-Around World Champion a cowboy must compete in two events. The top money winner who competed in two or more events is the world champion.

Average When a rodeo has more than one go-round in an event, the contestants are paid off for the best ride or time in each go-round and for the best average of all go-rounds. The winner of the average is the event winner.

Barrier In the timed events, the stock is given a predetermined head start depending on arena conditions, called the *score* and marked by the scoreline. A rope is stretched across the front of the box from which the contestant's horse will come out. The barrier rope is released by a measured length of twine that is pulled loose from the calf or steer as it crosses the scoreline.

Breaking the Barrier If the contestant rides through and breaks the barrier, before it is released, a penalty of 10 seconds is added to the contestant's total time.

Champion Many rodeos award buckles or saddles to the champions of each event and to the all-around winner of the rodeo. The top 15 money winners in each event after the year's-end rodeo, the Grand National at the Cow Palace in San Francisco, are eligible to compete in the National Finals Rodeo (NFR), held in Oklahoma City in early December.

Day Money The amount of prize money paid to the winners of each go-round.

Entry Fee The money paid by the contestant before he can compete at a rodeo. This is added to the prize money in that event. The size of the fee varies with the amount of the purse, ranging from $10 at the smallest rodeos to $200. Contestants must pay a separate entry fee for each event entered.

Hazer   The cowboy who rides on the opposite side of the steer to keep the steer from running away from the steer wrestler's horse.

Honda   The eye in one end of a rope through which the other end is passed to form the loop in a lariat.

Jackpot   An event for which no purse is put up by the rodeo. Winners split all or part of the entry fees.

No Time   When a flag fieldman waves "no time," it means that the contestant has not caught or thrown, animal properly and receives no time on that animal in that go-round, but is still entitled to compete in the next go-round.

Pickup Man   A mounted cowboy who helps the rider off a bronc when a ride is completed. The pickup man then pulls off the flank strap from the bronc and removes the bronc from the arena.

PRCA Champions   The top money winners in each event were called PRCA Champions during 1976–1978. (PRCA stands for Professional Rodeo Cowboys Association.)

Prize Money   Money paid to the winners of the various events in a rodeo. It is comprised of the purse and the entry fees paid by the contestants. The split of the prize money depends on the total monies in a go-round. If it is under $2,000, it is split four ways: 40, 30, 20, and 10 percent. There is a fixed ratio for the number of monies paid and a set percentage, depending on total go-rounds and amount of prize money.

Pulling Leather   When a saddle bronc rider touches any art of the saddle with a free hand, he or she is said to be *pulling leather* and is disqualified.

Reride   Another ride is given to a bronc or bull rider in the same go-round when the first ride was unsatisfactory for any of several reasons.

Rough Stock Events   The riding events: bareback riding, saddle bronc riding, and bull riding.

Score   The distance between the box opening and the scoreline, or head start given the stock in the timed events.

Timed Events   Calf roping, team roping, and steer wrestling.

World Champion   In 1976, under new championship regulations, the top 15 contestants in an event had an equal chance to become the World Champion of the event. The winner of the NFR was the World Champion. In 1979, the rules were changed, and the World Champion in each event is the largest money winner.

APPENDIX 14-1

SEMEN EVALUATION PROCEDURES

To prepare for the semen evaluation, turn on all equipment and allow it to equilibrate to temperature or voltage. Place all glassware and supplies that will come in contact with the semen in an incubator set at 100°F (38°C). Remove from the freezer extenders to be used during the evaluation and thaw them. Then place them in the incubator. Prepare colorimetric tubes for concentration determinations. If the semen is to be used to inseminate a mare, prewarm all necessary supplies to 100°F (38°C) in the incubator.

Volume

The first step in evaluating semen is to determine the gel-free volume. Gently pour the semen into a graduated cylinder (Appendix Figure 14-1). To determine the gel volume, turn the filter upside down in a beaker until it all drains out. Then measure it in a graduated cylinder. After reading the volume, immediately return the cylinder to the incubator and record the volume on the semen evaluation record (Appendix Figure 14-2).

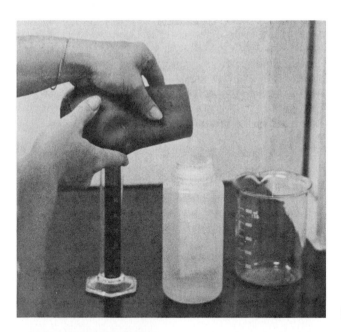

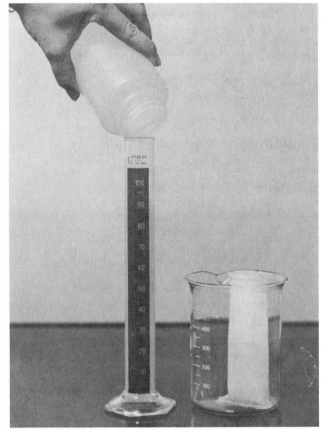

(a)

(b)

APPENDIX FIGURE 14-1

Semen volume determination.
(a) Semen is poured from the
collection bulb into the filter.
(b) Gel-free semen is poured into
a volumetric cylinder to determine
volume and filter is placed
in beaker to collect gel.
(Photos by B. Evans.)

Breeding Farm Name
Address
Phone Number

SEMEN EVALUATION RECORD

Stallion _____   Date _____

Collector _____   Time _____

Number of Mounts _____

Volume: gel-free semen   _____

gel   _____

semen (total volume) _____

Motililty: _____ % using extended semen

Concentration (gel-free):

_____ % transmittance = _____ $\times\ 10^6$/ml

_____ spermatozoa in hemocytometer $\times\ 10^6$ = _____ = 10^6/ml

Total spermatozoa:

_____ $\times\ 10^6$/ml $\times$ _____ ml (gel-free) = _____ $\times\ 10^9$

pH: _____

Motile sperm/ml gel-free: _____ concentration/ml

$\times$ _____ % motility

= _____ motile sperm

Morphology (number of spermatozoa per 100 observed):

abnormal head   _____   abnormal tail _____   normal _____

isolated head   _____   droplets   _____

abnormal mid piece _____   other   _____

APPENDIX FIGURE 14-2
"Semen Evaluation Record."

Motility

As soon as the volume is determined, estimate motility. Stallion spermatozoa usually move in a straight line, but because of their abaxial midpieces they may move in circles.

Motility estimates can be made by placing a drop of semen with a prewarmed pasteur pipette on a prewarmed slide, covering the drop with a prewarmed cover slip and observing the spermatozoa with a 100 or 400 $\times$ microscope (Appendix Figure 14-3). It is advisable to use a slide warmer to keep the spermatozoa from experiencing cold shock. To estimate motility, a series of yes–no decisions is made; for example, are 2

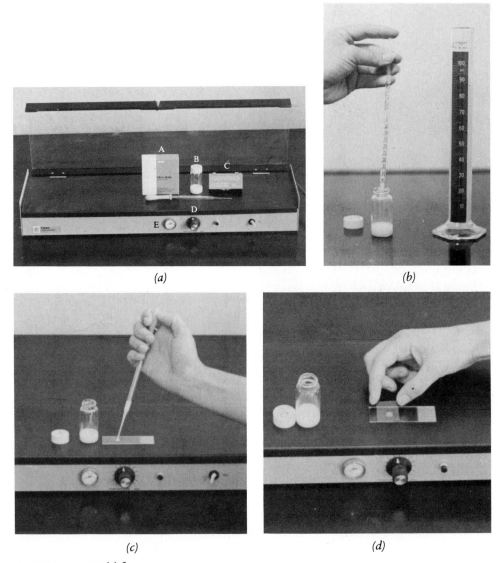

APPENDIX FIGURE 14-3

Semen motility evaluation. *(a)* Equipment: (A) Microscope slides. (B) Semen extender.
(C) Cover slips for microscope slides. (D) Pasteur pipette with rubber bulb. *(b)* Semen is pipetted
into warm extender with a warm pipette. *(c)* Drop of extended semen is placed on warm
microscope slide. *(d)* A warm cover slip is placed over the drop of extended semen.
(Photos by B. Evans.)

of 3, 3 of 4, 4 of 5, 5 of 6, or 9 of 10 motile? Three to five fields should be observed for
each drop. If the spermatozoa are clumping together, which prevents a good estimate,
the semen can be extended by pipetting 0.25 ml of semen into 4.75 ml of an extender
in a vial. Then observe a drop of the extended semen for motility. One of the most

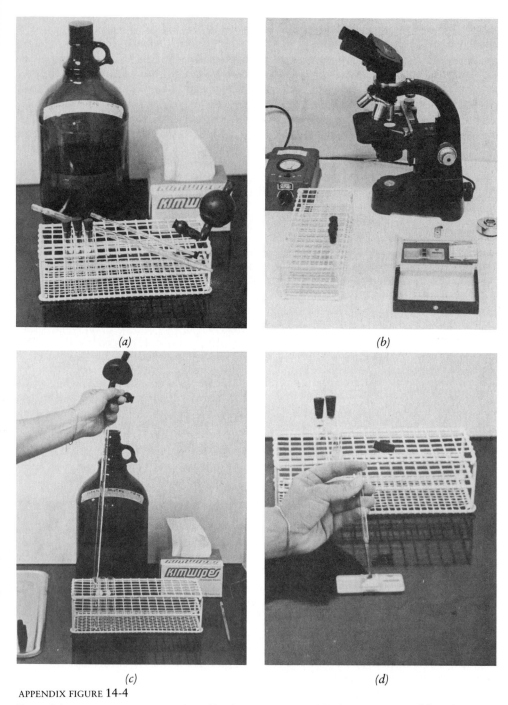

(a) *(b)*

(c) *(d)*

APPENDIX FIGURE 14-4

Determining spermatozoa concentration with a hemocytometer. *(a)* Equipment consists of formalin solution, paper wipes, pipette bulb, pipettes, test tubes, test tube stoppers, and a test tube rack. *(b)* Additional equipment consists of microscope, mechanical counter, and hemocytometer. *(c)* Formalin and semen are pipetted into test tubes. *(d)* Both sides of hemocytometer chamber are filled with diluted semen. (Photos by B. Evans.)

popular extenders is the dried skim milk extender. Prepare the extender by placing 50 ml of sterile water in a sterile 100-ml graduated cylinder. Then add 5 gm of nonfortified dried skim milk, 5 gm of glucose, 50,000 units of penicillin, and 50 mg of streptomycin to the cylinder. After they are dissolved, fill the cylinder to 100 ml with sterile distilled water. Freeze vials containing 4.5 ml of the extender until they are needed. This extender must be clearly labeled, because it is not suitable for extending semen for artificial insemination.

Concentration

There are two ways of determining concentration. The less expensive method is to use a hematocytometer (Appendix Figure 14-4). Prepare semen by pipetting 0.2 and 0.4 ml into 7.8 and 7.6 ml, respectively, of a 10-percent formalin solution, place a rubber stopper in the test tube, and slowly invert 8 to 10 times. If you shake the tubes, air bubbles will be trapped in the solution, which gives a false value. You can make 1 liter of the formalin solution, because it can be stored. Mix it by placing 500 ml of distilled water in a 1-liter volumetric flask or graduated cylinder. Then add 9.0 gm of sodium chloride and mix until it dissolves. Add 100 ml of a 37-percent formaldehyde solution (formalin) and mix thoroughly. Add distilled water, to a final volume of 1 liter. Store the solution in a bottle that prevents evaporation.

The tip of a pasteur pipette is filled with the diluted semen. It is used to fill the hemocytometer chambers (Appendix Figure 14-4). Both sides of the chamber are covered with a cover slip and filled by capillary action. The spermatozoa in the five squares (illustrated in Appendix Figure 14-5) are counted. For a 1:20 dilution (0.2 ml of semen

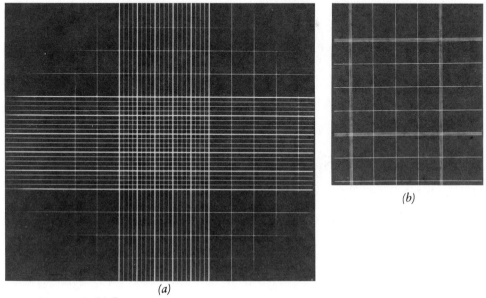

(b)

(a)

APPENDIX FIGURE 14-5

Hemocytometer field. (a) Entire field. (b) One of the five fields that are counted. (Photos by B. Evans.)

in a 7.8-ml formalin solution), the number of spermatozoa counted in the five squares is equivalent to that number of million spermatozoa per ml of undiluted semen. Count each square under 400×, and take care to count each spermatozoon once. Therefore, as you count the spermatozoa in the square that has 16 smaller squares, count the spermatozoa on the lines of the smaller squares at the top and right for each row of spermatozoa. Count those on the bottom and left lines when you count across the next row.

After use, wash the hemocytometer and cover slip by squirting them with distilled water and rinsing with ethanol. They must not be scratched.

If several stallions are being evaluated on a frequent basis, it is advisable to use a spectrophotometer (Appendix Figure 14-6). Set the spectrophotometer at a wavelength of 550 mμ. Set it to 0-percent transmittance without a tube. Place a colorimetric tube containing formalin solution in the spectrophotometer and adjust it to read 100-percent transmittance. Remove the tube and readjust the spectrophotometer to 0 percent. Return the formalin tube and readjust the spectrophotometer to 100 percent. Continue this until it reads 0 percent without a tube and 100 percent for the formalin tube without further adjustments. Prepare semen as for the hemocytometer method except prepare it in colorimeter tubes. Record the percent transmittance for the 1:20 and 1:40 dilution of semen. The concentration of spermatozoa per ml is obtained from a standard curve

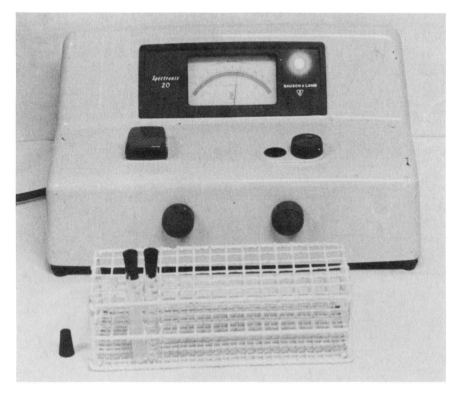

APPENDIX FIGURE 14-6

Spectrophotometer used to determine concentration of spermatozoa. (Photo by B. Evans.)

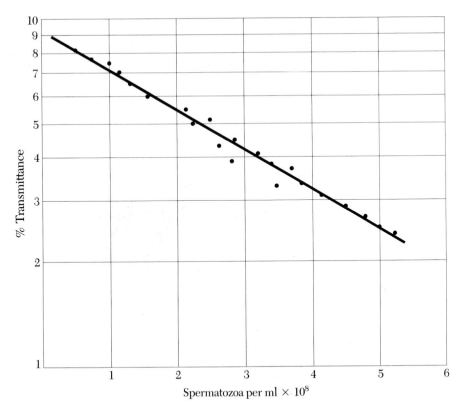

APPENDIX FIGURE 14-7
Standard curve for determining spermatozoa concentration using a spectrophotometer.

previously prepared (Appendix Figure 14-7). Prepare the standard curve by serially diluting a 1:10 dilution of spermatozoa in formalin solution, recording the percent transmittance, and determining the number of spermatozoa in a hemocytometer. The percent transmittance should be between 20 and 80 percent for accurate determination.

When a spectrophotometer is used, the colorimeter tubes must be clean and free from scratches. Wash them with a soft test tube brush and store them in a rubberized test tube rack. Immediately before insertion into the spectrophotometer, wipe the colorimeter tube with a tissue to remove any dust or fingerprints. Each time the light bulb is changed, a new standard curve must be made. Also the calibration should be checked every few months to maintain accuracy.

pH

Determine the pH with a pH meter (Appendix Figure 14-8). Digital pH meters are inexpensive and easy to use. Rinse the electrode with distilled water, dry with tissue paper, and submerge in a small portion of the semen.

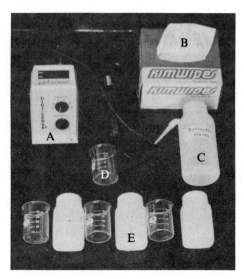

APPENDIX FIGURE 14-8

Equipment for determining pH of semen. (A) pH meter.
(B) Paper wipes. (C) Distilled water to wash electrodes
of pH meter. (D) Beaker to hold semen.
(E) pH standards. (Photo by B. Evans.)

Morphology

Prepare the slide for morphological examination carefully, because the spermatozoa can
be easily damaged, increasing the percentage of abnormal sperm. To prepare the slide,
place a small drop of gel-free semen and a small drop of Hancock's strain on one end of
a slide (Appendix Figure 14-9). Prepare Hancock's stain by dissolving 6 gm of nigrosin
in 60 ml of deionized water. After the nigrosin dissolves, add 1 gm of eosin. After
mixing, filter the solution. Mix the two drops with a small glass rod or the corner of
another slide. Holding the slide at a 45-degree angle, allow the edge of another slide to
touch the drop and then pull it across the slide with zigzag motions to form the smear.
Use very little pressure to pull the slide. A common mistake is to get the layer of stained
spermatozoa too thick, which makes it difficult to observe the spermatozoa. After making the smear, dry the stained slide on a slide warmer or in an incubator before using
it. Using the microscope at 400×, count and classify at least 100 spermatozoa.

Another procedure that stains the spermatozoa for morphological evaluation is as
follows:

1. Prepare a thin, uniform smear of fresh semen on a clean, dry microscope slide.

2. Dry the smear fast over an open flame (match).

3. Submerge the slide in a solution of one part ethyl alcohol and one part ethyl
ether for 3 minutes.

4. Remove the slide and dry it in air.

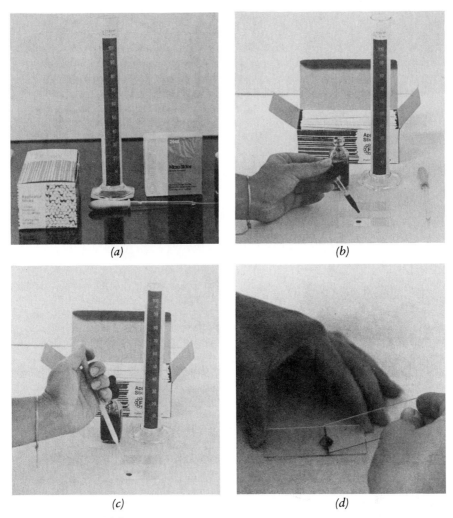

APPENDIX FIGURE 14-9

Morphological examination of spermatozoa. *(a)* Equipment: applicator sticks, semen, microscope slides, stain, and pasteur pipette with rubber bulb. *(b)* A drop of stain is placed on slide. *(c)* A drop of semen is placed adjacent to drop of stain and mixed. *(d)* Microscope slide is used to make a smear. (Photos by B. Evans.)

5. Add two or three drops of stain to the smear and place the slide over a steam bath for 6 minutes. Prepare the stain by mixing 200 ml (5 gm/100 ml) of aniline, 100 ml 5-percent solution (5 gm/100 ml) of Eosin B, and 100 ml of 1-percent solution (1 gm/100 ml) of phenol. After mixing, filter the strain through filter paper and store it in a bottle with an eyedropper.

6. Remove the slide and wash it with distilled water to remove excess stain.

7. Dry in air.

8. Examine in the microscope. The head of the spermatozoa will be pale bluish-gray, the neck outlined in dark blue and colorless inside, and the tail components dark blue.

APPENDIX 15-1

PREPARATION OF CREAM-GEL SEMEN EXTENDER

Prepare cream-gel semen extender by dissolving 1.3 gm of unflavored Knox gelatin in 10 ml of deionized water. Autoclave the solution and add half-and-half (light cream) to make 100 ml. Prepare the half-and-half by heating it in a double boiler at 92°C for 10 minutes. If a scum forms, remove it. Add 1 million units of crystalline penicillin and 1 gm of crystalline streptomycin. Freeze the 100 ml of extender in containers of 10-ml doses. Before collecting the stallion, thaw enough extender and warm it to 100°F (38°C). Extend semen to a ratio of 1:1 to 1:3 (semen:extender) in the extender storage vial.

INSTRUCTIONS FOR MIP-TEST KIT PREGNANCY DETERMINATION

Procedure

Determine pregnancy with an MIP-test kit as follows:

1. Place about 5 ml of the mare's blood in a clean tube.

2. Allow the blood to clot at room temperature.

3. Free the clot from the tube wall using a clean applicator stick.

4. Allow the blood to stand to obtain clear supernatant serum; centrifuge if necessary.

5. Place two of the clean test tubes supplied in the test tube holder. Mark one *control* (C) and the other *sample* (S). This may be done with a glass-marking pencil.

6. Fill a serum collection dropper to the line with mare serum and deliver it into Tube C. Deliver the same amount of serum into Tube S.

7. Place 1 drop of inhibitor solution (bottle with red label and dropper with red bulb) into Tube C and one drop into Tube S. (Make sure that the dropper is held in a vertical position and that the drop falls directly into the test tube.) Mix by shaking the tubes. Allow the tubes to stand at least 1 minute but not longer than 2 minutes before the next step.

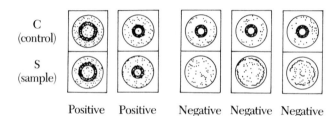

C (control)

S (sample)

Positive   Positive   Negative   Negative   Negative

APPENDIX FIGURE 15-1

Positive and negative reactions obtained with the MIP-test kit for pregnancy.

8. Add 1 drop of neutralizer solution (bottle with brown label and dropper with brown bulb) to Tube C and 1 drop to Tube S. (Make sure that the dropper is held in a vertical position and that the drop falls directly into the test tube.) Mix by shaking the tubes.

9. Using the calibrated dropper with the green bulb, fill the dropper to the line with control solution (bottle with green label), and add the contents to Tube C. Mix by shaking. *Do not add the control solution to Tube S.*

10. Using the calibrated dropper with the blue bulb, fill the dropper to the line with antiserum (bottle with blue label), and add the contents to Tube S. Mix by shaking. *Do not add antiserum to Tube C.*

11. Shake the cell suspension (bottle with black label and dropper with black bulb) until all cells are suspended. Then add 1 drop to Tube C and 1 drop to Tube S. Mix both tubes by shaking.

12. Allow the tubes to stand undisturbed in a test tube holder at room temperature for 2 hours.

Interpretation of Results

Read the cell patterns (Appendix Figure 15-1) in the mirror of the test tube holder and interpret as follows:

1. When both Tubes C and S show a doughnutlike ring, the test is interpreted as positive for pregnancy.

2. When the pattern in Tube S shows a clear mat of cells with or without a faint peripheral ring, the test is negative.

3. Occasionally, a definitely irregular-shaped ring will appear in Tube S; this also is interpreted as negative.

The diagrams in Appendix Figure 15-1 illustrate the patterns described.

Bibliography

Introduction: Owning a Horse

Baird, B. 1976. The remarkable rider: Horse-related activities for the handicapped. *Proceedings, First National Horsemen's Seminar*. Fredericksburg: Virginia Horse Council.

Horses and Land-Use Planning. 1973. Washington, D.C.: National Horse and Pony Youth Activities Council.

Howard, H. A. 1978. Horses help the handicapped. *Horseman* 23:13–20.

Jacobson, P., and M. Hayes. 1972. *A Horse Around the House*. New York: Crown.

Kohler, C. F. 1976. Livestock insurance—all breeds of horses. *Proceedings, First National Horsemen's Seminar*. Fredericksburg: Virginia Horse Council.

Lopez, J. 1974. Demographic survey. *American Horseman*.

McCowan, L. L. 1977. Riding for the handicapped. *Proceedings, Second National Horsemen's Seminar*. Fredericksburg: Virginia Horse Council.

Rhulen, P. 1977. Liability in respect to the horse business. *Proceedings, Second National Horsemen's Seminar*. Fredericksburg: Virginia Horse Council.

Wood, K. A. 1973. *The Business of Horses*. Chula Vista, Calif.: Wood.

Chapter 1: Selecting and Buying a Horse

Adams, O. R. 1974. *Lameness in Horses*. Philadelphia: Lea & Febiger.

Barbalace, R. C. 1977. *An Introduction to Light Horse Management*. Minneapolis: Burgess.

Bradley, M. 1971. *Determining the Age of Horses by Their Teeth*. Columbia: University of Missouri Science and Technology Guide 2842.

Ensminger, M. E. 1969. *Horses and Horsemanship*. Danville, Ill.: Interstate.

Getty, R. (Ed.). *Sisson and Grossman's The Anatomy of Domestic Animals*. Philadelphia: Saunders.

Gorman, J. A. 1967. *The Western Horse: Its Types and Training*. Danville, Ill.: Interstate.

Harrison, J. C. 1968. *Care and Training of the Trotter and Pacer*. Columbus, Ohio: U.S. Trotting Association.

Heglund, N. C., and R. C. Taylor. 1974. Scaling stride frequency and gait to animal size: Mice to horses. *Science* 186:1112–1113.

Hildebrand, M. 1965. Symmetrical gaits of horses. *Science* 150:701–708.

Littlejohn, A., and R. Munro. 1972. Equine recumbency. *Veterinary Record* 90:83–85.

Mellin, J. 1961. *The Morgan Horse*. Brattleboro, Vt.: Greene Press.

Messemer, G. 1978. Choosing your child's trainer. *Horseman* 22:14–19.

Miller, R. W. 1974. *Western Horse Behavior and Training*. New York: Doubleday.

Montgomery, G. G. 1957. Some aspects of the sociality of the domestic horse. *Transactions, Kansas Academy of Sciences* 60:419–424.

Official Guide for Determining the Age of the Horse. No date. American Association of Equine Practitioners, 14 Hillcrest Circle, Rt. 3, Golden, Colo. 80401.

Potter, G. D. 1976. Horse behavior and management. *Proceedings, First National Horsemen's Seminar*. Fredericksburg:Virginia Horse Council.

Pratt, G. W. 1977. Remarks on lameness and breakdown. *Thoroughbred Record* 209:1486–1489.

Pratt, G. W., and J. T. O'Conner. 1978. A relationship between gait and breakdown. *American J. Veterinary Research* 39:249–253.

Ruckebush, Y. 1972. The relevance of drowsiness in the circadian cycle of farm animals. *Animal Behavior* 20:637–643.

Stoneridge, M. A. 1968. *A Horse of Your Own*. New York: Doubleday.

Waring, G. H., S. Wierzbowski, and E. S. E. Hafez. 1975. The behavior of horses. In E. S. E. Hafez (Ed.), *The Behavior of Domestic Animals*. Baltimore: Williams & Wilkins.

Winkenwerder, V. 1978. 10 tips for trainer selection. *Horseman* 27:71–73.

Chapter 2: Evaluating Conformation and Soundness

Adams, O. R. 1974. *Lameness in Horses*. 3rd ed. Philadelphia: Lea & Febiger.

American Horse Shows Association Rule Book. Annual. New York: American Horse Shows Association.

Baum, J. L. 1974. Navicular disease. *Iowa State University Veterinarian* 36:34–36.

Beeman, M. 1972. The relationship of form to function. *Quarter Horse J.* 12:82–128.

Borton, A. 1977. Selection of the horse. In J. W. Evans, A. Borton, H. F. Hintz, and L. D. Van Vleck, *The Horse*. San Francisco: W. H. Freeman and Company.

Bradley, M. 1971. *Determining the Age of Horses by Their Teeth*. Columbia: University of Missouri Science and Technology Guide 2842.

Dial, S. E. 1977. Conformation is the key to Secretariat's greatness. *Horseman* 21:51–54.

Edwards, G. B. 1971. *Know the Arabian Horse*. Omaha, Neb.: Farnam Horse Library.

Edwards, G. B. 1973. *Anatomy and Conformation of the Horse*. Croton-on-Hudson, N.Y.: Dreenan Press.

4-H Horse Judging Guide. No date. University Park: Cooperative Extension, Pennsylvania State University.

Gay, C. W. 1932. *Lippincott's Farm Manuals: Productive Horse Husbandry*. Chicago: Lippincott.

Green, B. K. 1969. *Horse Conformation as to Soundness and Performance*. Greenville, Tex.: Green.

Harper, M. W. 1913. *Management and Breeding of Horses*. New York: Orange Judd.

Harrison, J. C. (Ed.). 1968. *Care and Training of the Trotter and Pacer*. Columbus, Ohio: U.S. Trotting Association.

Hayes, M. H. 1972. *Veterinary Notes for Quarter Horses*. New York: Arco.

Miller, R. M. 1970. Cancer of the eyelid and photosensitization. *Western Horseman* 35:48–49, 152.

Official Guide for Determining the Age of the Horse. No date. American Association of Equine Practitioners, 14 Hillcrest Circle, Rt. 3, Golden, Colo. 80401.

Roberts, S. J. 1969. Comments on equine leptospirosis. *J. American Veterinary Medical Association* 155:442–445.

Rossdale, P. D. 1975. *The Horse from Conception to Maturity*. London: Allen.

Runciman, B. 1940. Roaring and whistling in Thoroughbred horses. *Veterinary Record* 53:37.

Van Pelt, R. W. 1969. Inflammation of the tarsal synovial sheath (thoroughpin) in horses. *J. American Veterinary Medical Association* 155:1481–1488.

Van Vleck, L. D. 1977. Breeds in the United States. In J. W. Evans, A. Borton, H. F. Hintz, and L. D. Van Vleck, *The Horse*. San Francisco: W. H. Freeman and Company.

Chapter 3: Choosing the Breed

American Stud Book. Published every four years. New York: Jockey Club.

Briggs, H. M. 1969. *Modern Breeds of Livestock*. 3rd ed. New York: Macmillan.

Conn, G. H. 1957. *The Arabian Horse in America*. New York: Barnes.

Davenport, H. 1909. (Republished 1947.) My quest for the Arabian horse. In H. Davenport, *Arabian Horse Registry of America*. Denver, Colo.: Arabian Horse Registry of America.

Denhardt, R. M. 1948. *The Horse of the Americas*. Norman: University of Oklahoma Press.

Dinsmore, W. 1935. *Judging Horses and Mules*. Chicago: Horse and Mule Association of America.

Estes, J. A., and J. H. Palmer. 1942. *An Introduction to the Thoroughbred Horse*. 1st rev. ed. 1949, A. Bower (Ed.); 2nd rev. ed., 1972, C. H. Stone (Ed.). Lexington, Ky.: The Blood-Horse.

Fletcher, J. L. 1945. A genetic analysis of the American Quarter Horse. *J. Heredity* 36:346.

Fletcher, J. L. 1946. A study of the first fifty years of Tennessee Walking Horse breeding. *J. Heredity* 37:369.

Gazder, P. J. 1954. The genetic history of the Arabian horse in the United States. *J. Heredity* 45:95.

Gilbey, W. 1900. *Ponies: Past and Present*. London: Vinton.

Glyn, R. 1971. *The World's Finest Horses and Ponies*. New York: Doubleday.

Goodall, D. M. 1965. *Horses of the World*. London: Country Life.

Griffen, J. 1966. *The Pony Book*. New York: Doubleday.

Haines, F. 1963. *Appaloosa: The Spotted Horse in Art and History*. Fort Worth, Tex.: Amon Carter Museum of Western Art.

Haines, F., R. L. Peckinpah, and G. B. Hatley. 1957. *The Appaloosa Horse*. Lewiston, Idaho: Bailey.

Harper, M. W. 1913. *Management and Breeding of Horses.* New York: Orange Judd.

Hervey, J. 1947. *The American Trotter.* New York: Coward-McCann.

Miller, R. W. 1969. *Appaloosa Coat Color Inheritance.* Bozeman: Montana State University.

Nye, N. 1964. *The Complete Book of the Quarter Horse.* New York: Barnes.

Osborne, W. D. 1967. *The Quarter Horse.* New York: Grosset & Dunlap.

Patten, J. W. 1960. *The Light Horse Breeds.* New York: Barnes.

Reese, H. H. 1956. *Horses of Today: Their History, Breeds, and Qualifications.* Pasadena, Calif.: Wood and Jones.

Sanders, A. H., and W. Dinsmore. 1917. *A History of the Percheron Horse.* Chicago: Sanders.

Savitt, S. 1966. *American Horses.* New York: Doubleday.

Sires and Dams Book. Annual. Columbus, Ohio: U.S. Trotting Association.

Speelman, S. R. 1941. *Breeds of Light Horses.* USDA Farmers' Bulletin No. 952. Washington, D.C.: U.S. Department of Agriculture.

Taylor, L. 1961. *The Horse America Made: The Story of the American Saddle Horse.* New York: Harper & Row.

Telleen, M. 1972. The draft breeds. *Western Horseman* 37:80.

Trotting and Pacing Guide. Annual. Columbus, Ohio: U.S. Trotting Association.

Wentworth, L. 1945. *The Authentic Arabian Horse and His Descendants.* London: Allen & Unwin.

Widmer, J. 1959. *The American Quarter Horse.* New York: Scribner's.

Zahre, A. 1959. The genetic history of the Arabian horse in America. Unpublished doctoral thesis, Michigan State University, East Lansing.

Zarn, M., T. Keller, and K. Collins. 1977a. *Wild, Free-Roaming Horses, Status of Present Knowledge.* U.S. Departments of Interior and Agriculture Technical Note 294. Denver, Colo.: U.S. Department of Interior and U.S. Department of Agriculture.

Zarn, M., T. Keller, and K. Collins. 1977b. *Wild, Free-Roaming Burros, Status of Present Knowledge.* U.S. Departments of Interior and Agriculture Technical Note 296. Denver, Colo.: U.S. Department of Interior and U.S. Department of Agriculture.

Chapter 4: Providing Physical Facilities

Bingham, B. H., and A. Huff. 1968. *Horseman's Handbook: Barns and Related Equipment.* Publication No. 147. Blacksburg: Virginia Polytechnic Institute.

Blickle, J. D., R. L. Maddex, L. T. Windling, and J. H. Pedersen. 1971. *Horse Handbook Housing and Equipment.* Midwest Plan Service Series. Ames: Iowa State University.

Borton, A. 1977. Fences, buildings, and equipment for horses. In J. W. Evans, A. Borton, H. F. Hintz, and L. D. Van Vleck, *The Horse.* San Francisco: W. H. Freeman and Company.

Burton, D. W., and H. F. Hintz. 1974. Feeding facilities. *American Horseman* 5:14–16, 54–55.

Dunn, N. 1972. Horse handling facilities. In M. E. Ensminger (Ed.), *Stud Managers Handbook.* Vol. 8. Clovis, Calif.: Agriservices Foundation.

Ensminger, M. E. 1969. *Horses and Horsemanship.* Danville, Ill.: Interstate.

Evans, J. W. 1977. Management of a horse farm. In J. W. Evans, A. Borton, H. F. Hintz, and L. D. Van Vleck, *The Horse.* San Francisco. W. H. Freeman and Company.

Fallon, E. 1967. Breeding farm hygiene. *Lectures Stud Managers Course.* Lexington, Ky.: Grayson Foundation.

Gaines, D. 1977. Build that fence to last. *Horseman* 21:54–58.

Harvey, H. M. 1968. Stock farm management. In J. C. Harrison (Ed.), *Care and Training of the Trotter and Pacer.* Columbus, Ohio: U.S. Trotting Association.

Hollingsworth, K. 1971. New-design sheds serve very well. In *A Barn Well Filled.* Lexington, Ky.: *The Blood Horse.*

O'Dea, J. C. 1966. Veterinarian's view of farm layout. *The Blood Horse* 92:3790–3791.

Parsons, R. A. 1970. *Riding, Horse Barns, and Equipment.* M-25. Davis: University of California Cooperative Extension Service.

Smith, P. C. 1967. *The Design and Construction of Stables.* London: Allen.

Stone, C. H. 1971. Good fences make good neighbors and good horse farms. In *A Barn Well Filled.* Lexington, Ky.: The Blood-Horse.

Sutton, H. H. 1963. The veterinarian and horse farm management. In J. F. Bone and others (Eds.), *Equine Medicine and Surgery.* Wheaton, Ill.: American Veterinary Publications.

Umstadter, L. W. 1976. Do's and don'ts for horse farm facilities. *The Morgan Horse* 33:1409–1414.

Valliere, D. W. 1971. A barn for all seasons. In *A Barn Well Filled.* Lexington, Ky.: The Blood-Horse.

Willis, L. C. 1973. *The Horse Breeding Farm.* New York: Barnes.

Chapter 5: Feeding

Alexander, F., and M. E. Davies. 1969. Studies on vitamin B_{12} in the horse. *British Veterinary J.* 125:169.

Argenzio, R. A., J. E. Lowe, D. W. Pickard, and C. E. Stevens. 1974. Digesta passage and water exchange in the equine large intestine. *American J. Physiology* 226:1035.

Aronson, A. L. 1972. Lead poisoning of cattle and horses following long-term exposure to lead. *American J. Veterinary Research* 33:627.

Baker, H. J., and J. R. Lindsey. 1968. Equine goiter due to excess dietary iodine. *J. American Veterinary Medical Association* 153:1618.

Baker, J. P. 1971. Horse nutritive requirements. *Feed Management* 22:10–15.

Barth, K. M., J. W. Williams, and O. G. Brown. 1977. Digestible energy requirement of working and non-working ponies. *J. Animal Science* 44:585.

Bille, N. 1971. Vitamin D excess and calciphylaxis in domestic animals. *Nutrition Abstracts Rev.* 41:758.

Bradbury, P., and S. Werk. 1974. *Horse Nutrition Handbook.* Houston, Tex.: Cordovan.

Breuer, L. H. 1975. Effects of mare diet during late gestation and lactation, supplemental feeding of foal and early weaning on foal development. *Proceedings, Fourth Equine Nutrition Symposium.* Pomona: California State Polytechnic University.

Breuer, L. H., and D. L. Golden. 1971. Lysine requirement of the immature equine. *J. Animal Science* 33:227.

Caljuk, E. A. 1961. Water metabolism and water requirements of horses. *Tr. Vses. Inst. Konevodstvo.* 23:195. (As cited in *Nutrition Abstracts Rev.* 32:574, 1962.)

Carroll, F. D. 1950. B vitamin content in the skeletal muscle of the horse fed a B vitamin low diet. *J. Animal Science* 9:139.

Carroll, F. D., H. Goss, and C. E. Howell. 1949. The synthesis of B vitamins in the horse. *J. Animal Science* 8:290.

Cello, R. M. 1962. Recent findings in periodic ophthalmia. *Proceedings, Eighth Annual Convention of American Association of Equine Practitioners.*

Dawson, F. L. M. 1977. Recent advances in equine reproduction. *Equine Veterinary J.* 9: 4–11.

Drew, B., W. P. Barber, and D. G. Williams. 1975. The effect of excess dietary iodine on pregnant mares and foals. *Veterinary Record* 97:93.

Fonnesbeck, P. V. 1968. Consumption and excretion of water by horses receiving all hay and hay-grain diets. *J. Animal Science* 27:1350.

Fonnesbeck, P. V. 1969. Partitioning of the nutrients of forage for horses. *J. Animal Science* 28:624.

Fonnesbeck, P. V., and L. D. Symons. 1967. Utilization of carotene of hay by horses. *J. Animal Science* 26:1030.

Fowler, M. E. 1963. Poisonous plants. In J. F. Bone and others (Eds.), *Equine Medicine and Surgery.* Wheaton, Ill.: American Veterinary Publications.

Garton, C. L., G. W. Vander Noot, and P. V. Fonnesbeck. 1964. Seasonal variation in carotene and vitamin A concentration of the blood of brood mares in New Jersey. *J. Animal Science* 23:1233. (Abstract.)

Guenthner, H. R. 1974. *Horse Pastures.* Cooperative Extension Service Bull. C151. Reno: University of Nevada.

Hartley, W. J., and A. B. Grant. 1961. A review of selenium responsive diseases of New Zealand livestock. *Federation Proceedings* 20:678.

Hintz, H. F. 1968. Energy utilization in the horse. *Proceedings, Cornell Nutrition Conference.* Ithaca, N.Y.: Cornell University, pp. 47–50.

Hintz, H. F. 1975. Digestive physiology of the horse. *J. South African Veterinary Association* 46:13–16.

Hintz, H. F. 1977. Feeds and feeding. In J. W. Evans, A. Borton, H. F. Hintz, and L. D. Van Vleck, *The Horse.* San Francisco: W. H. Freeman and Company.

Hintz, H. F., J. E. Lowe, A. J. Clifford, and W. J. Visek. 1970. Ammonia intoxication resulting from urea ingestion by ponies. *J. American Veterinary Medical Association* 157:963.

Hintz, H. F., S. J. Roberts, S. Sabin, and H. F. Schryver. 1971. Energy requirements of light horses for various activities. *J. Animal Science* 32:100.

Hintz, H. F., and H. F. Schryver. 1972. Nitrogen utilization in ponies. *J. Animal Science* 34:592.

Hintz, H. F., H. F. Schryver, and J. E. Lowe. 1971. Comparison of a blend of milk products and linseed meal as protein supplements for young growing horses. *J. Animal Science* 33:1274.

Hintz, H. F., H. F. Schryver, J. E. Lowe, J. King, and L. Krook. 1973. Effect of vitamin D on Ca and P metabolism in ponies. *J. Animal Science* 37:282. (Abstract.)

Houpt, T. R., and K. A. Houpt. 1965. Urea utilization by the pony. *Federation Proceedings* 24:628.

Howell, C. E., G. H. Hart, and N. R. Ittner. 1941. Vitamin A deficiency in horses. *American J. Veterinary Research* 2:60.

Jefferies, N. E. 1970. *Better Pastures for Horses and Ponies.* Cooperative Extension Service Circular 294. Bozeman: Montana State University.

Johnson, R. J., and J. W. Hart. 1974. Utilization of nitrogen from soybean meal, biuret and urea by equine. *Nutrition Report International* 9:209.

Jordan, R. M., V. S. Myers, B. Yoho, and F. A. Spurrell. 1975. Effect of calcium and phosphorus levels on growth, reproduction and bone development of ponies. *J. Animal Science* 40:78.

Joyce, T. R., K. R. Pierce, W. M. Romain, and J. G. Baker. 1971. Clinical study of nutritional secondary hyperparathyroidism in horses. *J. American Veterinary Medical Association* 158:2033.

Kingsbury, J. 1964. *Poisonous Plants of the United States and Canada.* Englewood Cliffs, N.J.: Prentice-Hall.

Krook, L., and J. E. Lowe. 1964. Nutritional secondary hyperparathyroidism in the horse. *Veterinary Pathology* 1: Supplement 1.

Linerode, P. A. 1966. Studies on the synthesis and absorption of B complex vitamins in the equine. Unpublished doctoral thesis, Ohio State University, Columbus.

McNeely, G. H., and others. 1975. *Feed Requirements of the Light Horse.* Cooperative Extension Publication 75-FS/4005. Davis: University of California.

Martin, A. A. 1975. Nigro pallidal encephalomalacia in horses caused by chronic poisoning with yellow star thistle. *Nutrition Abstracts Rev.* 45:85.

Moxon, A. L. 1937. Alkali disease or selenium poisoning. *South Dakota Agricultural Experiment Station Technical Bull.* 311:1.

National Research Council. 1978. *Nutrient Requirements for Horses.* Washington, D.C.: National Academy of Sciences.

Nelson, A. W. 1961. *Nutrient Requirements of the Light Horse.* Amarillo, Tex.: American Quarter Horse Association.

Nelson, D. D., and W. J. Tyznik. 1971. Protein and non-protein nitrogen utilization in the horse. *J. Animal Science* 32:68.

Nickel, R., A. Schummer, E. Seiferle, and W. O. Sack. 1973. *The Viscera of Domestic Animals.* New York: Springer-Verlag.

Olsson, N., and A. Ruudvere. 1955. The nutrition of the horse. *Nutrition Abstracts Rev.* 25:1.

Ott, E. A. 1971. Energy and protein for reproduction in the horse. *Proceedings, Second Equine Nutrition Symposium.* Ithaca, N.Y.: Cornell University.

Potter, G. D., and J. D. Huchton. 1975. Growth of yearling horses fed different sources of protein with supplemental lysine. *Proceedings, Fourth Equine Nutrition Symposium.* Pomona: California State Polytechnic University.

Pratt, J. N. 1973. Let's grow better pastures. *Quarter Horse J.* 27:162, 192.

Prior, R. L., H. F. Hintz, J. E. Lowe, and W. J. Visek. 1974. Urea recycling and metabolism of ponies. *J. Animal Science* 38:565.

Quinn, C. R. 1975. Isobutylidene diurea as a protein substitute for horses. Unpublished doctoral thesis, Colorado State University, Fort Collins.

Romane, W. M., and others. 1966. Cystitis syndrome of the equine. *Southwest Veterinarian* 19:95–99.

Rusoff, L. L., R. B. Lank, T. E. Spillman, and H. B. Elliot. 1965. Non-toxicity of urea feeding to horses. *Veterinary Medicine* 60:1123.

Schryver, H. F., P. H. Craig, and H. F. Hintz. 1970. Calcium metabolism in ponies fed varying levels of calcium. *J. Nutrition* 100:955.

Schryver, H. F., H. F. Hintz, and P. H. Craig. 1971a. Calcium metabolism in ponies fed a high phosphorus diet. *J. Nutrition* 101:259.

Schryver, H. F., H. F. Hintz, and P. H. Craig. 1971b. Phosphorus metabolism in ponies fed varying levels of phosphorus. *J. Nutrition* 101:1257.

Slade, L. M., L. D. Lewis, C. R. Quinn, and M. L. Chandler. 1975. Nutritional adaptations of horses for endurance performance. *Proceedings, Fourth Equine Nutrition Symposium.* Pomona: California State Polytechnic University.

Slade, L. M., D. W. Robinson, and K. E. Casey. 1970. Nitrogen metabolism in nonruminant herbivores. I. The influence of non-protein nitrogen and protein quality on the nitrogen retention of adult mares. *J. Animal Science* 30:753.

Stillions, M. C., and W. E. Nelson. 1972. Digestible energy during maintenance of the light horse. *J. Animal Science* 34:981.

Stillions, M. C., S. M. Teeter, and W. E. Nelson. 1971. Ascorbic acid requirements of mature horses. *J. Animal Science* 32:249.

Stowe, H. D. 1967. Reproductive performance of barren mares following vitamins A and E supplementation. *Proceedings, Thirteenth Annual Convention of American Association of Equine Practitioners,* pp. 81–93.

Stowe, H. D. 1968. Alpha-tocopherol requirements for equine erythrocyte stability. *American J. Clinical Nutrition* 21:135.

Tyznik, W. J. 1975. Feeding the foal. Paper presented to May 1975 California Livestock Symposium, Fresno, Calif.

Ullrey, D. E., R. D. Struthers, D. G. Hendricks, and B. E. Brent. 1966. Composition of mare's milk. *J. Animal Science* 25:217.

Van Niekerk, C. H., and J. S. van Heerden. 1972. Nutrition and ovarian activity of mares early in the breeding season. *J. South African Veterinary Association* 43:351–360.

Wagoner, D. W. 1973. *Feeding to Win.* Grapevine, Tex.: Equine Research Publications.

Weeks, P. J. 1975. Good pastures. *Horseman* 19:21–27.

White, H. E. 1976. Establishing and maintaining horse pasture. *Proceedings, First National Horsemen's Seminar.* Fredericksburg: Virginia Horse Council.

White, H. E. 1977. Hay for horses. Cooperative Extension Service, Horse Facts, Vol. 2, No. 5. Blacksburg: Virginia Polytechnic Institute and State University.

Wooden, G., K. Knox, and C. L. Wild. 1970. Energy metabolism in light horses. *J. Animal Science* 30:544.

Wysocki, A. A., and J. P. Baker. 1975. Utilization of bacterial protein from the lower gut of the equine. *Proceedings, Fourth Equine Nutrition Symposium.* Pomona: California State Polytechnic University.

Chapter 6: Handling

Barbalace, R. C. 1974. *Light Horse Management.* Minneapolis: Burgess.

Beck, D., and S. Sidles. 1975. Trailering. *The Morgan Horse* 35:1400–1403; 35:1631–1633.

Carlson, T. 1975. Taking the burden out of trailer loading. *Horseman* 29:72–82.

Gorman, J. A. 1967. *The Western Horse.* Danville, Ill.: Interstate.

Kaplan, S. 1976. Trailering tips. *Horseman* 22:36–39.

Leahy, J. R., and P. Barrow. 1953. *Restraint of Animals.* Ithaca, N.Y.: Cornell Campus Store.

Miller, R. W. 1975. *Western Horse Behavior and Training.* New York: Doubleday.

Rondum, K. 1973. Teach your horse to load. *Practical Horseman* 74:41–45.

Trailer Towing Guide. 1975. Wausaw, Wis.: Hammer Blow Company.

Chapter 7: Grooming

Grooming. 1973. *Practical Horseman* 74:19–24, 37.
How to Clip Your Horse. 1976. Temple, N.H.: Derbyshire.
Jennings, J., and S. Vaughn. 1976. Body clipping. *Quarter Horse J.* 28:172–180.
McHam, D. 1975. Groom your horse if you can't ride him. *Horseman* 43:60–63.
Meek, N. 1978. Grooming: How professionals do it. *Horse Lovers National* 74:42–43, 46.

Chapter 8: Caring for the Feet

Adams, O. R. (Ed.). 1974. *Lameness in Horses*. 3rd ed. Philadelphia: Lea & Febiger.
Butler, K. D. 1977a. The hoof's role in competitive trail riding. *Saddle Action* 1:18–21, 50.
Butler, K. D. 1977b. Practical hoof physiology. *Proceedings, Second National Horsemen's Seminar*. Fredericksburg: Virginia Horse Council.
Butler, K. D. 1974. *The Principles of Horseshoeing*. Ithaca, N.Y.: Butler.
Canfield, D. M. *Elements of Farrier Science*. Albert Lea, Minn.: Enderes Tool Co.
Emery, L., J. Miller, and N. Van Hoosen. 1977. *Horseshoeing Theory and Hoof Care*. Philadelphia: Lea & Febiger.
Evans, J. W. 1977. Anatomy and care of the foot. In J. W. Evans, A. Borton, H. F. Hintz, and L. D. Van Vleck, *The Horse*. San Francisco: W. H. Freeman and Company.
Getty, R. (Ed.). 1953. *Sisson and Grossman's The Anatomy of Domestic Animals*. Philadelphia: Saunders.
Greeley, R. G. 1970. *The Art and Science of Horseshoeing*. Philadelphia: Lippincott.
Harrison, J. C. 1968. *Care and Training of the Trotter and Pacer*. Columbus, Ohio: U.S. Trotting Association.
Kays, D. J. 1969. *How to Save a Horse*. New York: Barnes.
O'Connor, J. T., and J. C. Briggs. 1971. Feet: Conformation and motion. *The Morgan Horse* 364:27–32.
Springhall, J. A. 1964. *Elements of Horseshoeing*. Brisbane, Australia: University of Queensland Press.
The Horseshoer. 1941. Technical Manual TMA-220. Washington, D.C.: War Department.
Wiseman, R. F. 1968. *The Complete Horseshoeing Guide*. Norman: University of Oklahoma Press.

Chapter 9: Recognizing Illness and Giving First Aid

Brown, C. M. 1978. Bandage the forearm. *Horseman* 23:34.
Hewett, D. K. 1973. Leg wraps. *Quarter Horse J.* 25:170–171.
Hoeppner, G. 1976. Keeping your horse sound: Therapy for lameness. *Horse of Course* 5:36–39.
McHam, D. 1976. Emergency first aid. *Horseman* 21:35–38.

McHam, D. 1978. First aid: Do it right. *Horseman* 23:31–34.

MacKay-Smith, M., and L. DeButts. 1977. Locating lameness. *Equus* 12:47–55.

Naviaux, J. L. 1974. First aid. In J. L. Naviaux, *Horses in Health and Disease*. New York: Arco.

Nesbitt, C. 1977. A bandaging primer: The right way to bandage a horse. *Horse of Course* 6:26–28; 6:26–27.

Rossdale, P. D. 1971. Diagnosis of lameness. *Thoroughbred of California* 62:780–792.

The wrap session I. 1978. *Equus* 4:51–55.

The wrap session II. 1978. *Equus* 9:29–33.

Chapter 10: Knowing About Parasites and Diseases

Bushnell, R. B., D. P. Furman, E. C. Loomis, and L. A. Riehl. 1976. *Control of External Parasites of Livestock*. Division of Agricultural Sciences, Leaflet 2854. Davis: University of California.

Catcott, E. J., and J. F. Smithcors (Eds.). 1972. *Equine Medicine and Surgery*. Wheaton, Ill.: American Veterinary Publications.

Cernick, S. L. 1977. *Preventative Medicine and Management for the Horse*. New York: Barnes.

Deal, A. S., E. C. Loomis, G. G. Gurtle, and W. C. Fairbanks. 1969. *Fly Control on the Horse Ranch*. Cooperative Extension Service Bull. AXT-307. Davis: University of California.

Donawick, W. J. Colics of the horse. Unpublished. Veterinary Clinic, New Bolton, Pa.

Equine Infectious Anemia. 1977. Gaithersburg, Md.: Equine Health Publications.

Georgi, J. R. 1977. Parasites of the horse. In J. W. Evans, A. Borton, H. F. Hintz, and L. D. Van Vleck, *The Horse*. San Francisco: W. H. Freeman and Company.

Hayes, M. H. 1972. *Veterinary Notes for Horse Owners*. New York: Arco.

Hintz, H. F., and J. E. Lowe. 1977. Diseases of the horse. In J. W. Evans, A. Borton, H. F. Hintz, and L. D. Van Vleck, *The Horse*. San Francisco: W. H. Freeman and Company.

Loomis, E. C., J. P. Hughes, and E. L. Bramhall. 1975. *The Common Parasites of Horses*. Cooperative Extension Service Publication 4006. Davis: University of California.

Proctor, D. L. 1969. Internal parasites of the horse. *Proceedings, Fifteenth Annual Convention of American Association of Equine Practitioners*, pp. 351–363.

Rossdale, P. D. 1971. Colic. *Thoroughbred of California* 61:770–779.

Rossdale, P. D. 1976. *Inside the Horse*. Arcadia: California Thoroughbred Breeders Association.

Schroeder, W. G. 1969. Suggestions for handling horses exposed to rabies. *J. American Veterinary Medical Association* 155:1842–1843.

The first totally complete worming chart. 1978. *Horse Lovers National* 43:52–53.

Wise, W. E., J. H. Drudge, and E. T. Lyons. 1972. *Controlling Internal Parasites of the Horse*. Lexington: University of Kentucky Cooperative Extension Booklet Vet-1.

Chapter 11: Riding and Driving

About bits. 1973. *Practical Horseman* 74:30–33, 50–51.

Barbalace, R. C. 1974. *Light Horse Management*. Minneapolis: Burgess.

Bradley, M. 1973. *Western Equitation: Mounting, Correct Seat, Dismounting*. Science and Technology Guide 2876. Columbia: University of Missouri Cooperative Extension.

Bradley, M., and S. D. Hardwicke. 1971. *English Equitation: Mounting, Correct Seat, Dismounting*. Science and Technology Guide 2875. Columbia: University of Missouri Cooperative Extension.

Chamberlain, H. D. 1937. *Training Hunters, Jumpers, and Hacks*. New York: Surrydale Press.

Christopher, R. 1978. Saddle seat: Who says it's English. *Horse Lovers National* 43:8–15.

Close, P. 1977. Les Corbett: Saddle seat equitation. *Western Horseman* 42:32–34.

Davis, J. A. *The Reins of Life*. Plant City, Fla.: Kimbal Horse Books.

DeRomasskan, G. 1964. *Fundamentals of Riding*. New York: Doubleday.

DeRomasskan, G. 1968. *Riding Problems*. Brattleboro, Vt.: Greene Press.

Dillon, J. 1960. *A School for Young Riders*. New York: Arco.

Dillon, J. 1961. *Form over Fences*. New York: Arco.

DuPont, J. 1973. From egg butt to billet strap. *Quarter Horse J.* 25:84.

Edwards, R. H. 1965. *Saddlery*. New York: Barnes.

Ensminger, M. E. 1972. *Breeding and Raising Horses*. U.S. Department of Agriculture Handbook No. 394. Washington, D.C.: Agricultural Research Service.

Evans, J. W. 1977. The recreational use of the horse. In J. W. Evans, A. Borton, H. F. Hintz, and L. D. Van Vleck, *The Horse*. San Francisco: W. H. Freeman and Company.

Evans, J. W. 1977. The recreational use of the horse. In J. W. Evans, A. Borton, H. F. Hintz, and L. D. Van Vleck, *The Horse*. San Francisco: W. H. Freeman and Company.

Felton, S. W. 1967. *The Literature of Equitation*. London: Allen.

Ferguson, B. 1971. An explanation of the aids as used at the walk, trot, and canter. *The Morgan Horse* 31:32.

Floyd, J. 1979a. Mounting and dismounting. *Side-Saddle News* 6:3.

Floyd, J. 1979b. Ready to ride. *Side-Saddle News* 6:3.

Forbes, J. 1972. *So Your Kids Want a Pony*. Brattleboro, Vt.: Greene Press.

Foreman, M. *Horse Handling Science*. Vols. 1, 2, and 3. Fort Worth, Tex.: Horse Handling Science (P.O. Box 9371).

Gianoli, L. 1969. *Horses and Horsemanship Through the Ages*. New York: Crown.

Gorman, J. A. 1967. *The Western Horse: Its Types and Training*. Danville, Ill.: Interstate.

Greenall, J. 1976. Hitching a single horse to an antique carriage. *The Morgan Horse* 36:509–512.

Greenall, J. 1977. Appointments for driving antique carriages. *The Morgan Horse* 37:554–566.

Hamilton, S. 1977. Before you saddle up. *Equus* 11:36–42.

Hyland, A. 1971. *Beginner's Guide to Western Riding*. London: Pelham Books.

Jones, S. N. 1966. *The Art of Western Riding*. Portales, N.M.: Bishop.

Kaiser, N. W. 1975. Sidesaddle—the past revisited. *The Morgan Horse* 35:346–347.

Kiser, J. J. 1972. *What Judges Look for in Equitation Classes*. Science and Technology Guide 2838. Columbia: University of Missouri Cooperative Extension.

Kulesza, S. 1965. *Modern Riding*. New York: Barnes.

Levings, N. P. 1968. *Training the Quarter Horse Jumper*. New York: Barnes.

Littauer, V. S. 1963. *Common Sense Horsemanship*. Rev. ed. New York: Arco.

Littauer, V. S. 1967. *The Development of Modern Riding*. London: Allen.

Littauer, V. S. 1973. *How the Horse Jumps*. London: Allen.

McCowan, L. L. *It Is Ability That Counts*. Augusta, Mich.: Cheff Center for the Handicapped (Rural Route 1, P.O. Box 171).

Mason, B. S. 1940. *Roping*. New York: Ronald Press.

Morris, G. 1971. *Hunter Seat Equitation*. New York: Doubleday.

Powell, M. D. 1970. Who owns horses? *Arabian Horse World* 11:66.

Santini, P. 1967. *The Caprilli Papers*. London: Allen.

Saunders, G. C. 1966. *Your Horse*. Rev. ed. New York: Van Nostrand Reinhold.

Shannon, E. 1970. *Manual for Teaching Western Riding*. Washington, D.C.: American Association for Health, Physical Education, and Recreation.

Stoneridge, M. A. 1968. *A Horse of Your Own*. New York: Doubleday.

Tree talk. 1969. *Appaloosa News* 26:8–9.

Trench, C. C. 1970. *A History of Horsemanship*. Norwich, England: Jarrold.

Tuke, D. 1965. *Bit by Bit*. New York: Barnes.

Williamson, C. O. 1965. *Breaking and Training for Stock Horses*. 5th ed. Hamilton, Neb.: Williamson School of Horsemanship (P.O. Box 506).

Wright, G. 1966. *Learning to Ride, Hunt and Show*. New York: Doubleday.

Yeates, B. F., and M. Bradley. 1972. *Pre-Bit Hackamore Training*. Science and Technology Guide 2864. Columbia: University of Missouri Cooperative Extension.

Young, J. R. 1954. *The Schooling of the Western Horse*. Norman: University of Oklahoma Press.

Chapter 12: Showing and Racing

Abby, H. C. 1970. *Show Your Horse*. New York: Barnes.

American Horse Shows Association Rule Book. Annual. New York: American Horse Shows Association.

American Racing Manual. Annual. Lexington, Ky.: Daily Racing Form, Inc.

Essary, D. 1974. *Quarter Horse Racing*. Amarillo, Texas: American Quarter Horse Racing Association.

Evans, J. W. 1977. The recreational use of the horse. In J. W. Evans, A. Borton, H. F. Hintz, and L. D. Van Vleck, *The Horse*. San Francisco: W. H. Freeman and Company.

Evans, L. T. 1964. *Standardbred Sport*. Columbus, Ohio: U.S. Trotting Association.

Mercer, T. (Ed.). *Quarter Running Horse Chart Book*. Annual. Amarillo, Tex.: American Quarter Horse Racing Association.

Racing's leading earners. 1979. *Thoroughbred Record* 211:64.

Third Level Test 1 (Revised). 1977. New York: American Horse Shows Association.

Chapter 13: Rodeo, Trail, and Sports Events

Back, J. 1959. *Horses, Hitches and Rocky Trails.* Chicago: Swallow Press.

Barsaleau, R. B. 1977. Initiating the schedule. *Saddle Action* 1:39, 58.

Beacon, N. 1970. Competitive trail riding. *The Morgan Horse* 30:25.

Bradley, M. 1973. *Intermediate Trail Riding.* Science and Technology Guide 2883. Columbia: University of Missouri Cooperative Extension.

Casewell, L. F. 1976. *Packing and Outfitting.* Cooperative Extension Service Publication No. 622. Blacksburg: Virginia Polytechnic Institute.

Clevenger, B. 1975. Cowboy polo. *Horseman* 19:26–28.

Davis, F. W. 1975. *Horse Packing in Pictures.* New York: Scribner's.

Distance Riding Manual. 1972. San Martin, Calif.: North American Trail Ride Conference.

Evans, J. W. 1977. The recreational use of the horse. In J. W. Evans, A. Borton, H. F. Hintz, and L. D. Van Vleck, *The Horse.* San Francisco: W. H. Freeman and Company.

For Our Fans. 1972. Special Events Directory for Golden Gate Fields and Bay Meadows Race Track. Albany, Calif.: Tanforan Racing Association.

Hannan, L. T. 1977. Menu for a horse camper's feast. *Horseman* 21:64–68.

Hill, O. C. 1976. *Packing and Outfitting Field Manual.* Cooperative Extension Service Bull. 636. Laramie: University of Wyoming.

Jennings, J. 1978. Three-day eventing. *Quarter Horse J.* 30:162–171.

Jones, C. 1973. Fox hunting in America. *American Heritage* 24:62–68, 101.

Miles, H. T. 1974. The three-day event. *American Horseman* 5:57–66.

Official Pro Rodeo Media Guide. 1978. Denver, Colo.: Professional Rodeo Cowboys Association.

Rider's Manual. 1976. Gilroy, Calif.: North American Trail Ride Conference.

Saare, S. 1972. *Endurance Riding and Management.* Santa Rosa, Calif.: Northwest Printing.

Saare, S. 1973. *Distance Riding Manual.* Moscow, Idaho: Appaloosa Horse Club.

Wiggins, W. 1978. *The Great American Speedhorse.* New York: Sovereign Books.

Yearbook of the United States Polo Association. 1978. Oak Brook, Ill.: U.S. Polo Association.

Chapter 14: Breeding the Stallion

Asbury, A. C., and J. P. Hughes. 1964. Use of the artificial vagina for equine semen collection. *J. American Veterinary Medical Association* 144:879–882.

Berndtson, W. E., B. W. Pickett, and T. M. Nett. 1974. Reproductive physiology of the stallion. IV: Seasonal changes in the testosterone. *J. Reproduction & Fertility* 39:115–118.

Bielanski, W. 1960. Reproduction in horses. Vol. 1: Stallions, Instytut Bull. 116. Kracow, Poland: Instytut Zootechniki.

Chenoweth, P. J., R. R. R. Pascoe, H. L. McDougal, and P. J. McCosker. 1970. An abnormality of the spermatozoa of a stallion *(Equus caballus). British Veterinary J.* 126:476–481.

Eckstein, P., and S. Zuckerman. 1956. Morphology of the reproductive tract. In A. S. Parkes (Ed.), *Marshall's Physiology of Reproduction.* 3rd ed. Boston: Little, Brown.

El Wishy, A. B. 1975. Morphology of epididymal spermatozoa in the ass *(Equus asinus)* and stallion *(Equus caballus). Zeitschrift fur Tierzuchtung und Zuechtungsbiologie* 92:67–72.

Fraser, A. F. 1968. *Reproductive Behavior in Ungulates.* New York: Academic Press.

Gebauer, M. R., B. W. Pickett, and E. E. Swierstra. 1974a. Reproductive physiology of the stallion. II: Daily production and output of sperm. *J. Animal Science* 39:732–736.

Gebauer, M. R., B. W. Pickett, and E. E. Swierstra. 1974b. Reproductive physiology of the stallion. III: Extragonadal transit time and sperm reserves. *J. Animal Science* 39:737–742.

Gebauer, M. R., B. W. Pickett, J. L. Voss, and E. E. Swierstra. 1974. Reproductive physiology of the stallion: Daily sperm output and testicular measurements. *J. American Veterinary Medical Association* 165:711–713.

Getty, R. (Ed.). 1953. *Sisson and Grossman's The Anatomy of Domestic Animals.* Philadelphia: Saunders.

Hafez, E. S. E. 1968. *Reproduction in Farm Animals.* Philadelphia: Lea & Febiger.

Haner, E. P., H. C. Dellgren, S. E. McCraine, and C. K. Vincent. 1970. Pubertal characteristics of Quarter Horse stallions. *J. Animal Science* 30:321.

Hansen, J. C. 1965. Artificial insemination. *The Blood Horse* 90:3368.

Heinze, C. D. 1966. Methods of equine castration. *J. American Veterinary Medical Association* 148:428.

Hughes, J. P., and R. G. Loy. 1970. Artificial insemination in the equine: A comparison of natural breeding and artificial insemination of mares using semen from six stallions. *Cornell Veterinarian* 60:463.

Julian, L. M., and W. S. Tyler. 1959. Anatomy of the male reproductive organs. In H. H. Cole and P. T. Cupps (Eds.), *Reproduction in Domestic Animals.* Vol. 1. New York: Academic Press.

Kenny, R. M., and W. L. Cooper. 1974. Therapeutic use of a phantom for semen collection from a stallion. *J. American Veterinary Medical Association* 165:706–707.

Kenny, R. M., R. S. Kingston, A. H. Rajamannon, and C. F. Ramberg. 1971. Stallion semen characteristics for predicting fertility. *Proceedings, Seventeenth Annual Convention of American Association of Equine Practitioners* 17:53.

Kosniak, K. Characteristics of the successive jets of ejaculated semen of stallions. *J. Reproduction & Fertility,* Supplement 23, pp. 59–61.

Lowe, J. E., and R. Dougherty. 1972. Castration of horses and ponies by a primary closure method. *J. American Veterinary Medical Association* 160:183.

Nishikawa, Y. 1959. *Studies on Reproduction in Horses.* Tokyo: Japan Racing Association.

Pickett, B. W. 1974. Evaluation of stallion semen. In O. R. Adams (Ed.), *Lameness in Horses.,* 3rd ed. Philadelphia: Lea & Febiger.

Pickett, B. W. 1976. Stallion management with special reference to semen collection, evaluation and artificial insemination. *Proceedings, First National Horsemen's Seminar.* Fredericksburg: Virginia Horse Council.

Pickett, B. W., and D. G. Back. 1973. Procedures for preparation, collection, evaluation, and insemination of stallion semen. General Series 935. Fort Collins: Colorado State University, Experimental Station.

Pickett, B. W., D. G. Back, L. D. Burwash, and J. L. Voss. 1971. The effect of extenders, spermatozoal number and rectal palpation on equine fertility. *Proceedings, Fifth Technology Conference on Artificial Insemination and Reproduction.*

Pickett, B. W., and G. Marvin Beeman. 1975. The arguments for artificial insemination. *The Morgan Horse* 35:38–40.

Pickett, B. W., L. C. Faulkner, and T. M. Sutherland. 1970. Effect of month and stallion on seminal characteristics and sexual behavior. *J. Animal Science* 31:713.

Pickett, B. W., and J. L. Voss. 1973. Reproductive management of the stallion. General Series 934. Fort Collins: Colorado State University, Experimental Station.

Pickett, B. W., J. L. Voss, and E. W. Anderson. 1976. Are you managing your stallion for maximum reproductive efficiency? *The Morgan Horse* 36:145–151.

Pickett, B. W., J. L. Voss, and E. L. Squires. 1977. Impotence and abnormal sexual behavior in the stallion. *Theriogenology* 8:329–347.

Skinner, J. D., and J. Bowen. 1968. Puberty in the Welsh stallion. *J. Reproduction & Fertility* 16:133.

Swierstra, E. E., M. R. Gebauer, and B. W. Pickett. 1974. Reproductive physiology of the stallion. I: Spermatogenesis and testis composition. *J. Reproduction & Fertility* 40:113–123.

Tischner, M., K. Kosniak, and W. Bielanski. 1974. Analysis of the patterns of ejaculation in stallions. *J. Reproduction & Fertility* 39:329–335.

Voss, J. L., and B. W. Pickett. 1976. Reproductive management of the broodmare. General Series 961. Fort Collins: Colorado State University, Experimental Station.

Wierzbowski, S. 1958. Ejaculatory reflexes in stallions following natural stimulation and the use of the artificial vagina. *Animal Breeding Abstracts* 26:367.

Chapter 15: Breeding the Mare

Allen, W. R. 1977. *Ova Transfer: Techniques and Results in Horses.* Canadian Agricultural Monograph No. 16.

Allen, W. R., and L. E. A. Rowson. 1973. Control of the mare's oestrous cycle by prostaglandins. *J. Reproduction & Fertility* 33:539.

Allen, W. R., and L. E. A. Rowson. 1975. Surgical and nonsurgical egg transfer in horses. *J. Reproduction & Fertility*, Supplement 23, pp. 525–530.

Arthur, G. H. 1970. The induction of oestrus in mares by uterine infusion of saline. *Veterinary Record* 86:584.

Arthur, G. H., and W. E. Allen. 1972. Clinical observations on reproduction in a pony stud. *Equine Veterinary J.* 4:109.

Berliner, V. R. 1959. The estrous cycle of the mare. In H. H. Cole and P. T. Cupps (Eds.), *Reproduction in Domestic Animals.* Vol. 1. New York: Academic Press.

Bitteridge, K. J., and J. A. Laing. 1970. The diagnosis of pregnancy. In J. A. Laing (Ed.), *Fertility and Infertility in Domestic Animals.* Baltimore: Williams & Wilkins.

Braselton, W. E., and W. H. McShan. 1970. Purification and properties of follicle-stimulating and luteinizing hormones from horse pituitary glands. *Archives of Biochemical Biophysics* 139:56.

Burkhardt, J. 1949–1950. Sperm survival in the genital tract of the mare. *J. Agricultural Science* 39–40:201–203.

Burkhardt, J. 1947. Transition from anoestrus in the mare and the effects of artificial lighting. *J. Agricultural Science* 37:64.

Cole, H. H., and P. T. Cupps (Eds.). 1959. *Reproduction in Domestic Animals*. New York: Academic Press.

Eckstein, P., and S. Zuckerman. 1956. Morphology of the reproductive tract. In A. S. Parkes (Ed.), *Marshall's Physiology of Reproduction*. 3rd ed. Boston: Little, Brown.

Evans, J. W., J. P. Hughes, D. P. Neely, G. H. Stabenfeldt, and C. M. Winget. 1979. Episodic LH secretion patterns in the mare during the oestrus cycle. *J. Reproduction & Fertility*, Supplement 27, pp. 143–150.

Frandson, R. D. (Ed.). 1974. *Anatomy and Physiology of Farm Animals*. Philadelphia: Lea & Febiger.

Geschwind, I. I., R. Dewey, J. P. Hughes, J. W. Evans, and G. H. Stabenfeldt. 1975. Circulating luteinizing hormone levels in the mare during oestrus cycle. *J. Reproduction & Fertility*, Supplement 23, pp. 207–212.

Getty, R. (Ed.). 1953. *Sisson and Grossman's The Anatomy of Domestic Animals*. Philadelphia: Saunders.

Ginther, O. J. 1979. *Reproductive Biology of the Mare*. Ann Arbor, Mich.: McNaughton and Gunn.

Ginther, O. J., and N. L. First. 1971. Maintenance of the corpus luteum in hysterectomized mares. *American J. Veterinary Research* 32:1687.

Ginther, O. J., M. C. Garcia, E. L. Squires, and W. P. Steffenhagen. 1972. Anatomy of vasculature of uterus and ovaries in the mare. *American J. Veterinary Research* 33:1561.

Hammond, J., and K. Wodzicki. 1941. Anatomical and histological changes during the oestrus cycle in the mare. *Proceedings of the Royal Society* 130(B):1.

Hansen, J. C. 1965. Artificial insemination. *The Blood-Horse* 90:3368–3370.

Hansen, J. C. 1966. Artificial insemination. *Modern Vet Practice* 47:45.

Harrison, R. J. 1946. The early development of the corpus luteum in the mare. *J. Anatomy* 80:160.

Harvey, E. B. 1959. Implantation, development of the fetus and fetal membranes. In H. H. Cole and P. T. Cupps (Eds.), *Reproduction in Domestic Animals*. Vol. 1. New York: Academic Press.

Householder, D. D., J. L. Fleeger, Y. H. Clamann, and A. M. Sorensen. 1974. Criteria for detecting ovulation in the mare. *J. Animal Science* 39:212.

Hughes, J. P., G. H. Stabenfeldt, and J. W. Evans. 1972a. Clinical and endocrine aspects of the estrous cycle in the mare. *Proceedings, American Association of Equine Practitioners* 1:119.

Hughes, J. P., G. H. Stabenfeldt, and J. W. Evans. 1972b. Estrous cycle and ovulation in the mare. *J. American Veterinary Medical Association* 161:1367.

Irvine, D. S., B. R. Downey, W. G. Parker, and J. J. Sullivan. 1975. Duration of oestrus and time of ovulation in mares treated with synthetic Gn-Rh (AY-24,031). *J. Reproduction & Fertility*, Supplement 23, pp. 279–283.

Kiefer, B. L., J. F. Roser, J. W. Evans, D. P. Neely, and C. A. Pacheco. 1979. Progesterone patterns observed with multiple injections of a $PGF_{2\alpha}$ analogue in the cyclic mare. *J. Reproduction & Fertility*, Supplement 27, pp. 237–244.

Loy, R. G. 1970. The reproductive cycle of the mare. *Lectures of the Stud Managers Course* 9:20.

Loy, R. G., and J. P. Hughes. 1966. The effects of human chorionic gonadotropin on ovulation, length of estrus and fertility in the mare. *Cornell Veterinarian* 61:41.

McGee, W. R., and H. D. McGee. *Veterinary Notes for the Standardbred Breeder*. Columbus, Ohio: U.S. Trotting Association.

Matthews, R. G., R. T. Ropiha, and R. M. Butterfield. 1967. The phenomenon of foal heat in mares. *Australian Veterinary J.* 43:579–582.

Meacham, T., and C. Hutton. 1968. Reproductive efficiency on 14 horse farms. *J. Animal Science* 27:434.

Nishikawa, Y. 1959. *Studies on Reproduction in Horses.* Tokyo: Japan Racing Association.

Noden, P. A., H. D. Hafs, and W. D. Oxender, 1973. Progesterone, estrus and ovulation after prostaglandin F-2-alpha in horses. *Federation Proceedings* 32:229.

Nuti, L. C., H. J. Grimek, W. E. Braselton, and W. H. McShan. 1972. Chemical properties of equine pituitary follicle-stimulating hormone. *Endocrinology* 91:1418.

Pattison, M. L., C. L. Chen, and S. L. King. 1972. Determination of LH and estradiol-17_β surge with reference to the time of ovulation in mares. *Biology of Reproduction* 7:136.

Penida, M. H., O. J. Ginther, and W. H. McShan. 1972. Regression of the corpus luteum in mares treated with an antiserum against equine pituitary fraction. *American J. Veterinary Research* 33:1767.

Rogers, T. 1971. *Mare Owner's Handbook.* Houston: Cordovan.

Rosborough, J. P. 1960. Breeding hygiene, procedures and practices. *Proceedings, Illinois Short Courses for Horse Breeders,* p. 30.

Roser, J. F., B. L. Kiefer, J. W. Evans, D. P. Neely, and C. A. Pacheco. 1979. The development of antibodies to human chorionic gonadotropin following its repeated injection in the cyclic mare. *J. Reproduction & Fertility,* Supplement 27, pp. 173–179.

Rossdale, P. D. 1973. Mares in breeding season. *Thoroughbred of California* 61:42.

Ryan, R. J., and R. V. Short. 1965. Formation of estradiol-17_β by granulosa and theca cells of the equine ovarian follicle. *Endocrinology* 76:108.

Stabenfeldt, G. H., J. P. Hughes, and J. W. Evans. 1970. Ovarian activity during the estrous cycle of the mare. *Endocrinology* 90:1379.

Stabenfeldt, G. H., J. P. Hughes, J. D. Wheat, J. W. Evans, P. C. Kennedy, and P. T. Cupps. 1974. The role of the uterus in ovarian control in the mare. *J. Reproduction & Fertility* 37:343.

Turner, C. W. 1952. *The Mammary Gland.* Vol. 1: *The Anatomy of the Udder of Cattle and Domestic Animals.* Columbia, Mo.: Lucas.

Warszawsky, L. F., W. G. Parker, N. L. First, and O. J. Ginther. 1972. Gross changes of internal genitalia during the estrous cycle in the mare. *American J. Veterinary Research* 33:19.

Witchi, E. 1956. *Development of Vertebrates.* Philadelphia: Saunders.

Wood, K. A. 1973. *The Business of Horses.* Chula Vista, Calif.: Wood.

Chapter 16: Care During and After Pregnancy

Allen, W. R. 1970. Endocrinology of early pregnancy in the mare. *Equine Veterinary J.* 2:64.

Allen, W. R. 1970. Equine gonadotropins. Unpublished doctoral thesis, University of Cambridge, England.

Amoroso, E. C. 1955. Endocrinology of pregnancy. *British Medical Bull.* 11:117.

Andrist, F. 1962. *Mares, Foals and Foaling.* London: Allen.

Bain, A. M. 1969. Foetal losses during pregnancy in the Thoroughbred mare: A record of 2,562 pregnancies. *New Zealand Veterinary J.* 17:155.

Bitteridge, K. J., and J. A. Laing. 1970. The diagnosis of pregnancy. In J. A. Laing (Ed.), *Fertility and Infertility in Domestic Animals.* Baltimore: Williams & Wilkins.

Brown, S. 1978. Raising an orphan foal. *Horse of Course* 7:29–32.

Cole, H. H., and H. Goss. 1943. The source of equine gonadotropin. In *Essays in Biology.* No. 107. Berkeley: University of California Press.

Cole, H. H., and G. H. Hart. 1942. Diagnosis of pregnancy in the mare by hormonal means. *J. American Veterinary Medical Association* 101:124.

Fowler, M. E. 1973. Diarrhea in foals. *Quarter Horse of the Pacific Coast* 10:13.

Frandson, R. D. (Ed.). 1974. *Anatomy and Physiology of Farm Animals.* Philadelphia: Lea & Febiger.

Harvey, E. B. 1959. Implantation, development of the fetus, and fetal membranes. In H. H. Cole and P. T. Cupps (Eds.). *Reproduction in Domestic Animals.* Vol. 1. New York: Academic Press.

Jeffcott, L. B., J. G. Atherton, and J. Mingay. 1969. Equine pregnancy diagnosis. *Veterinary Record* 84:80.

Jeffcott, L. B., and P. D. Rossdale. 1977. A critical review of current methods for induction of parturition in the mare. *Equine Veterinarian J.* 9:208–215.

Laing, J. A. (Ed.). 1970. *Fertility and Infertility in Domestic Animals.* Baltimore: Williams & Wilkins.

Lovell, J. D., G. H. Stabenfeldt, J. P. Hughes, and J. W. Evans. 1975. Endocrine patterns of the mare at term. *J. Reproduction & Fertility,* Supplement 23, pp. 449–456.

McMullen, W. C. 1975. Foal doctoring. *Horseman* 19:22–24.

Mahaffey, Leo W. 1968. Abortion in mares. *Veterinary Record* 82:681.

Nett, T. M., D. W. Holtan, and V. L. Estergreen. 1972. Plasma estrogens in pregnant mares. *Proceedings, Western Section of the American Society of Animal Science* 23:509.

Papkoff, H. 1969. Chemistry of the gonadotropins. In H. H. Cole and P. T. Cupps (Eds.), *Reproduction in Domestic Animals.* Vol. 2. New York: Academic Press.

Paul, R. R. 1973. Foaling date. *The Morgan Horse* 33:40.

Rogers, T. 1971. *Mare Owner's Handbook.* Houston: Cordovan.

Rollins, W. C., and C. E. Howell. 1951. Genetic sources of variation in the gestation length of the horse. *J. Animal Science* 10:797.

Ropiha, R. T., R. G. Matthews, R. M. Butterworth, F. M. Moss, and W. J. McFadden. 1969. The duration of pregnancy in Thoroughbred mares. *Veterinary Record* 84:552.

Sisson, S., and J. D. Grossman. 1953. *The Anatomy of Domestic Animals.* Philadelphia: Saunders.

Zemjanis, R. 1970. Examination of the mare. In R. Zemjanis, *Diagnostic and Therapeutic Techniques in Animal Reproduction.* Baltimore: Williams & Wilkins.

Chapter 17: Understanding Basic Genetics

Adalsteinsson, S. 1974. Inheritance of the palomino color in Icelandic horses. *J. Heredity* 65:15.

Adalsteinsson, S. 1977. Inheritance of yellow dun and blue dun in the Icelandic Toelter horse. *J. Heredity* 69:146–148.

Archer, R. K. 1972. True haemophilia in horses. *Veterinary Record* 91:655.

Aurich, R. 1959. A contribution to the inheritance of umbilical hernia in the horse. *Berler und Munchener Tierartzliche Wochenschrift* 72:420.

Basrur, P. K., H. Kanagawa, and J. P. W. Gilman. 1969. An equine intersex with unilateral gonadal agenesis. *Canadian J. Comparative Medicine* 33:297–305.

Basrur, P. K., H. Kanagawa, and L. Podliachouk. 1970. Further studies on the cell populations on an intersex horse. *Canadian J. Comparative Medicine* 34:294.

Bengtsson, S., B. Gahne, and J. Rendel. 1968. Genetic studies on transferrins, albumins, prealbumins, and esterases in Swedish horses. *Acta Agriculturae Scandinavia* 18:60.

Bengtsson, S., and K. Sandberg. 1972. Phosphoglucomutase polymorphism in Swedish horses. *Animal Blood Groups and Biochemical Genetics* 3:115.

Benirschke, K., L. E. Brownhill, and M. M. Beath. 1962. Somatic chromosomes of the horse, the donkey, and their hybrids, the mule and hinny. *J. Reproduction & Fertility* 4:319.

Benirschke, K., and N. Malouf. 1957. Chromosome studies of equidae. In H. Dathe (Ed.), *Equus*. Vol. 1, No. 2. Berlin: Tierpark.

Bielanski, W. 1946. The inheritance of shortening of the lower jaw (brachygnathia inferior) in the horse. *Przeglad Hodowlany* 14:24.

Blakeslee, L. H., R. S. Hudson, and R. H. Hunt. 1943. Curly coat of horses. *J. Heredity* 34:115–118.

Blunn, C. T., and C. E. Howell. 1936. The inheritance of white facial markings in Arabian horses. *J. Heredity* 27:393.

Bornstein, S. 1967. The genetic sex of two intersexual horses and some notes on the karotypes of normal horses. *Acta Veterinaria Scandinavia* 8:291.

Bouters, R., M. Vandeplassche, and A. De Moor. 1972. An intersex (male pseudo-hermaphrodite) horse with 64,XX/65,XXY mosaicism. *Equine Veterinary J.* 4:150.

Braend, M. 1964. Serum types of Norwegian horses. *Nordisk Veterinary Medicine* 16:363.

Braend, M. 1967. Genetic variation of horse hemoglobin. *J. Heredity* 58:385.

Braend, M. 1970. Genetics of horse acidic prealbumins. *Genetics* 65:495.

Braend, M., and C. Stormont. 1964. Studies on hemoglobin and transferrin types of horses. *Nordisk Veterinary Medicine* 16:31.

Briquet, R., Jr. 1957. So-called "albino horses." *Boletin de Industria Animal* 16:243.

Briquet, R., Jr. 1959. Investigations on the relationship between markings on face and limbs in horses. *Revista de Ramonta Veterinaria* 19 (entire issue).

Britton, J. W. 1945. An equine hermaphrodite. *Cornell Veterinarian* 35:373.

Britton, J. W. 1962. Birth defects in foals. *Thoroughbred* 34:288.

Buckingham, J. 1936. Hermaphrodite horses. *Veterinary Record* 48:218.

Butz, H., and H. Meyer. 1957. Epitheliogenesis imperfecta neonatorium equi (incomplete skin formation in the foal). *Deutsche Tierarztliche Wochenschrift* 64:555.

Castle, W. E. 1946. Genetics of the Palomino horse. *J. Heredity* 37:35.

Castle, W. E. 1948. The ABC of color inheritance in horses. *Genetics* 33:22.

Castle, W. E. 1951a. Dominant and recessive black in mammals. *J. Heredity* 42:48.

Castle, W. E. 1951b. Genetics of the color variety of horses. *J. Heredity* 42:297.

Castle, W. E. 1953. Note on the silver dapple mutation of Shetland ponies. *J. Heredity* 44:224.

Castle, W. E. 1954. Color coat inheritance in horses and other mammals. *Genetics* 39:35.

Castle, W. E. 1960. Fashion in the color of Shetland ponies and its genetic basis. *J. Heredity* 51:247.

Castle, W. E. 1961. Genetics of the Claybank dun horse. *J. Heredity* 51:121.

Castle, W. E., and W. R. Singleton. 1961. The Palomino horse. *Genetics* 46:1143.

Castle, W. E., and W. H. Smith. 1953. Silver dapple, a unique color variety among Shetland ponies. *J. Heredity* 44:139.

Clegg, M. T., J. M. Boda, and H. H. Cole. 1954. The endometrial cups and allantochorionic pouches in the mare with emphasis on the source of equine gonadotropin. *Endocrinology* 54:448.

Cradock-Watson, J. E. 1967. Immunological similarity of horse, donkey, and mule haemoglobins. *Nature* 215:630.

Cronin, N. T. I. 1956. Hemolytic disease of newborn foals. *Veterinary Record* 67:474.

Cross, R. S. N. 1966. Equine periodic ophthalmia. *Veterinary Record* 78:8.

Dimock, W. W. 1950. "Wobbles"—an hereditary disease in horses. *J. Heredity* 41:319.

Dreux, P. 1970. The degree of expression on limited piebaldness in the domestic horse. *Annales de Genetique et de Selection Animale* 2:119.

Dungsworth, D. I., and M. E. Fowler. 1966. Cerebral hypoplasia and degeneration in a foal. *Cornell Veterinarian* 56:17.

Dunn, H. O., J. T. Vaughan, and K. McEntee. 1974. Bilaterally cryptorchid stallion with female karyotype. *Cornell Veterinarian* 64:265.

Dusek, J. 1965. The heritability of some characters in the horse. *Zivocisna Vyroba* 10:449.

Dusek, J. 1970. Heritability of conformation and gait in the horse. *Zeitschrift fur Tierzuchtung und Zuchtbiologie* 87:14.

Dusek, J. 1971. Some biological factors and factors of performance in the study of heredity in horse breeding. *Scientia Agriculturae Bohemoslovaca* 3:199.

Emik, L. O., and C. E. Terrill. 1949. Systematic procedures for calculating inbreeding coefficients. *J. Heredity* 40:51.

Eriksson, K. 1955. Hereditary aniridia with secondary cataract in horses. *Nordisk Veterinary Medicine* 7:773.

Fischer, H., and K. Helbig. 1951. A contribution to the question of the inheritance of patella dislocation in the horse. *Tierzuchter* 5:105.

Flechsig, J. 1950. Hereditary cryptorchidism in a depot stallion. *Tierzuchter* 4:208.

Foye, D. B., H. C. Dickey, and C. J. Sniffen. 1972. Heritability of racing performance and a selection index for breeding potential in the Thoroughbred horse. *J. Animal Science* 35:1141.

Franks, D. 1962a. Differences in red-cell antigen strength in the horse due to gene interaction. *Nature* 195:580.

Franks, D. 1962b. Horse blood groups and hemolytic disease of the newborn foal. *Annals of the New York Academy of Sciences* 97:235.

Fraser, W. 1966. Two dissimilar types of cerebellar disorders in the horses. *Veterinary Record* 78:608.

Gahne, B. 1966. Studies of the inheritance of electrophoretic forms of transferrins, albumins, prealbumins, and plasma esterases of horses. *Genetics* 53:681.

Gahne, B., S. Bengtsson, and K. Sandberg. 1970. Genetic control of cholinesterase activity in horse serum. *Animal Blood Groups and Biochemical Genetics* 1:207.

Gluhovski, N., H. Bistriceanu, A. Sucui, and M. Bratu. 1970. A case of intersexuality in the horse with type 2A + XXXY chromosome formula. *British Veterinary J.* 126:522.

Hadwen, S. 1931. The melanomata of gray and white horses. *Canadian Medical Association J.* 21:519.

Hamori, D. 1940. Inheritance of the tendency to hernia in horses. *Allatorvosi Lapok* 63:136.

Hamori, D. 1941. Parrot mouth and hog mouth as inherited deformities. *Allatorvosi Lapok* 64:57.

Hartwig, W., and U. Reichardt. 1958. The heritability of fertility in horse breeding. *Zuchtungskunde* 30:205.

Hitenkov, G. G. 1941. Stringhalt in horses and its inheritance. *Vestnik Sel'jsko-hozyaistvennoi Nauki Zivotn* 2:64.

Hsu, T. C., and K. Benirschke. 1957. *An Atlas of Mammalian Chromosomes*. New York: Springer-Verlag.

Hughes, J. P., and A. Trommershausen-Smith. 1977. Infertility in the horse associated with chromosomal abnormalities. *Australian Veterinary J.* 53:253–257.

Huston, R., G. Saperstein, and H. W. Leipold. 1977. Congenital defects in foals. *J. Equine Medical Surgery* 1:146–161.

Hutchins, D. R., E. E. Lepherd, and I. G. Crook. 1967. A case of equine haemophilia. *Australian Veterinary J.* 43:83.

Jeffcott, L. B., and K. E. Whitehall. 1973. Twinning as a cause of foetal and neonatal loss in the Thoroughbred mare. *J. Comparative Pathology* 73:83.

Jones, T. C., and F. D. Maurer. 1942. Heredity in periodic ophthalmia. *J. American Veterinary Medical Association* 101:248.

Jones, W. E. 1970. Cytogenetics of the horse. *Quarter Horse J.* 22:38–46, 78.

Jones, W. E., and R. Bogart. 1971. *Genetics of the Horse*. East Lansing, Mich.: Caballus.

Kieffer, N. 1973. Inheritance of racing ability in the Thoroughbred. *Thoroughbred Record* 204:50.

Kieffer, N. M., N. Judge, and S. Burns. 1971. Some cytogenetic aspects of an *Equus caballus* intersex. *Mammalian Chromosome News Letter* 12:18.

King, J. M., R. V. Short, O. E. Mutton, and J. L. Hammerton. 1965. The reproductive physiology of male zebra–horse and zebra–donkey hybrids. *J. Reproduction & Fertility* 9:391.

Lasley, J. F. 1970. Genetic principles in horse breeding. (Available from J. F. Lasley, 2207 Bushnell Drive, Columbia, MO 65201.)

Lasley, J. F. 1972. *Genetics of Livestock Improvement*. Englewood Cliffs, N.J.: Prentice-Hall.

McGuire, T. C., M. J. Poppie, and K. L. Banks. 1974. Combined (B- and T-lymphocyte) immunodeficiency: A fatal genetic disease in Arabian foals. *J. American Veterinary Medical Association* 164:70.

Miller, R. W. *Appaloosa Coat Color Inheritance*. Moscow, Idaho: Montana State University, Bozeman and Appaloosa Horse Club.

Miser, M. 1951. Observation and discussion of variation and inheritance of white markings in horses. *Japanese J. Zootechnical Science* 22:23.

Niece, R. L., and D. W. Kracht. Genetics of transferrins in burros *(Equus asinus)*. *Genetics* 57:837.

Odriozola, M. 1948. Agouti color in horses: Change of dominance in equine hybrids. *Proceedings, Eighth International Congress of Genetics*, p. 635.

Odriozola, M. 1951. *A Los Colores del Caballo*. Madrid: National Syndicate of Livestock.

O'Ferral, G. J., and E. P. Cunningham. 1974. Heritability of racing performance in Thoroughbred horses. *Livestock Production Science* 1:87.

Osterhoff, D. R. 1967. Haemoglobin, transferrin, and albumin types in equidae (horses, mules, donkeys, and zebras). *Proceedings, Tenth European Conference on Animal Blood Groups and Biochemical Polymorphism, Paris, 1966,* p. 345.

Osterhoff, D. R., and I. S. Ward-Cox: 1972. Quantitative studies on horse hemoglobins. *Proceedings, Twelfth European Conference on Animal Blood Groups and Biochemical Polymorphism, Budapest, 1970,* p. 541.

Osterhoff, D. R., D. O. Schmid, and I. S. Ward-Cox. 1970. Blood group and serum type studies in Basuto ponies. *Proceedings, Eleventh European Conference on Animal Blood Groups and Biochemical Polymorphism, Warsaw, 1968,* p. 453.

Payne, H. W. 1968. Aneuploidy in an infertile mare. *J. American Veterinary Medical Association* 153:1293–1299.

Pern, E. M. 1970. The heritability of speed in Thoroughbred horses. *Genetika Moskva* 6:110.

Pirri, J., and D. G. Steele. 1952. The heritability of racing capacity. *The Blood Horse* 63:976.

Poppie, M. J., and T. C. McGuire. 1977. Combined immunodeficiency in foals of Arabian breeding: Evaluation of mode of inheritance and estimation of prevalence of affected foals and carrier mares and stallions. *J. American Veterinary Medical Association* 170:31–33.

Prawochenski, R. T. 1936. A case of lethal genes in the horse. *Nature* 137:689.

Pulos, W., and F. B. Hutt. 1969. Lethal dominant white in horses. *J. Heredity* 60:59.

Rooney, F. R., and M. E. Prickett. 1966. Foreleg splints in horses. *Cornell Veterinarian* 53:411.

Sandberg, K. 1970. Blood group factors and erythrocytic protein polymorphism in Swedish horses. *Proceedings, Eleventh European Conference on Animal Blood Groups and Biochemical Polymorphism, Warsaw, 1968,* p. 447.

Sandberg, K. 1974. Blood typing of horses: Current status and application to identification problems. *Proceedings, First World Congress Applied to Livestock Production* 1:253.

Sangor, V. L., R. E. Mairs, and A. L. Trapp. 1964. Hemophilia in a foal. *J. American Veterinary Medical Association* 142:259.

Scekin, V. A. 1973. The inheritance of stringhalt (cock gait) in the horse. *Konevodstru* 2:20.

Schwark, H. J., and E. Neisser. 1971. Breeding of English Thoroughbred horses in the G.D.R. Vol. 2: Results of estimates of heritability and breeding value. *Archiv Tierzuchter* 14:69.

Scott, A. M. 1972. Improved separation of polymorphic esterases in horse. *Proceedings, Twelfth European Conference on Animal Blood Groups and Biochemical Polymorphism, Budapest, 1970,* p. 527.

Searle, A. G. 1968. *Comparative Genetics of Coat Colour in Mammals.* London: Logo Press.

Severson, B. D. 1917. Cloven hoof in Percherons. *J. Heredity* 8:466.

Severson, B. D. 1918. Extra toes in horse and steers. *J. Heredity* 9:39.

Singleton, W. R., and Q. C. Bond. 1966. An allele necessary for dilute coat color in horses. *J. Heredity* 57:75.

Smith, A. T. 1972. Inheritance of chin spot markings in horses. *J. Heredity* 63:100.

Speed, J. G. 1958. A cause of malformation of the limbs of Shetland ponies with a note on its phylogenic significance. *British Veterinary J.* 114:18.

Spence, S. E. 1967. Can you tell tobiano from overo? *Paint Horse J.* 1:17.

Sponseller, J. L. 1967. Equine cerebellar hypoplasia and degeneration. *Proceedings, Annual Meeting of American Association of Equine Practitioners,* pp. 123–126.

Stormont, C. 1972. Genetic markers in the blood of horses. *Proceedings, Horse Identification Seminar.* Pullman: Washington State University.

Stormont, C., and Y. Suzuki. 1963. Genetic control of albumin phenotypes in horses. *Proceedings, Society for Experimental Biology and Medicine* 114:673.

Stormont, C., and Y. Suzuki. 1964. Genetic systems of blood groups in horses. *Genetics* 50:915.

Stormont, C., and Y. Suzuki. 1965. Paternity tests in horses. *Cornell Veterinarian* 55:365.

Stormont, C., Y. Suzuki, and J. Rendel. 1965. Application of blood typing and protein tests in horses. *Proceedings, Ninth European Conference on Animal Blood Groups and Biochemical Polymorphism, Prague, 1964,* p. 221.

Stormont, C., Y. Suzuki, and E. A. Rhode. Serology of horse blood groups. *Cornell Veterinarian* 54:439.

Theile, H. 1958. Polydactyly in a foal. *Monatshefte Veterinary Medicine* 13:342.

Tolksdorff, E., W. Jochle, D. R. Lamond, E. Klug, and H. Merkt. 1976. Induction of ovulation during the post partum period in the Thoroughbred mare with a prostaglandin analogue, Synchrocept. *Theriogenology* 6:403–412.

Trommershausen-Smith, A. 1972. Inheritance of chin spot markings in horses. *J. Heredity* 63:100.

Trommershausen-Smith, A. 1977. Lethal white foals in matings of overo spotted horses. *Theriogenology* 8:303–311.

Trujillo, J. M., C. Stenius, L. Christian, and S. Ohno. 1962. Chromosomes of the horse, the donkey, and the mule. *Chromosoma* 13:243.

Trujillo, J. M., B. Walden, P. O'Neil, and H. B. Anstall. 1965. Sex-linkage of glucose-6-phosphate dehydrogenase in the horse and donkey. *Science* 148:1603.

Trujillo, J. M., B. Walden, P. O'Neil, and H. B. Anstall. 1967. Inheritance and sub-unit composition of haemoglobin in the horse, donkey, and their hybrids. *Nature* 213:88.

Tuff, P. 1948. The inheritance of a number of defects in the joints, bones, and ligaments of the foot of the horse. *Norske Veterinaertidsskrift* 60:385.

Van Vleck, L. D. 1977a. Principles of Mendelian inheritance. In J. W. Evans, A. Borton, H. F. Hintz, and L. D. Van Vleck, *The Horse.* San Francisco: W. H. Freeman and Company.

Van Vleck. L. D. 1977b. Principles of selection for quantitative traits. In J. W. Evans, A. Borton, H. F. Hintz, and L. D. Van Vleck, *The Horse.* San Francisco: W. H. Freeman and Company.

Van Vleck, L. D. 1977c. Relationships and inbreeding. In J. W. Evans, A. Borton, H. F. Hintz, and L. D. Van Vleck, *The Horse.* San Francisco: W. H. Freeman and Company.

Van Vleck, L. D. 1977d. Some Mendelian traits: Blood factors, colors, lethals. In J. W. Evans, A. Borton, H. F. Hintz, and L. D. Van Vleck, *The Horse.* San Francisco: W. H. Freeman and Company.

Van Vleck, L. D., and M. Davitt. 1977. Confirmation of a gene for dominant dilution of horse colors. *J. Heredity* 68:280–282.

Varo, M. 1965a. On the relationship between characters for selection in horses. *Annales Agriculturae Fenniae* 4:38.

Varo, M. 1965b. Some coefficients of heritability in horses. *Annales Agriculturae Fenniae* 4:223.

Weber, W. 1947a. Congenital cataract, a recessive mutation in the horse. *Schweizerische Archiv Tierheilkunde* 89:397.

Weber, W. 1947b. Schistosoma reflexum in the horse, with a contribution on its origin. *Schweizerische Archiv Tierheilkunde* 89:244.

Weisner, E. 1955. The importance of inherited eye defects in horse breeding. *Tierzuchter* 4:310.

Wheat, J. D., and P. C. Kennedy. 1953. Cerebellar hypoplasia and its sequelae in a horse. *J. American Veterinary Medical Association* 131:241.

White, D. J., and D. A. Farebrother. 1969. A case of intersexuality in the horse. *Veterinary Record* 85:203.

Index

Abortion, 536–537
Abrasions, treatment, 293–294
Accessory sex glands, 454
Afterbirth, 530–531, 545–546, 549
Age:
  determining, 31–38
  feeding aged horses, 202–203
  value of horse, 13, 30–31
Agonistic behavior, 39–40
Albino, 121, 563–565, 570–572
Alfalfa hay, 175
Allantois, 530
Alleles, 558, 561
Allowance race, 398
American Albino (breed), 121
American Bashkir Curly (breed), 120
American Buckskin (breed), 122
American Carriage (breed), 137
American Horse Shows Association, divisions:
  dressage, 391–398
  equitation, 376
  gymkhana, 388
  halter, 376
  hunter, 383
  jumper, 383
  roadster, 386

  saddle horse, 385
  three-day event, 387
  western, 379
American Quarter Horse (breed), 102–105
  See also Quarter Horse
American Saddlebred (breed), 112
Amnion, 530
Anestrus, 482, 484, 486, 490–492, 499–500, 505
Anglo-Arab (breed), 107
Anthrax, 321
Antibodies:
  blood groups, 555
  passive transfer, 538
Appaloosa:
  breed, 126
  characteristics, 56
  color patterns, 571–572
Appetite, 204
Arabian (breed), 101–102, 105–107, 556
Arm, 70
Arteritis, viral, 326–327
Artificial insemination, 503, 523–525
Artificial vagina, 456–458
Ascarids, 302
Atresia coli, 555
Atretic follicle, 473

Attacking (vice), 48
Auction sales, 12
Azoturia, 327

Backing up, 22
Bandages, leg, 288–293
Bareback, bronc riding, 421–422
Barley:
    hay, 175
    grain, 180
Barns:
    design, 149
    exterior, 156
    fire safety, 165
    interior, 158
    layout, 150
    portable, 162
Barrel racing, 388, 424
Barren mare, 505, 525–527
Base-narrow fault, 73, 75, 82, 275, 277
Base-wide fault, 73, 75, 82, 275, 277–278
Basket hitch, 443
Bathing, 235
Bay (color), 563–569
"Bean" in penis, 454
Bedding, 164
Behavior:
    agonistic, 39–40
    dominance, 40–41
    elimination, 43–44
    epimeletic, 41–42
    estrous cycle, 476–480
    et-epimeletic, 41–42
    grooming, 43
    ingestive, 44
    investigative, 42–43
    learning, 45
    mimicry, 45
    play, 44
    sexual, 457, 463, 466–468, 476–482, 522
    sleeping, 45–46
    temperament, 38–39
    vices, 47–52
    waking, 45–46
Belgian (breed), 134
Bench knees, 70
Bermudagrass hay, 175
"Big head" disease, 194
Biting (vice), 47–48
Bits, 343–352, 354–355
    action, 350–353
Black (color), 54, 563–564
Blankets, 236
Blemish, definition, 83
Blindness, 84
Blind spavin, 97
Blisters, leg, 297
Blood groups, 560–561
    incompatible, 555
Blood spavin, 97

Body color: See Color, body
Body temperature, 280, 549
Bog spavin, 82, 96
Bolting (vice), 48
Bone spavin, 82, 97
Borium, 267
Bots, 307–308
Bowed tendon, 88
Bowlegs, 73, 76, 82–83
Braces, leg, 297
Breed associations, 581–584
Breeds, 12–13
    American Albino, 121
    American Bashkir Curly, 120
    American Buckskin, 122
    American Carriage, 137
    American Quarter Horse, 102–105
    American Saddlebred, 112
    Appaloosa, 126
    Arabian, 101–102, 105–107, 556
    Belgian, 134
    Chickasaw, 129
    Clydesdale, 133
    Connemara, 139
    feral, 129
    French Coach, 137
    Galiceño, 119
    German Coach, 137
    Gotland, 116
    Hackney, 137
    miniature, 142
    Missouri Fox Trotting Horse, 117
    Morab, 117
    Morgan, 111
    Morocco Spotted Horse, 126
    Paint, 124
    Palomino, 123
    Paso Fino, 118
    Percheron, 131
    Peruvian Paso, 118
    Pinto, 124–125
    Pony of the Americas, 140
    Racking Horse, 116
    Russian Orloff, 137
    Shetland, 141
    Shire, 135
    Spanish-Barb, 128
    Standardbred, 109
    Suffolk, 132
    Tennessee Walking Horse, 114
    Thoroughbred, 107
    Welsh Mountain Pony, 138
Bridles, 336, 341, 343, 354
Bridle teeth, 33
Bridling, 220, 357–258
Brown (color), 563–564
Bruised sole, 92
Bruises, treatment, 295
Brushing (gait defect), 25
Bucked shins, 86

Buck knees, 73, 76
Buckskin (color), 55, 565–569
Bulbourethral gland, 454
Bull riding, 423–424
Burns, treatment, 293–294
Business associations, 584
Byerly Turk, 101–102

Calcium, 193
Calcium–phosphorus ratio, 194
Calf knees, 73, 76–77
Calf roping, 419–420
Camped-out fault, 73, 77
Camped-under fault, 73, 77, 277
Cannon bones, 71, 81
Canter, 18–22
Capillary refill time, 281–282
Capped elbow, 89
Capped hock, 97
Carbohydrate, digestion, 171
Caslick's operation, 526
Castration, 467–468
Catching, 206–208, 219
Cattle grubs, 309
Cervix, 469, 474, 483–484, 500, 533
Change of leads, 18–22
Chest, 66, 77
Chestnut (color), 55, 564–569
Chewing wood (vice), 52, 205
Chickasaw (breed), 129
Chiggers, 319
Chilling cold, 296
Chopped knee, 71
Chorion, 530
Chromosome, 558–560
CID (combined immunodeficiency disease), 556
Claiming race, 399
Cleft palate, 554–555
Clitoris, 469, 475
Clover hay, 175
Clydesdale (breed), 133
Coat color inheritance, 561–572
Coggins' test, 323
Colic, 328, 549
Color, body:
  Appaloosa, 56, 571–572
  bay, 54, 563–569
  black, 54, 563–569
  brown, 563–564
  buckskin, 55, 565–569
  chestnut, 55, 564–569
  dun, 55, 565–569
  gray, 55, 565, 570
  grulla, 55, 566–569
  overo, 56, 124, 570
  paint, 56, 565, 570
  palomino, 55, 565–569
  piebald, 56, 124, 571
  pinto, 56, 570–571
  roan, 55, 570
  seal-brown, 54–55, 563–569
  skewbald, 56, 124, 571
  sorrel, 55, 564–569
  tobiano, 56, 124, 570
  white, 55, 563, 565, 570
Color, head markings, 53, 571–572
Color, leg markings, 54, 571–572
Colostrum, 538, 548, 555, 557
Combined immunodeficiency disease (CID), 556
Competitive trial riding, 429–430
Concentration, semen, 461, 640–643
Conformation, 14
  defects, 25–28, 82, 277
Congenital defects, 554–557
Connemara (breed), 139
Contracted heels, 91
Coprophagy, 205
Corium, 254
Corn (unsoundness), 92
Corn grain, 181
Coronary band, 251
Corpus albicans, 473
Corpus luteum, 471–473, 482, 484, 486
  prolonged, 490, 500–501, 527
Corrective trimming, 273–278
Coupling, 79
Cow-hocked fault, 82, 277
Cracks: toe, quarter, and heel, 94, 278
Creep feeding, 198, 203
Cremello (color), 565–569
Cresty neck, 66
Cribbing (vice), 49, 64
Cross-firing (gait defect), 23–24, 77
Croup, 79–81
Cryptorchid, 468–555
Cup in tooth, 33, 36
Curb, 82, 98
Cuts, treatment, 286
Cutting horses, 425

Darley Arabian, 101–102
Defecation, 43–44
Dehydration, 228, 281
Dental star, 36
Derby race, 399
Diamond hitch, 441
Diarrhea, 549
Diestrus, 475, 484–486, 491–492
Digestible energy, 185
Digestive tract, anatomy, 168–173
Digital cushion, 255
Diseases:
  anthrax, 321
  control, 322, 329
  encephalomyelitis, 321–323
  equine infectious anemia, 323
  influenza, 323–324
  melanoma, 64, 329

Diseases *(continued)*
  metabolic, 327–329
  piroplasmosis, 324
  pneumonia, 324
  prevention, 329–330
  rabies, 324–325
  rhinopneumonitis, 325
  ringworm, 325
  strangles, 325–326
  tetanus, 326, 538, 548
  viral arteritis, 326–327
  warts, 328
Dishing (gait defect), 26
Disinfectant, 539
Dominance (behavior), 40–44
Donkey, 560
Double diamond hitch, 442
Draft horses, 131–137
Dressage, 391–398
Driving, 226, 371–374
Dun (color), 55, 565–569
Dystocia, 546–547

Electric shock, first-aid treatment, 294
Elimination (behavior), 43–44
Embryo transfer, 527
Encephalomyelitis, 321–323
Endometrial cups, 531
Endometritis, 527
Endurance riding, 427–430
Energy, 185–188
  requirement table, 593–597
Entropion, 549
Epididymis, 452–453
Epimeletic behavior, 41–42
Equine infectious anemia (EIA, or swamp
    fever), 323
Ergot, 60
Esophagus, 171
Estrogen, 482
  pregnancy, 529, 536
Estrous cycle:
  behavior, 476–480
  diestrus, 475
  estrus, 475
  hormone changes, 481–482
  manipulation, 500–505
  reproductive-tract changes, 482–485
  variability, 486–500
Estrus, 475, 484–486, 491–493, 499–501
  behavior changes, 476–480
Et-epimeletic behavior, 41–42
Ewe neck, 65–66
Exercise, 205, 223–226
  conditioning, 226–228
  energy requirements, 186
  pregnant mare, 538, 539
Eye injury, 295

Face markings, 53, 571–572
Fallopian tubes, 469, 474, 483

Fat:
  in diet, 185
  digestion, 171
Feces, 282, 313–314, 548
Feeding:
  foals, 187, 551
  management practices, 201–205
  mares, 187, 202–203
  mature horses, 186
  orphans, 551
  working horses, 186–187
  yearlings, 187
Feeds:
  composition tables, 598–604
  vitamin-content table, 605
Feet:
  anatomy, 251–256
  bruised sole, 92
  cleaning, 247–248
  conformation, 72–73
  contracted heels, 91
  corns, 92
  founder, 92
  growth, 248
  handling, 242–247
  laminitis, 92
  navicular, 93
  shape, 72
  shoeing, 256–278
  trimming, 248–250
Femur, 81
Fences, 162
Feral horses, 129
Fetal growth, 532–536
Fever, 560
Fighting (vice), 49
Figure-eight stake race, 390–391
Fire safety, 165
Firing, leg treatment, 297
First aid:
  abrasions, 293–294
  bandages, 288–293
  bruises, 295
  burns, 293–294
  chilling cold, 296
  cuts, 286
  electric shock, 294
  eye, 295
  fractures, 296
  heat stroke, 295
  kit, 284–285
  lameness, 296–297
  punctures, 293–294
  saddle sores, 295
  snake bites, 296
  sprains, 295
  strains, 295
  tears, 286
  treatments, 286–288
Fistulous withers, 86
Flat-foot walk, 22

Flies, 320–321
Floating teeth, 298
Foal:
  care, 547–549
  congenital defects, 554–557
  diseases, 549–550
  orphan, 550–551
  training, 552–553
  trimming feet, 273–275
  weaning, 551–552
Foal heat, 500
Foaling, 538
  dystocia, 546–547
  inducing, 547
  kit, 539–540
  process, 540–545
  signs, 540
Follicle, 471–473, 482, 484, 501
  atretic, 473
Follicle stimulating hormone (FSH), 454,
    482, 505
Foot flight arc, 73
Forge, blacksmith, 77
Forging (gait defect), 24–25, 82, 277
Forward riding, 363–366
Founder, 92
Fox hunting, 426–427
Fox trot, 23
Fractures, 296
French Coach (breed), 137
Frog, 250, 254
FSH (follicle stimulating hormone), 454,
    482, 505

Gait defects:
  brushing, 25
  cross-firing, 23–24
  dishing, 26
  forging, 24–25, 82, 277
  hock hitting, 28
  overreaching, 24–25, 77, 82, 277
  paddling, 27–28
  pounding, 28
  scalping, 28, 277
  shin hitting, 28
  speedy cutting, 28
  trappy, 29
  winding, 29
  winging, 25–26
Gaits, 14–29
  backing up, 22
  canter, 18–22
  defects, 23–29, 77, 82, 277
  flat-foot walk, 22
  fox trot, 23
  leads and lead changes, 18–22
  lope, 18–22
  pace, 18
  pasi-trote, 23
  paso llano, 23
  rack, 23

run, 22
running walk, 22
slow pace, 23
sobreandando, 23
stepping pace, 23
trot, 18
walk, 15
Galiceño (breed), 119
Gall bladder, 171
Galvayne's groove, 36, 37
Gaskin muscle, 81
Gates, 162–163
Gelding, 57, 467–468
Gene, 558–559
Genetics:
  blood factors, 555–556, 560–562
  coat color, 561–572
  congenital defects, 554–555
  genes and chromosomes, 558–559
  inbreeding, 576–577
  infertility, 560
  linebreeding, 576–577
  sex determination, 559–560
  qualitative inheritance, 573–574
  quantitative inheritance, 574–575
German Coach (breed), 137
Gestation, 528
Girls' Rodeo Association, 418
Gnats, 319–320
Goiter, 194
Goose rump, 81
Godolphin Barb, 101–102
Gotland (breed), 116
Grain:
  barley, 180
  corn, 181
  milo, 182
  oat, 180
  wheat, 182
Granulosa cell tumor, 527
Gravel (disease), 95
Gray (color), 55, 565–570
Grooming, 233–236
Grooming (horse behavior), 43
Grubs, cattle, 309
Grulla (color), 55, 566–569
Gymkhana, 388–391

Hackney (breed),137
Hair, curly, 572
Halter classes, 376
Haltering, 206–208
Halter pulling (vice), 49–50
Hand (height measurement), 13
Handicap race, 380
Harness, 371–373
Harness racing, 403–405
Hay:
  alfalfa, 175
  barley, 175
  bermudagrass, 175

Hay *(continued)*
  clover, 175
  oat, 175
  prairie, 175
  sudangrass, 175
  timothy, 175
  wild native, 175
HCG (human chorionic gonadotropin), 501, 505
Head, 63–65, 169
  color markings, 53, 571–572
Heart rate, 280–281
Heat, 475, 476–480, 499
  silent, 493
  foal, 500
Heat stroke, 295
Heaves, 98, 328–329
Heel crack, 94, 278
Height, 13
Hemolytic icterus, 555–556
Heritability estimates, 574–575
Hernia, 98
  umbilical, 555
  scrotal, 555
Hinny, 560
Hitches, packing, 441–443
Hobbles, 211–212
Hock:
  sickle, 82, 277
  cow, 82, 277
Hock hitting (gait defect), 28
Hoof handling, 242–247
Hoof wall, 251
"Hooks" on teeth, 298
Hormones, 452, 454–456, 481–482, 490–491, 501–503, 505, 528–539, 526, 547
Horseshoes, Horseshoeing: *See* Shoes, Shoeing
Hot walker, 223
Human chorionic gonadotropin (HCG), 501, 505
Hyperparathyroidism, nutritional secondary, 194

Illness, recognition, 279–281
Immunity, failures, 557
Immunizations, 322
Inbreeding, 576–577
Incisive arcade (tooth), 36–37
Infertility:
  genetic, 560
  mare, 525–527
  stallion, 465
Influenza, 323–324
Ingestive behavior, 43–44
Inheritance:
  qualitative, 573–574
  quantitative, 574–575
  See also Genetics
Insecticides, 315–318

Insurance policies, 2–4
Intestine, 171–173
Investigative behavior, 42–43
Iodine, 194
Isoerythrolysis, neonatal, 555
Isohemolytic icterus, 555

Jaquima, 346, 348–349, 381
Jaundice foal, 555
Jaw:
  overshot, 64
  undershot, 64
Justin Morgan, 109, 111

Keyhole racing, 390
Kicking (vice), 50
Knees:
  buck, 73, 76
  calf, 73, 76
  chopped, 71
  knock, 73, 76
  offset, 70
  open, 70
  popped, 90
  tied-in, 71
Knock knees, 76
Knots, 210–211

Lactation, energy requirement, 187
Lameness:
  first aid, 296–297
  recognition, 282–284
Laminae, 251
Laminitis, 92
Large intestine, 172
Lateral cartilage, 255
Lead changes, 18–22
Leading, 208–209, 219
Lead limbs, 18–19
Learning (behavior), 45
Leg markings, 53, 571–572
Lethal genes, 554–555
LH (luteinizing hormone), 455, 481–482, 501, 505
Lice, 318
Lights, artificial, for breeding, 505
Linebreeding, 576–577
Lockjaw, 326, 538, 548
Longeing, 225–226
Lope, 18–22
Luteinizing hormone (LH), 455, 481–482, 501, 505

Magazines, addresses, 585–586
Magnesium, 194
Maiden race, 399
Mammary gland, 469–475
Mane, 238–240
Mare:
  foaling, 538–546
  physiology of reproduction, 475–500

preparation for breeding, 516–520
reproductive tract, 469–475
Martingales, 353
Masturbation, 465–466
Mecate, 346–347
Meconium, 548
Melanoma, 64, 329
Metabolic diseases, 327–329
Milo grain, 182
Mimicry (behavior), 45
Minerals, 184
   calcium, 193
   iodine, 194
   magnesium, 194
   phosphorus, 193
   potassium, 195
   requirement tables, 593–597
   selenium, 194
   sodium chloride, 193
Miniature horses, 142
MIP-test kit, 536, 647–648
Missouri Fox Trotting Horse (breed), 117
Mites, 319
Molasses, 182
Moon blindness, 84
Morab (breed), 117
Morgan (breed), 111
Morocco Spotted Horse (breed), 126
Morphology of spermatozoa, 462, 644–645
Mosquitos, 320
Motility of spermatozoa, 461, 638–639,
   641
Mounting, 221, 358–359, 363,
   366–369
Mud fever, 91
Mule, 560
Muscular system, 69
Mutton withers, 67

National High School Rodeo Association,
   418
National Intercollegiate Rodeo Association,
   415–418
Navel cord, 547–548
Navicular disease, 93, 278
Neck:
   conformation, 65
   cresty, 66
   ewe, 66
   swan, 65
   upside-down, 66
Neonatal isoerythrolysis, 555
Night blindness, 84
North American Trail Ride Conference,
   428–429
Nutrients:
   analysis of feed, 199
   requirement tables, 593–597
Nutritional secondary hyperparathyroidism,
   194

Oat:
   grain, 180
   hay, 175
Offset knees, 70
Open knees, 70
Orphan foals, 550–551
Osselets, 90
Ovary, 469, 471–473, 482–484,
   527–528
   anatomy, 471–473
   corpus albicans, 473
   corpus luteum, 471–473, 482, 484, 486,
      490, 500
   follicles, 471–473, 482, 484
   ovulation fossa, 471–472
   pregnancy, 528
Overo (color), 56, 124, 570
Overreaching (gait defect), 77, 82, 277
Overshot jaw, 64, 85
Oviducts, 469, 474, 483
Ovulation, 492–493, 497, 499–501, 503
   during pregnancy, 529
Ovulation fossa, 471–472
Ownership:
   benefits, 5–6
   costs, 2–4
   obligations, 1–2
Oxytocin, 547

Pace, 18
Packing and outfitting, 430–444
   equipment, 431–436
   packing loads, 439–444
   saddling, 437–439
Paddling (gait defect), 27–28
Pads, saddle, 336, 340, 343, 354
Paint:
   breed, 124
   color, 56, 565, 570
Palomino:
   breed, 123
   color, 55, 565–569
Palpation:
   ovaries, 483
   pregnancy determination, 532–535
   stallion reproductive tract, 465
Parasites, external:
   chiggers, 319
   control, 315–318
   flies, 320–321
   gnats, 319–320
   lice, 318
   mites, 319
   ticks, 314
Parasites, internal:
   Ascarids, 302
   bots, 307–308
   control, 309–313
   fecal examination, 313–314
   grubs, cattle, 309

Parasites, internal *(continued)*
  stomach worms, 306–307
  strongyles, 303–305
  pinworms, 305
  threadworms, 306
Parentage disputes, 560–561
Parimutuel betting, 400
Parrot mouth, 554
Parturition, 538–547
Pasi-trote, 23
Paso Fino (breed), 118
Paso llano, 23
Passive transfer, immunity, 538
Pasterns, 71, 72
Pasture, 177
Pawing (vice), 50
Pedigree, 57–58
Penis, 453, 517, 520–522
Percheron (breed), 131
Periodic ophthalmia, 84
Periople, 251
Peruvian Paso (breed), 118
pH, semen, 462, 643
Phalanx, 255
Pharynx, 170
Phosphorus, 193
Photosensitization, 99–100
Physical facilities, 145
  barns, 149–150, 156, 158, 162, 165
  breeding, 162
  codes and regulations, 147
  fences, 162
  fire safety, 165
  planning, 146
  site selection, 147
Piebald (color), 56, 124, 570
Pigeon toes, 75–76
Pig eyes, 64
Pinto:
  breed, 124–125
  color, 56, 570–571
Pinworms, 305
Piroplasmosis, 324
Placenta, 530–531, 545–546, 549
Play (behavior), 44
Pleasure horse, 381
PMSG (pregnant mare serum
    gonadotropin), 505, 528–529, 536
Pneumonia, 324
Pneumovagina, 525–526
Poisonous plants, 178–179
Pole bending, 389
Poll evil, 84
Polo, 444–446
Pony of the Americas (breed), 140
Popped knees, 90
Potassium, 195
Pounding (gait defect), 28
Prairie hay, 175

Pregnancy:
  abortions, 536–537
  duration, 528
  examination, 532–536
  fetal growth, 532
  hormone changes, 528–529
  placentation, 530
Pregnant mare serum gonadotropin
    (PMSG), 505, 528–529, 536
Prepuce, 453–454, 517, 523
Professional Rodeo Cowboys Association,
    413–415
Progesterone, 481, 490, 505, 528
Prolonged corpus luteum, 527
Prostaglandin, 482, 490–491, 501, 503,
    505, 547
Prostate gland, 454
Protein, 182–183, 188–189
  requirement tables, 593–597
Puberty:
  colt, 454
  filly, 475
Pulse rate, 228, 280–281
Puncture wounds, 293–294

Qualitative inheritance, 573–574
Quantitative inheritance, 574–575
Quarter crack, 94, 278
Quarter Horse:
  breed, 102
  racing, 405–409
Quittor, 94

Rabies, 324–325
Race, Racing:
  allowance, 298
  barrel, 388, 424
  claiming, 399
  derby, 399
  figure-eight stake, 390–391
  glossary of terms, 628–633
  handicap, 398
  harness, 403
  maiden, 399
  Quarter Horse, 405–409
  stakes, 398
  statistics, 608–627
  supervision, 398–399
  Thoroughbred, 409–410
  types, 298, 388, 390–391, 398–399, 403,
    405–410, 424
Rack, 23
Racking Horse (breed), 116
Ration:
  analysis, 199–200
  change, 203–204
  examples, 197–199

formulation, 195–197, 587–592
nutrient concentrations, 190–191
Rearing (vice), 51
Records:
  breeding, 505–516
  racing, 58, 608–627
Respiration rate, 228, 280
Restraint:
  foal, 552–554
  hobbles, 211
  tranquilizers, 218
  twitches, 215
  tying, 211, 219
  war bridles, 217
Rhinopneumonitis, 325
Ribs, 77
Ridgeling, 57
Riding:
  bridling, 357–358
  equipment, 333–355
  exercise, 226
  mounting, 358–359, 363, 366–369
  position, 359–363, 367, 369–370
  saddling, 355–357
  safety, 221–222
Ring bone, 72, 75, 88
Ringworm, 325
Roach back, 77
Roan (color), 55, 570
Roaring (unsoundness), 99
Rodeo, 411–424
  associations, 413–418
  events, 418–424
  terms, 634–635
Roughages, 173–178
  hays, 173–177
  pasture, 177–178
Roundworms, 302
Run, 22
Running walk, 22
Russian Orloff (breed), 137

Saddlebred (breed), 109
Saddle bronc riding, 422
Saddle pads, 336, 340, 343, 354
Saddles, 333–335, 338–340, 341–342,
  353–354
Saddle seat equitation, 366–367
Saddle sores, 295
Saddling, 220–221, 355–357
Safety, 165, 218–222
Salt, 193
Sand crack, 94
Scalping (gait defect), 28, 277
Scratches, 94
Scrotal hernia, 555
Scrotum, 449
Seal-brown (color), 54–55, 563–569
Seedy foot, 95

Selection:
  basic factors, 9–58
  breeds, 101–142
  conformation, 59–100
Selenium, 194
Semen:
  characteristics, 455
  collection, 455, 457–458
  concentration, 461, 640–643
  evaluation, 455, 459–463, 636–645
  extender, 646
  pH, 462, 643
  processing. 455, 459
  production, 454
  volume, 461
Seminal vesicles, 454
Sesamoiditis, 91
Sex, 13
  accessory glands, 454
  determination, genetic, 559, 560
  geldings, 57
  value when purchasing, 56
Sexual behavior:
  mare, 478–481
  stallion, 457, 463, 466–468, 522
Sheath, 453
Shetland (breed), 141
Shin hitting (gait defect), 28
Shire (breed), 135
Shoe boil, 89
Shoes, Shoeing, 256, 268
  corrective, 273–278
  equipment, 256–260
  evaluation, 272–273
  procedure, 268–271
  shoes, 260–268, 275–278
  nails, 268
Shoulder, 85
  conformation, 67–69
  slope, 68
  sweeny, 86
Shying (vice), 51
Sickle hock, 82, 277
Side bones, 72, 94
Sidesaddle, riding, 367–370
Silent heat, 493
Size, 13
Skeleton, 61
Skewbald (color), 56, 124, 571
Sleeping (behavior), 45–46
Sleeping sickness, 321–323
Sling hitch, 443
Slow pace, 23
Small intestine, 171
Smegma, 453
Snake bites, 296
Sobreandando, 23
Sodium chloride, 193
Sole, 254

Sorrel (color), 55, 564–569
Spanish-Barb (breed), 128
Speedy cutting (gait defect), 28
Spermatozoa:
    morphology, 462, 644–645
    motility, 461, 638–639, 641
Splay feet, 76
Splints, 89–90
Split estrus, 493
Sprains and strains, 295
Spavins, 96–97
Stakes race, 398
Stallion:
    breeding soundness, 463–464
    castration, 467–468
    infertility, 465–466
    preparation for breeding, 517, 521–522
    reproductive tract, 449–454
    semen, 454, 457–462
    training young, 467
Stalls, 158–159, 166
Stall vices, 47
Stall walking (vice), 51
Standardbred (breed), 109
Steer wrestling, 418–419
Stepping pace, 23
Stifle, 95
Stock horse, 380
Stomach, 171
Stomach worms, 306–307
Strains and sprains, 295
Strangles (disease), 325–326
Stratum tectorium, 252
Striking (vice), 51
Stringhalt, 96
Strongyles, 303–305
Sudangrass hay, 175
Suffolk (breed), 132
Swamp fever (EIA, equine infectious
    anemia), 323
Swan neck, 65
Swayback, 77
Sweat, leg, 297
Sweeny, 86

Tail, 240–241
Tail rubbing (vice), 51–52
TDN (total digestible nutrients), 186
Team roping, 420–421
Teasing mares, 476–480
Teeth:
    aging, 31–38
    bridle, 33
    care, 298–300
    cup, 33
    dental star, 36
    floating, 298
    Galvayne's groove, 36–37
    incisive arcade, 36

    table surface, 33
    temporary, 33
    wolf, 298
Temperament, 38–39
Temperature, body, 280, 549
Tennessee Walking Horse (breed), 114
Testicle, 449–452
Testosterone, 452, 455–456
Tetanus, 326
    foal, 538, 548
    pregnant mare, 538
Thick wind, 99
Thoroughbred:
    breed, 107
    racing, 409–410
Thoroughpin, 96
Threadworms, 306
Three-day event, 387
Thrush (disease), 95, 247
Tibia, 61, 81
Ticks, 314
Tied-in knees, 71
Timothy hay, 175
Tobiano (color), 56, 124, 570
Toe crack, 94, 278
Toe-narrow, toe-in fault, 25–28, 75, 275,
    277
Toe-wide, toe-out fault, 25–28, 76,
    277–278
Total digestible nutrients (TDN), 186
Trace minerals, 195
Trailer, 229–232
Trail horses, 381
Trail riding, 427–430
Trainer, 10–11
Training, 13
    influencing value, 29–30
    stallions, 467
Tranquilizers, 218
Trappy (gait defect), 29
Treadmill, 224
Trimming:
    body, 237
    feet, 248, 273
    mane, 238
    tail, 240
Trot, 23
Twitches, 215–216
Tying, 210–211, 219
Tying-up syndrome, 327

Umbilical hernia, 555
Undershot jaw, 64, 85
Unsoundness:
    definition, 83
    recognition, 282, 284
Unsoundnesses:
    blindness, 84
    blind spavin, 97

blood spavin, 97
bog spavin, 82, 96
bone spavin, 82, 97
bowed tendon, 88
bruised sole, 92
bucked shins, 86
capped elbow, 89
capped hock, 97
contracted heels, 91
corn, 92
curb, 82, 98
fistulous withers, 86
founder, 92
gravel, 95
heaves, 98, 328–329
hernia, 98, 555
laminitis, 92
moon blindness, 84
mud fever, 91
navicular disease, 93, 278
night blindness, 84
osselets, 90
overshot jaw, 64, 85
periodic ophthalmia, 84
photosensitization, 99–100
poll evil, 84
popped knees, 90
quarter crack, 94, 278
quittor, 94
ring bone, 72, 75, 88
roaring, 99
scratches, 94
seedy foot, 95
sesamoiditis, 91
shoe boil, 89
side bones, 72, 94
splints, 89–90
stifle, 95
stringhalt, 96
thoroughpin, 96
thrush, 95, 247
undershot jaw, 64, 85
wind broke, 98
wind puffs, 90
wobbler, 85, 555
Upside-down neck, 66
Urachus, 547
Urination, 44
Urine:
  characteristics, 282
  pooling in mares, 526
Uterus, 469, 474, 483–484
  infections, 527
  during pregnancy, 533–536

Vaccination program, 322, 329–330
Vagina, 469, 475, 483, 485
Vas deferens, 453
Veterinary assistance, 284

Vices, 47–52
  attacking, 48
  biting, 47–48
  bolting, 48
  cribbing, 49, 64
  definition, 47
  fighting, 49
  halter pulling, 49–50
  kicking, 50
  pawing, 50
  rearing, 51
  shying, 51
  stall, 47
  stall walking, 51
  striking, 51
  tail rubbing, 51–52
  weaving, 52
  wood chewing, 52, 205
Viral arteritis, 326–327
Vitamins, 183–184
  A, 189
  B, 192
  C, 192
  D, 190
  E, 191
  K, 192
  requirement tables, 593–597
Volume, semen, 461
Vulva, 475, 527, 538

Ysabella, 124

Waking (behavior), 45–46
Walk, 15
War bridles, 217–218
Warts, 328
Water, requirement, 185
Weaving (vice), 52
Weight conversion factors, 606
Welsh Mountain Pony (breed), 138
Western riding, 358–363
Weymouth bridle, 341
Wheat grain, 182
White (color), 55, 563, 565, 570
Wild native hays, 175
Wind broke, 98
Winding (gait defect), 29
Wind puffs, 90
"Windsucker," 525–526
Winging (gait defect), 25–26
Withers, 66
  fistulous, 86
Wobbler, 85, 555
Wolf teeth, 298–300
Wood chewing (vice), 52, 205
Worm medicine, 310–313

Zebra, 559